Complete Fuel Systems and Emission Control

by Robert Scharff & the Editors of MOTOR SERVICE

WE ENCOURAGE
PROFESSIONALISM

THROUGH TECHNICIAN
CERTIFICATION

Fuel modes!!!

 Delmar Publishers Inc.®

NOTICE TO THE READER

Text, Design, and Production by
Scharff Associates, Ltd.
RD 1 Box 276
New Ringgold, PA 17960

Scharff Staff
Production Manager: Marilyn Hauptly
Editor: Lois Breiner
Text Design: Barbara Gould
Cover Design: Eric Schreader

Delmar Staff
Editor-in-Chief: Mark W. Huth
Associate Editor: Joan Gill

For more information, address Delmar Publishers Inc.
3 Columbia Circle, Box 15-015
Albany, New York 12212-5015

Printed in the United States of America
Published simultaneously in Canada
by Nelson Canada,
a division of International Thomson Limited

10 9 8 7 6 5 4 3 2

Library of Congress Cataloging-in-Publication Data

Scharff, Robert.
    Complete fuel systems and emission control / by Robert Scharff & the editors of Motor service.
        p.   cm.
    Includes index.
    ISBN 0-8273-3576-8.—ISBN 0-8273-3577-6 (instructor's guide).—
ISBN 0-8273-3578-4 (shop manual).
    1. Automobiles—Fuel systems.   2. Automobiles—Fuel systems—
Maintenance and repair.   3. Automobiles—Pollution·control devices.
4. Automobiles—Pollution control devices—Maintenance and repair.
I. Motor Service.   II. Title
TL214.F8S34   1989   89-7672
629.25'3—dc20   CIP

[Handwritten notes:]

1. Verify complaint
2. Determine Related symptoms
3. analize the Symptoms
4. Isolate the Problem
5. Fix the Problem
6. check for Proper Repairs.

CONTENTS

PREFACE

Although emission controls have been on vehicles for only about 20 years, fuel systems have been with us since the very first cars. However, in the last few years, a great number of changes have taken place in automotive fuel systems. Many modern vehicle models have such items as electronically controlled carburetors and fuel injection systems, various emission control devices, engine sensors, computers, and so on. With the addition of these components, the once simple fuel system has developed into a complex function of automotive servicing. As a result, the technician must have a complete knowledge of fuel systems and emission controls. As an aid in helping to accomplish this task, we have broken down subject matter into twelve chapters plus appendices.

Chapter 1 is an overview of the basic components of the fuel and emission control system. It also includes safety procedures, how to keep shop records, and the importance of ASE certification.

Chapter 2 examines in detail the various fuels used in vehicles today, including gasoline, no-lead gasoline, diesel fuel, LP gas, and the various additives employed.

Chapter 3 explains how the vacuum and air intake systems work in conjunction with fuel operation. It also gives a complete diagnosis of vacuum, air, and fuel delivery system problems.

Chapter 4 describes the operation and service of both conventional and feedback (electronic) carburetors. Although conventional carburetors are being phased out in new cars, the technician must have a knowledge of their operation and servicing, since they will still be around for many years.

Chapters 5, 6, and 7 cover fuel injection, including the basic principles of electronic fuel injection, both throttle body and port fuel injection, as well as their servicing. Chapter 8 features an overview of diesel injection engines.

Chapter 9 explains the importance of the exhaust and tells how it operates. A portion of the chapter is devoted to the operation and servicing of turbochargers.

Chapters 10, 11, and 12 detail the theory, operation, and servicing of various components found in both conventional and electronic operated vehicles.

Each chapter opens with a list of objectives so that the student knows what he or she can expect to learn. At the end of each chapter, there are review questions to check the student's understanding of the text. The glossary and list of abbreviations at the back of the book give the most commonly used fuel and emission terms.

Complete Fuel Systems and Emission Control contains various cautions and warnings that should be carefully noted to avoid personal injury or vehicle damage when servicing. However, it is important to understand that these cautions and warnings could not possibly cover all conceivable service procedure variations and their particular hazards, nor can the contributors, the publisher, or the editors and their staff possibly know or investigate all such variations. It is, therefore, the responsibility of anyone using the service procedures or tools (whether or not recommended in this book) to ensure to their own satisfaction that neither personal safety nor vehicle safety will be jeopardized.

ACKNOWLEDGMENTS

To organize a book requires the help of a great number of sources of information and the aid of many people and organizations. The cooperation of the staff of *Motor Service* and its editorial director, Jim Halloran, played a great part in the preparation of the text and furnishing of illustrations. In addition we would like to thank the big three automotive manufacturers—Chrysler, Ford, and General Motors—for permission to use information and illustrations from their training programs and service manuals.

Grateful acknowledgment is also made to the following companies, which provided fuel system and emission controls reference material and illustrations used in this book.

AC-Delco (General Motors Corporation)
Allied Aftermarket Division (Allied Signal Inc.)
American Honda Motor Company, Inc.
Arrow Automotive Industries
Buick Division (General Motors Corporation)
BWD Automotive Corporation
Cadillac Motor Division (General Motors Corporation)
Car-Quest Inc.
Carter Automotive (Federal Mogul Corporation)
Chevrolet Motor Division (General Motors Corporation)
Chrysler Motor Division (General Motors Corporation)
CR Industries, Inc.
Cummins Engine Company
Dana Corporation
Echlin Inc.
Fel-Pro Inc.
Filko Automotive Products
Ford Motor Company
Ford Parts and Service Division (Ford Motor Company)
Garrett Turbocharger Division (Allied Signal Inc.)
Hennessy Industries

Holley Parts Division (Colt Industries)
Huth Manufacturing, Inc.
Isuzu Motors, Ltd.
Kem Manufacturing Company, Inc.
Lincoln
Maremont Corporation
Mazda Motors of America, Inc.
Moog Automotive, Inc.
Motorcraft Products (Ford Motor Company)
National Automotive Parts Association (NAPA)
National Institute for Automotive Service Excellence (ASE)
Oldsmobile Motor Division (General Motors Corporation)
Perfection Automotive Company
Pontiac Motor Division (General Motors Corporation)
Robert Bosch Corporation
Rochester Division (General Motors Corporation)
Subaru of America, Inc.
Toyola Motor Sales, Inc.
TRW Automotive Aftermarket Inc.
Volkswagen of America, Inc.
Wix Corporation

We would like to thank the following for reviewing the manuscript and for their helpful comments:

Mr. Richard Neet
Spokane Community College
North Bessie Road
Spokane, WA 99212

Mr. Robert Nudd
Northwestern Business College
1441 North Cable Road
Lima, OH 45805

Mr. Fred Raadsheer
BCIT
Barnuady Campus
3700 Willingdon Avenue
Burnaby, British Columbia
Canada, V5G3H2

Mr. Santiago Urdiales
Alamo Community College District
Automotive Technology Department
St. Phillips College
2111 Nevada Street
San Antonio, TX 78203

CHAPTER ONE

INTRODUCTION TO FUEL SYSTEMS AND EMISSION CONTROLS

Objectives

After reading this chapter, you should be able to:
- Explain how a gasoline engine operates, including the four-stroke cycle.
- Describe the three basic design characteristics of an engine.
- Understand the function of the systems that are related to engine operation.
- Name the components of a fuel system and explain their operation.
- Name the components of an exhaust system and explain their operation.
- Name the components of an emission control system and explain their operation.

The automotive engine has been changing since its invention in the nineteenth century. Automobile manufacturers have tried different designs and used different materials in the effort to build engines that offer better performance, improved fuel economy, and reduced exhaust emissions. Many design features that have become popular recently were actually developed long ago. However, automotive engine (Figure 1-1) changes have come faster in recent years because, while government regulations have required better fuel economy and lower exhaust emissions, consumers have continued to demand good performance.

To understand the fuel and emission control systems, the service technician must have an understanding of the operation of a gasoline engine.

GASOLINE ENGINE OPERATION

In a passenger car or truck, the engine provides the rotating power to drive the wheels through the transmission and driving axle. All automobile engines, both gasoline and diesel, are classified as "internal combustion engines" because the combustion or burning that creates heat energy takes place inside the engine. These systems require an air/fuel mixture that arrives in the combustion chamber with exact timing and an engine constructed to withstand the temperatures and pressures created by thousands of explosions.

The combustion chamber is the space between the top of the piston and cylinder head. It is an enclosed area in which the gasoline and air mixture is exploded. The piston is a hollow metal tube with one end closed that moves up and down in the cylinder. This reciprocating motion is produced by the burning of fuel in the cylinder.

The reciprocating motion must be converted to rotary motion before it can drive the wheels of a vehicle. As shown in Figure 1-2, this conversion is achieved by linking the piston to a crankshaft with a connecting rod. The upper end of the connecting rod moves with the piston as it moves up and down in the cylinder. The lower end of the connecting rod is attached to the crankshaft and moves in a circle. The

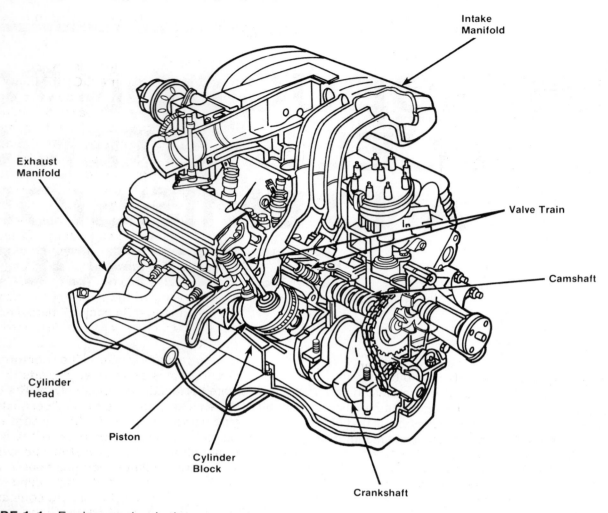

FIGURE 1-1 Engine mechanical components

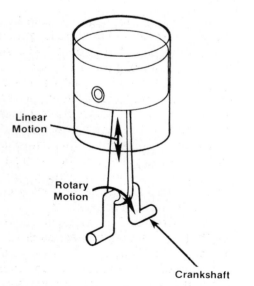

FIGURE 1-2 Crankshaft changes linear motion to rotary motion.

end of the crankshaft is connected to the transmission to continue the power flow through the drivetrain and to the wheels.

For the explosion in the cylinder to take place completely and efficiently, precisely measured amounts of air and fuel must be combined in the right proportions. The carburetor (or in some cases, a fuel injection system) makes sure that the engine gets exactly as much fuel and air as it needs for the many different conditions under which the vehicle must operate: starting, idling, accelerating, or cruising.

There are usually two valves at the top of the cylinder. The air/fuel mixture enters the combustion chamber through an intake valve and leaves (after having been burned) through an exhaust valve. The valves are accurately machined plugs that fit into machined openings. A valve is said to be seated or closed when it rests in its opening. When the valve is pushed off its seat, it opens.

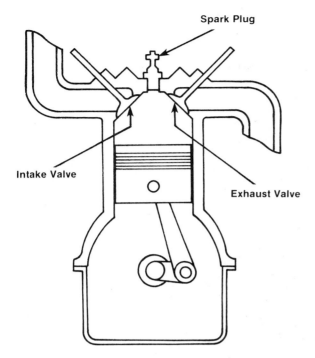

Spark Plug

Intake Valve

Exhaust Valve

FIGURE 1-3 Valves and spark plug control the combustion process.

A rotating camshaft, connected to the crankshaft, opens and closes the intake and exhaust valves (Figure 1-3). Cams are raised sections of the shaft, or collars, with high spots called *lobes*. As the camshaft rotates, the lobes rotate and push away a spring-loaded valve tappet. The tappet transfers the motion to a pushrod and perhaps a rocker arm to open the valve by lifting it off its seat. Once the lobe on the cam rotates out of the way, the valve, forced by a spring, moves down and reseats.

In summary, the essentials for the complete combustion process include the following:

- Admit a proper mixture of air and fuel into the cylinder.
- Compress (squeeze) the mixture so it will burn better and deliver more power.
- Ignite and burn the mixture.
- Remove the burned gases from the cylinder so that the process can be completed and repeated.

With the proper timing of the action of the valves and spark plug to the movement of the piston, the combustion cycle takes place in four strokes of the piston. The basis of automotive gasoline engine operation is the four-stroke cycle. A stroke is the full travel of the piston up or down. There are four strokes in this cycle: the intake stroke, the compres-

sion stroke, the power stroke, and the exhaust stroke.

- *Intake Stroke.* As the piston moves away from top dead center (TDC), the intake valve opens (Figure 1-4A). The downward movement of the piston increases the volume of the cylinder above it. This, in turn, reduces the pressure in the cylinder below atmospheric pressure. The reduced pressure causes atmospheric pressure to push a mixture of air and fuel through the open intake valve. As the piston reaches the bottom of its stroke, the reduction in pressure stops and the intake of air/fuel mixture nearly ceases. But due to the weight and movement of the air/fuel mixture, it will continue to enter the cylinder until the intake valve closes. The delayed closing of the intake valve increases the volumetric efficiency of the cylinder by packing as much air and fuel into it as possible.
- *Compression Stroke.* The compression stroke begins as the piston starts to move from bottom dead center (BDC). The intake valve closes, trapping the air/fuel mixture in the cylinder (Figure 1-4B). Upward movement of the piston compresses the air/fuel mixture. At TDC, the piston and cylinder walls form a combustion chamber in which the fuel will be burned. The volume of the cylinder with the piston at BDC compared to the volume of the cylinder with the piston at TDC determines the compression ratio of the engine.
- *Power Stroke.* The power stroke begins as the compressed fuel mixture is ignited in the combustion chamber (Figure 1-4C). An electrical spark across the electrode of a spark plug ignites the air/fuel mixture. The burning fuel rapidly expands, creating a very high pressure against the top of the piston. This drives the piston down toward BDC. The downward movement of the piston is transmitted through the connecting rod to the crankshaft. Up-and-down movement of the piston on all four strokes is converted to rotary motion of the crankshaft.
- *Exhaust Stroke.* The exhaust valve opens just before the piston reaches BDC on the power stroke (Figure 1-4D). Pressure within the cylinder when the valve opens causes the exhaust gas to rush past the valve and into the exhaust system. Movement of the piston from BDC pushes most of the remaining ex-

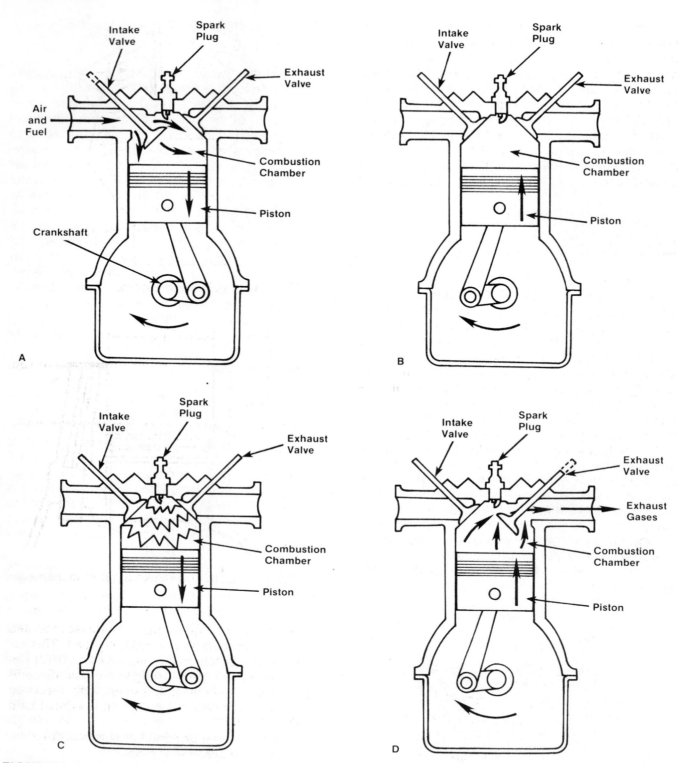

FIGURE 1-4 (A) Intake stroke; (B) compression stroke; (C) power stroke; (D) exhaust stroke

haust gas from the cylinder. As the piston nears TDC, the exhaust valve begins to close as the intake valve starts to open. The exhaust stroke completes the four-stroke cy-

cle. The opening of the intake valve begins the cycle again. This cycle occurs in each cylinder and is repeated over and over, as long as the engine is running.

The four strokes of the cycle require two full revolutions of the crankshaft. Also, the piston is being acted on by combustion pressure during only about half of one stroke, or about one-quarter of one revolution. This makes it easier to understand the function of the flywheel. Even if the engine has multiple cylinders, a certain amount of power it produces has to be stored momentarily in the flywheel. From there, it is used to keep the piston in motion during about seven-eighths of the total cycle and to compress the fuel mixture just before combustion.

CHARACTERISTICS OF ENGINE DESIGN

There are many variations in engine design. Generally, modern automotive engines have four, six, or eight cylinders arranged either in-line or with two banks of in-line cylinders in a "V" or "Y."

In-line means that the cylinders are all in one straight row around the crankshaft. Most four- and six-cylinder engines are built with the cylinders in line. Figure 1-5 shows a typical arrangement of four- or six-cylinder engines. In the past there have been eight-cylinder in-line engines on automobiles, but the modern eight-cylinder engine is a V-8 (Figure 1-6), which has two in-line banks of four cylinders, each arranged in a "V" pattern with a common crankshaft. A V-6 engine on some units has the same basic arrangement, but has only three cylinders in each bank. The V-4 has two banks of two cylinders each.

Opposite air-cooled engines have been used in compact vehicles in the past. In such an arrangement, the two rows of cylinders are located opposite the crankshaft.

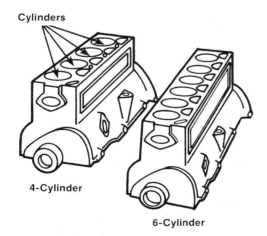

FIGURE 1-5   Typical in-line engines

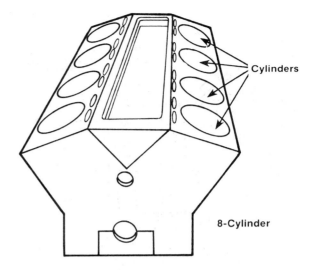

FIGURE 1-6   Typical V-8 engine

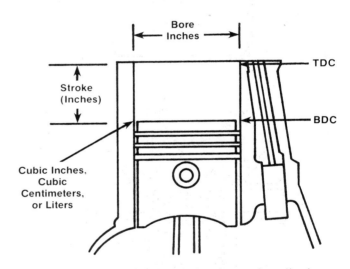

FIGURE 1-7   Bore and stroke determine displacement.

Most modern engines have the valves mounted above the cylinders in the cylinder head. This arrangement is called an overhead valve (OHV) design. A variation of this arrangement has the camshaft, which operates the valves, also mounted above the cylinders. It is called an overhead cam (OHC) design.

There are three design characteristics that are the basis of the gasoline engine.

1. *Bore and Stroke.* The bore of a cylinder is simply its diameter (Figure 1-7). The stroke is the length of the piston travel between TDC and BDC. Between them, bore and stroke determine the displacement of the cylinders.

2. *Displacement.* Displacement is the volume the cylinder holds between TDC and BDC positions of the piston. It is usually measured in cubic inches, cubic centimeters, or liters. The total displacement of an engine (including all cylinders) is a rough indicator of its power output. Displacement can be increased by opening the bore to a larger diameter or by increasing the length of the stroke.

3. *Compression Ratio.* Compression ratio is the amount the air/fuel mixture is squeezed or compressed, based on the cylinder and combustion chamber volume at BDC and TBC. In the example shown in Figure 1-8, the volume before compression is eight times the volume after compression, so the compression ratio is 8 to 1. Increasing compression ratio usually increases the power output of the engine. However, high-compression engines require premium gasoline, which might be higher in lead content. Modern compression ratios are in the range of 8.5 to 1 or less to take advantage of the lower lead content of regular and lead-free gasoline.

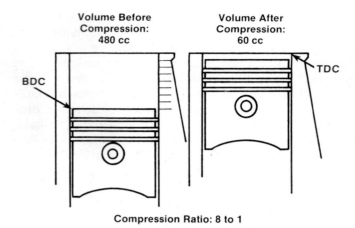

Compression Ratio: 8 to 1

FIGURE 1-8 Compression ratio is a measure of volume to volume.

ENGINE SYSTEMS

Besides the major power-generating system in the engine, there are several other systems essential to engine operation.

- *Air/Fuel System.* This system ensures that the engine gets the right amount of both air and fuel needed for efficient operation. These two systems join in the carburetor, which blends the fuel and air and supplies the resulting mixture to the cylinder. Some automobiles have a fuel injector system that replaces the carburetor.
- *Ignition System.* This system supplies a precisely timed spark to ignite the compressed air/fuel mixture in the cylinder at the end of the compression stroke.
- *Lubrication System.* The system supplies oil to the various moving parts in the engine. The oil lubricates all parts that slide in or on other parts, such as the piston, bearings, crankshaft, and valve stems. The oil enables the parts to move easily so that little power is lost and wear is kept to a minimum.
- *Cooling System.* This system is also extremely important. Water circulates in jackets around the cylinder and in the cylinder head. This water removes part of the heat produced by the combustion of the air/fuel mixture and prevents the engine from being damaged by overheating.
- *Exhaust System.* This system efficiently removes the burned gases and limits noise produced by the engine.
- *Emission Control System.* Several control devices, which are designed to reduce emission levels of combusted fuel, have been added to the engine. Engine design changes, such as reshaped combustion chambers and altered tune-up specs, have also been implemented to help control the auto's smog-producing byproducts. These devices and adjustments have reduced emissions considerably but have changed automotive engine servicing to a great extent.

In this book, the major concern is to train automotive technicians in the operation and servicing of the air/fuel system, exhaust system, and emission control system. To understand these systems, it is important to have an overview of their functions and components.

THE FUEL SYSTEM

The fuel system supplies to the engine cylinders a combustible mixture of gasoline and air. To do this, it must store the fuel and deliver it to the fuel metering and atomization system, where it is mixed with air to provide the combustible mixture that is delivered in a manner that meets the varying load requirements of the engine.

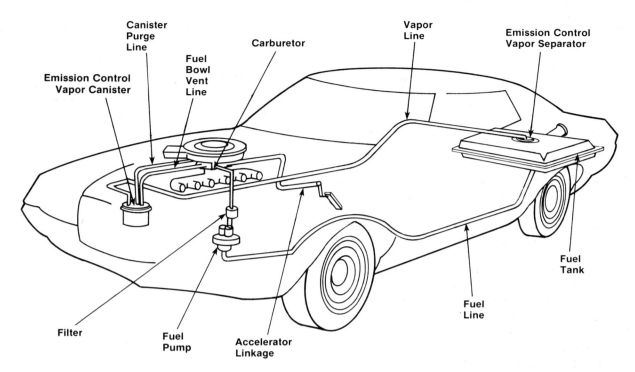

FIGURE 1–9 Components of a fuel system

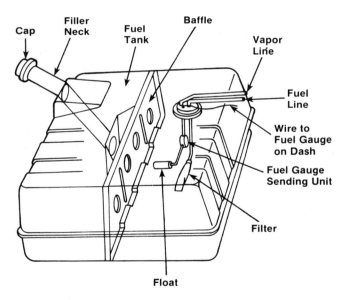

FIGURE 1–10 Parts of a fuel tank

The system uses several components to accomplish this task: a fuel tank to store the gasoline in liquid form, fuel lines to carry the liquid from the tank to the other parts of the system, a pump to move the gasoline from the tank, a filter to remove dirt or other harmful particles that might be in the fuel, a carburetor or electronic fuel injector to mix the liquid gasoline with air for delivery to the cylinders after the air

has passed through an air filter, and an intake manifold through which the air/fuel mixture from the carburetor or fuel injector is directed to each of the engine cylinders.

FUEL SYSTEM COMPONENTS

The components of a fuel system, shown in Figure 1–9, are as follows:

- Fuel tank
- Fuel lines
- Fuel pump
- Fuel filter
- Air filter
- Fuel metering and atomization system (carburetor or electronic fuel injector)

Fuel Tank

The fuel tank is usually located at the rear of the vehicle, aft of the rear axle. It is mounted by straps or brackets to keep it in place. It is constructed of either pressed sheet metal with welded seams and special coating to prevent corrosion or fiberglass-reinforced plastic material. The latter tanks conform better to unusual underbody contours.

The fuel tank has several openings (Figure 1–10), the largest being the fill pipe. This is the pipe

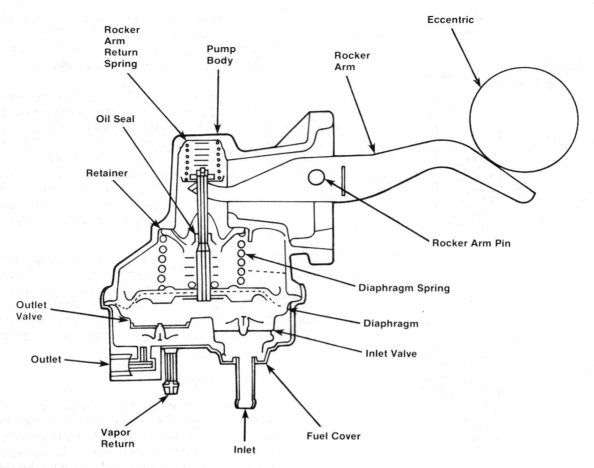

FIGURE 1-11 Mechanical fuel pump

through which the gasoline is pumped into the tank from an external source. Adjacent to it is a vent line to permit air vapors in the tank to escape while it is being filled. The gasoline exits the tank through an outlet pipe that usually enters from the top of the tank. The outlet pipe extends down into the tank to about 1/2 inch from the bottom. This prevents sediment and debris at the bottom of the tank from being pulled into the outlet line. Some outlet lines also have a screen to prevent particles from being pulled in.

Prior to 1971, self-venting filler caps were used to vent the tank. Since the introduction of emission controls, several different internal venting devices have been used to extract vapors from the tank. In the majority of these devices, the vapors are carried by a tube to a charcoal canister. Here they are absorbed by the charcoal and later drawn with the air/fuel mixture into the combustion chambers of the engine.

Fuel tanks usually have some type of device to indicate the amount of fuel that is in the tank. The indicating fuel gauge is located on the vehicle's dashboard, while the fuel information sensing unit can be found in the tank.

Fuel Lines

The fuel lines are small-diameter steel tubing and neoprene hoses. Usually the steel line runs along the chassis from the tank to the fuel pump and has flexible neoprene attachments to the fuel tank pickup at one end and to the fuel pump at the other. This arrangement absorbs the vibrations when the engine moves and the frame does not thus preventing the metal tubing from cracking or breaking.

Fuel Pump

The fuel metering and atomization system is located higher in the vehicle than the fuel tank. A mechanical or electrical pump is employed to draw the fuel from the tank and deliver it to the carburetor or fuel injectors. It must have sufficient capacity to supply the engine with fuel under all operating con-

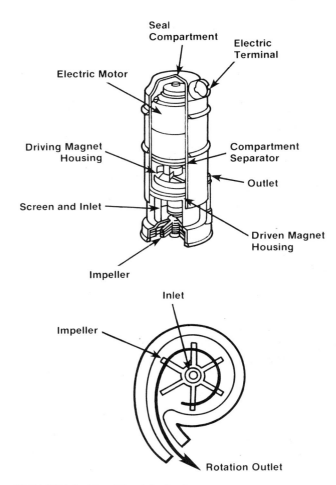

FIGURE 1-12   Electric fuel pump

ditions. It should also maintain sufficient pressure in the line to the carburetor to keep the fuel from boiling and causing vapor lock due to high engine temperatures under the hood.

Mechanical Fuel Pump. The most common type of fuel pump is the universal mechanical diaphragm-type pump (Figure 1-11). The rocker arm rides against an eccentric lobe on the camshaft. The amount of fuel pumped increases as the camshaft rotates faster. The force of the diaphragm spring establishes the maximum working pressure of the fuel pump. It limits the amount of fuel according to engine requirements.

Mechanical pumps in the engine compartment are subject to heat. A low-pressure area is also created in the fuel line during fuel intake. Both of these conditions can lead to vapor lock.

Electric Fuel Pumps. Electric fuel pumps come in four basic types.

1. *Diaphragm.* This type works the same way as a mechanical pump, with the exception that an electromagnet moves the diaphragm.
2. *Plunger.* An electromagnetic switch controls a plunger or piston that moves up and down in this type.
3. *Impeller.* This pump has no input and outlet valves. Instead it has a revolving armature that works much the same way as a fan blows air (Figure 1-12). The impeller creates suction that draws fuel into the pump and pushes it out to the carburetor.
4. *Bellows.* This pump is similar to the diaphragm type. However, a metal accordian-pleated bellows is used instead of a diaphragm.

The distinct advantage of the electric fuel pump is its capability of being mounted anywhere between the fuel tank and the carburetor. When installed close to the fuel tank or in the fuel tank, it eliminates vapor lock because with fuel under pressure, it is difficult for vapor to form even if the fuel lines get hot.

Fuel Filters

Fuel filters are in several locations in the fuel system between the tank and the fuel metering and atomization system intake. Moisture and contaminants can collect in the fuel tank. Consequently, particles of rust from the tank can clog the fuel lines and small passages in the carburetor or fuel injectors. There are three basic types of fuel filters.

1. *Strainer.* These fuel filters consist of small fibers molded into a suitable shape for installation in a see-through glass bowl or metal canister. This type of filter is usually mounted in the fuel pump (Figure 1-13).

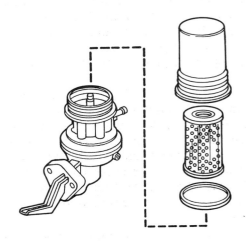

FIGURE 1-13   Fuel filter mounted on fuel pump

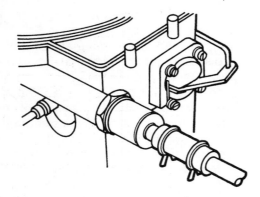

FIGURE 1-14 Fuel filter on carburetor fuel inlet

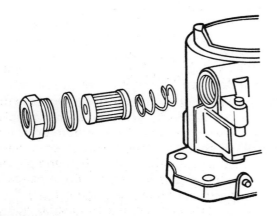

FIGURE 1-15 Fuel filter in carburetor

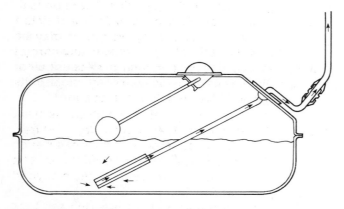

FIGURE 1-16 Fuel filter in tank

2. *Screen.* These fuel filters are usually mounted in the carburetor fuel inlet fitting. The fuel line must be disengaged to replace the filter. Screen filters are usually made of porous bronze or woven brass wire. The filter openings are small enough to stop contaminants from entering the carburetor.

3. *Paper.* Although these filters are sometimes found in other parts of the system, they are usually mounted in the fuel line between the pump and carburetor. They remove smaller sized particles and larger amounts of them than it is possible to remove with the strainer or screen types.

Some fuel filters screw into the fuel inlet fitting of the carburetor and connect to the fuel line with a hose and clamps (Figure 1-14). Others are located inside the carburetor itself, as shown in Figure 1-15. Often these are small pleated paper filters with a built-on gasket to provide a positive seal. Sometimes, however, they are sintered bronze, which would be considered a strainer type filter. Some of the sintered bronze filters contain a built-in check valve to limit fuel leakage if the vehicle overturns.

Some cars and light trucks use two gasoline filters. The first filter, made of fine woven fabric, is found in the gasoline tank (Figure 1-16). Besides preventing large pieces of dirt or other contaminants from damaging the fuel pump, this filter also prevents most water from going to the carburetor. The tank filter will not require servicing or replacement under normal circumstances.

The second filter can be found in a number of different locations in the engine compartment. This filter must be serviced regularly. Figure 1-17 illustrates one type, called an in-line fuel filter.

Air Filter

An internal combustion engine typically "breathes in" about 10,000 gallons of air for every 1 gallon of fuel it consumes. To avoid damage to the engine, this air brought into the system must be very clean. The most common contaminants that an air filter must remove are dirt, exhaust soot, vegetable matter, and insects. Naturally, geographic location will affect the type of contaminants encountered. An automobile used in a wet, humid climate will have different filtration problems than one used in a dry, dusty climate. An off-road vehicle will encounter

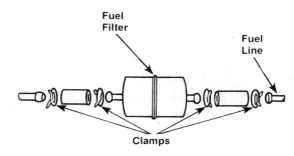

FIGURE 1-17 In-line fuel filter

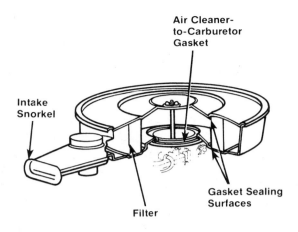

FIGURE 1-18   Light-duty air filter

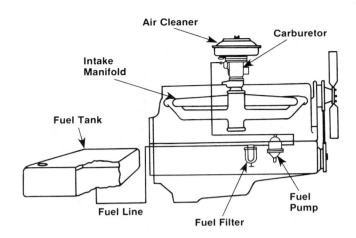

FIGURE 1-20   Fuel and air system

more dirt than an automobile used only on the highway.

The oil bath air cleaner was used for many years to clear the intake air of the engine. When the airstream was forced to make an abrupt 180-degree turn, dirt particles were trapped on the surface of a pool of oil. However, the efficiency of such an air cleaner (that is, the percentage of contaminants it removed from the air) was approximately only 95 percent. At low engine speed, efficiency dropped by as much as 25 percent. Thus, the invention of the dry air filter was an important step toward prolonging engine life.

There are both light- and heavy-duty dry types of air filters. The light-duty air filters (Figure 1-18) are used most often on passenger car and pickup truck engines. Because of the limitations of space under the hood, these filters are usually small. Nevertheless, these filters allow the free flow of air for even the largest gasoline engines while providing a minimum efficiency of 98 percent.

Heavy-duty air filters (Figure 1-19) have a minimum efficiency rating of 99.9 percent. However, they are much larger and bulkier than light-duty types. Therefore, they are seldom found in passenger vehicles.

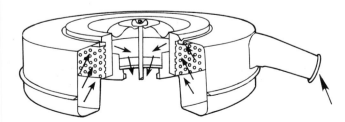

FIGURE 1-19   Heavy-duty air filter

Fuel Metering and Atomization System

Technicially, it is inaccurate to say that cars run on gasoline. A vehicle actually runs on a precise mixture of gasoline and air. The carburetor or fuel injector can be considered the "chemist" of the vehicle. It mixes the proper amount of fuel and air and releases it to the engine according to the needs of the engine in various situations (Figure 1-20).

The carburetor and fuel injectors subdivide or "atomize" the fuel and mix these fine particles of fuel with air. This air/fuel mixture is expressed as a number of parts of air to 1 part of gasoline. The chemically correct mixture (at sea level) is 14.7 parts of air to 1 part of gasoline, measured by weight. It may be expressed by the ratio 14.7:1. A chemically correct fuel mixture is called *stoichiometric*. Because air at higher elevations contains less oxygen, more parts of air are required for complete combustion.

Air/fuel ratios having high concentrations of air, such as 16:1, 18:1, or 20:1, are called *lean mixtures*. Air/fuel ratios having low concentrations of air, such as 12:1, 10:1, or 8:1, are called *rich mixtures*. The engine will not run if the air/fuel ratio goes beyond these rich and lean limits.

The engine requires a mixture as rich as 11.5:1 during warm-up and on heavy acceleration. But a mixture as lean as 18:1 can be used after warm-up and during low-load cruising. An air/fuel ratio of 14.7:1 (at sea level) provides the most power for the amount of fuel consumed.

Carburetor. The purpose of a carburetor on a gasoline engine is to meter, atomize, and distribute the fuel throughout the air flowing into the engine. These functions are designed into the carburetor and are carried out by the carburetor automatically

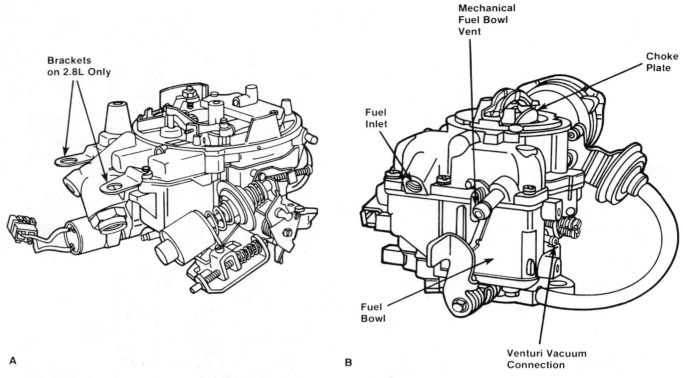

FIGURE 1-21 (A) Two-barrel and (B) single-barrel carburetors

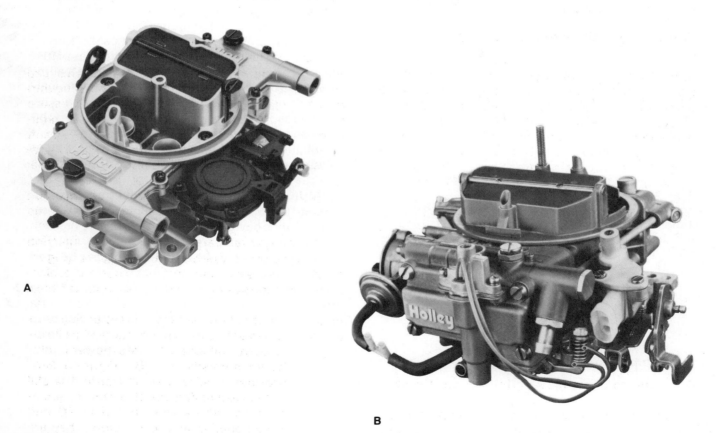

FIGURE 1-22 (A) Typical conventional carburetor and (B) feed back carburetor

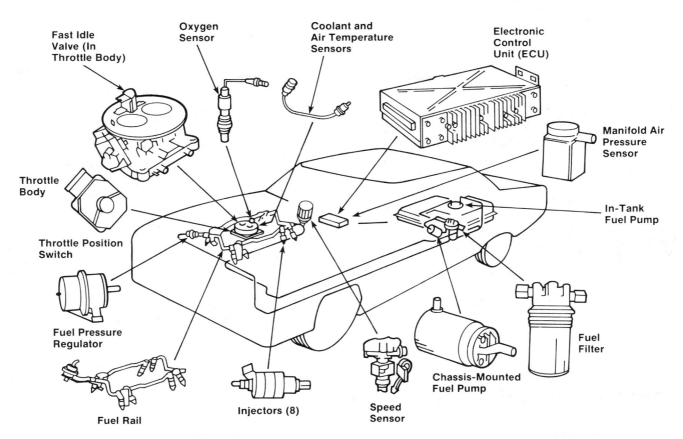

FIGURE 1-23 Electronic fuel injection

over a wide range of engine operating conditions such as varying engine speeds, load, and operating temperature. The carburetor also regulates the amount of air/fuel mixture that flows to the engine. It is this mixture flow regulation that gives the driver control of the engine speed. Typical two-barrel and single-barrel carburetors are shown in Figure 1-21.

Although the carburetor performs a comparatively simple job, it does so under such varied conditions that it is necessary to have several systems to alter its functions so it can adjust to various situations. Most carburetors contain the following basic systems:

- Float system
- Main system
- Choke system
- Power system
- Acceleration system
- Idle system

Components of a typical carburetor are shown in Figure 1-22.

Full details on the operation of these six carburetor systems as well as their servicing are given in Chapter 3.

Fuel Injection. Fuel injection systems use pressure to force fuel through a small opening into the air. Unlike a carburetor, however, the pressure forcing the fuel into the air is not atmospheric pressure. Instead, the pressure is produced by a pump in the injection system itself. The benefits of fuel injection include a more precise control of the air/fuel ratio, improved emission control, fuel efficiency, and driveability. On the other hand, fuel injection systems are more expensive, usually considered harder to service, and require clean fuel.

In late-model cars, electronic fuel injection (EFI) is the most widely used fuel system (Figure 1-23). There are currently two major designs of electronic injection systems used in passenger vehicles.

1. Throttle body injection (TBI) uses one or two injectors centrally mounted in a throttle body unit located where the carburetor would normally be. TBI (Figure 1-24A) provides precise fuel control, but is still subject to manifold problems that are common to a carbureted system. If the TBI unit is equipped with two injectors, they are opened alternately to prevent the fuel pres-

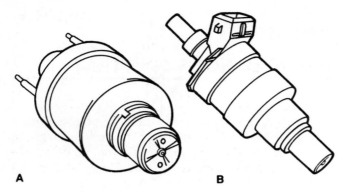

FIGURE 1-24 (A) Throttle body injector; (B) port fuel injector

sure from dropping too low. TBI systems are also known as central fuel injection (CFI), single-point fuel injection (SPFI), and digital fuel injection (DEFI).

2. Port fuel injection (PFI) uses one injector for each cylinder in the engine, which is located in the intake runner near the intake valve port (Figure 1-24B). By positioning the injector in the intake port, manifold problems associated with the TBI or carbureted systems are eliminated. Both TBI and PFI use similarly designed, electronically controlled injectors. The difference between the two systems is the number of injectors used and where they are located. PFI systems are also known as multiport (MFI), rail, and sequential fuel injection (SFI).

The electronic fuel injector is a solenoid-operated valve that is controlled by the car's computer. The amount of fuel that is delivered to the engine is controlled by the following factors:

- The size of the orifice in the injector is determined when the system is designed. This factor is fixed; it cannot be changed or modified.
- The amount of fuel pressure is also predetermined by system design (Figure 1-25). The fuel pump and fuel pressure regulator set the system's operating pressure. Pressure regulators on some systems can vary the pressure setting depending upon the engine load.
- The length of time the injector is held open is controlled by the car's computer and can be altered. The computer varies this amount of time the injector is held open in accordance with various computer inputs and specifications from the Programmable Read Only Memory (PROM).

The typical EFI system provides a pulse of fuel to get the system underway. The length of the pulse, that is, the length of time the injector is held open, is controlled by the electronic system (Figure 1-26).

An electrical fuel pump pressurizes the fuel. Fuel pressure is regulated according to manifold absolute pressure, with excess fuel returned to the tank through a return line. The fuel damper prevents pulsations in the line and maintains fuel pressure in the system. Because extremely clean fuel is required for fuel injection, the fuel filter is extremely important.

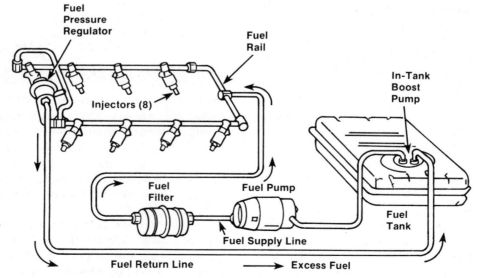

FIGURE 1-25 Electronic fuel injector fuel delivery subsystem

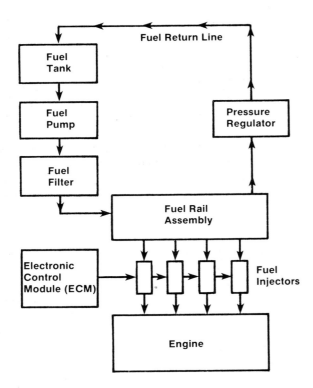

FIGURE 1-26   Electronic fuel supply system

grammed signals for warm-up, idle, and full-load enrichment.

THE EXHAUST SYSTEM

The exhaust system is a relatively simple part of a vehicle engine system and because of this, there is a tendency on the part of many car owners to ignore it until there is an obvious problem. However, remember that exhaust performs the following important functions:

1. Allows the engine's spent combustion gases to escape
2. Carries these gases safely away from the passenger compartment and bodywork
3. Muffles engine noise
4. Cleans up the exhaust before it is released into the atmosphere

Each of these functions of the exhaust system is critically important. If the system fails, the results can range from mere inconvenience to disaster.

EXHAUST SYSTEM COMPONENTS

The various components of the typical exhaust system are shown in Figure 1-27: They include the following:

- Exhaust manifold and gasket
- Exhaust pipe, seal, and connector pipe
- Catalytic converter
- Intermediate pipes
- Muffler
- Resonator
- Tail pipe
- Heat shields
- Clamps, gaskets, and hangers

Exhaust Manifold and Gasket

The exhaust manifold (Figure 1-28) is a bank of pipes that collects the burned gases as they are expelled from the engine cylinders and directs them to the exhaust pipe. Exhaust manifolds for most vehicles are made of cast iron, but some newer, smaller passenger cars have stamped, heavy-gauge sheet metal exhaust units. The manifold unit has one port for each cylinder.

The operation of the exhaust manifold is simple, but its design might not be. Near the end of the power stroke, the exhaust valve or port opens and exhaust gases begin to leave the cylinder. The piston continues on its exhaust stroke, pushing the

Air flowing into the system is measured by an airflow sensor. The signal from the airflow sensor is called the *base pulse.* The base pulse provides the minimum amount of fuel to be injected. A solid state electronic control unit (ECU) modifies the base pulse to increase the time the injector stays open. The ECU receives and interprets signals from other sensors that monitor coolant temperature, throttle positions, engine speed, and engine load to adjust the injectors to engine operating conditions.

Electronic injectors use a solenoid to open the nozzle valve. The ECU controls the pulse width at the injector by the length of time the solenoid receives a voltage signal.

A cold start injector valve, a thermo-time switch, and an auxiliary air device are used to make provision for cold starting and warm-ups. Extra fuel is delivered by the cold start injector while the engine is cranking. During starting and warm-up, the auxiliary air device provides extra air to overcome cold engine friction.

A feedback or closed loop fuel control system can meter the air/fuel mixture more closely. Because a closed loop system continually monitors itself and adjusts the inject pulse rate, the air/fuel ratio can be kept as close to ideal as possible. The ECU in an open loop operation sends prepro-

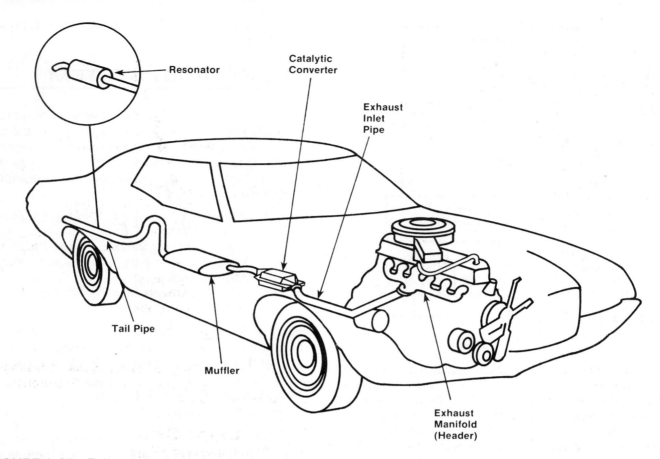

FIGURE 1-27 Exhaust system components

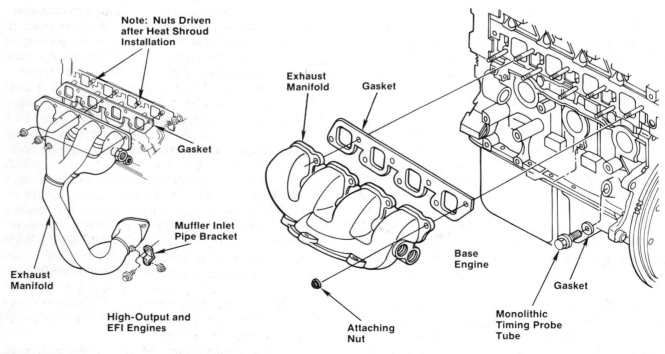

FIGURE 1-28 Exhaust manifold

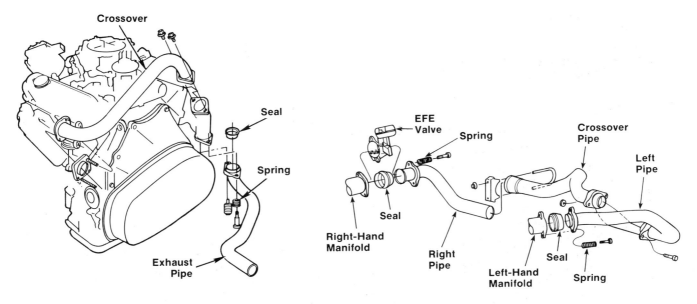

FIGURE 1-29   Crossover pipe arrangement

burned gases into the exhaust manifold. But each time the exhaust valve opens, waves of high-pressure gases are forced into the exhaust manifold. Sound, or noise, is produced by pressure waves.

Exhaust noises are reduced by several design factors. For example, one important factor is the length of passages within the manifold. If an exhaust manifold is designed correctly, pulses from the cylinders do not interfere with one another. Thus, exhaust gases can flow easily through the manifold.

To assure a good flow of gases with minimum back pressure, the exhaust manifold has large tubular sections. In addition, automotive design engineers try to avoid sharp bends that can slow the passages of gases. However, the exhaust manifold often has to fit into narrow, cramped spaces between the engine and body of the vehicle. Access to spark plugs and other parts also helps determine the shape of the exhaust manifold. Thus, exhaust manifold shapes or designs in smaller cars are often compromises between good gas flow and limited mounting space.

There is a heat riser valve (HRV) in most exhaust manifolds to help cold starting. This valve provides extra heat under the carburetor as the engine is warming up to improve performance. It is located at the outlet on one side of the exhaust manifold, usually on the right.

The manifold itself rarely causes any problems. The gasket that seals the manifold to the engine block and the gasket that seals the exhaust pipe to the manifold occasionally fail. A leak in these areas causes a ticking sound when the engine is running.

Visually inspect the gaskets for signs of leakage, looking at the gasket and its surrounding area for a grayish-white deposit.

Exhaust, Pipe Seal, and Connector Pipe

The exhaust pipe carries collected gases and vapor from the exhaust manifold to the next component in the exhaust system. A "Y" pipe or a crossover pipe arrangement (Figure 1-29) is often used to connect both exhaust manifolds of a V-type engine to form a single exhaust system. The crossover pipe can be rotated either below or behind the engine.

In a dual or V-type engine setup with a dual exhaust system, an "H" pipe is often used. This consists of right and left exhaust pipes connected by a balance pipe. In other words, each bank of engine valves has its own separate exhaust system. Some car buffs believe that a dual exhaust system, which consists of two complete exhaust systems, is a worthwhile way to cut down back pressure and increase gasoline mileage. Dual exhaust systems are excellent when designed into the vehicle by the manufacturer, who usually specifies them only for large high-performance engines. The dual system is seldom very useful on smaller engines. The single system is capable of handling the car's exhaust at speeds from idle to the top of the speedometer and has no difficulty coping with today's *maximum* 65 miles per hour speed limits.

A typical turbocharged engine exhaust pipe connection configuration is shown in Figure 1-30.

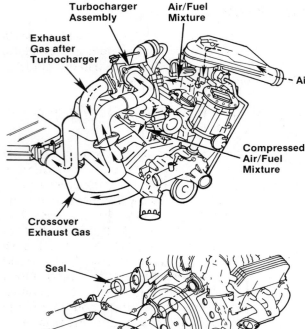

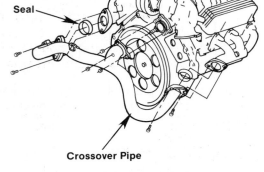

Crossover Pipe

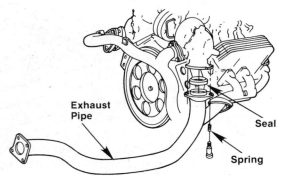

FIGURE 1-30 Typical turbocharged exhaust pipe configurations

The exhaust gases that drive a turbocharger must be routed from the exhaust pipe into and out of the exhaust turbine housing.

Catalytic Converter

The catalytic converter (Figure 1-31) is an emission control device added to the exhaust system to reduce pollutants. On some vehicles, the catalytic converter is located under the floor between the exhaust manifold and the muffler. On some later model vehicles, the catalytic converter is incorporated in the exhaust manifold. One or two catalytic converters can be used in an exhaust system. In a system using two converters, one can be a "light-off" converter that operates during warm-up. Such a converter pretreats and heats the exhaust gases for treatment by the main converter.

The operation of catalytc converters is explained later in this chapter.

Intermediate Pipes

Also known as an extension or connecting pipe, the intermediate pipe connects the exhaust pipe with the muffler or resonator. These intermediate pipes can be shaped to fit around axles and suspension components, away from shock absorbers and fuel tanks, and so on.

Mufflers

The primary task of the muffler is to deaden the roar of the exhaust gases. Most quality mufflers are galvanized or treated to protect them from corrosion by acid or rust. A good muffler is "tuned" to the requirements of the vehicle with the proper number of baffles (Figure 1-32), each designed to dampen sound without unnecessarily obstructing the flow of gases.

Specific mufflers are designed for specific cars and engines. Installing a muffler on a car that fits the space but is not designed for that car or engine can hurt the car's performance and will not silence as well. It can also cause damage if it develops too much back pressure.

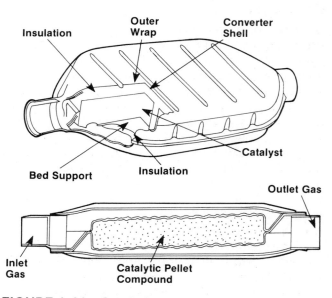

FIGURE 1-31 Catalytic converter

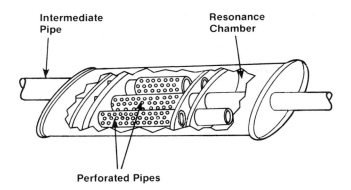

FIGURE 1-32 Typical muffler

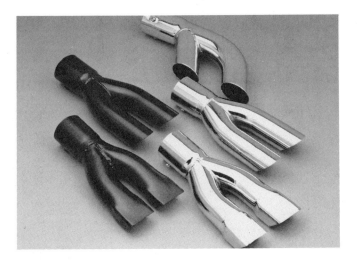

FIGURE 1-33 Typical tail pipe extension

Resonator

The resonator is a second, usually smaller silencing element in some exhaust systems where the space available under the car does not permit installation of a muffler that can completely silence the exhaust. The resonator smooths out any loudness or roughness from the first muffler. While most resonators are positioned toward the rear of the muffler, some import models have the unit located ahead of the muffler. However, not every car is equipped with a resonator. It is frequently used on vehicles with large engines.

Tail Pipe

Like the exhaust pipe, the tail pipe is nothing more than a pipe and/or the final extension of the exhaust system (Figure 1-33). The purpose of a tail pipe(s) is to direct the exhaust gases and vapor into the air away from the passenger compartment.

Heat Shields

Heat shields are used to protect vehicle parts from the heat of the exhaust system and catalytic converter (Figure 1-34). They are usually made of pressed or perforated sheet metal.

Clamps, Gaskets, and Hangers (Figure 1-35)

Clamps provide leak-free connections at joints in the muffler system. Gaskets, usually made from asbestos (decreasing in use), pressed steel, or sintered iron, are used to ensure tighter connections between the exhaust manifold and the exhaust pipe.

Hangers hold the clamps, pipes, and muffler to the underside of the car. They are flexible to absorb road vibration, although if they are too flexible, road vibration can break them. On the other hand, if they are too firm, vibration and noise from the exhaust system are transferred to the underbody of the car and can be heard in the passenger compartment.

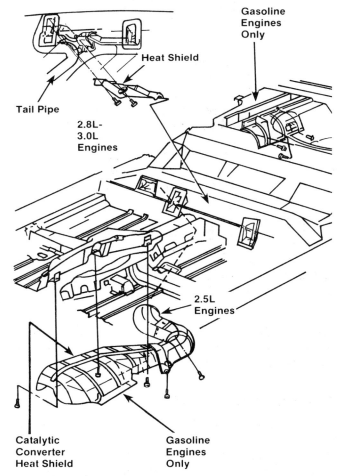

FIGURE 1-34 Typical heat shield

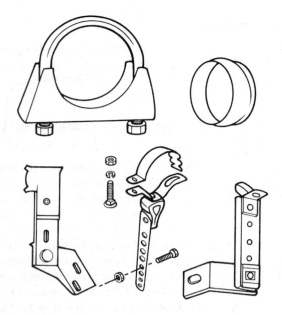

FIGURE 1-35 Clamps, gaskets, and hangers

THE EMISSION CONTROL SYSTEM

Beginning as early as the 1960s, the federal government's Environmental Protection Agency (EPA) began to enforce strict controls over the pollutants emitted from automobiles. To understand how these pollutants originate, one must first understand how an internal combustion engine operates. Being viewed as the chemical equation "whatever goes in must come out" in the internal combustion process, all elements that have been entered into the system are also in some form altered and expelled (Figure 1-36).

The two items fed into an engine during combustion are fuel and oxygen. Secondary elements, including pollutants in the atmosphere and additives in fuel, are also consumed. When combustion is complete, there are more than 280 compounds expelled from the automobile's tail pipe. Some of these compounds are hazardous to a person's health and the environment and become even more harmful when combined with outside elements such as rain and sunlight.

Three main pollutants are manufactured during internal combustion.

1. *Carbon monoxide (CO).* Carbon monoxide is formed whenever there is not enough oxygen to completely burn the fuel—in other words, wherever the air/fuel mixture is too rich. The richer the mixture, the

greater the quantity of CO produced. Actually, carbon monoxide is the worst pollutant of the three because it is deadly. It is invisible and odorless, yet it only takes about a 5 percent concentration to render a person unconscious. CO emissions are reduced by keeping the air/fuel ratio lean and by converting the remaining CO into harmless carbon dioxide in the catalytic converter.

2. *Hydrocarbon (HC).* Hydrocarbon emissions are unburned gasoline. HC is not directly harmful, but it contributes to smog formation. A fouled spark plug, leaky exhaust valve, or a fuel mixture so lean it will not ignite (lean misfire) can all allow unburned fuel to enter the exhaust. HC emissions can also come from oil in an engine that burns oil. HC is reduced by maintaining a balanced air/fuel mixture that is neither excessively lean nor rich, by making sure compression and ignition func-

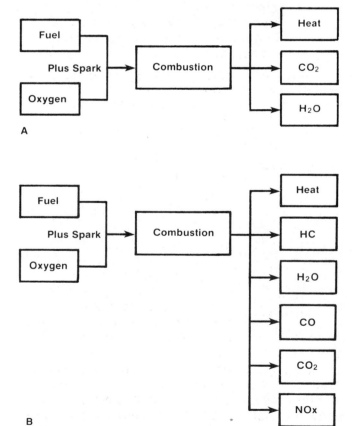

FIGURE 1-36 (A) Perfect combustion; (B) by-products of imperfect combustion

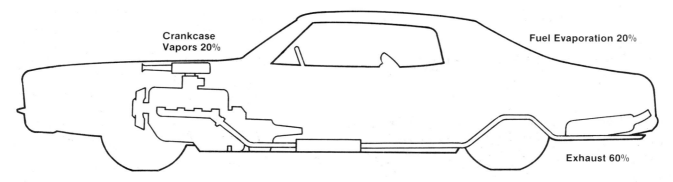

FIGURE 1-37 Major sources of automotive pollutants

tion properly, and by reburning any HC that is left in the catalytic converter.

3. *Nitrous oxide (NO$_x$).* Nitrous oxide is formed in the combustion chamber when temperatures rise above 2500 degrees Fahrenheit and nitrogen begins to react with oxygen creating various compounds. Lean air/fuel mixtures can increase NO$_x$ formation because of the higher combustion temperatures. Although not as poisonous as carbon monoxide, NO$_x$ irritates the eyes, nose, and lungs and combines with atmospheric oxygen to form ozone and acid rain.

These automotive-generated pollutants come from three major sources (Figure 1-37):

1. Tail pipe
2. Crankcase blowby vapors inside the engine
3. Fuel vapors that evaporate from the fuel tank and carburetor

In order to conform to the strict regulations issued by the EPA, various emission cleaning devices were added onto the engine, fuel, and exhaust systems. Most of the devices, however, did little to improve gas mileage and in some cases actually lowered fuel economy and engine performance. Driveability problems, including surging, detonation, and poor power, also began to develop due to smaller engine designs, leaner fuel mixtures, lower octane fuels, and excessive power draws from emission control devices.

In the mid-1970s, the gas shortage caused by the Arab oil boycott resulted in an additional problem. The government, concerned with our growing dependency on foreign fuel supplies, issued the Corporate Average Fuel Economy (CAFE) regulations. CAFE set strict requirements for automobile manufacturers to improve fuel economy. To meet the new

demands for better gas mileage, the need for emission control devices to operate more efficiently through the use of feed-back carburetors, onboard computers, and fuel injection became apparent.

Now that the emission control problems have been covered, look at the various systems and parts (Figure 1-38) that were designed to correct them. The systems and parts will appear in the approximate order in which they were introduced by car manufacturers. Keep in mind that not all systems or parts are on all applications, only those required to meet EPA and CAFE requirements.

EMISSION CONTROL SYSTEM COMPONENTS

The components of a vehicle's emission control system can be subdivided into six basic systems.

1. Positive crankcase ventilation (PCV)
2. Evaporative emission controls
3. Heated air intake (includes early fuel evaporation)
4. Air injection
5. Catalytic converter
6. Exhaust gas recirculation (EGR)

Positive Crankcase Ventilation (PCV)

The PCV system prevents crankcase blowby emissions from entering the atmosphere by siphoning the vapors back into the intake manifold through the PCV valve (Figure 1-39).

The PCV valve is a spring-loaded valve designed to restrict the amount of air passing from the crankcase to the intake manifold. In essence, the valve is a calibrated vacuum leak because it allows air to be drawn into the engine below the throttle plates. To prevent the air/fuel ratio from being upset, the spring inside the PCV valve closes the valve partially

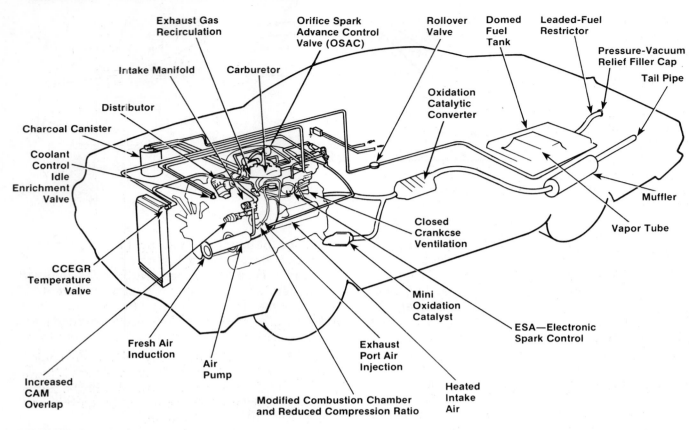

FIGURE 1-38   Emissions systems parts

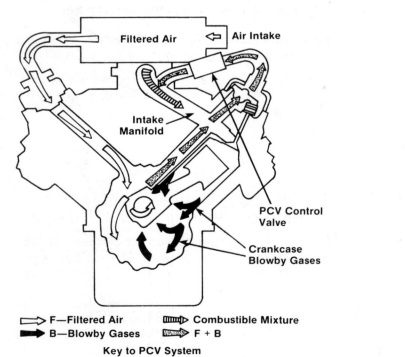

F—Filtered Air   Combustible Mixture
B—Blowby Gases   F + B
Key to PCV System

FIGURE 1-39   PCV system

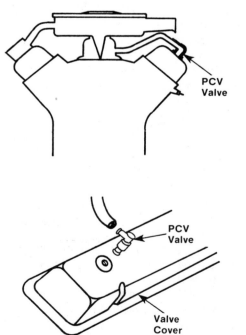

FIGURE 1-40   Location and replacement of a PCV valve

at idle to reduce airflow. Under cruise conditions, the valve opens to allow maximum airflow. During wide-open throttle acceleration or when the engine is off, the valve is closed.

The PCV valve can have an important effect on engine performance. Vacuum leaks in the valve or hose can lean out the air/fuel ratio, resulting in lean misfire, hesitation, hard starting, and stalling. An accumulation of deposits in either valve or hose can restrict or block the flow of air, resulting in a rich air/fuel mixture. Or more seriously, oil leaks can develop caused by a buildup of pressure in the crankcase. Rapid oil contamination can also occur because of moisture accumulation in the crankcase.

The PCV valve should be replaced if found to be defective or replaced every two to three years as a preventative maintenance item (Figure 1-40). One item that is often overlooked when replacing the PCV valve is the PCV filter located inside the air cleaner. It should always be changed when replacing the PCV valve. More information on servicing the PCV system as well as the rest of the emission control system components can be found in Chapter 11.

Evaporative Emission Controls

Evaporative emissions are controlled by sealing the fuel tank and carburetor bowl. The gas tank is sealed with a cap that contains a spring-loaded pressure relief valve. The tank itself is also vented to the charcoal canister that traps and stores fuel vapor until the engine is started and the vapors can be siphoned into the engine and reburned (Figure 1-41).

The charcoal canister is filled with activated charcoal that soaks up fuel vapors like a sponge

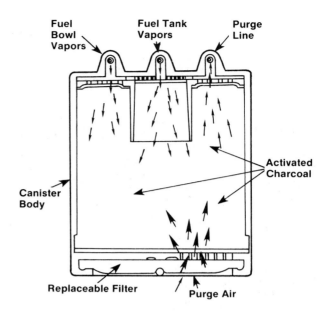

FIGURE 1-42   Typical charcoal canister

soaks up water (Figure 1-42). When the engine starts, a purge valve on the canister opens (Figure 1-43). Manifold vacuum draws air through the canister, which pulls the fuel vapors out of the charcoal and flushes them into the engine.

The carburetor bowl is likewise vented to the charcoal canister so when the engine is shut off, no fuel vapors can escape into the atmosphere.

Heated Air Intake and Early Fuel Evaporation (EFE)

To aid the process of fuel vaporization when a cold engine is started, one of several devices may be

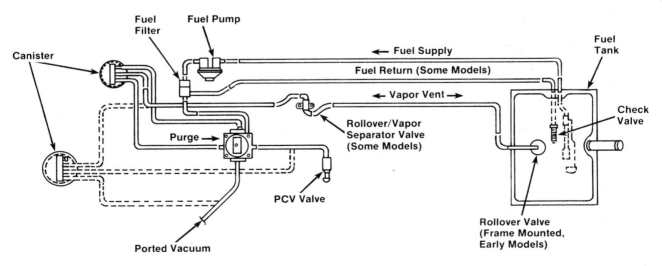

FIGURE 1-41   Typical evaporative control system

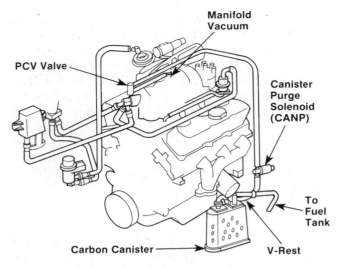

FIGURE 1-43 Canister purge system

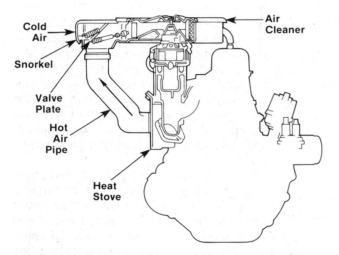

FIGURE 1-44 Heated air intake system

used. Most engines have a heated air intake system that draws warm air from a stove around the exhaust manifold into the air cleaner (Figure 1–44).

A temperature sensor (Figure 1–45) inside the air cleaner controls vacuum to a control valve in the air cleaner inlet. When the engine is cold, the thermostat passes vacuum to the control valve, which closes a flap to outside air allowing heated air to be drawn into the air cleaner. This aids fuel vaporization and makes it easier for the carburetor to maintain a more consistent air/fuel mixture. As the engine warms up, the thermostat begins to bleed air, allowing the control door to open to the outside air.

An early fuel evaporation aid is the heat riser valve found on many older V-6 and V-8 engines (Figure 1–46). A thermostatically controlled valve on one exhaust manifold blocks the flow of exhaust when the engine is cold, forcing the hot exhaust gases to flow back through a special crossover passage in the intake manifold directly under the carburetor. The hot exhaust heats the manifold to speed fuel vaporization and engine warm-up. Once the engine starts to warm up, the heat riser valve opens.

If the heat riser valve sticks open, no heating of the intake manifold will result, causing slow warm-up and hesitation and stumbling when the engine is cold. If it sticks shut, it causes a restriction and excessive heating of the manifold. The result can be detonation and a loss of high-speed power.

Air Injection

Air injection works in conjunction with the catalytic converter, although some vehicles have air pumps but no converter. The air pump forces air into the exhaust manifold so oxygen will combine with the hot exhaust gases to reduce HC and CO (Figure

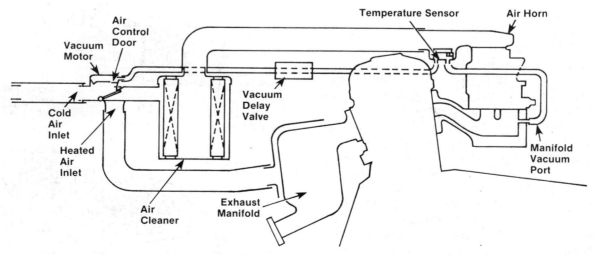

FIGURE 1-45 A sensor in the air cleaner controls a vacuum signal to the air control door. The system supplies air at the proper temperature to assure good driveability.

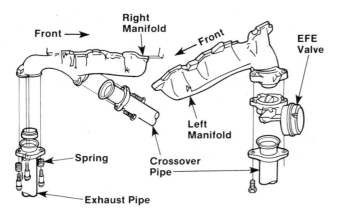

FIGURE 1–46   EFE mounting

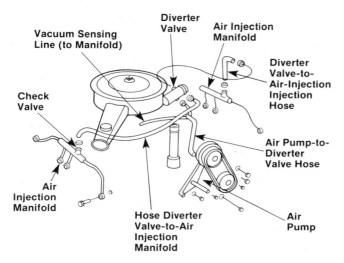

FIGURE 1–47   Air injection system is another vital part of late-model emission control systems.

When computerized engine controls (CEC) and three-way catalytic converters were added, the air pump gained yet another control valve. When the engine is cold, air is routed to the exhaust manifold to help reduce the initial HC and CO emissions. NO_x is not a problem when the engine is cold. But as the engine warms up and NO_x starts to rise, the flow of air is diverted from the exhaust manifold directly to the converters where it enters a chamber between the two catalysts. This is necessary because too much oxygen upstream of the converter interferes with the reduction of NO_x.

Catalytic Converter

The catalytic converter plays an important role in the emission control system. It contains a ceramic honeycomb or ceramic pellets coated with a thin layer of catalyst (Figure 1–48). In an oxidizing converter (one that reburns HC and CO), the catalysts are platinum and palladium metal. In a three-way converter, a reducing catalyst containing rhodium is added to break NO_x down into oxygen and nitrogen. The catalyst inside the converter is not used up in the chemical reactions but serves as a kind of chemical spark that helps trigger the reactions in the first place. For the converter to function efficiently, the engine must maintain a balanced air/fuel mixture. The converter also needs extra oxygen provided by the air pump.

The converter is normally a trouble-free emission control device, but two things can damage it. One is leaded gasoline. Lead coats the catalyst and renders it useless. The other is overheating. If raw fuel enters the exhaust because of a fouled spark plug or leaky exhaust valve, the temperature of the

1–47). The converter adds the extra kick needed to reburn the pollutants. The result is significantly lower HC and CO emissions.

The air pump is belt driven and feeds air to the exhaust manifold through a diverter valve and check valve. The diverter valve dumps excess air back into the atmosphere when it is not needed (during deceleration, for example). The check valve prevents backfiring through the air pump plumbing.

On some engines, a gulp valve is another part of the plumbing. The gulp valve diverts air from the pump into the intake manifold to momentarily lean out the mixture during deceleration to prevent backfiring in the exhaust from too much fuel.

On some engines, an aspirator is used in place of an air pump. An aspirator is a one-way valve that allows air to be siphoned into the exhaust system between exhaust pulses.

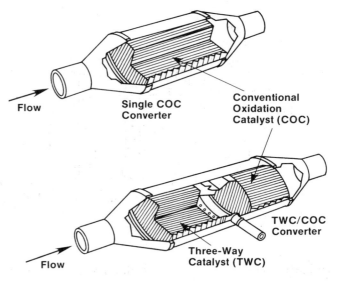

FIGURE 1–48   Typical catalytic converters

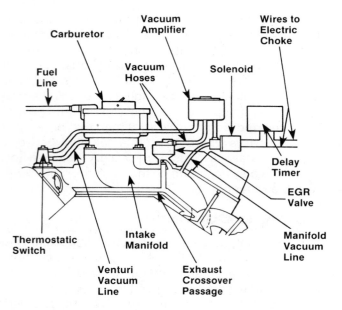

FIGURE 1–49 Exhaust gas recirculation

converter will soar. This can melt the ceramic honeycomb or pellets inside, causing a severe or complete exhaust blockage.

 SHOP TALK

Under no circumstances should the converter be removed and replaced with a straight piece of pipe (a test pipe). Do-it-yourselfers can still get away with it (until the vehicle undergoes an emission test, then the converter will have to be put on), but this practice is illegal for the professional auto mechanic (see Chapter 11).

Exhaust Gas Recirculation (EGR)

EGR reduces NO_x by diluting the air/fuel mixture and lowering combustion temperatures. Recirculating a small amount of exhaust back into the intake manifold keeps combustion temperatures below the 2,500 degree Fahrenheit NO_x formation threshold.

The heart of the system is the EGR valve, which opens to allow engine vacuum to siphon exhaust into the intake manifold (Figure 1–49). The EGR valve consists of a poppet valve and a vacuum actuated diaphragm. When ported vacuum is applied to the diaphragm, it lifts the valve off its seat. Intake

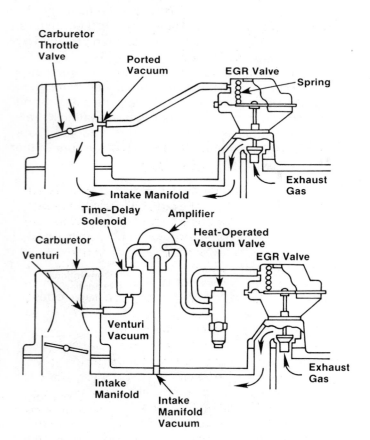

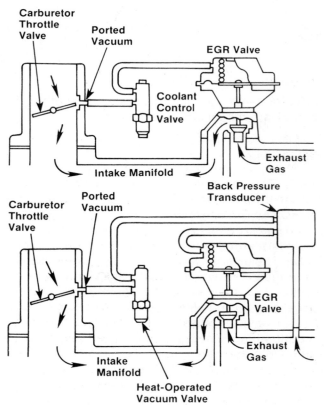

FIGURE 1–50 EGR variations

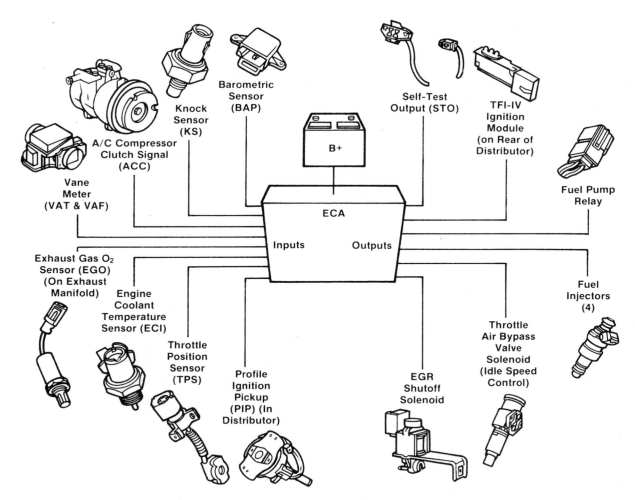

FIGURE 1-51 Typical EEC system

vacuum then siphons exhaust into the engine. Like a PCV valve, the EGR valve is a kind of calibrated vacuum leak.

On many late-model vehicles, a positive back pressure or negative back pressure EGR valve is often used (Figure 1-50). This type of valve requires a certain level of back pressure in the exhaust before it will open when vacuum is applied.

The EGR valve uses ported vacuum as its primary vacuum source. Ported vacuum is used because EGR is not needed at idle. (NO_x levels are low at idle.) If the EGR valve sticks open at idle or if the vacuum line is mistakenly connected to a source of manifold vacuum, it will cause a rough idle.

The vacuum control plumbing to the EGR valve usually includes a temperature vaccum switch (TVS) or solenoid to block or bleed vacuum until the engine warms up. On some late-model vehicles with computerized engine controls, the computer actuates the solenoid to further modify the opening of the EGR valve.

ELECTRONIC ENGINE CONTROLS (EEC)

Electronic engine controls have given the automotive manufacturer the capability to meet federal government regulations by controlling various engine systems accurately. In addition, electronic control systems have fewer moving parts than old style mechanical and vacuum controls. Therefore, the engine can maintain its calibration almost indefinitely. As an added advantage, the EEC system is very flexible. Because it uses microcomputers, it can be modified through programming changes to meet a variety of different vehicle engine combinations or calibrations. Critical quantities that describe an engine can be changed easily by altering data that is stored in the system's computer memory. In other words, the EEC system (Figure 1-51) is an assembly of electronic and electromechanical components that monitor engine operation to meet emission, fuel economy, and driveability requirements.

TOOLS AND SAFETY

There is no substitute for getting the job done right the first time. For parts personnel, that means being able to provide both the correct part(s) and helpful information. For the technicians, it means doing the job as quickly and safely as possible.

HAND AND POWER TOOLS

Many of the tools an auto technician uses every day are general-purpose hand tools (Figure 1–52). For instance, a complete collection of wrenches is indispensable. A variety of auto body parts, accessories, and related parts, not to mention shop equipment, use common bolt and nut fasteners as well as special hex screws and fasteners. Depending on the make and model of the vehicle, the fasteners can be standard SAE or metric size fasteners. So, a well-equipped auto body technician will have both metric and SAE wrenches in a variety of sizes and styles. The proper use of the appropriate hand tools by the technician is very important for performing quality auto service.

This is also true of power tools, which make the job a great deal easier. Although electric drills, wrenches, grinders, chisels, drill presses, and various other tools are found in shops, pneumatic (air) tools are used more frequently. Pneumatic tools have four major advantages over electrically powered equipment in the auto repair shop:

FIGURE 1–52   A professional set of hand tools

- *Flexibility.* Air tools run cooler and have the advantage of variable speed and torque; damage from overload or stalling is eliminated. They can fit in tight spaces.
- *Lightweight.* The air tool is lighter in weight and lends itself to a higher rate of production with less fatigue.
- *Safety.* Air equipment reduces the danger of fire hazard in some environments where the sparking of electric power tools can be a problem.
- *Low-cost operation and maintenance.* Due to fewer parts, air tools require fewer repairs and less preventive maintenance. Also, the original cost of air-driven tools is usually less than the equivalent electric type.

The automotive industry was one of the first industries to see the advantages of air-powered tools. Today they are known as the tools of the professional auto technician.

Of all the electric power tools, the MIG welder is used most by the emission technician. It can be used to cut away and weld mufflers, exhaust pipes, and so on.

Some special tools are used by emission and fuel service technicians. These include digital amp-volt-ohmmeters, dwell-tachometers, oscilloscopes, diagnostic testers, and exhaust gas analyzers (Figure 1–53).

REPAIR MANUALS

Some of the most important tools of the trade are repair or service manuals. Publications that should be available to fuel and emission technicians include the following:

FIGURE 1–53 Diagnostic tester

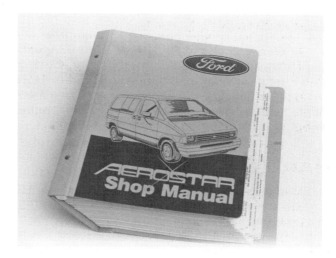

FIGURE 1–54 Typical service manual

- *Auto Manufacturers' Service Manuals.* The main source of repair and specification information for any car, van, or truck is the manufacturer. Service manuals are published each year for every vehicle built (Figure 1–54). The manuals are written in technical language for professional technicians working in the dealerships. Technicians must be able to understand the language, procedures, and specifications. Because of the enormous amount of information, some manufacturers publish more than one manual per year per car model. They can be separated into sections such as chassis, suspension, steering, emission controls, fuel systems, brakes, basic maintenance and tune-up, engine, transmission, body, and so on.

  When complete information with step-by-step testing and repair procedures is desired, nothing can beat the auto manufacturers' repair manuals. They are the most reliable because they cover all repairs, adjustments, specs, detailed diagnostic procedures, and special tools required. They are distributed to technicians in the dealerships who, in turn, are continually notified of all pertinent changes.

  Automobile manufacturers also publish a series of so-called "technician reference books." The publications provide general instructions on all their current vehicles for accomplishing service and repair with their tested, effective techniques.
- *Aftermarket Suppliers' Guides and Catalogs.* Many of the larger manufacturers have excellent guides on the various parts that they

manufacture or supply. They also provide updated service bulletins on their products.

- *General and Specialty Repair Manuals.* These are published by independent companies rather than the manufacturers. However, they pay for and get most of their information from the carmaker. They contain component information, diagnostic steps, repair procedures, and specs for several car makes in one book. Information is usually condensed and is more general in nature, depending on which manual is used. The condensed format allows for more coverage areas in less space and therefore is not always specific. They also can contain several years of models as well as several car makes in one book.

One of the best sources of up-to-date fuel and emission control information is the trade magazine. Most are published monthly.

GENERAL SHOP SAFETY

The most important considerations in any automotive repair shop should be accident prevention and safety. Carelessness and the lack of safety habits cause accidents. Accidents have a far-reaching effect, not only on the victim, but also on the victim's family and society in general. More important, accidents can cause serious injury, temporary or permanent, or even death. Therefore, it is the obligation of all shop employees and the employer to develop a safety program to protect the health and welfare of those involved.

For example, extreme caution should be used while working with any of the components of the fuel system. Gasoline is a very volatile substance. Do not expose it to an open flame, spark, or high heat. Disconnect the negative terminal of the battery before doing any task that will release gasoline from any part of the system. Use containers to catch the gasoline and cloths to wipe up minor spills. Use a flashlight rather than a trouble light or droplight. Gasoline spilled on a hot bulb could cause the bulb to explode and ignite the gasoline. Always keep a Class B fire extinguisher or one capable of fighting type A, B, and C fires (Table 1–1) nearby to deal with problems that may occur. The Class B type is specially intended for use on gasoline fires.

In the following chapters of this book, the text contains special notations labeled **SHOP TALK, CAUTION,** and **WARNING.** Each one is there for a specific purpose. **SHOP TALK** gives added information that will help the technician to complete a particular procedure or make a task easier. **CAU-** **TION** is given to prevent the technician from making an error that could damage the vehicle. **WARNING** reminds the technician to be especially careful of those areas where carelessness can cause personal injury. The following text contains some general **WARNINGs** that should be followed when working in an automobile repair shop on fuel inspection and emission control systems:

- Always wear safety glasses, protective clothing, respirator, or other protective equipment whenever required.
- Use safety stands (Figure 1–55) whenever a procedure requires getting under the vehicle. Never trust jacks alone.
- Be sure that the ignition is always in the OFF position, unless otherwise required by the procedure.
- Set the parking brake when working on the vehicle. If it is an automatic transmission, set it in PARK unless instructed otherwise for a specific service operation. If it is a manual transmission, it should be in REVERSE (engine off) or NEUTRAL (engine on) unless instructed otherwise for a specific service operation.
- Operate the engine only in a well-ventilated area to avoid the danger of carbon monoxide. Most shop exhaust systems carry the exhaust away from a vehicle directly to the outside of the shop.
- Keep clothing away from moving parts when the engine is running, especially the fan and belts.
- To prevent serious burns, avoid contact with hot metal parts such as the radiator, exhaust manifold, tail pipe, catalytic converter, and muffler.

FIGURE 1–55 Use safety stands whenever working under a vehicle.

TABLE 1-1:   GUIDE TO EXTINGUISHER SELECTION

	Class of Fire	Typical Fuel Involved	Type of Extinguisher
Class A Fires (green)	**For Ordinary Combustibles** Put out a class A fire by lowering its temperature or by coating the burning combustibles.	Wood Paper Cloth Rubber Plastics Rubbish Upholstery	Water*[1] Foam* Multipurpose dry chemical[4]
Class B Fires (red)	**For Flammable Liquids** Put out a class B fire by smothering it. Use an extinguisher that gives a blanketing, flame-interrupting effect; cover whole flaming liquid surface.	Gasoline Oil Grease Paint Lighter fluid	Foam* Carbon dixoide[5] Halogenated agent[6] Standard dry chemical[2] Purple K dry chemical[3] Multipurpose dry chemical[4]
Class C Fires (blue)	**For Electrical Equipment** Put out a class C fire by shutting off power as quickly as possible and by always using a nonconducting extinguishing agent to prevent electric shock.	Motors Appliances Wiring Fuse boxes Switchboards	Carbon dioxide[5] Halogenated agent[6] Standard dry chemical[2] Purple K dry chemical[3] Multipurpose dry chemical[4]
Class D Fires (yellow)	**For Combustible Metals** Put out a class D fire of metal chips, turnings, or shavings by smothering or coating with a specially designed extinguishing agent.	Aluminum Magnesium Potassium Sodium Titanium Zirconium	Dry power extinguishers and agents only

*Cartridge-operated water, foam, and soda-acid types of extinguishers are no longer manufactured. These extinguishers should be removed from service when they become due for their next hydrostatic pressure test.

Notes:

(1) Freeze in low temperatures unless treated with antifreeze solution, usually weighs over 20 pounds, and is heavier than any other extinguisher mentioned.

(2) Also called ordinary or regular dry chemical. (sodium bicarbonate)

(3) Has the greatest initial fire-stopping power of the extinguishers mentioned for class B fires. Be sure to clean residue immediately after using the extinguisher so sprayed surfaces will not be damaged. (potassium bicarbonate)

(4) The only extinguishers that fight A, B, and C classes of fires. However, they should not be used on fires in liquefied fat or oil of appreciable depth. Be sure to clean residue immediately after using the extinguisher so sprayed surfaces will not be damaged. (ammonium phosphates)

(5) Use with caution in unventilated, confined spaces.

(6) May cause injury to the operator if the extinguishing agent (a gas) or the gases produced when the agent is applied to a fire is inhaled.

- Do not smoke while working on the vehicle.
- To avoid injury, always remove rings, watches, loose hanging jewelry, and loose clothing before beginning to work on a vehicle. Tie long hair securely behind the head.
- Keep hands and objects clear of the radiator fan blades. Electric cooling fans can start to operate at any time by an increase in underhood temperatures, even though the ignition is in the OFF position. Therefore, care should be taken to ensure that the electric cooling fan is completely disconnected when working under the hood.
- Whenever draining lubricant, extreme caution should be used. Lubricant can be hot enough to cause injury.
- All bolts, nuts, lock rings, and other fastening components mentioned in the manufacturer's service manual are crucial to the safe operation of the car. Failure to use those specific items could cause extensive damage. Manufacturer's torque specifications must be followed.
- Be careful when using sharp or pointed tools that can slip and cause injury. If a tool is to be sharp, make sure it is sharp.

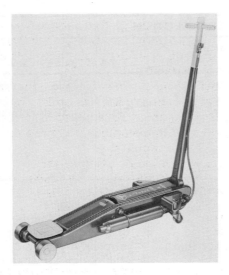

FIGURE 1-56 End lift jack

FIGURE 1-57 Raising a vehicle on a hoist

- Never leave a power tool unattended when it is running. Before leaving, turn off the machine. Anyone passing an unattended machine can be hurt seriously.
- Do not substitue metric wrenches or sockets for standard, or vice versa.
- Store oily rags and all other combustibles in a safe place, such as covered metal containers.
- Remove the ground (B–) cable from the battery to prevent accidental starting of the engine.
- Keep aisles and walkways free of tools, creepers, and any material that might cause a person or fellow worker to trip or stumble.
- Do not attempt repairs that are not fully understood.

Lift Safety

Some undercar work is necessary to raise the car to repair fuel, emission, and exhaust systems. Raising a vehicle on a lift (Figure 1-56) or a hoist (Figure 1-57) requires specific care. For example, it is important to make sure that vehicles equipped with a catalytic converter have enough clearance between the hoist and exhaust system components before driving the vehicle onto the ramps.

Adapters and hoist plates must be positioned correctly on twin post and rail type lifts to prevent damage to the underbody of the vehicle. The tie-rod, rod bracket, and shock absorbers are some of the undercar components that could be damaged if the adpaters and hoist plates are incorrectly placed.

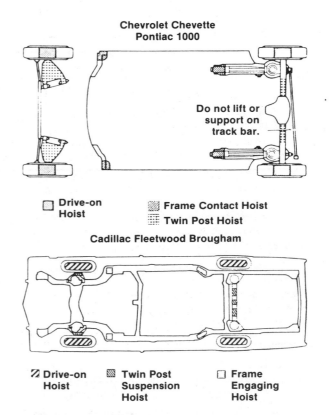

FIGURE 1-58 Lift points for unibody and frame cars

There are specific contact points to use where the weight of the vehicle is evenly supported by the adapters or hoist plates. The correct lifting points can be found in the vehicle's service manual. Figure 1-58 shows typical locations for both frame and unibody cars. These diagrams are for illustration only;

always use the manufacturer's instructions as to the safe lifting procedure.

- Before operating any lift or hoist, carefully read the owner's manual and understand all the operating and maintenance instructions.

WARNING: Always use the rated tonnage of a lift jack for tons specified. If a jack is rated for 2 tons, do not attempt to use it for a job requiring 20 tons. It is dangerous for both the under-car technician and the vehicle.

- Before driving a vehicle over a lift, position the arms and supports to provide unobstructed clearance. Do not hit or run over lift arms, adapters, or axle supports. This could damage the lift or vehicle.
- Load the vehicle on the lift carefully. Check to make sure adpaters or axle supports are in secure contact with the vehicle, according to manual instructions, before raising it to the desired working height. Remember that unsecured loads can be dangerous.
- Make sure the vehicle's doors, hood, and trunk are closed prior to raising the vehicle. Never raise a car with passengers inside.
- Position the lift supports to contact at the vehicle manufacturer's recommended lifting points. Raise the lift until the supports contact the vehicle. Check supports for secure contact with the vehicle and raise the lift to the desired working height.

WARNING: When working under a car, the lift should be raised high enough for the locking device to be engaged.

- After lifting a vehicle to the desired height, be sure to always lower the unit onto mechanical safeties.
- Note that with some vehicles, the removal (or installation) of components can cause a critical shift in the center of gravity and result in raised vehicle instability. Refer to the vehicle manufacturer's service manual for recommended procedures when vehicle components are removed.
- Make sure tool trays, stands, and so forth are removed from under the vehicle. Release locking devices as per instructions before attempting to lower the lift.

- Before removing the vehicle from the lift area, position the arms, adapters, or axle supports to assure that the vehicle or lift will not be damaged.
- Inspect the lift daily. Never operate it if it malfunctions or if it has broken or damaged parts. It should be removed from service and repaired immediately. A lift requires immediate attention if it
  —Jerks or jumps when raised
  —Slowly settles down after being raised
  —Slowly rises when not in use
  —Slowly rises when in use
  —Comes down very slowly
  —Blows oil out of the exhaust line
  —Leaks oil at the packing gland
- Repairs should be made with original equipment parts only.

SHOP RECORDS AND PROCEDURES

Automotive servicing, like any other business, is made smoother when the paperwork is done correctly. The forms used most often are covered here.

- *Repair Order.* Once the job is in the shop, either the service manager or an assistant fills out a repair order. As shown in Figure 1–59, the repair order includes information about the customer, the car (service description and parts replaced), and billing.

  Many states have information disclosure laws to protect consumers from fraud. Written job estimates must be furnished to the customer by the garage. If the cost will be higher than the original written estimate, additional approval must be received by the customer. Also, any replaced part is returned to the customer.
- *Work Order.* When the customer arrives with the vehicle, a service writer notes on a work order any customer complaints or labor instructions. The repair order and vehicle are dispatched to the mechanic who is assigned to do the servicing. The mechanic services the vehicle following the instructions on the work order.
- *Parts Requisition.* To order new parts, the mechanic writes the names of what is needed in the parts requisition section of the repair order. Then the mechanic takes the form to the parts department where the counter at-

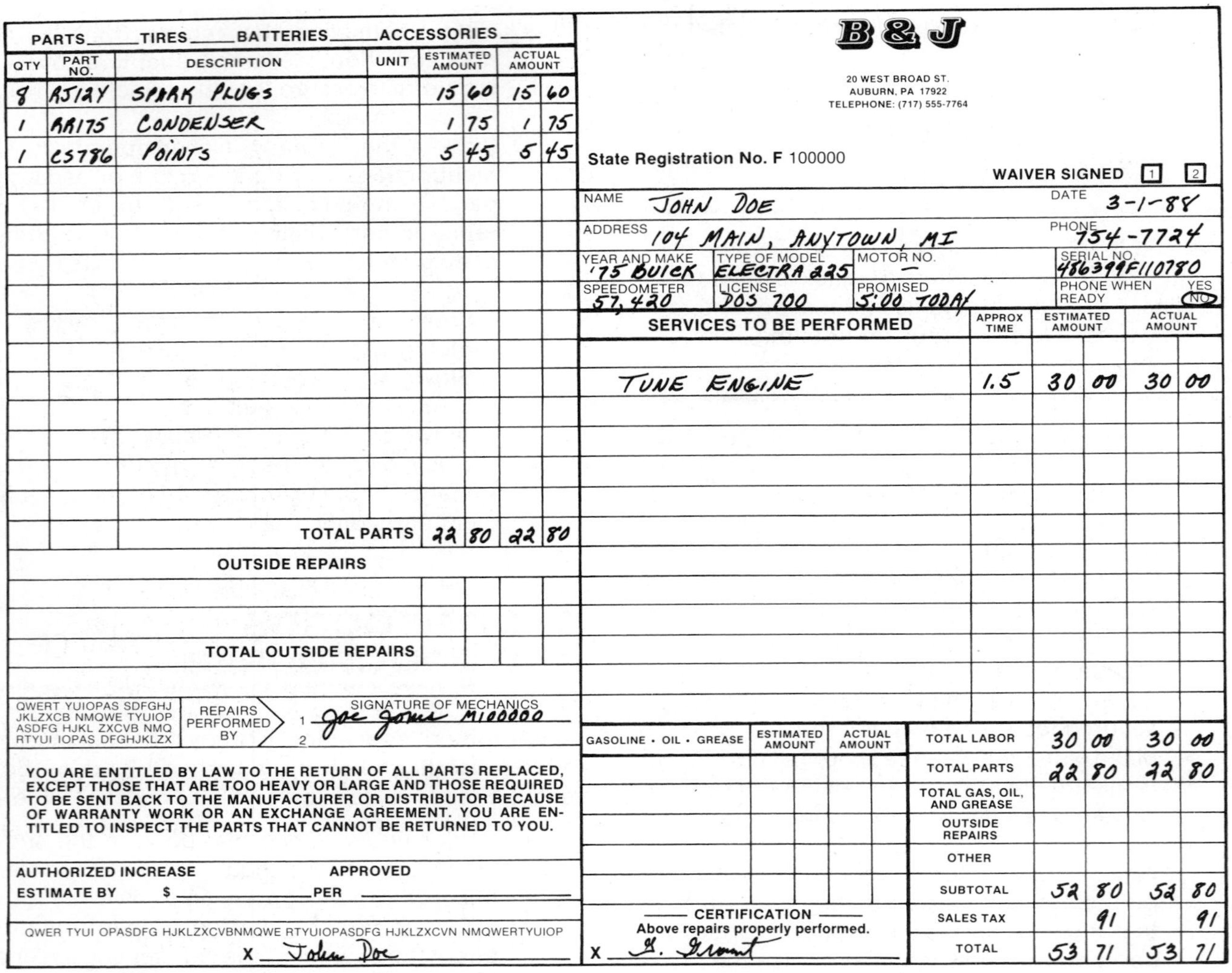

B & J

20 WEST BROAD ST.
AUBURN, PA 17922
TELEPHONE: (717) 555-7764

PARTS_____ TIRES_____ BATTERIES_____ ACCESSORIES_____

QTY	PART NO.	DESCRIPTION	UNIT	ESTIMATED AMOUNT	ACTUAL AMOUNT
8	AJ12Y	SPARK PLUGS		15 60	15 60
1	RR175	CONDENSER		1 75	1 75
1	CS786	POINTS		5 45	5 45
		TOTAL PARTS		22 80	22 80
		OUTSIDE REPAIRS			
		TOTAL OUTSIDE REPAIRS			

State Registration No. F 100000

WAIVER SIGNED ☐1 ☐2

NAME JOHN DOE   DATE 3-1-88
ADDRESS 104 MAIN, ANYTOWN, MI   PHONE 754-7724
YEAR AND MAKE '75 BUICK   TYPE OF MODEL ELECTRA 225   MOTOR NO. —   SERIAL NO. 486399F110780
SPEEDOMETER 57,420   LICENSE DOS 700   PROMISED 5:00 TODAY   PHONE WHEN READY YES (NO)

SERVICES TO BE PERFORMED	APPROX TIME	ESTIMATED AMOUNT	ACTUAL AMOUNT
TUNE ENGINE	1.5	30 00	30 00

QWERT YUIOPAS SDFGHJ JKLZXCB NMQWE TYUIOP ASDFG HJKL ZXCVB NMQ RTYUI IOPAS DFGHJKLZX

REPAIRS PERFORMED BY
1 SIGNATURE OF MECHANICS Joe Jones M100000
2

YOU ARE ENTITLED BY LAW TO THE RETURN OF ALL PARTS REPLACED, EXCEPT THOSE THAT ARE TOO HEAVY OR LARGE AND THOSE REQUIRED TO BE SENT BACK TO THE MANUFACTURER OR DISTRIBUTOR BECAUSE OF WARRANTY WORK OR AN EXCHANGE AGREEMENT. YOU ARE ENTITLED TO INSPECT THE PARTS THAT CANNOT BE RETURNED TO YOU.

AUTHORIZED INCREASE   APPROVED
ESTIMATE BY $ _____ PER _____

QWER TYUI OPASDFG HJKLZXCVBNMQWE RTYUIOPASDFG HJKLZXCVN NMQWERTYUIOP
X John Doe

GASOLINE · OIL · GREASE	ESTIMATED AMOUNT	ACTUAL AMOUNT			
			TOTAL LABOR	30 00	30 00
			TOTAL PARTS	22 80	22 80
			TOTAL GAS, OIL, AND GREASE		
			OUTSIDE REPAIRS		
			OTHER		
			SUBTOTAL	52 80	52 80
			SALES TAX	91	91
			TOTAL	53 71	53 71

——— CERTIFICATION ———
Above repairs properly performed.
X S. Grant

FIGURE 1-59 Example of a repair order

tendant records prices of the parts requested.

Parts managers must fill out forms to keep an adequate number of parts in stock. When the stock gets low, new parts are ordered. Occasionally it is necessary to order special parts. To ensure customer and shop satisfaction, these special orders must always be coordinated between the service and parts department. Parts can be ordered by using standard catalogs or computerized catalogs.

- *Dispatch Sheet.* A dispatch sheet, or work schedule, keeps track of appointments. Whenever a customer calls, the dispatch sheet is consulted so that a suitable service appointment is made. The dispatch sheet is usually posted in the office of the garage or wherever it is convenient for the service manager.

- *Labor Charges.* Once the work on the automobile is completed, the mechanic turns in the repair order to the service manager, or service waiter, who adds the labor charges.

- *Billing.* Billing is the total cost of servicing. It includes all labor charges, parts charges, sales tax, and any other charges. The billing department totals the amount and receives payment from the customer.

ASE CERTIFICATION

Just as doctors, nurses, accountants, electricians, and other professionals are licensed or certi-

FIGURE 1-60 Automobile Technician shoulder patch

FIGURE 1-61 Master Automobile Technician shoulder patch

fied to practice their profession, a mechanic can also be certified. Certification protects the general public and the practitioner or professional. It assures the general public and the prospective employer that certain minimum standards of performance have been met. Many employers now expect their mechanics or technicians to be certified. The certified technician is recognized as a professional by employers, peers, and the public. For this reason, the certified technician usually receives higher pay than the noncertified operator.

Mechanics can get certified in one or more technical areas by taking and passing a mechanic certification test. The National Institute for Automotive Service Excellence (ASE) offers a voluntary certified program that is recommended by the major vehicle manufacturers in the United States. This program has certification tests that fall into eight categories or specialties as follows:

1. Engine repairs
2. Automatic transmissions/transaxle
3. Manual drivetrain and rear axle
4. Suspension and steering
5. Brakes
6. Electrical systems
7. Heating and air-conditioning
8. Engine performance

A technician who passes one examination receives an Automobile Technician shoulder patch (Figure 1-60). A Master Automobile Technician meeting minimum standards in all eight categories receives the shoulder patch shown in Figure 1-61.

To help prepare for the Engine Repairs and the Engine Performance programs, the test questions at the end of each chapter are similar in design and content to those used by the ASE. For further information on the ASE certification program, write: National Institute for Auto Service Excellence, 1920 Association Drive, Reston, Virginia 22091.

REVIEW QUESTIONS

1. In a gasoline-operated engine, the space between the top of the piston and cylinder head is called the _____ .
   a. combustion chamber
   b. crankshaft

c. rotator cuff
d. camshaft

2. The basis of automotive gasoline engine operation is the _____.
   a. crankshaft motion
   b. bore and stroke
   c. four-stroke cycle
   d. compression ratio

3. The fuel system supplies a _____ to the engine cylinders.
   a. vapor lock
   b. combustible mix of gas and air
   c. fuel injector device
   d. mechanical fuel pump

4. A chemically correct fuel mixture is called

   _____.
   a. lean
   b. contaminant
   c. stoichiometric
   d. rich

5. Air/fuel ratios with high concentrations of air are called _____ mixtures.
   a. lean
   b. rich
   c. gasoline
   d. poor

6. Technician A wants to replace a two-year-old PVC valve and filter as a preventative maintenance measure. Technician B recommends replacing only the PCV valve. Who is correct?
   a. Technician A
   b. Technician B
   c. Both A and B
   d. Neither A nor B

7. _____ is formed in the combustion chamber when temperatures rise above 2500 degrees Fahrenheit and nitrogen reacts with oxygen.
   a. carbon monoxide
   b. hydrocarbon
   c. carbon dioxide
   d. nitrous oxide

8. Technician A wants to replace a catalytic converter with a straight piece of pipe. Technician B refuses, saying it is illegal. Who is correct?
   a. Technician A

b. Technician B
c. Both A and B
d. Neither A nor B

9. The most important considerations in an automotive repair shop is _____.
   a. proper ventilation
   b. safety
   c. accident prevention
   d. all of the above

10. ASE certification gives _____.
    a. protection to the practitioner and public
    b. assurance of performance standards
    c. recognition of professionalism
    d. all of the above

11. Which of the following is an essential part of the combustion process?
    a. charging the cylinders with fuel and air
    b. compressing the air/fuel mixture
    c. expelling the burned gasses from the cylinders
    d. all of the above

12. The volume of the cylinder with the piston at BDC compared to the volume of the cylinder with the piston at TDC determines the _____ of the engine.
    a. compression ratio
    b. bore and stroke
    c. displacement
    d. carburetion

13. You are most likely to find a filter made of paper in the _____.
    a. carburetor's fuel inlet
    b. air cleaner
    c. fuel line
    d. all of the above

14. Which of the following is not another name for port fuel injection?
    a. sequential fuel injection
    b. multiport fuel injection
    c. throttle body injection
    d. all of the above

15. Which of the following emissions is not controlled by the emission control system?
    a. sulfur
    b. carbon monoxide
    c. hydrocarbons
    d. oxides of nitrogen

CHAPTER TWO

AUTOMOTIVE FUELS

Objectives

After reading this chapter, you should be able to:
- Identify the services that a good motor fuel must provide.
- Explain how compression ratio and detonation affect a gasoline engine.
- Identify and explain the major factors that affect fuel performance.
- Name the performance characteristics and driving conditions affected by the volatility of gasoline, and describe how each one is affected.
- Explain the differences between leaded and unleaded gasoline.
- Identify commonly used gasoline additives and the function they perform.
- Explain the differences between gasoline and diesel engines.
- Name the characteristics of diesel fuel.
- Explain the principles behind diesel fuel's cetane rating.
- Describe the relationship of carbon residue and sulfur content to diesel engines.
- Explain the use of LP-gas.

Before anyone can fully understand automotive fuel and emission control systems, a knowledge of the performance of petroleum-based material used to power the engine is necessary. Gasoline and diesel fuels are the most commonly used motor fuels in modern vehicles.

Crude oil, as removed from the earth, is a mixture of hydrocarbon compounds ranging from gases to heavy tars and waxes. The crude oil can be refined into products such as lubricating oils, greases, asphalts, kerosene, diesel fuel, gasoline, and natural gas (Figure 2–1). Before its widespread use in the internal combustion engine of automobiles, gasoline was an unwanted byproduct of refining for oils and kerosene.

The liquid components of the hydrocarbons that are blended together in gasoline have a boiling range of approximately +80 to +400 degrees Fahrenheit. On the other hand, conventional diesel fuels are obtained by the distillation of crude oil. These distillates have a boiling range of about +300 to +700 degrees Fahrenheit. Diesel fuels are predominantly straight-run fractions that contain the largest amount of normal paraffins and naphthenes and the least amount of isoparaffins and aromatics. While normal paraffins and naphthenes have superior diesel ignition qualities, they also have the disadvan-

tage of higher pour points than isoparaffins and aromatics.

The automotive technician must remember that a good motor fuel must provide the following:

- Quick starts
- Fast warm-up
- Rapid acceleration
- Smooth performance
- Minimum engine maintenance
- Good mileage under various driving conditions

GASOLINE

Two important factors affect the power and efficiency of a gasoline engine.

1. *Compression Ratio.* The higher the compression ratio, the greater the power output and efficiency. The better the efficiency, the less fuel consumed to produce a given power output. To have a high compression ratio requires an engine of greater structural integrity in pistons, crankshafts, connecting rods, and heads. The 1915 Ford Model T had a low compression

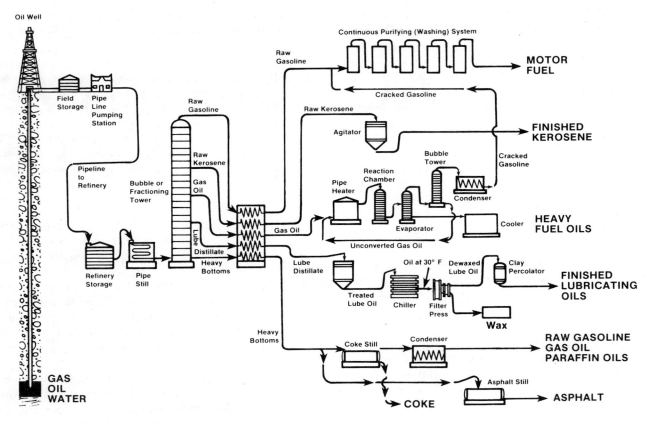

FIGURE 2-1 Flow chart traces crude oil from well to finished product

ratio of 3.6 to 1. As improvements were made to the engines, compression ratios slowly increased. A 1927 Model A was 4.5 to 1; and in the late 1960s it had approached 12.5 to 1 on some optional high-performance engines. Due to the use of low-octane unleaded gasoline in post-1975 models, the compression ratio generally ranges from 8.2:1 to 10.2:0.

2. *Detonation (Abnormal Combustion).* As described in Chapter 1, normal combustion occurs gradually in each cylinder. The flame front (the edge of the burning area) advances smoothly across the combustion chamber until all the air/fuel mixture has been burned (Figure 2–2). Detonation occurs when the flame front fails to reach a pocket of mixture before the temperature

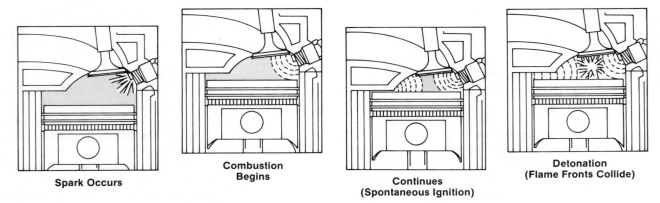

Spark Occurs   Combustion Begins   Continues (Spontaneous Ignition)   Detonation (Flame Fronts Collide)

FIGURE 2-2 Normal combustion and detonation in a spark ignition engine

in that area reaches the point of self-ignition. Normal burning at the start of the combustion cycle raises the temperature and pressure of everything in the cylinder. The last part of the air/fuel mixture will be both heated and pressurized, and the combination of those two factors can raise it to the self-ignition point. At that moment, the remaining mixture burns almost instantaneously. The two flame fronts create a destructive pressure wave between them that can destroy cylinder head gaskets, break piston rings, and burn pistons and exhaust valves. When detonation occurs, a hammering, pinging, or knocking sound is heard. But, when the engine is operating at high speed, these sounds cannot be heard because of motor and road noise.

FUEL PERFORMANCE

Many of the performance characteristics of gasoline can be controlled in refining and blending to provide proper engine function and vehicle driveability. The major factors affecting fuel performance are:

- Antiknock Quality
- Volatility
- Sulfur content
- Deposit control

Antiknock Quality

An octane number or rating was developed by the petroleum industry so the antiknock quality of a gasoline could be rated. It is a measure of burning quality; the tendency not to produce knock in an engine. The higher the octane number, the lesser the tendency to knock. By itself, antiknock rating has nothing to do with the fuel economy or efficiency of an engine.

Two commonly used methods for determining the octane number of motor gasoline are the motor octane number (MON) method and the research octane number (RON) method. Both use a laboratory single-cylinder engine equipped with a variable head and knock meter to indicate knock intensity. The test sample is used as fuel, and the engine compression ratio and air/fuel mixture are adjusted to develop a specified knock intensity. There are two primary standard reference fuels, isooctane and heptane. Isooctane will not knock in an engine, but is not used in gasoline because of its expense. Heptane knocks severely in an engine. Isooctane has an octane number of 100; heptane has an octane number of zero.

A fuel of unknown octane value is run in the special test engine and the severity of knock is measured. Then various proportions of heptane and isooctane are run in the test engine to duplicate the severity of the knock of the fuel being tested. When the knock caused by the heptane/isooctane mix is identical to the test fuel, the octane number is established by the percentage of isooctane in the mixture. For example, if 85 percent isooctane and 15 percent heptane produce the same severity of knock as the fuel in question, the fuel is assigned an octane number of 85.

- *Lean Fuel Mixture.* A lean mixture burns slower than a rich mixture. The heat of combustion is higher, which promotes the tendency for unburned fuel in front of the spark-ignited flame to detonate.
- *Ignition Timing Overadvance.* Advancing the ignition timing induces knock. Slowing ignition timing suppresses knock because lower cylinder pressures result when the piston is on its way down as combustion nears completion.
- *Compression Ratio.* Compression ratio affects knock because cylinder pressures are increased with the increase in compression ratio. The higher the compression ratio, the more sensitive an engine is to knock.
- *Valve Timing.* The timing of intake and exhaust valves affects the cylinder pressure and knock. A valve timing that fills the cylinder with more air/fuel mixture promotes higher cylinder pressures, increasing the chances for detonation.
- *Turbocharging.* Turbocharging or supercharging forces additional fuel and air into the cylinder (see Chapter 3). This induces higher cylinder pressures and promotes knock. It also raises the temperature of the air coming in.
- *Coolant Temperature.* Hot spots in the cylinder or combustion chamber because of inefficient cooling or a damaged cooling system raise combustion chamber temperatures and promote knock.
- *Cylinder-to-Cylinder Distribution.* If an engine has poor distribution of the air/fuel mixture from cylinder to cylinder, the leaner cylinders could promote knock. The distributor spark advance curve must be set for the advance required for the leanest.
- *Excessive Carbon Deposits.* The accumulation of carbon deposits on the piston, valves, and combustion chamber causes poor heat transfer from the combustion chamber. Car-

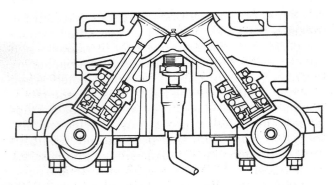

FIGURE 2-3 Cross section of a hemispherical combustion chamber

bon accumulation also artificially increases the compression ratio. Both conditions cause knock.

- *Air Inlet Temperature.* The higher the air temperature when it enters the cylinder, the greater the tendency to knock. The higher air temperature increases the temperature of combustion.

- *Combustion Chamber Shape.* The optimum combustion chamber shape for reduced knocking is hemispherical with a spark plug located in the center (Figure 2–3). This hemi-head allows for faster combustion, allowing less time for detonation to occur ahead of the flame front.

- *Octane Number.* Recommendations about the best octane fuel to use in any given engine can only be approximate. Many factors determine octane ratings of fuel and octane

requirements of engines. It is important to keep in mind that little is gained or lost by using gasoline of higher octane quality than required by the engine under a complete range of operating conditions. Only when an engine is designed and adjusted to take advantage of the higher octane gasoline can the fuel value be obtained. Most modern engines are designed to operate efficiently with regular grade gasoline and do not require high-octane premium grade (Table 2-1). Those designs will provide better fuel economy and performance when converted to a higher-octane gasoline.

Volatility

Volatility is the ability of gasoline to evaporate so its vapor will adequately mix with air for combustion. Only vaporized fuel will support combustion. To insure complete combustion, complete vaporization must occur.

The volatility of gasoline affects the following performance characteristics or driving conditions:

- *Cold Starting and Warm-Up.* Vaporization characteristics affect cold starting and warm-up. A fuel can cause hard starting, hesitation, and stumbling during warm-up if it does not vaporize readily. A fuel that vaporizes too easily in hot weather can form vapor bubbles in the fuel line and fuel pump resulting in vapor lock or loss of performance. The volatility of a fuel must be seasonally blended at the refinery to insure proper operation of the engine.

	EXXON 2000 (UNLEADED)	EXXON (LEADED)	EXXON EXTRA UNLEADED
TABLE 2-1: PROPERTIES OF EXXON® GASOLINES			
Research octane number[a]	93	94	97
Motor octane number[a]	84	84	87
TEL content	Nil[b]	Approx. 2 grams/gal	Nil[b]
Approximate compression ratio served	8.2:1	8.2:1	8.2:1 and higher
Carburetor icing control		Adequate	
Volatility adjustments		Continually with season for each market area	
Detergent		For required carburetor deposit control	

[a] Lower for higher-altitude market areas.
[b] Trace lead contents are due to pickup in the distribution system.

- *Temperature.* Because a highly volatile fuel vaporizes at a lower temperature than a less volatile fuel, winter grade gasoline is more volatile than summer grade gasoline. Winter grade fuels, however, give poorer fuel economy than low volatility fuels. Seasonal blending provides a more volatile fuel for the northern states during the winter months than a fuel sold in southern states.
- *Altitude.* Gasoline vaporizes more easily at high altitudes, so volatility is controlled in blending according to the elevation of the place where fuel is sold.
- *Carburetor Icing Protection.* Although carburetor icing is not as common in modern engines as in those of the past, it is most likely to occur at ambient temperatures between 28 to 55 degrees Fahrenheit when the relative humidity rises above 65 percent. The humid air enters the carburetor and is mixed with drops of fuel. When the fuel evaporates, it removes heat from the air and surrounding metal parts. When this occurs, the throttle temperature is rapidly lowered to below 32 degrees Fahrenheit (if the ambient temperature is within the range indicated), and condensing water vapor forms ice. The ice will cause the engine to stall if it is idling during this phase. Gasoline with higher mid-volatility (that is, a higher percentage of evaporation at 212 degrees Fahrenheit) creates greater evaporative cooling. Therefore, the percentage of gasoline evaporated at 212 degrees Fahrenheit is the significant factor in minimizing carburetor icing.
- *Crankcase Oil Dilution.* A fuel must vaporize well to prevent diluting the crankcase oil with liquid fuel. If parts of the gasoline do not vaporize, droplets of liquid break down the oil film on the cylinder wall, causing scuffing or scoring. The liquid eventually enters the crankcase oil and results in the formation of sludge, gum, and varnish accumulation as well as decreasing the lubrication properties of the oil.
- *Driveability.* Poor vaporization can also affect the distribution of fuel from cylinder to cylinder since vaporized fuel will travel farther and faster in the manifold. This results in driveability problems and increased exhaust emissions.

Sulfur Content

Gasoline can contain some of the sulfur present in the crude oil. Sulfur content is reduced at the refinery to limit the amount of corrosion of the engine and exhaust system.

When the hydrogen in the hydrocarbon of the fuel is burned with air, one of the products of combustion is water. Water leaves the combustion chamber as steam but can condense back to water when passing through a cool exhaust system. When the engine is shut off and cools, steam condenses back to a liquid and forms water droplets in the exhaust system and on engine parts. Steam present in crankcase blowby also condenses to water.

When the sulfur in the fuel is burned, it combines with oxygen to form sulfur dioxide. This sulfur dioxide can then combine with water to form highly corrosive sulfuric acid. This corrosion is the leading cause of exhaust valve pitting and exhaust system deterioration. With catalysts, the sulfur dioxide can cause the obnoxious odor of rotten eggs during vehicle warm-up. To reduce corrosion caused by sulfuric acid, the sulfur content in gasoline is limited to less than 0.01 percent.

Deposit Control

Several additives are put in gasoline to control harmful deposits. These are described later in this chapter.

BASIC FUELS

Two basic motor fuels are used in modern vehicles: leaded and unleaded gasolines.

Leaded Gasoline

The two most commonly used additives to regular and premium lead fuels are tetraethyl lead (TEL) and tetramethyl lead (TML). These additives are used to improve the antiknock quality. For example, if 3 grams of TEL are added per gallon of gasoline, the octane rating will increase by up to 15 numbers. Two factors determine the amount of TEL used: the response of the gasoline base stock to TEL and the economics of obtaining octane by refining processes versus the cost of TEL. Currently, there are federal restrictions that limit TEL or TML content to 3.5 grams of lead per gallon. Scavengers are often used with the TEL and TML additives to promote the removal of lead salts formed in the combustion chamber after combustion.

Unleaded Gasoline

Unleaded gasolines are the key to emission control. Since 1975, as previously stated, most domestic cars have used catalytic converters to treat the exhaust of the engine. Because of the deactivating or

FIGURE 2-4 *Unleaded fuel only* labels

poisoning effect lead has on the catalyst, these gasolines are limited to a lead content of 0.06 grams per gallon. Since TEL or TML is not added to unleaded gasolines, the required octane number is obtained by blending compounds of the required octane quality. Methylcylopentadienyl manganese tricarbonyl (MMT) is a catalyst-compatible octane improver. It may also be used to further improve the octane number of unleaded gasolines.

Vehicles with catalytic converters are labeled at both the fuel gauge and fuel filler *unleaded fuel only* (Figure 2-4). The fuel filler neck also has a restrictor plate (Figure 2-5). The smaller hole in the plate pre-

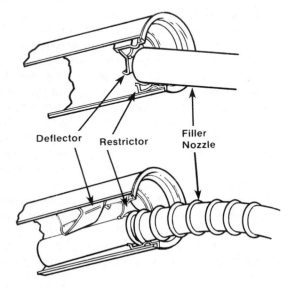

Deflector   Restrictor   Filler Nozzle

FIGURE 2-5 Fuel filler restrictor used on cars with catalytic converters to prevent the use of leaded gasoline

vents using the larger nozzle on all fuel pumps delivering leaded fuel.

OTHER GASOLINE ADDITIVES

At one time, all a gasoline-producing company had to do to produce their product was pump the crude from the ground, run it through the refinery to separate the fractions, dump in a couple grams of lead per gallon, and deliver the finished product to the service station. Of course, automobiles were much simpler then and what they burned was not very critical. As long as gasoline vaporized easily and did not cause the low compression engines to knock, everything was fine.

Times have changed. Today, refiners are under constant pressure to ensure that their product passes a series of rigorous tests for seasonal volatility, minimum octane, existent gum and oxidation stability as well as add the correct de-icers and detergents to make the product competitive in the price-sensitive marketplace.

Gasoline additives—primarily used in unleaded types—have different properties and a variety of uses.

Anti-Icing or De-Icer

Isopropyl alcohol is seasonally added to gasoline as an anti-icing agent to prevent gas line freeze in cold climates.

Gum or Oxidation Inhibitors

During storage, harmful gum deposits can form due to the reaction of some gasoline components with each other and with oxygen. Oxidation inhibitors are added to promote gasoline stability. They help control gum, deposit formation, and staleness. Stale gasoline becomes cloudy and smells like paint thinner.

Gasoline must not leave excessive deposits, gums, or varnishes on engine parts exposed to fuel before and after combustion. An undesirable residue is gum, a sticky substance, which can harden into varnish as it absorbs heat and oxygen. Gum content is also influenced by the age of the gasoline and its exposure to oxygen and certain metals such as copper. If gasoline is allowed to evaporate, the residue left can form gum and varnish.

Gasoline must have as little gum residue as possible to prevent gum formation in the intake manifold, carburetor, and on intake valve stems. Some gasoline contains aromatic amines and phenols to prevent formation of gum and varnish.

Metal Deactivators and Rust Inhibitors

These additives are used to inhibit reactions between the fuel and the metals in the fuel system that can form abrasive and filter-plugging substances.

Detergents

The use of detergent additives in gasoline has been the subject of some confusion among the motoring public. Detergent additives are designed to do only what their name implies—clean certain critical components inside the engine. They do not affect octane.

Aromatics

In general, the aromatics are defined as those hydrocarbon-based compounds that contain the characteristic benzene ring of molecules. In gasolines, the aromatics consist of the benzene, toluene, and xylene (BTX) group. Although all gasolines contain some level of aromatics in combination with the long-chain aliphatics (olefins and paraffins), the BTX group is used chiefly as an octane enhancer.

Aromatics must be carefully blended due to their inherent volatility—too much and vapor pressure rises sufficiently to cause vapor lock and driveability problems. Between 25 and 40 percent of all gasolines on the market today contain some level of additional aromatic compounds, according to industry experts.

Since the aromatics do not contain any oxygen, they can be blended at the refinery or at the terminal. Aromatics will increase octane by two to three octane points. They cause some swelling of fuel system elastomers in older cars, but should not affect newer vehicles.

Ethanol

By far the most widely used gasoline additive today is ethanol, or grain alcohol. Ethanol's value as an octane enhancer becomes more apparent when considered in the context of the government-mandated phasedown of tetraethyl lead. Blending 10 percent ethanol into gasoline is seen as a comparatively inexpensive octane booster that results in an increase of 2.5 to 3 road octane points.

One of the biggest arguments against the use of ethanol in gasoline is that it increases gasoline volatility. The addition of 10 percent ethanol to gasoline results in an increase of 4.5 to 5 percent, or about 0.5 psi, in final reid vapor pressure (RVP).

The level of RVP change from blending ethanol depends entirely on the volatility of the base gasoline. In other words, if the base gasoline had an unusually high RVP to start with, the addition of ethanol could contribute to vapor lock or driveability problems during unseasonably warm weather. For gasoline with a normal volatility for the geographic region and the season, a 10 percent ethanol blend should not contribute to any driveability problems.

In addition to octane enhancement, some of the other benefits of ethanol blending include the following:

- Keeping the carburetor or fuel injectors clean due to detergent additive packages found in most of the ethanol marketed
- Inhibiting fuel system and injection corrosion due to additive packages
- Decreasing carbon monoxide emissions at the tail pipe due to the higher oxygen content of blended fuel

While ethanol blends help reduce CO emissions, HC and NOx emissions are higher where ethanol blends are marketed. It should be noted that the increase in HC levels can be attributed more to evaporate (refueling and fuel system) emissions than tail pipe emissions.

It should also be noted that running an ethanol blend in older, carbureted vehicles will cause the car to run leaner than on straight gasoline. Unless the car is already set to run on the lean side, the driver should not notice any difference. On computer-controlled vehicles, the on-board system should offset the higher oxygen content of the fuel whenever operating in closed-loop mode.

Ethanol blends can also have an effect on fuel economy. This is due to the lower energy content (measured in Btu's) of ethanol. However, the decrease in fuel economy will not likely exceed 4.5 percent on average, which is virtually unnoticeable to most drivers. Remember, ethanol-blended motor fuels are not, according to ethanol industry spokesmen, to be referred to as "gasohol." That term stopped being used in the late 1970s because it conveys a negative image of what is intended as a modern octane enhancer and fuel extender.

When ethanol first appeared on the market, problems associated with its use often focused on its effect on plastic parts within the carburetor—particularly accelerator pumps—or fuel lines. Its use resulted in swelling, loss of elasticity, and reduced tensile strength of the material. Within a few years, however, automakers were able to identify those parts that were susceptible to the effects of ethanol blends and substitute more resistant compounds.

Methanol

Methanol is the lightest and simplest of the alcohols and is also known as wood alcohol. It can be distilled from coal, but most of what is used today is derived from natural gas.

In the early 1980s, a few companies began selling a blend of 5 percent methanol in gasoline in the Northeast and Midwest. This showed that methanol plus anticorrosion co-solvents could compete in the motor fuel market as an octane enhancer. Today, both GM and Ford have working prototypes on the road that can burn any combination of fuel from nearly neat methanol (85 percent methanol with 15 percent gasoline) to straight gasoline. The key is a fuel line sensor that detects methanol concentration and adjusts spark timing and air/fuel mixture accordingly. The federal government is pushing for methanol acceptance through industry incentives and fleet-sponsored programs.

For now, however, many automakers continue to warn motorists about using a fuel that contains more than 10 percent methanol and co-solvents by volume. Methanol is recognized as being far more corrosive to fuel system components than ethanol, and it is this corrosion that has automakers concerned.

As a result, many motorists appear confused when considering an aftermarket fuel additive that contains methanol. Generally, these are sold in 10-ounce bottles for the specific purpose of keeping the fuel system free of ice during winter cold snaps. At least one additive manufacturer has issued a statement assuring consumers its gas line de-icer will not lead to fuel system corrosion nor will it affect their new car warranties.

MTBE

Methyl tertiary butyl ether (MTBE)—a member of the BTX group—has become very popular in the past few years as an octance enhancer and supply extender because of excellent compatibility with gasoline.

For the most part, MTBE is manufactured from a forced chemical reaction of methanol and isobutylene. Current U.S. EPA restrictions on oxygenates limit MTBE in unleaded gasoline to 11 percent of volume. At that level, it increases pump octane (RM/2) by 2.5 points. However, it is usually found in concentrations of 7 to 8 percent of volume.

MTBE has found favor among refiner/suppliers and independent marketers alike because it is not nearly as sensitive to moisture as the other oxygenates, has virtually no effect on fuel volatility, and can be shipped through product pipelines whereas ethanol and methanol blends cannot. As a result, ethanol and methanol must be blended at the terminal or loading rack, while MTBE blending can be closely controlled by the refiner.

MTBE's prime asset in today's market is that it increases octane while reducing carbon monoxide emissions at the tail pipe and does it at a cost that makes it very attractive to gasoline marketers across the country. MTBE has also become the oxygenate additive of choice for those marketing fuel in the Denver, Colo., area where a first-of-its-kind state clean air program requires gasoline with 1.5 percent oxygen content by weight.

Possibly the only major drawback to MTBE is its effect on fuel system elastomers and plastics. Studies have shown that gasoline containing MTBE (like those containing methanol, ethanol, or higher levels of aromatics) can lead to some swelling of elastomers but, in general, most cars manufactured since 1980 should not be affected. The same applies to plastic seals and connectors on the fuel handling systems at most service stations. Although oxygenated blends and high aromatic gasolines caused some problems with these materials when they first hit the market, manufacturers were quick to respond with new compounds that resist the effects of these additives.

DIESEL FUEL

Gasoline and diesel engines are similar in basic construction. The main difference is the fuel system and the means of initiating combustion. The two types are different in combustion; what is desirable in a diesel is totally undesirable in a gasoline engine.

With a gasoline engine, vaporized fuel and air are mixed before going into the cylinder for combustion. Combustion is started by a spark ignition at the spark plug.

The diesel engine brings air into the cylinder, and it is compressed to increase its temperature. This heat of compression is significantly higher than in a gasoline engine because of the high compression ratios commonly used on diesels. Fuel is then injected under high pressure into the heated air of the cylinder. The vaporized fuel ignites because of the high temperature of the compressed air. Diesel combustion is started by the same conditions causing knock in a gasoline engine.

Diesel engines are called compression ignition (CI); gasoline engines are called spark ignition (SI). In the CI engine, knock is desirable. In contrast, in an SI engine, knock is undesirable and leads to destruction of the engine. Even though both engines are

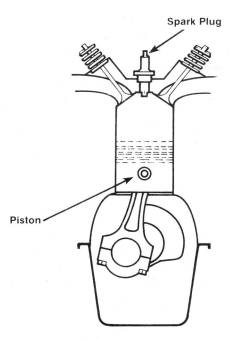

FIGURE 2-6   Spark ignition

constructed essentially the same, this difference in the effect of knock is not contradictory. Knock in an SI engine occurs at the end of combustion. Knock in a CI engine occurs at the beginning of combustion. This is the key difference.

In an SI engine (Figure 2-6), combustion is completed while the piston is at the upper end of travel. This means the volume of the mixture stays about the same during most of the burning process. When the piston moves down and the volume increases, there is little additional combustion to maintain pressure.

In a CI engine (Figure 2-7), there is continuous combustion during most of the power stroke. The expansion of gases increases at the same rate the volume of the cylinder increases as the piston is forced down. The pressure from combustion remains approximately constant throughout the power stroke.

Diesels operate efficiently because of the injection and autoignition of the fuel. More engineering has been devoted to control of injection and autoignition than any other factor in a diesel engine.

COMPRESSION RATIOS

CI engines have compression ratios ranging from 18:1 to 23:1. The volume of air drawn into the cylinder is compressed to a small fraction depending on the ratio. When air is compressed its temperature increases. In CI engines the heat of compression can range from 1100 to 1500 degrees Fahren-

heit. This high temperature in the cylinder ignites the diesel fuel when it is injected.

INJECTION TIME

With this elevated temperature, the injection system must be able to inject the fuel into the cylinder within 0.001 to 0.006 seconds. The liquid fuel sprayed into the cylinder quickly vaporizes, helped by the intense heat.

At the beginning of the injection cycle, the fuel instantly vaporizes then autoignites, resulting in audible knock. With this combustion started, fuel injection continues to carry on combustion as the piston travels down. At the beginning of combustion, temperatures of 3500 degrees Fahrenheit are reached. This further vaporizes and burns the injected fuel. The shape of the fuel spray, turbulence in the combustion chamber, beginning and duration of injection, and the chemical properties of the diesel fuel all affect power output.

ADVANTAGES AND DISADVANTAGES OF DIESEL ENGINES

The basic advantages of the diesel engine are low fuel consumption (greater thermal efficiency), reduced fire hazard, and lower emission levels.

- *Low Fuel Consumption.* The three primary factors for the diesel engine's lower fuel consumption are air/fuel ratio, compression ra-

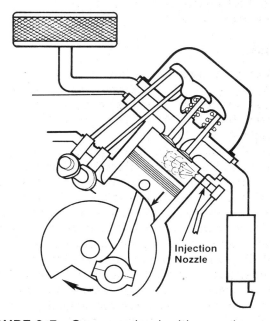

FIGURE 2-7   Compression ignition engine

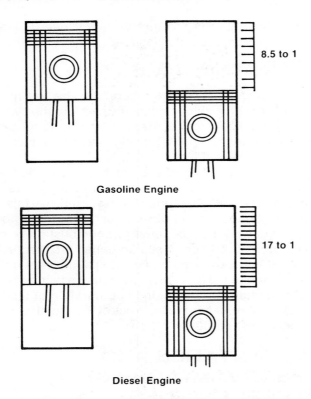

Gasoline Engine

8.5 to 1

17 to 1

Diesel Engine

FIGURE 2-8 The diesel engine operates on a much higher compression ratio.

tio (Figure 2-8), and low pumping losses. Figure 2-9 shows a comparison of the consumption of various fuels.

• *Reduced Fire Hazard.* While not as great a fire hazard as gasoline, there is no reason to handle diesel fuel any less carefully than gasoline.

• *Lower Emission Levels.* The high air/fuel ratio lowers HC and CO emissions (Figure 2-10). The diesel also produces less NO_x.

The following disadvantages of the diesel engine can be attributed to the diesel's characteristics or to the subjective opinion of the owner:

• *Diesel Engine Construction Costs Are Higher.* Because the diesel puts more stress on its compounds, the engine must be constructed from special materials. The quality of the material must be exact, and the parts must be fitted together with very little tolerance for error. This means assembly costs are higher.

• *Different Maintenance Procedures.* Because the diesel engine's design, service, and maintenance procedures are different, and the technician must be more precise in making repairs.

• *Cold Weather Starting.* Diesels use the heat of compression to ignite the fuel. The colder the air temperature, the harder it is to build up enough heat for ignition. To aid cold weather starting, manufacturers have added special starting help packages.

• *Engine Noise.* Diesel engines produce a knock, particularly at idle, that is very noticeable.

• *Exhaust Smoke and Odor.* Anyone who has followed a diesel, particularly one with a malfunction, knows that the diesel produces smoke and an odor all its own.

DIESEL FUEL CHARACTERISTICS

Diesel fuel, like gasoline, is made from petroleum. However, at the refinery, the petroleum is sepa-

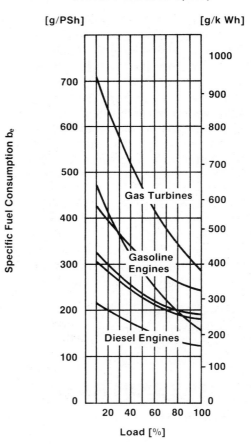

Fuel Consumption
(Compared with Load at
Constant Rotational Speed)

FIGURE 2-9 Compared to the gas turbine and gasoline engine, the diesel engine has the lowest fuel consumption.

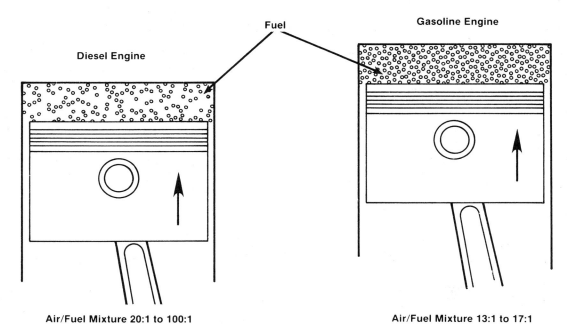

Air/Fuel Mixture 20:1 to 100:1 Air/Fuel Mixture 13:1 to 17:1

FIGURE 2-10 Reason for low HC and CO emission

rated into three major components: gasoline, middle distillates, and all remaining substances. Diesel fuel comes from the middle distillate group, which has properties and characteristics different from gasoline.

Wax Appearance Point and Pour Point

Temperature affects diesel fuel more than it affects gasoline. This is because diesel fuels contain paraffin, a wax substance common among middle distillate fuels. As temperatures drop past a certain point, wax crystals begin to form in the fuel. The point where the wax crystals appear is the wax appearance point (WAP) or cloud point. WAP can change as a result of the origin of the crude oil and the quality of the fuel. The better the quality, the lower the WAP. As temperatures drop, the wax crystals grow larger and restrict the flow of fuel through the filters and lines. Eventually, the fuel, which may still be liquid, stops flowing because the wax crystals plug a filter or line. As the temperature continues to drop, the fuel reaches a point where it solidifies and no longer flows. This is called the *pour point*. In cold climates it is recommended that a low-temperature pour point fuel be used.

Specific Gravity

The specific gravity of a liquid is a measurement of the liquid's weight compared to water. Water is

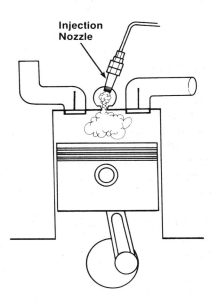

FIGURE 2-11 Combustion is not spread evenly through the combustion chamber.

assigned a value of 1. Diesel fuel is lighter than water but heavier than gasoline and can change if it is mixed with other fuels. The specific gravity of diesel fuel is important to engine operation. The fuel must be heavy enough to achieve adequate penetration into the combustion chamber. If the specific gravity is too low, all the fuel immediately burns upon entering the combustion chamber. This puts all the force of combustion on one small area of the piston instead of equal force across the dome (Figure 2-11).

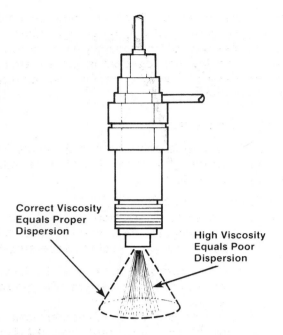

FIGURE 2-12 Fuel viscosity affects spray pattern.

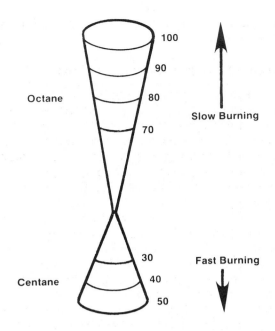

FIGURE 2-13 The octane scale is the opposite of the cetane scale.

As a result, performance suffers, engine noise increases, and the piston could eventually be damaged.

Viscosity

Viscosity is a measure of a fluid's resistance to flow. Low viscosity fluids flow easily; high viscosity fluids do not. The viscosity of diesel fuel must be controlled when produced at the refinery. Like most fluids, diesel fuel viscosity varies with temperature.

The viscosity of diesel fuel directly affects the spray pattern of the fuel into the combustion chamber and the fuel system components. Fuel with a high viscosity produces large droplets that are hard to burn. Fuel with a low viscosity sprays in a fine, easily burned mist (Figure 2-12). If the viscosity is too low, it does not adequately lubricate and cool the injection pump and nozzles.

Volatility

Volatility, as stated earlier, is the ability of a liquid to change into a vapor. Gasoline is extremely volatile compared to diesel fuel. For instance, if diesel fuel and gasoline are exposed to the atmosphere at room temperature, the gasoline vaporizes and the diesel fuel does not.

Flash Point

Flash point is the lowest temperature at which the fuel burns when ignited by an external source.

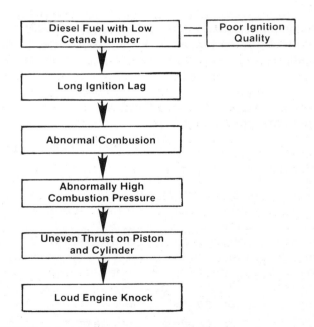

FIGURE 2-14 Note the effect of using the wrong cetane-rated fuel

The flash point has little bearing on engine performance, but is important in fuel storage safety. (The temperature at which the flash point occurs is regulated.) If the flash point of diesel fuel were lower than specified, it would have the right combination of air and fumes that would ignite too easily, making the handling of it hazardous. Gasoline evaporates at a very low temperature, filling the tank with fumes that are potentially explosive.

Cetane Number or Rating

Diesel fuel's ignition quality is measured by the cetane rating. Much like the octane number, the cetane number is measured in a single-cylinder test engine with a variable compression ratio. The diesel fuel to be tested is compared to cetane, a colorless, liquid hydrocarbon that has excellent ignition qualities. Cetane is rated at 100. The higher the cetane number, the shorter the ignition lag time (delay time) from the point the fuel enters the combustion chamber until it ignites. The exact rating is determined by mixing the cetane with a chemical called methylnapthalene, which is rated at zero because it does not ignite. The percentage of cetane mixed with methylnapthalene that produces a similar ignition quality to the fuel being tested is the cetane number rating. Ignition quality and flash point should not be confused. Flash point is the lowest temperature at which the fuel burns when ignited by an external source.

In those fuels that are readily available, the cetane number ranges from 40 to 55 with values of 40 to 50 being most common. These cetane values are satisfactory for medium-speed engines whose rated speeds are from 500 to 1200 rpm and for high-speed engines rated over 1200 rpm. Low-speed engines related below 500 rpm can use fuels in the above 30 cetane number range. The cetane number improves with the addition of certain compounds such as ethyl nitrate, acetone peroxide, and amyl nitrate. Amyl nitrate is commercially available for this purpose.

As previously mentioned, the octane number indicates the resistance of a fuel to self-ignite (knock). Premium gasoline has poor ignition quality because it burns slower than regular gasoline and has more resistance to preignition and detonation. The higher an octane number, the more resistance a fuel has to knocking. Diesel fuel cetane ratings are the opposite of gasoline octane ratings (Figure 2–13). It is important that the cetane rating is correct for the vehicle (Figure 2–14).

Carbon Residue

Carbon residue is the material left in the combustion chamber after burning. It is found not only in diesel engines, but also in other engines that burn hydrocarbon fuel.

The amount of carbon residue left by diesel fuel depends on the quality and the volatility of that fuel. Fuel that has a low volatility is much more prone to leaving carbon residue. The small, high-speed diesels found in automobiles require a high-quality fuel because they cannot tolerate excessive carbon deposits. Large, low-speed industrial diesels are relatively unaffected by carbon deposits and can run on low-quality fuel.

Sulfur Content

Sulfur content is common in fuels made from low-quality crude oil. Sulfur in diesel fuel can cause the following problems:

- Chamber deposits
- Exhaust system corrosion
- Wear on pistons, rings, and cylinders (particularly at low water-jacket temperature)

There are two factors that determine the level of sulfur tolerance by an engine: the type of engine and the type of service.

A fuel sulfur content below 0.4 percent is generally considered low; sulfur content above 0.4 is considered medium or high. Summer grade diesel fuels that are commercialy available commonly run in the 0.2 to 0.5 percent sulfur range. Winter grades usually contain less than 0.2 percent sulfur. Heavy fuels that have sulfur contents up to 1.25 percent or even higher are used in some medium-speed engines in stationary service.

Bacteria Content

Diesel fuel is attacked by various fungi and bacteria. They ingest the diesel fuel as food, changing it to their waste products—a slimy, gelatin-type growth. This growth not only plugs the fuel system but also produces an acid that is corrosive to fuel system components. Because the fuel may contain harmful organisms, any wound exposed to diesel fuel should be cleaned immediately. Fungicides and bactericides, which prevent their formation and growth, are available.

Water and Dirt Content

Water in diesel fuel is a major problem because water and diesel fuel readily mix. Careless storage and distribution of diesel fuel invites problems. Diesel fuel that appears cloudy often contains water. Some of the problems that water causes include

- Corrosion of the fuel system. This can cause the fuel filter to plug with rust particles.
- Icing of the fuel system. Wherever the water collects and the temperature is low enough, ice forms, causing severe damage to the fuel system components.

- Inadequate lubrication of the injection pump and nozzles. Water does not have good lubricating qualities.
- Bacteria growth in diesel fuel.

Most diesel engines have a water separator to catch the moisture before it reaches the injection system. Some water separators are integral with the fuel filter (see Chapter 8). To prevent diesel fuel line filters and injector filters from becoming plugged with dirt and emulsions, fuel tank caps, hoses, and nozzles must be kept clean, and water must be kept to a minimum.

DIESEL FUEL STANDARD

Minimum quality standards for diesel fuel grades have been set by the American Society for Testing Materials (ASTM) as a guide for engine operators. As pointed out in Table 2-2, there are three grades of diesel fuel for automotive use: 1-D, 2-D, and 4-D. At one time there was a grade 3-D, but it has been discontinued.

Grade 1-D is a kerosene-type fuel that has a lower viscosity, lower wax content, and lower Btu per gallon than grade 2-D. It is also more volatile than 2-D. Grade 2-D is the fuel recommended for automotive and some industrial applications. Grade 4-D is a fuel for low- and medium-speed engines.

CAUTION: Heating fuel, which is similar to grade 2-D fuel, should not be used in automotive applications. Heating fuel does not meet the strict standards or have the needed additives for automotive use.

In cold climates it is often necessary to run on a blended fuel. A blended fuel reduces the WAP and pour point, allowing the fuel to flow at low temperatures. Typically, grade 1-D fuel is used to lower the WAP and pour point of grade 2-D fuel. Each manufacturer has specific instructions on what blend should be used at certain temperatures. Usually, a 10 percent increase of grade 1-D to grade 2-D lowers the WAP by 2 degrees Fahrenheit. However, since grade 1-D has a lower heat energy content, fuel economy also decreases.

Additives are chemicals put in at the refinery to lower the WAP and pour point. At the refinery, the composition of the oil and wax content is known. The proper additives are blended with the fuel to give it the desired properties. Additives used in the aftermarket by owners and technicians may or may not work because of variations in oil composition. Furthermore, using additives might violate the manufacturer's warranty.

WARNING: Never blend gasoline with diesel fuel. This can create a powerful explosive. Diesel fuel alone in the tank emits very little vapor. Gasoline fills the tank with fumes that are too rich to burn. When mixed together, the combination of fuel vapor and air is potentially explosive. This mixture can be ignited in a variety of ways: A spark created by a static charge can occur merely by filling the tank. A person performing mechanical work on the vehicle can create a spark with tools or a lighted cigarette. Also, if the vehicle is in an accident, the fuel tank can explode.

TABLE 2-2: COMPARISON OF GRADES 1-D, 2-D, AND 4-D DIESEL FUEL

	1-D	2-D	4-D
Minimum flash point, °F	100 or Legal	125 or Legal	130 or Legal
Viscosity, 100°F			
Minimum	1.4	2.0	5.8
Maximum	2.5	5.8	26.4
Carbon residue weight percent maximum	0.15	0.35	—
Ash, weight percent maximum	0.01	0.02	0.10
Sulphur, weight percent maximum	0.50	1.0	2.0
Ignition quality, cetane no. minimum	40	40	30
Distillation temperature °F, 90% evaporated			
Minimum	—	540	—
Maximum	550	576	—

LP-GAS

LP-gas ("LP" stands for "liquefied petroleum") is a byproduct of crude oil refining and it is also found in natural gas wells. Fuel grade LP-gas is almost pure propane with a little butane and propylene usually present. Because of its high propane content, many people simply refer to LP-gas as propane.

Propane's main attraction has always been its low price relative to gasoline. But over the past several years, the price advantage has narrowed and consequently the number of conversions has noticeably been reduced. Propane conversions are generally made by specialty engine rebuilding shops.

Propane burns clean in the engine, precisely controlled. Because it vaporizes at atmospheric temperatures and pressures, it does not puddle in the intake manifold. This means it emits less hydrocarbons and carbon monoxide. Emission controls on the engine can be simpler. Cold starting is easy, down to much below zero. At normal cold temperatures, the propane engine fires easily and takes power without surge or stumble.

One of the most noticeable differences between propane and gasoline is that propane is a "dry" fuel. It enters the engine as vapor. Gasoline, on the other hand, enters the engine as tiny droplets of liquid, whether it flows through a carburetor or is sprayed in through a fuel injector. This is an important difference because it has a pronounced affect on cooling the intake valves as well as the combustion chamber. Contrary to a popular misconception, LP-gas does not burn hotter than gasoline.

The propane fuel system is a completely closed system that contains a supply of pressurized liquid LP-gas. Since the fuel is already under pressure, no fuel pump is needed. From the pressurized fuel tanks, the fuel flows to a vacuum filter fuel lock. This serves as a filter and a control allowing fuel to flow to the engine. Fuel flows to a converter or heat exchanger where it changes from a liquid to a gas. When the propane flows through the converter, it expands as it changes into a gas. The carburetor mixes gaseous propane with the gaseous air. Airflow into the engine is controlled by a butterfly valve in the venturi. Mixture is controlled by a fuel metering valve operated by a diaphragm, which is controlled by intake manifold pressure. The idle system is an air bleed, similar to a gasoline engine. In fact, except for the fact that the propane carburetor type does not require a fuel bowl, the two carburetor types are basically the same.

REVIEW QUESTIONS

1. Crude oil is a mix of _____ .
   a. gasoline
   b. diesel fuel
   c. hydrocarbon compounds
   d. petroleum

2. Which of the following is not a desirable ingredient in gasoline?
   a. isooctane
   b. sulfur
   c. hydrocarbons
   d. detergents

3. Technician A says that adding isopropyl alcohol to fuel decreases the possibility of gas line freeze. Technician B says ethanol is added to dissolve gum and varnish. Who is correct?
   a. Technician A
   b. Technician B
   c. Both A and B
   d. Neither A nor B

4. Gasoline's ability to evaporate and mix with air for combustion is known as _____ .
   a. temperature
   b. volatility
   c. combustibility
   d. driveability

5. Which of the following does not contribute to knock or detonation?
   a. increased compression ratios
   b. high octane fuel
   c. overadvanced ignition timing
   d. turbocharging

6. Advantages of a diesel engine include _____ .
   a. low fuel consumption
   b. reduced fire hazard
   c. low emission levels
   d. all of the above

7. The specific gravity of a liquid measures a liquid's _____ .
   a. weight compared to water
   b. volatility

c. viscosity
d. temperature compared to water

8. Technician A says that water must be period- ically drained from a water separator in a die- sel fuel system. Technician B says that gaso- line should be added to diesel fuel to lower the viscosity. Who is correct?
   a. Technician A
   b. Technician B
   c. Both A and B
   d. Neither A not B

9. Which of the following is not an advantage of propane gas over gasoline?
   a. burns cleaner
   b. keeps valves and combustion chambers cooler

c. easy cold starting
d. does not puddle in the intake manifold

10. Which of the following fuel systems does not require a fuel pump?
    a. gasoline injection
    b. diesel
    c. propane
    d. methanol

11. During cold weather, winter conditions, diesel fuel should have a _____ .
    a. low cetane rating
    b. high wax point
    c. high viscosity
    d. low pour point

CHAPTER THREE

VACUUM, AIR INTAKE, AND FUEL SYSTEMS

Objectives

After reading this chapter, you should be able to:
- Define *driveability* and symptoms of driveability problems.
- Define *vacuum* and explain how it affects and assists vehicle operation and driveability.
- Define the air intake system components and their functions, including air intake ductwork, air cleaner/filter assembly, and intake air temperature controls.
- Define the fuel transport system components and their functions, including the fuel tank, fuel lines, filters, and gauges.
- Conduct inspection and servicing procedures for the vacuum system, air intake system, and fuel system.
- Define the components and their functions in an electronic engine control (EEC) fuel delivery system.

Driveability is one of the major service concerns in modern vehicles. Driveability includes everything that the vehicle's owner expects from a properly operating engine, such as easy starting (hot or cold engine), acceptable smoothness at idle, good acceleration in all speed ranges, instant response, and full power. Driveability can also apply to how the transmission operates, the way the vehicle handles, and to owner expectations in regard to noise, vibration, and harshness. In this book, the driveability concerns relate only to engine performance. The rest of the driveability terms consist of the various symptoms that can occur when the vehicle has a driveability problem.

Table 3–1 gives the correct term to use for each symptom, the definition of the symptom, and the engine systems that might cause the symptom. Notice how one engine system can cause many symptoms and, on the other hand, how most symptoms can have more than one system as their cause.

VACUUM SYSTEMS

The term *vacuum* refers to a pressure level that is lower than the earth's atmospheric pressure at any given altitude. The higher the altitude, the lower the atmospheric pressure.

Vacuum pressure can be measured in relation to atmospheric pressure. Atmospheric pressure is the pressure exerted on every object on earth and is caused by the weight of the surrounding air. At sea level, the pressure exerted by the atmosphere is 14.7 psi (absolute). Most pressure gauges ignore atmospheric pressure and read zero under normal conditions. For service purposes, atmospheric pressure is zero psi gauge. But, the usual measure of vacuum is in inches of mercury instead of psi. Other units of vacuum measure seen on automotive service gauges are kilopascals and the manometer, which measures pressure in inches of water.

Vacuum in any four-stroke engine is created by the downward movement of the piston during the intake stroke. With the intake valve open and the piston moving downward, a partial vacuum is created within the cylinder and intake manifold (Figure 3–1). The automotive engine creates a partial vacuum that is relatively continuous. In a typical four-cylinder engine, one cylinder is always at some stage of the intake cycle. The amount of vacuum created is partially related to the positioning of the choke (on carbureted vehicles) and to the throttle

Complete Fuel Systems and Emission Control

TABLE 3-1: DRIVEABILITY SYMPTOMS AND DEFINITIONS

Term	Definition	Probable System Cause
Cranks normally, but will not start.	The starter will crank the engine at normal speed, but the engine shows no indication of "firing" or will not continue running when the key is moved from the start to the RUN position.	C D E F
Starts normally, but will not run (stalls).	The starter will crank the engine at normal speed, the engine will start, but will not continue to run.	A D E F
Cranks normally, but slow to start.	The starter will crank the engine at normal speed, but the engine requires excessive cranking time before starting. If an engine will start, but will not stay running when the key is returned to the ON position, the symptom "starts normally, but will not run (stalls)" should be used.	A B C D E F
Rough idle	Engine vibrates excessively during idle.	A C D E F
Misses under load.	One or more cylinders do not fire under load (acceleration, climbing hills, pulling a trailer). Usually accompanied by lack of power.	C D E
Stalls on decel or quick stop.	Engine stalls when decelerating or during a quick stop.	C E F
Hesitates or stalls on acceleration.	A lag between the time the throttle is depressed and acceleration begins.	A B C D E F
Backfires (induction or exhaust).	Abnormal combustion in the induction or exhaust system, which produces a loud noise.	A C D E F
Lack of power	Performance does not meet design intent or customer expectations.	B C D E F
Surge at steady speed	An unexpected change in engine speed, which usually has a repetitive pattern or a repetitive occurrence of one or more cylinders not firing.	A C D E F
Engine diesels (runs on).	Engine continues to run after the key is returned to the OFF position.	A C F
Engine noise	Any unusual engine noise.	A B C D E F
Poor fuel economy	Fuel economy does not meet design intent or published information.	B C D E F
High oil consumption	Oil consumption, which exceeds one quart per 900 miles after break-in.	B E F
Spark knock/pinging	Abnormal combustion accompanied by an audible "pinging" noise.	A B D E F
Engine vibrates at normal speeds.	During a steady cruise, the engine vibrates excessively	F
Engine runs cold.	An abnormal condition under which the coolant temperature does not reach operating temperature. Symptoms include poor fuel economy and no heat from the vehicle's heater.	F
Engine runs hot.	An abnormal condition under which the coolant temperature becomes excessive. Symptoms include temperature warning lamp coming on, temperature gauge in the overheat range, or coolant overflowing the overflow tank.	A D E F

TABLE 3-1:   DRIVEABILITY SYMPTOMS AND DEFINITIONS

Term	Definition	Probable System Cause
Exhaust smoke	An abnormal condition where the exhaust is noticeable or any unusual exhaust smell, such as rotten eggs (hydrogen sulfide).	B C E F
Gas smell	Any gasoline fuel smell or visible leaks.	C E

Key:
A: Vacuum
B: Air intake
C: Fuel
D: Ignition
E: Emissions
*F: Mechanical

*Engine mechanical system servicing is not covered in this book. Information on engine mechanical systems can be found in vehicle service manuals and other textbooks.

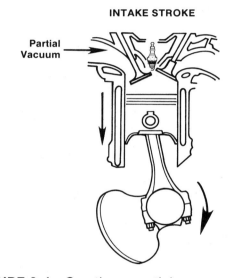

INTAKE STROKE

Partial Vacuum

FIGURE 3-1   Creating a partial vacuum

plates(s). The throttle plate not only admits air or air/fuel into the intake manifold, it also helps to control the amount of vacuum available during engine operation. At closed throttle idle, the vacuum available is usually between 15 to 22 inches. At a wide-open throttle acceleration, the vacuum can drop to zero.

 SHOP TALK _____

A perfect vacuum—29.92 inches mercury—is the complete absence of pressure (zero psi absolute). It is interesting to note that at 32 degrees Fahrenheit water will boil in a "perfect vacuum." Remember that a perfect vacuum would prevent the engine from running. If the engine were airtight, air and fuel would not be able to enter it for combustion.

TYPICAL VACUUM-CONTROLLED SYSTEMS

During the first fifty years of automotive vehicle production, the controls for gasoline engines were relatively distinct. Early electrical, ignition, fuel, and vacuum systems were easy to service because they had a limited operational interrelationship. For example, the effect and application of the vacuum system were quite apparent when operating windshield wipers; under heavy acceleration, the wipers would stop. Spark advance was vacuum controlled and as technology progressed to meet improved fuel economy and lower emissions, vacuum applications were increased.

Engine vacuum is used for emission controls, improved fuel economy, driveability, and to operate accessories. Electronic controls and vacuum applications have now made previously distinct systems interdependent. The following describe the engine systems that have vacuum inputs and controls:

- *Fuel Induction System.* Certain vacuum-operated devices are added to carburetors and some central fuel-injected throttle bodies to ease engine start-up, warm-and-cold engine driveaway, and to compensate for an air conditioner load on the engine. Chapters 4 through 8 describe the use of vacuum in the fuel induction system operation and servicing.

- *Emission Control System.* Most emission control output devices operate on vacuum. This vacuum is usually controlled by solenoids that are opened or closed, depending on electrical signals received from the electronic control assembly (ECA). Other systems use switches that are controlled by engine coolant temperature, such as a ported vacuum switch (PVS), or by ambient air such as a temperature vacuum switch (TVS). The complete use of engine vacuum in emission control operation and servicing is covered in Chapters 10 to 12.

- *Accessory Controls.* Engine vacuum is used to control operation of certain accessories, such as air conditioner/heater systems, power brake boosters, speed-control components, automatic transmission vacuum modulators, and so on.
- *Air Intake and Fuel System.* Vacuum is used to draw filtered air into the engine and to control vacuum-operated full system devices which, in turn, are used to control the temperature of the air allowed into the engine. This improves fuel vaporization so that fuel will burn more efficiently. This chapter covers the operation and servicing of air intake and fuel systems.

To operate these various automotive systems, there are basic groups of vacuum controls generally employed. They are:

1. *Vacuum Switches and Valves.* These devices control vacuum flow and are usually operated either electrically or thermostatically (Figure 3–2A).
2. *Vacuum Delay Valves.* These valves, often called *restrictors,* are typically located in the vacuum line between the vacuum source and device to be acted upon (Figure 3–2B). They are used to delay vacuum flow.
3. *Vacuum Diaphragm.* These controls, often called *actuators,* are used to operate various vehicle mechanical parts (Figure 3–2C).

Vacuum Schematic

An engine emissions vacuum schematic is located inside the engine compartment on all Ford vehicles. The following schematic shows the vacuum hose routings and vacuum source for all emissions-related equipment. The vacuum schematic illustrated in Figure 3–3A is the new type that has just been introduced in some vehicles. It shows the relationship of the components as they are mounted on the engine. It is different from the computer-generated types that have been used over the previous years (Figure 3–3B). Graphic symbols that are used on later vacuum schematic drawings can be found in Appendix C.

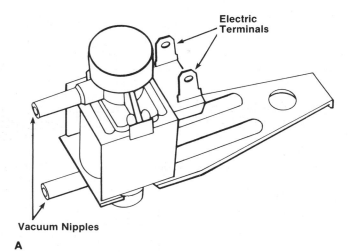

Electric Terminals

Vacuum Nipples

A

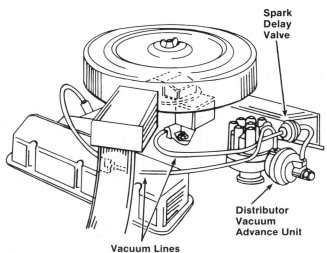

Spark Delay Valve

Distributor Vacuum Advance Unit

Vacuum Lines

B

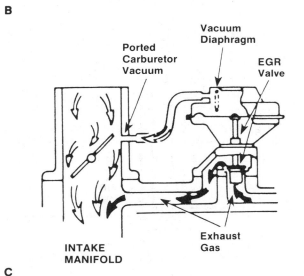

Vacuum Diaphragm

Ported Carburetor Vacuum

EGR Valve

Exhaust Gas

INTAKE MANIFOLD

C

FIGURE 3–2 Vacuum control groups: (A) solenoid vacuum valve; (B) delay valve; (C) vacuum diaphragm

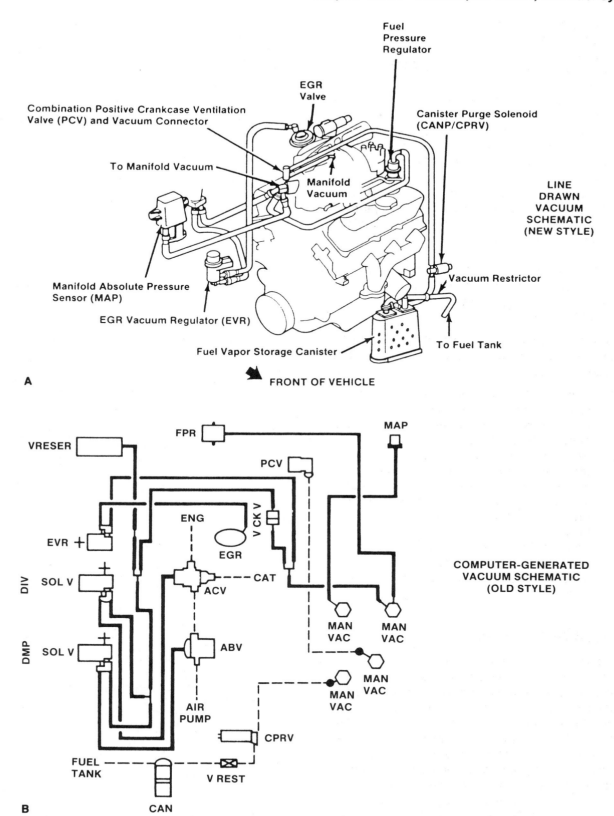

FIGURE 3-3   (A) New style line-drawn vacuum schematic; (B) old style computer-generated vacuum schematic

VACUUM SYSTEM SERVICING

Vacuum system problems can produce or contribute to driveabilty symptoms, such as:

- Stalls
- No start (cold)
- Hard start (floods hot)
- Backfire (deceleration)
- Rough idle
- Poor acceleration
- Rich or lean stumble
- Overheating
- Knock or pinging
- Detonation
- Rotten eggs odor
- Poor fuel economy

As a routine part of problem diagnosis, the service technician suspecting a vacuum problem should first

- Inspect vacuum hoses for improper routing or disconnects (engine decal identifies hose routing).
- Look for kinks, tears, or cuts in vacuum lines.
- Check vacuum hose routing and wear near hot spots, such as the exhaust manifold or the EGR tubes.
- Make sure there is no evidence of oil or transmission fluid on vacuum hose connections (valves can become contaminated by oil getting inside).
- Inspect vacuum system devices for damage (dents in cans; bypass valves; broken nipples on VCV or PVS valves; broken "tees" in vacuum lines, and so on).

If there is reason to believe a vacuum leak exists, this schematic helps to locate the vacuum hose routing for emission-related vacuum components only. If there is a vacuum leak somewhere in the vacuum system, it could be in a nonemission-related component. Vacuum is used to control speed control, heater or A/C controls, and the like. If there is a crack or broken vacuum line in these areas, the vacuum system is affected in the same manner as a cracked or broken vacuum line in the emission system vacuum hoses.

The following vacuum system problems are the most common:

- *Leaking Vacuum Hose (Rough Idle).* A broken vacuum line allows a vacuum leak. A vacuum leak admits more air into the intake manifold than the engine is calibrated for. Then the engine runs roughly due to the leaner air/fuel mixture.
- *Kinked Vacuum Hose (Spark Knock/Pinging).* If the vacuum hose to the exhaust gas recirculation (EGR) valve is kinked, the exhaust gas recirculation (EGR) valve will not open when required. The engine is calibrated to allow a certain amount of exhaust gas to enter the combustion chamber. This not only reduces the amount of NO_x, but also cools down the combustion chamber to prevent spark knock and pinging.

To check vacuum controls, refer to the vehicle manufacturer's service manual for the correct location and identification of components. Typical locations of vacuum-controlled components are shown in Figure 3-4.

Vacuum Test Equipment

The two most common vacuum diagnostic tools are the vacuum gauge (Figure 3-5) and the vacuum pump (Figure 3-6). Until the introduction of the computerized engine analyzer (Figure 3-7) the vacuum gauge was one of the most important engine diagnostic tools used by service technicians. With the gauge installed according to manufacturer's instruction and the engine warm and idling properly, getting an instant engine diagnosis is easy. Simply watch the action of the gauge's needle. A healthy engine will give a steady, constant vacuum reading between 17 and 22 inches. (A vacuum gauge measures vacuum in inches of mercury.) On some four- and six-cylinder engines, however, a reading of 15 inches is considered acceptable. With high-performance engines, a slight flicker of the needle can also be expected. The operator should discount for the fact that the gauge reading will drop about 1 inch for each 1,000 feet above sea level. Figure 3-8 shows some of the common readings and what engine malfunctions they indicate.

A vacuum gauge and hand-held vacuum pump are used to check many of the vehicle's vacuum control devices in the following manner:

- *Vacuum Switches and Valves.* Apply the source of vacuum to the inlet or other specified part. Then the switch or valve must be operated (either electrically or by temperature), and the vacuum flow through the device can be measured (Figure 3-9A).
- *Vacuum Delay Valves.* These valves are checked by applying vacuum to the inlet side of the valve and connecting a second vacuum gauge to the outlet side. The length of

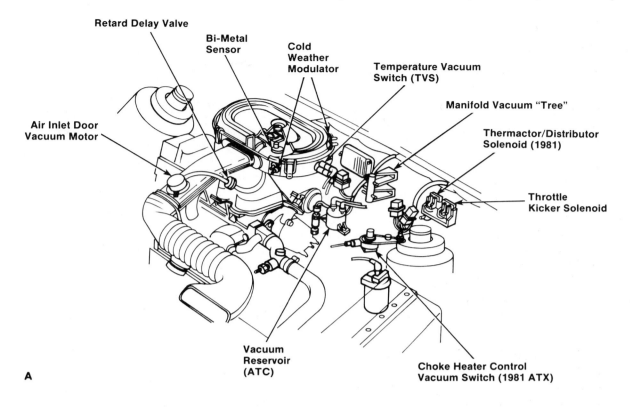

Retard Delay Valve

Bi-Metal Sensor

Cold Weather Modulator

Temperature Vacuum Switch (TVS)

Manifold Vacuum "Tree"

Thermactor/Distributor Solenoid (1981)

Air Inlet Door Vacuum Motor

Throttle Kicker Solenoid

Vacuum Reservoir (ATC)

Choke Heater Control Vacuum Switch (1981 ATX)

A

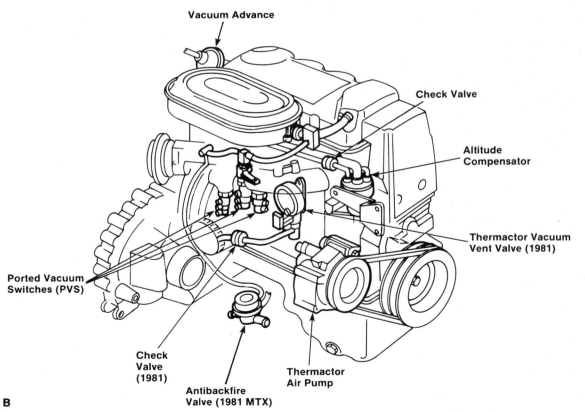

Vacuum Advance

Check Valve

Altitude Compensator

Thermactor Vacuum Vent Valve (1981)

Ported Vacuum Switches (PVS)

Check Valve (1981)

Antibackfire Valve (1981 MTX)

Thermactor Air Pump

B

FIGURE 3-4   Vacuum devices and controls: (A) in dash panel inside engine compartment; (B) in a 1.6-liter engine with PVS; (C) in a 1.6-liter engine with an EGR valve

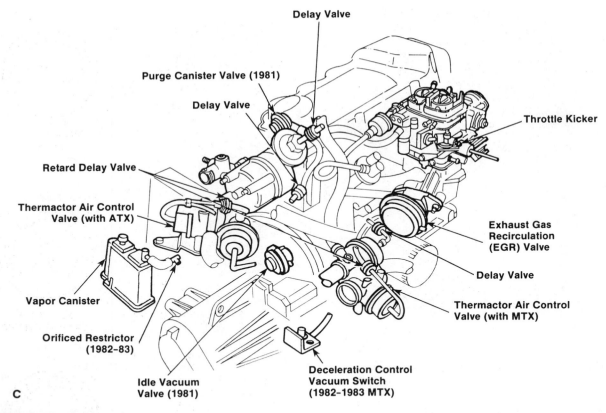

Delay Valve

Purge Canister Valve (1981)

Delay Valve

Throttle Kicker

Retard Delay Valve

Thermactor Air Control Valve (with ATX)

Exhaust Gas Recirculation (EGR) Valve

Delay Valve

Vapor Canister

Thermactor Air Control Valve (with MTX)

Orificed Restrictor (1982–83)

Idle Vacuum Valve (1981)

Deceleration Control Vacuum Switch (1982–1983 MTX)

C

FIGURE 3-4 continued

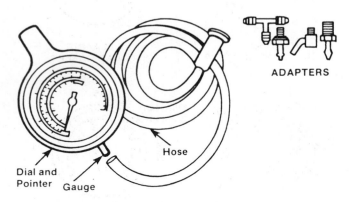

ADAPTERS

Dial and Pointer

Gauge

Hose

FIGURE 3-5 Vacuum gauge and adapters

time needed for the vacuum gauge readings to stabilize on both sides of the valve should be compared to the specifications given in the manufacturer's service manual to determine the valve's performance (Figure 3-9B).

• *Vacuum Diaphragms.* Control devices of this type are generally checked by applying vacuum to the inlet port and noting the diaphragm movement.

Additional test procedures for various vacuum emission controls are covered in Chapter 11.

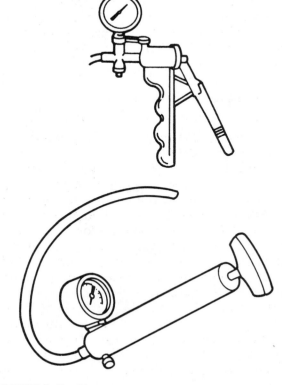

FIGURE 3-6 Vacuum pumps

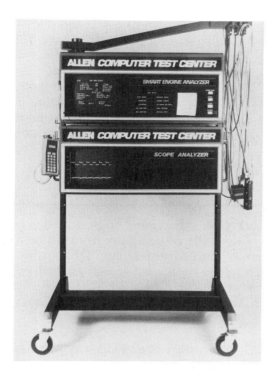

FIGURE 3-7   Computerized engine analyzer

 SHOP TALK _____

When a vacuum valve is closed, vacuum will be blocked from passing through the valve. When a valve opens, vacuum will pass through. Refer to the vehicle's service manual for correct valve operating conditions and test procedures.

Vacuum Hose Lines

The major cause of problems in a vacuum system occur in its hose lines. Check all vacuum hoses carefully for cracks, breaks, kinks, hardening, and loose or disconnected connections (Figure 3-10). Any defective hoses should be replaced one at a time to avoid misrouting. Refer to the vacuum schematic located inside the engine compartment or check the vehicle's service manual for correct hose routing if more than one hose is disconnected.

Most original equipment manufacturers (OEM) equipped vacuum lines are installed in a harness consisting of 1/8-inch outer diameter and 1/16-inch inner diameter nylon hose with bonded nylon or rubber connectors. Occasionally, a rubber hose might be connected to the harness. The nylon connectors have rubber inserts to provide a seal between the nylon connector and the component connection (nipple).

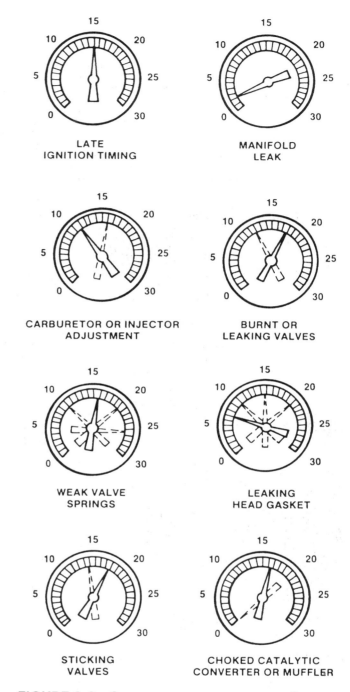

LATE
IGNITION TIMING

MANIFOLD
LEAK

CARBURETOR OR INJECTOR
ADJUSTMENT

BURNT OR
LEAKING VALVES

WEAK VALVE
SPRINGS

LEAKING
HEAD GASKET

STICKING
VALVES

CHOKED CATALYTIC
CONVERTER OR MUFFLER

FIGURE 3-8   Common vacuum gauge readings

The following is a typical vacuum hose replacement procedure:

 1.   If a nylon hose is broken or kinked and the damaged area is 1/2 inch or more from a connector, the hose can be repaired by cutting out the damaged section (not more than 1/2 inch) and then installing rubber union (Figure 3-11A).

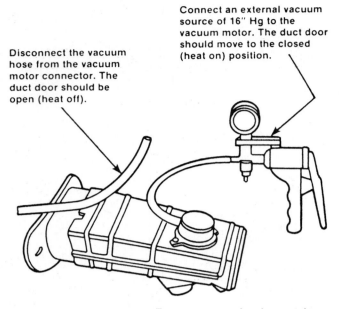

Disconnect the vacuum hose from the vacuum motor connector. The duct door should be open (heat off).

Connect an external vacuum source of 16" Hg to the vacuum motor. The duct door should move to the closed (heat on) position.

Trap vacuum and make sure door remains closed for 60 seconds. If not, replace vacuum motor.

A

B

FIGURE 3–9 (A) Hand-operated vacuum pump being used to test an air cleaner duct vacuum motor; (B) two-port/four-port VCV or PVS test

2. If the remaining hose is too short or the damaged portion is more than 1/2 inch, replace the entire hose and connectors with rubber vacuum hoses and a tee. Use existing service stock of 5/32-inch hose, 7/32-inch hose, and tees (Figure 3–11B). The circled number in these illustrations identifies the same connection points on both the original and the repaired harnesses.

CAUTION: Care must be exercised to keep the nylon parts routed away from hot components. In addition, holes might be worn into the nylon hoses if allowed to rub against rough surfaces.

---

3. If only part of a nylon connector is damaged or broken, cut the connector apart (as illustrated in Figure 3–11C) and discard the damaged half of the harness. Replace it with rubber vacuum hoses and a tee from service stock.
4. Figure 3–11D shows how to replace a damaged harness with available service stock—rubber hoses (5/32-inch and 7/32-inch), tees, and elbows. Cut the hoses to the required lengths but never less than 1 1/2 inches. Be sure to properly route the hoses, using the identification numbers provided for this purpose.

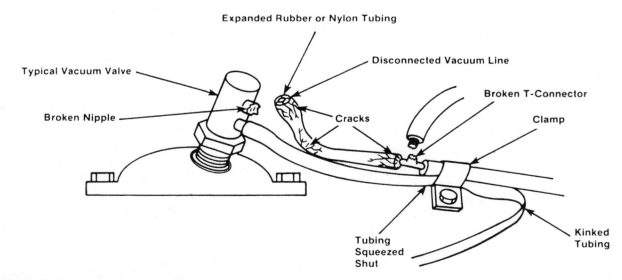

Expanded Rubber or Nylon Tubing

Disconnected Vacuum Line

Typical Vacuum Valve

Broken T-Connector

Broken Nipple

Cracks

Clamp

Tubing Squeezed Shut

Kinked Tubing

FIGURE 3–10 Vacuum line defects

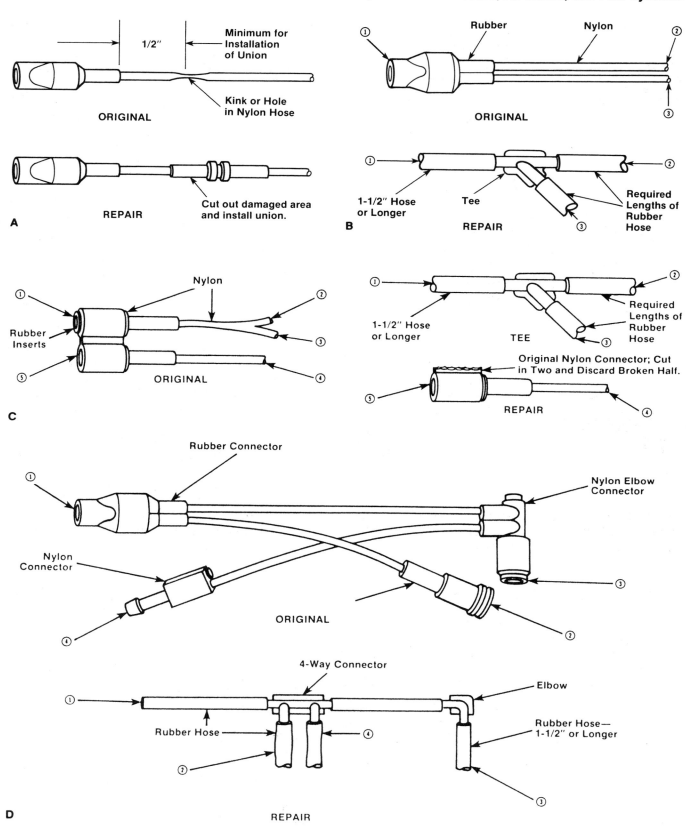

FIGURE 3-11   (A) Rubber union installation; (B) hose replacement procedure; (C) damaged nylon connector repair; (D) replacement of damaged harness

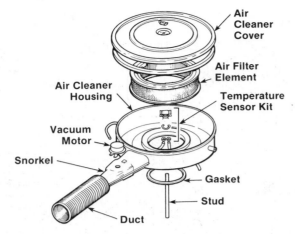

FIGURE 3-12   Air cleaner assembly

AIR SYSTEMS

This chapter will be concerned with the use of air in the automotive engine, dwelling primarily on the intake of fresh air into the air cleaner system. The expulsion air from the engine along with harmful pollutants go through the vehicle's exhaust system (see Chapter 9).

AIR INTAKE SYSTEMS

The air intake or induction system has the following main functions:

- Provide the air that the engine needs to operate

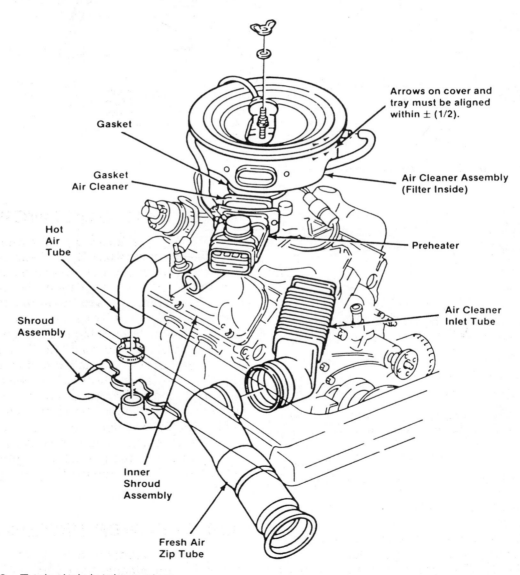

FIGURE 3-13   Typical air intake system

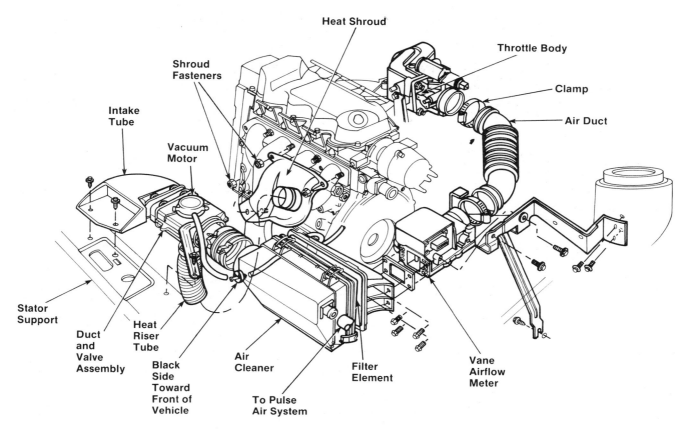

FIGURE 3-14   Air intake system for a fuel injected engine

- Filter the air to protect the engine from wear
- Monitor air temperature for more efficient combustion and reduction of hydrocarbon (HC) and carbon monoxide (CO) emissions
- Control the amount of air flowing into the engine.
- Operate in conjunction with the positive crankcase ventilation (PCV) system to burn the crankcase fumes in the engine.

To perform these tasks, the typical carburetor intake air cleaner filter assembly (Figure 3-12) contains the following parts:

- *Air Cleaner Housing* (metal or plastic container). Holds components and supporting parts.
- *Air Ducts.* Carry air into the cleaner housing.
- *Filtering Element.* Cleans inkeeping engine air.
- *Air Cleaner Cover.* Allows service of filtering element.
- *Air Cleaner Gaskets.* Prevent air leakage between components of the air cleaner.
- *Snorkel.* Rigid air inlet that extends out from the air cleaner housing.

AIR INTAKE DUCTWORK

Figure 3-13 shows a typical air intake system. Cool air is drawn through the fresh air tube duct from the leading edge of the engine compartment and routed through a preheat assembly, the air cleaner, and into the carburetor. In a fuel injection system the air from the cleaner is usually sent to the air sensor on to the throttle body (Figure 3-14).

It is important to make sure that the intake ductwork is properly installed and all connections are airtight—especially those between an airflow sensor or remote air cleaner and the carburetor or fuel injection manifold. Generally metal or plastic clamps are used to hold the ducts together. Plastic ducts are used where engine heat is a problem. Special paper-metal ducts or all-metal ducts are used when they will be exposed to very high temperatures.

AIR CLEANER/FILTER

The primary function of the air filter is to prevent airborne contaminants and abrasives from entering the engine through the carburetor with the air/fuel

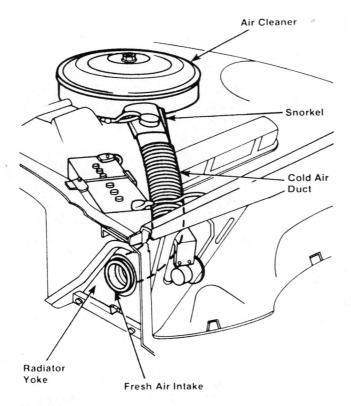

FIGURE 3-15   Air cleaner with fresh air intake mounted on the radiator yoke

el cars, the air cleaner assembly provides a control and mixing function. Incoming air is maintained at a constant temperature of about 100 degrees Fahrenheit to help reduce emissions and improve engine performance. The air cleaner also provides filtered air to the positive crankcase ventilation (PCV) emission control system.

Types of Air Filters

For years, the oil bath air cleaner was used to clean the intake air of engines. Dirt particles were

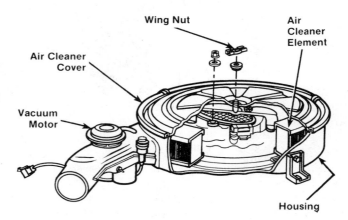

FIGURE 3-16   Typical air cleaner design for a carbureted engine

mixture. That is, the air portion of the air/fuel mixture can carry minute particles of foreign matter such as dust, dirt, and exhaust carbons. Without proper filtration, these contaminants can cause serious damage and appreciably shorten engine life. Remember that all incoming air must pass through the filter element before entering the engine (Figure 3-15).

The air filter unit used on most gasoline-driven cars and light-duty trucks is located inside the air cleaner housing (Figure 3-16), which can be mounted directly to a flange on the carburetor air horn, the fuel injection throttle body, or the intake manifold (on diesel engines). On some automobiles and trucks (especially diesel vehicles), it can also be mounted in a remote location on the engine or engine compartment to reduce overall engine height (Figure 3-17). In such cases, the air cleaner unit is connected to the carburetor (or injector system), air horn, throttle body, or manifold by an air transfer tube or duct.

The air cleaner provides other services to the air intake system. For example, airflow through the carburetor and intake valve operation creates "noise" from the intake manifold. The air cleaner assembly helps to eliminate or reduce this noise. On late mod-

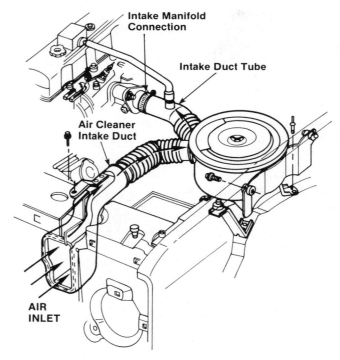

FIGURE 3-17   Ducts are used for remote air cleaner on a diesel engine

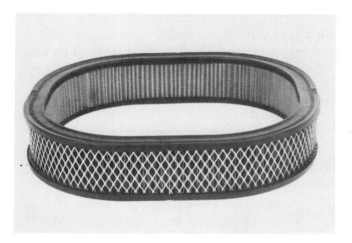

FIGURE 3-18   Dry air filter

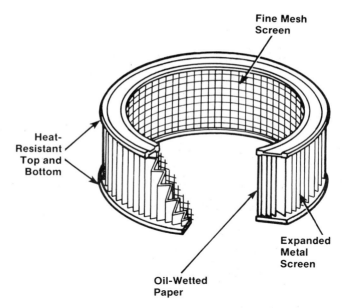

FIGURE 3-20   AC automotive air filter design features

trapped on the surface of a pool of oil when the airstream was forced to make an abrupt 180 degree turn. However, the efficiency of such an air cleaner (the percentage of contaminants it removed from the air) was only around 95 percent, and at low engine speeds it dropped by as much as 25 percent. So the invention of the dry air filter (Figure 3-18) was a significant step in prolonging engine life.

Air filters for today's automobiles and trucks are available in two basic air filter classifications:

1. *Light-Duty (Standard).* These are usually made of pleated paper or oil-wetted polyurethane.
2. *Heavy-Duty.* Better heavy-duty air filters have an oil-wetted polyurethane outer cover over a dry-type paper element (Figure 3-19). The polyurethane effectively traps

larger dirt particles on its surface. The finer particles that pass through the polyurethane are trapped by the paper media. The polyurethane outer cover can be removed, cleaned, and reinstalled, thus extending filter life.

When selecting a new air filter keep these design features in mind (Figure 3-20):

- Fine mesh screen on the inside reduces the possibility of fire hazards caused by engine backfire.
- Heat-resistant plastisol on the top and bottom with special sealing beads provides a positive dust seal over the wide range of automotive operating conditions.
- Oil-wetted resin-impregnated paper provides long-lasting filter efficiency.
- Wire or expanded metal outer screen adds construction strength and protects against accidental paper damage during handling and shipping.

Air Filter Servicing

The engine manufacturer's recommended air filter replacement interval is usually specified in the vehicle owner's service manual. However, replacement of the air filter might be required on a more frequent schedule if the vehicle is subjected to continuous operation in an extremely dusty or severe off-the-road environment.

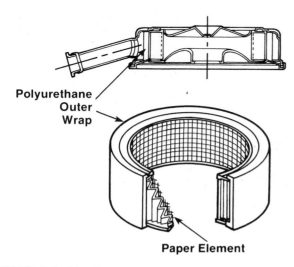

FIGURE 3-19   Polyurethane outer wrap and paper element

The air filter should be replaced on a regular basis because, if damaged, it can accelerate wear of engine cylinder walls, pistons and piston rings; an extremely dirty air filter can act as a choke, affecting engine performance and fuel consumption. To replace an air filter, proceed as follows:

1. Identify the correct replacement air filter for the engine.
2. Remove the wing nut or other fasteners from the engine air cleaner assembly; then remove the cover. Some unitized air cleaners with nonremovable covers must be replaced as a total assembly.
3. The air filter should easily lift out of the air cleaner assembly.
4. Visually inspect the air filter for damage and dirt accumulation. An air filter that shows signs of damage or heavy dirt accumulation should be replaced with a new air filter of the specified type.
5. Reassemble the air cleaner cover and wing nut or other fasteners. Do not overtighten the wing nut.

Certain types of heavy-duty air filters can be cleaned and reused. These filters have a heavy paper media encased in an element with metal end caps. Generally, reusable heavy-duty air filter elements should be replaced with a new element after six cleanings, or once each year.

WARNING: When working on pressurized gasoline systems, be sure to wear safety glasses. Shop rags should be used to wipe up any fuel leaks from the filter during service.

Before attempting to clean a reusable element, always examine it carefully. Look for a torn or punctured paper media, bent end covers, or pinholes. If end gaskets are damaged or missing, replace the element. Air filter elements that have precleaner fins or tubes should be inspected for dirt accumulations. A stiff nylon or fiber bristle brush can be used to remove light coatings of dust.

If dust is the main contaminant, compressed air can often clean the element to near "new service" capability. Heavy-duty air filter elements that are lightly coated with dust can be cleaned by directing a stream of air up and down the pleats on the clean air side of the element (Figure 3–21).

CAUTION: Do not allow air pressure at the nozzle to exceed 30 psi and maintain a 6-inch minimum distance from the nozzle to the paper. Air pressure must be applied only on the inner surface since dirt has collected on the outer surface.

Heavy-duty elements that are extremely dirty or have soot or oil vapor deposits require a more thorough cleaning. Clean these elements as follows:

1. Remove all loose dust with a flow of water or compressed air.
2. Immerse the element in a solution of approved filter cleaner for 15 minutes or more. Do not allow the temperature of the solution to exceed 140 degrees Fahrenheit. Occasionally agitate the element to loosen particles.
3. Remove the element and rinse it thoroughly by running clean water through it from the clean air side to the dirty side until the water comes out clean. Use a hose without a nozzle and do not permit water pressure to exceed 40 psi.
4. Allow the element to air dry completely before reinstalling. A flow of warm air (temperature not to exceed 140 degrees Fahrenheit) can be used to shorten drying time.

FIGURE 3–21 Low-pressure shop air can be used to blow dirt from the element. Air pressure must be applied on the inner surface only since dirt has collected on the outer surface.

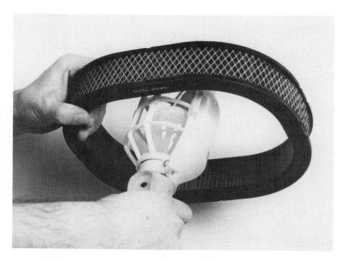

FIGURE 3-22   Inspect element with light.

Inspect the element with a light bulb after cleaning with compressed air or washing (Figure 3-22). Pinholes or slight ruptures will admit enough airborne dirt to render the element unfit for further

service. Use of a damaged filter can cause rapid failure of piston rings.

After cleaning, check the gaskets on the ends for condition and proper location. If they are damaged or missing, replace the element.

Some 4-cylinder engines use disposable air cleaners (check the vehicle's service manual). The air cleaner is a one-piece unit that contains a built-in filter (Figure 3-23). The entire assembly is replaced at 50,000-mile intervals.

AIR INTAKE TEMPERATURE CONTROLS

Air temperature is controlled by the heated or thermostatic air cleaner intake system. The heated air intake system aids vaporization of the fuel on cold start-up by warming the intake air at the exhaust manifold. The air valve in the snorkel (cold air intake) should close off outside (ambient) air when a cold engine starts. At 75 to 105 degrees Fahrenheit in the air cleaner (depending on temperature setting of the bimetal sensor), the bimetal sensor causes the air valve to open to outside air and shut off heated air (Figure 3-24).

When a vacuum signal is applied to the motor, it closes the door to cold air and allows warm air to enter the air cleaner. When a vacuum signal is not applied, the door opens to cold air.

A bimetal sensor is installed in the air cleaner housing tray. When the sensor is below a specified temperature, it allows a vacuum signal to flow through it to the air cleaner vacuum door motor. At a given increase in temperature, the sensor bleeds off vacuum, permitting the vacuum motor to open the duct door to allow fresh air to enter the engine. The air cleaner sensor can be checked with a magnetic base thermometer (Figure 3-25). The tape will hold the thermometer beside the sensor and keep it from being sucked into the air horn.

The cold weather modulator is clipped to holes in the air cleaner housing assembly and senses the temperature of the inlet air. A bimetal disc actuates a check valve inside the modulator, depending on the air temperature.

- *Open.* The modulator passes vacuum freely either way.
- *Closed.* The modulator can trap vacuum at an actuator or block vacuum from being applied, depending on how the hoses are connected.

When the modulator is used to trap vacuum, it will trap the vacuum signal to the air cleaner vacuum

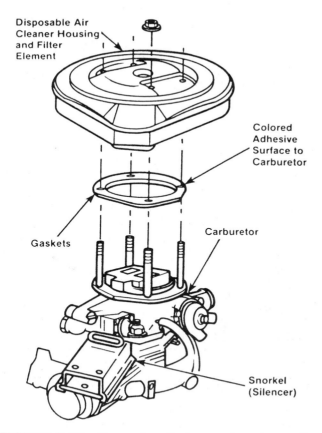

FIGURE 3-23   Some 4-cylinder engines have disposable air cleaners.

Disposable Air Cleaner Housing and Filter Element

Colored Adhesive Surface to Carburetor

Gaskets

Carburetor

Snorkel (Silencer)

MANIFOLD-HEATED AIR ON
(FRESH AIR OFF)

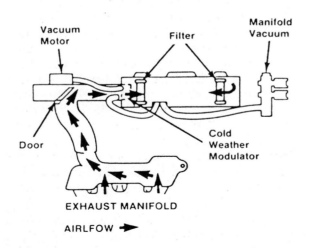

FRESH AIR ON
(HEATED AIR OFF)

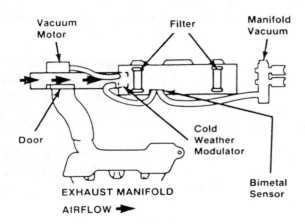

FIGURE 3-24 Heated air intake system operation

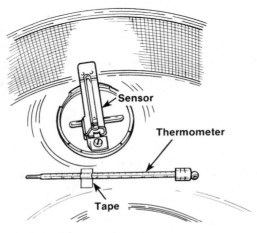

FIGURE 3-25 Checking the air cleaner sensor with a magnetic base thermometer

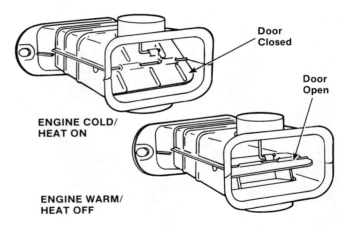

ENGINE COLD/
HEAT ON

ENGINE WARM/
HEAT OFF

FIGURE 3-26 Air cleaner thermostatic door positions: (A) when engine is cold, door should be closed; (B) when engine is at operating temperature, door should be open.

door motor. This prevents the door from switching to cold inlet air when the vacuum drops during acceleration (Figure 3-26A). After the engine warms up, there is no need to trap the vacuum to the door (Figure 3-26B). Some intake air temperature control systems use a vacuum retard delay valve (VRDV) instead of the cold weather modulator. Further details on the operaton of air intake systems can be found in Chapter 10.

If the air filter becomes very dirty, the dirt will partially block the flow of air into the fuel charging assembly (Figure 3-27). Without enough air, the engine will constantly burn a rich air/fuel mixture, using up more fuel than it normally would. The use of extra fuel means poor fuel economy, and the reduced airflow can cause lack of power. Also, if the heated air inlet door sticks open, heated air cannot reach the engine, even if it is cold. The cold air that does enter the engine will not allow the fuel to vaporize completely. Liquid fuel will not ignite as easily as fuel that is vaporized.

Common air intake problems are shown in Figure 3-28, and their servicing is covered in Chapter 11.

FUEL SYSTEMS

The automobile's fuel system must serve several major functions:

- Store fuel
- Move fuel from the storage container to the engine
- Blend the fuel with the air so it can be used by the engine

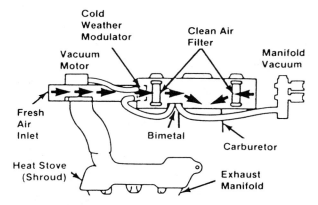

CLEAN AIR FILTER ALLOWS THE REQUIRED
AMOUNT OF AIR TO ENTER THE ENGINE.

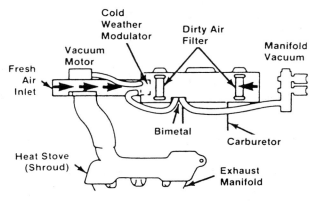

DIRTY AIR FILTER PREVENTS THE REQUIRED
AMOUNT OF AIR FROM ENTERING THE ENGINE.

FIGURE 3-27   Restricted airflow—dirty air filter

WARNING: Extreme caution should be used while working with any of the components of the fuel system. Gasoline is a very volatile and flammable substance. Do not expose it to an open flame, spark, or high heat. Disconnect the negative terminal of the battery before doing any task that will release gasoline from any part of the system. Use an approved safety-type gasoline can (never an open container) to catch the gasoline and cloths to wipe up the minor spills. Use a flashlight rather than a trouble light or drop-light. Gasoline spilled on a hot bulb could cause the bulb to explode and ignite the gasoline. Never attempt to weld a fuel tank. Even when a tank is empty of fuel, it can still contain enough gas fumes to cause an explosion if they come in contact with a flame or excessive heat. Always keep a Class B fire extinguisher nearby. It is specially intended for use on gasoline fires.

The automobile's fuel system is both simple and complicated: simple in the systems that transfer fuel to the engine and complex in the carburetor or fuel injector system that mix that fuel with air in the correct amounts and proportion to meet all needs of the engine.

This chapter will focus on the simple part of the fuel system. It will cover the basic parts of the fuel

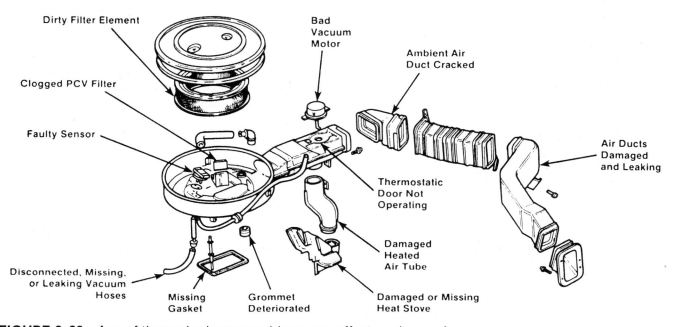

FIGURE 3-28   Any of these air cleaner problems can affect engine performance.

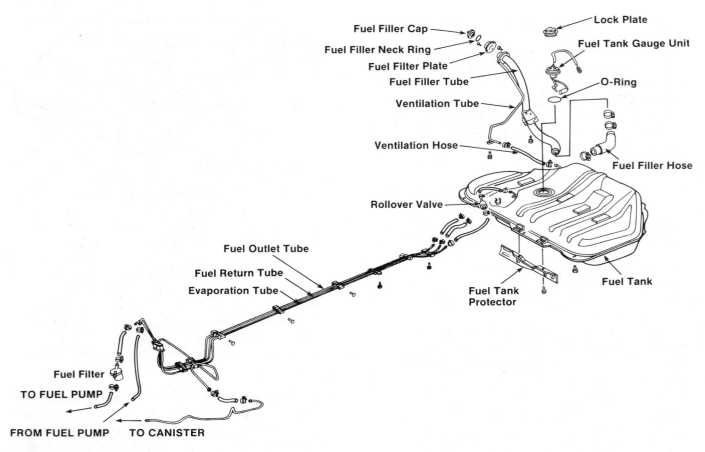

FIGURE 3–29 Typical fuel line tubing that should be visually checked for any leak or damage and looseness of clamps

transport system, including the fuel tank, lines, filters, and pump (Figure 3–29). Chapters 4 to 8 examine the more complex principles and functions of the fuel metering and atomization portion of the system. Remember, however, that the fuel transport portion of the system is very important in the operation of either a gasoline or diesel vehicle.

FUEL TANKS

Modern fuel tanks have several features that differ from those of tanks in older cars. On early model cars, gas tanks were vented, meaning vapors escaped into the atmosphere, leading to both fire dangers and pollution. Modern fuel tanks include devices that prevent vapors from leaving the tank. For example, to contain vapors and allow for expansion, contraction, and overflow that results from changes in the temperature, the fuel tank has a separate air chamber dome at the top. Another way to contain vapors is to use a separate internal expansion tank within the main tank (Figure 3–30). All fuel tank designs provide some control of fuel height

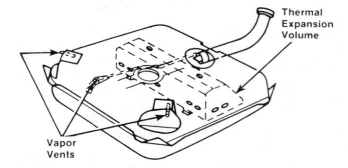

FIGURE 3–30 Fuel tank with an internal expansion tank to allow for changes in fuel volume due to changes in temperature.

when the tank is filled. Frequently, this is achieved by using vent lines within the filter tube or tank (Figure 3–31). With tank designs such as this, only 90 percent of the tank is ever filled, leaving 10 percent for expansion in hot weather. Some vehicles have an overfill limiting valve to prevent overfilling of the tank. If a tank is filled to capacity, it will overflow whenever the temperature of the fuel increases.

From the fuel tank, the vapors travel through a line, or continuous tube, to a charcoal-filled canister. When the engine starts, fresh air is drawn through the charcoal canister where it mixes with the vapors from the fuel tank. From there, these vapors are drawn into the intake manifold. The vapors then pass through the intake manifold and into the cylinders, where they are burned along with the normal air/fuel mixture. For more information on evaporative emission control, see Chapters 10 and 11.

As mentioned in Chapter 1, the fuel tank is usually made of pressed corrosion-resistant steel halves, which are ribbed for additional strength. Often, exposed portions of the tank are made of heavy-gauge steel for protection from road damage. The seams of the tank are welded. In an attempt by vehicle manufacturers to reduce overall car weight, some are now using aluminum or molded reinforced polyethylene fuel tanks.

Most metal tanks have slosh baffles or surge plates to prevent the fuel from splashing around inside the unit. In addition to slowing down fuel movement, the plates tend to keep the fuel pickup or sending assembly immersed in the fuel during hard braking and acceleration. The sheet metal plates or baffles are usually welded in place inside the tank and have holes or slots in them to permit the fuel to move from one end of the tank to the other.

Except for rear engine vehicles, the fuel tank in a passenger car is located in the rear of the vehicle for improved safety. Figure 3-32 illustrates the more common locations. To increase the driving range of a vehicle, some cars install an auxiliary or saddle tank, generally in the trunk. But when making such an installation, extreme care must be taken to be sure it is tightly secured in a safe location and that all emission regulations are met. Some states have laws against the use of auxiliary fuel tanks.

The fuel tank is provided with an inlet filler tube and cap. The location of the fuel inlet filler tube depends on the tank design and tube placement. It is usually positioned behind the filler cap or a hinged

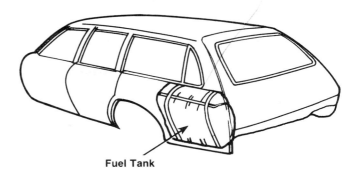

Fuel Tank

FUEL TANK IN REAR QUARTER PANEL

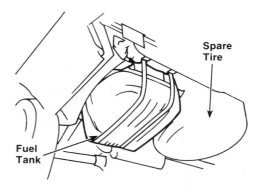

Spare Tire

Fuel Tank

FUEL TANK UNDER TRUNK ON ONE SIDE

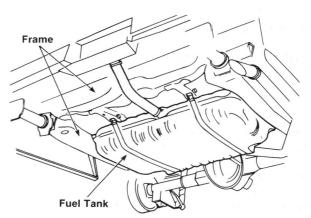

Frame

Fuel Tank

FUEL TANK INSIDE FRAME

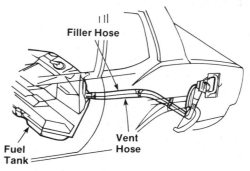

Filler Hose

Fuel Tank

Vent Hose

FUEL TANK UNDER SEAT

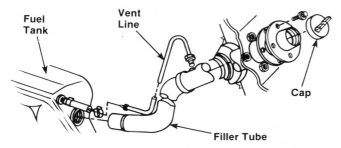

Fuel Tank

Vent Line

Cap

Filler Tube

FIGURE 3-31 Ford EEC fill control vent system

FIGURE 3-32 Gasoline tanks can be located almost anywhere that is safe from damage.

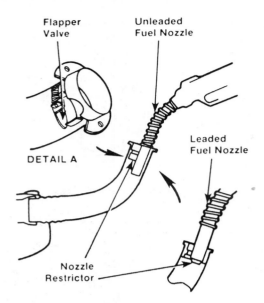

FIGURE 3-33 Cars that require unleaded fuel have restrictors in the filler tubes to allow only the entry of smaller unleaded fuel pump nozzles

quired a control on gasoline leakage from passenger cars and certain light trucks and buses, after they were subjected to barrier impacts and rolled over. Tests conducted under these severe conditions showed the most common gasoline leak path to be

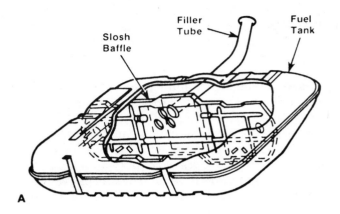

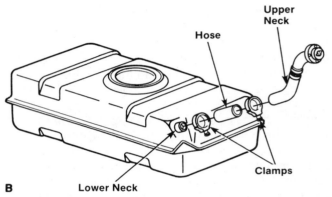

FIGURE 3-34 Filler tubes: (A) one piece; (B) three piece

door in the center of the rear panel or in the outer side of either rear fender panel. Vehicles designed for unleaded fuel use have a restrictor in the filler tube that prevents the entry of the larger leaded fuel delivery nozzle at the gas pumps (Figure 3-33). The filler pipe can be a rigid one-piece tube soldered to the tank (Figure 3-34A) or a three-piece unit (Figure 3-34B). The three-piece unit has a lower neck soldered to the tank and an upper neck fastened to the inside of the body sheet metal panel.

From 1971 on, filler tube caps are the nonventing type and usually have some type of pressure-vacuum relief valve arrangement (Figure 3-35). Under normal operating conditions the valve closed, but whenever pressure or vacuum is more than the calibration of the cap, the valve opens. Once the pressure or vacuum has been relieved, the valve closes. Most pressure caps have four anti-surge tangs that lock onto the filler neck to prevent the delivery system's pressure from pushing fuel out of the tank. By turning such a cap one-half turn, the tank pressure is released slowly and keeps pressure from blasting out of the tank all at once. Then, with another quarter turn, the cap can be removed.

It is important that the service technician inspects the pressure to be sure that the seal is not torn, split, cracked, or hardened. Always check the service manual for the proper filler cap when replacement is necessary.

Starting with the 1976 model year, a Federal Motor Vehicle Safety Standard (FMVSS 301) re-

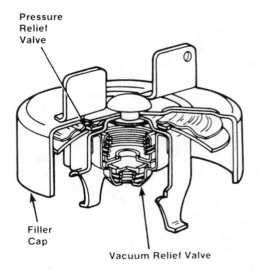

FIGURE 3-35 Pressure-vacuum relief filler cap

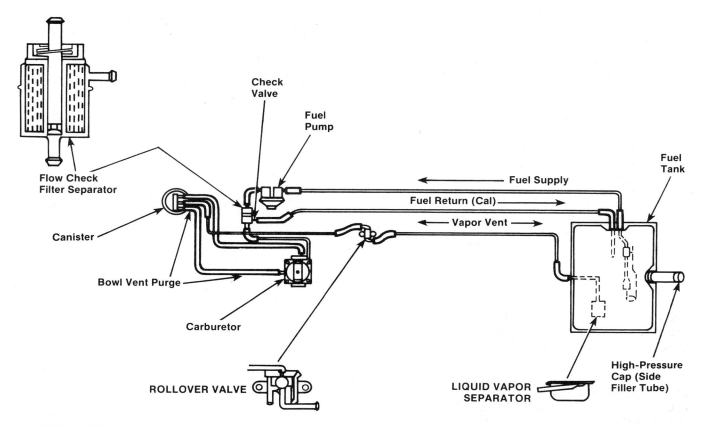

FIGURE 3-36   Rollover leakage protection devices

the gasoline supply line from the fuel tank to the carburetor. Under normal operation, fuel pump pressure is sufficient to open the valve and supply gasoline to the carburetor. If the vehicle rolls over, the fuel in the carburetor spills out, the engine stops, and the fuel pump ceases to operate. Most rollover leakage protection devices are variations of a basic one-way check valve. They are usually installed in the fuel vapor vent line between the tank and the vapor canister and at the carburetor fuel feed or return line fitting (Figure 3-36). In some protection systems the check valve is incorporated into the carburetor inlet fuel filter (Figure 3-37). A check valve might also be fitted in the fuel tank filler cap (Figure 3-38), and most caps' pressure-vacuum relief valve settings have been increased so that fuel pressure cannot open them in a rollover.

Some fuel line systems contain a fuel return arrangement which aids in keeping the gasoline cool, thus reducing chances of vapor lock. The return system (Figure 3-39) consists of a special fuel pump equipped with an extra outlet fitting and necessary fuel line. The fuel return line generally runs next to the conventional fuel line (Figure 3-40), except that flow is in the opposite direction. That is, the

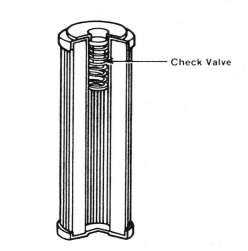

FIGURE 3-37   Gasoline filter with check valve

flow is through the return line toward the fuel tank, not toward the engine. The fuel return system allows a metered amount of cool fuel to circulate through the tank and fuel pump, thus reducing vapor bubbles caused by overheated fuel from upsetting the engine's operation by changing its fuel mixture.

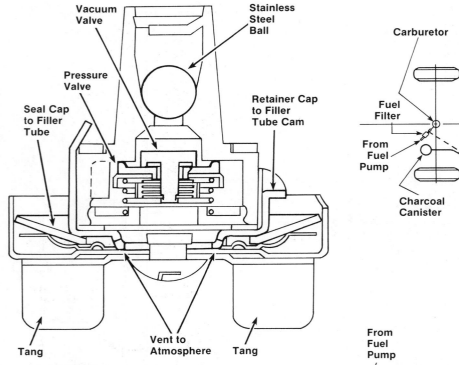

FIGURE 3–38 Cap with rollover check valve. Gravity causes steel ball to close off vent in cap.

Some form of liquid vapor separator is incorporated into most modern vehicles to stop liquid fuel or bubbles from reaching the vapor storage canister or the engine crankcase (Figure 3–41). It can be located inside the tank, on the tank (Figure 3–42), or in

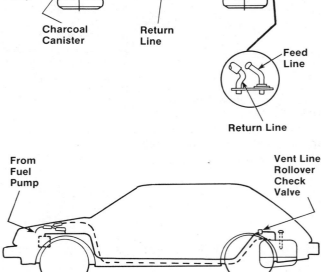

FIGURE 3–40 Parallel routing of fuel return line and vapor line

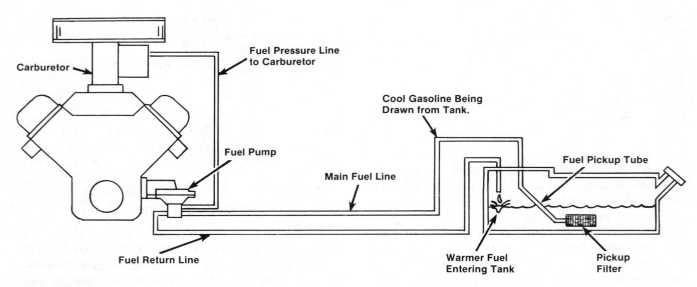

FIGURE 3–39 Gasoline is drawn from the tank into the fuel pump. From the pump, gasoline flows to the carburetor as well as back to the fuel tank. This tends to cool the gasoline in the pump and reduces the change of vapor lock.

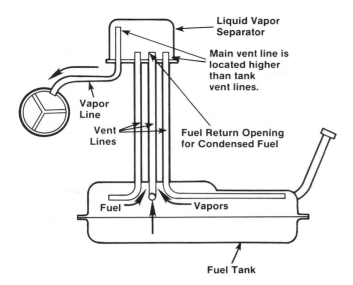

FIGURE 3-41 Fuel tank vapor separator allows some of the fuel vapors to condense back into liquid and return to the tank. Only vapors can normally enter the higher main vent tube opening.

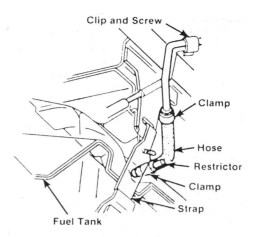

FIGURE 3-43 Vapor separator in fuel vent lines

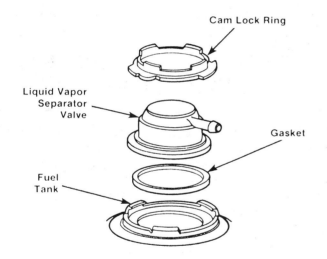

FIGURE 3-42 Vapor separator attached to the fuel tank

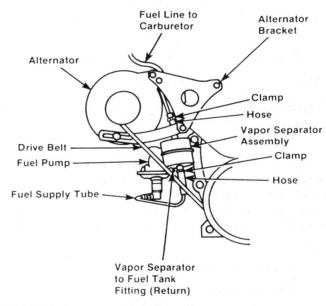

FIGURE 3-44 Vapor separator located near the fuel pump

fuel vent lines (Figure 3-43). It can also be located near the fuel pump (Figure 3-44). Check the service manual for the exact location of the liquid vapor separator and line routing.

Inside the fuel tank there is also a sending unit which includes a pickup tube and float-operated fuel gauge (Figure 3-45). The fuel tank pickup tube is connected to the fuel pump by the fuel line. Some electric fuel pumps are combined with the sending unit (Figure 3-46). The pickup tub extends nearly, but not completely, all the way to the bottom of the tank. Rust, dirt, sediment, and water cannot be drawn up into the fuel tank filter which can cause clogging. The ground wire is often attached to the fuel tank unit.

Servicing the Fuel Tank

Leaks in the metal fuel tank can be caused by a weak seam, rust, or road damage. The best method of permanently solving this problem is to replace the tank. Another method is to remove the tank and steam clean or boil in a caustic solution to remove the gasoline residue. After this has been done, the leak can be soldered or brazed.

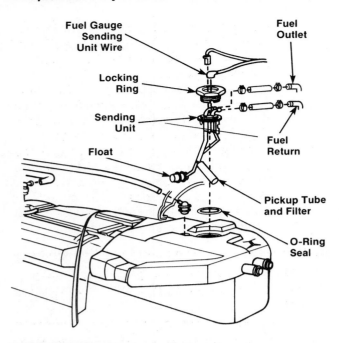

FIGURE 3-45 Tank sending unit

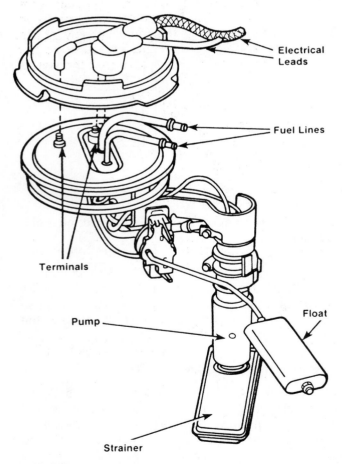

FIGURE 3-46 Electric fuel pump located inside a fuel tank

WARNING: Extreme care must be taken when using steam cleaning equipment or when washing with a caustic solution. Follow the manufacturer's instructions exactly. If the shop does not have steam cleaning or washing facilities, these services can be done by a radiator specialty shop. Do not use a steam cleaning procedure on a plastic tank. Never attempt to repair a fuel tank by soldering or brazing without first cleaning the inside of the tank. Also be sure that the inside of the tank is thoroughly dry and no liquid is present.

Although there are several methods of patching small holes in tanks other than welding, none are industry approved. Holes in a plastic tank can sometimes be repaired by using a special tank repair kit. Be sure to follow manufacturer's instructions when doing the repair.

WARNING: When raising a car or truck to remove a fuel tank, never use a vehicle jack to support the vehicle if you are going to work under it. Safety stands should be used when you do service or maintenance work under your vehicle.

Replacing the Fuel Tank. When a fuel tank is leaking dirty water or has water in it, the tank must be cleaned, repaired, or replaced. To do this, disconnect the negative terminal from the battery, remove the tank filler cap, and drain the tank of all fuel. Then proceed as follows:

 1. Disconnect the fuel line at the tank that runs to the fuel pump. Then connect a syphon hose or similar device to the tank and draw the fuel into a clean, approved safety can (Figure 3-47). Operate the syphon device as directed by its manufacturer.

CAUTION: Abide by local laws for the disposal of contaminated fuels. Also be sure to wear eye protection when working under the vehicle.

2. Attach a piece of masking tape identifying tag to each tank line to insure correct reinstallation. Disconnect the vent lines and the plugged wire connected to the sending

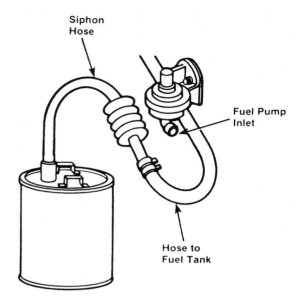

FIGURE 3-47 Proper equipment should be used to safely handle and store highly flammable and explosive gasoline

unit. This wire is usually mounted on the upper front or top section of the fuel tank.

3. Unfasten the filler from the tank. If it is a rigid one-piece tube, remove the screws around the outside of the filler neck near the filler cap. If it is a three-piece unit, remove the neoprene hoses after the clamp has been loosened.

4. Loosen the bolts holding the fuel tank straps to the vehicle (Figure 3-48) until they are about two threads from the end. Holding the tank securely against the underchassis with one hand, remove the strap bolts and lower the tank to the ground. When lowering the tank, make sure all wires and tubes are unhooked. Also keep in mind that small amounts of fuel might still remain in the tank.

CAUTION: Use a drain pan to catch any spilled fuel or be sure to clean it up immediately.

---

5. To reinstall the new or repaired fuel tank, reverse the removal procedure. Be sure that all the rubber or felt tank insulators are in place. Then, with the tank straps in place, position the tank. Loosely fit the tank straps around the tank, but do not tighten them. Make sure that the hoses,

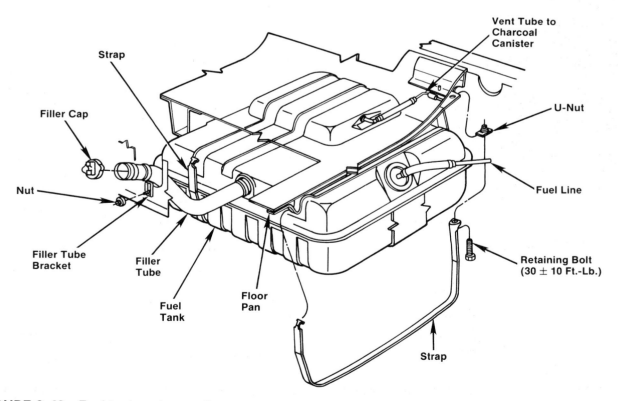

FIGURE 3-48 Fuel tank and mounting components

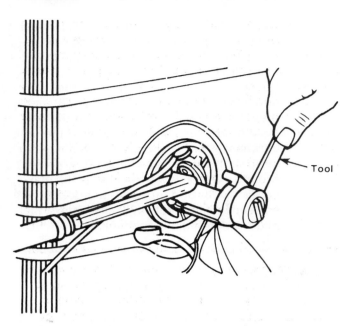

FIGURE 3-49 A special tool will help to remove a sending unit retaining ring

wires and vent tubes are connected properly. Also check the filler neck for alignment and for insertion into the tank. Tighten the strap bolts and secure the tank to the car. Install all of the tank accessories (vent line, sending unit wires, ground wire, fill tube, and so on). Fill the tank with fuel and check it for leaks, especially around the filler neck and the pickup assembly. Reconnect the battery and check the fuel gauge for proper operation.

Removing and Replacing the Fuel Gauge Sending Unit. The sending unit is held in the tank by either a retaining ring or screws. The easiest way to remove a sending unit retaining ring is to use a special tool designed for this purpose (Figure 3-49). This tool fits over the metal tabs on the retaining ring, and after about a quarter turn, the ring will come loose and the sender unit can be returned. If the special tool is not available, a drift punch or screwdriver and ball peen hammer will usually do the job (Figure 3-50).

When removing the sending unit from the tank, be very careful not to damage the float arm, the float, or the fuel gauge sender. Check the unit carefully for any damaged components. Shake the float, and if fuel can be heard inside it, replace the float. Make sure the float arm is not bent. It is usually wise to replace the filter and O-ring before replacing the unit (Figure 3-51). Check the fuel gauge as de-

scribed in the service manual. When reinstalling the pickup pipe-sending, be very careful not to damage any of the components.

FUEL LINES

Fuel lines can be made of either metal tubing or flexible nylon or synthetic rubber hose. The latter must be able to resist gasoline. It must also be non-permeable, so gas and gas vapors cannot evaporate

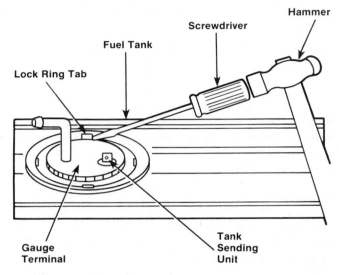

FIGURE 3-50 A drift punch or screwdriver and ball peen hammer can be used to remove the tank unit.

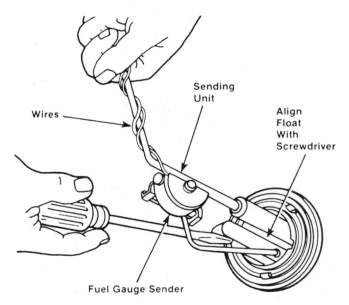

FIGURE 3-51 Be extremely careful not to damage the float, float arm, or screen when installing a tank sending unit.

through the hose. Ordinary rubber hose, such as that used for vacuum lines, deteriorates when exposed to gasoline. Only hoses made for fuel systems should be used for replacement. Similarly, vapor vent lines must be made of material that will resist attack by fuel vapors. Replacement vent hoses are usually marked with the designation EVAP to indicate their intended use. The inside diameter of a fuel delivery hose is generally larger (5/16 to 3/8 of an inch) than that of a fuel return hose (1/4 of an inch).

The fuel line system actually starts at the fuel tank. Here a fuel filler neck tube or hose connects with the tank. This short connection is made by a straight or curved tubing or hose that is roughly 1 1/2 to 2 1/2 inches in diameter. Where the latter is employed, there is usually an internal wire coil that resists collapsing.

Many fuel tanks have vent hoses to allow air in the fuel tank to escape when the tank is being filled with fuel. Vent hoses are usually installed alongside the filler neck hose.

The fuel lines carry fuel from the fuel tank to the fuel pump, the fuel filter, and to the carburetor or fuel metering pump. These lines are usually made of rigid metal, although some sections are constructed of rubber hose to allow for car vibrations. This fuel line, unlike filler neck or vent hoses, must work under pressure or vacuum. Because of this, the flexible synthetic hoses must be stronger. This is especially true for the hoses on fuel injection systems, where pressures reach 50 psi or more. For this reason flexible fuel line hose must also have special resistance properties. In fuel injection systems, unused gas is returned to the fuel tank where it becomes "sour gas," which ages when hydroperoxide molecules form. Hydroperoxides can cause some fuel line hoses to crack and disintegrate. Many auto manufacturers recommend that flexible hose should only be used as a delivery hose to the fuel metering unit in fuel injection systems. It should *not* be used on the pressure side of the injector systems. This application requires a special high-pressure hose.

All fuel lines should occasionally be inspected for holes, cracks, leaks, kinks, or dents. Many fuel system troubles that occur in the lines are blamed on the fuel pump or carburetor. For instance, a small hole in the fuel line will admit air but will not necessarily show any drip marks under the car. This usually occurs if the hole is high up in the line, especially where it curves up over the rear axle. Air can then enter the fuel line, allowing the fuel to gravitate back into the tank. Then, instead of drawing fuel from the tank, the fuel pump draws only air through the hole in the fuel line. When this condition exists, the fuel pump is frequently tested, and if there is insufficient

fuel, it is considered faulty, when in fact there is nothing wrong with it. If the hole is suspected, remove the coupling at the tank and the pump and pressurize the line with air. The leaking air is easily spotted.

Since the fuel is under pressure, leaks in the line between the pump and carburetor and injectors are relatively easy to recognize. When a damaged fuel line is found, replace it with one of similar construction—steel with steel, and the flexible with nylon or synthetic rubber. When installing flexible tubing, always use new clamps. The old ones lose some of their tension when they are removed and will not provide an effective seal when used on the new line.

CAUTION: Do not substitute aluminum or copper tubing for steel tubing. Never use hose within 4 inches of any hot engine or exhaust system component. A metal line must be installed.

Fuel supply lines from the tank to the carburetor or injectors are routed to follow the frame along the underchassis of vehicles. Generally, rigid lines are used extending from near the tank to a point near the fuel pump. To absorb engine vibrations, the gaps between the frame and tank or fuel pump are joined by short lengths of flexible hose (Figure 3-52).

Any damaged or leaking fuel line—either a portion or the entire length—must be replaced. To fabricate a new fuel line, select the correct tube and fitting dimension and start with a standard length that is slightly longer than the old line. With the old line as a reference, use a tubing bender to form the same bends in the new line as those that exist in the old. Although steel tubing can be bent by hand to obtain a gentle curve, any attempt to bend a tight curve by hand will usually kink the tubing. Since a kink in a brake line will weaken the line, a kinked line should never be used. To avoid kinking, always use a bending tool like the one shown in Figure 3-53.

The two most-used tubing fittings are either the compression type or the double-flare type (Figure 3-54). The double flare—which is the most common—is made with a special tool that has an anvil and a cone (Figure 3-55). The double flaring process is performed in two steps.

1. The anvil begins to fold over the end of the tubing.
2. The cone is used to finish the flare by folding the tubing back on itself. This action will double the thickness and create two sealing surfaces.

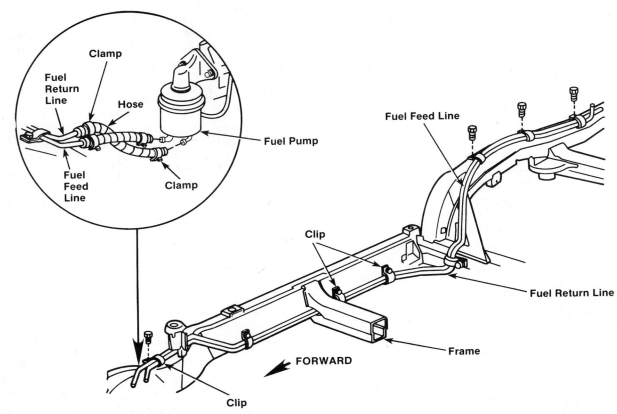

FIGURE 3-52 Gaps between the frame and tank or fuel pump are joined by short lengths of flexible hose.

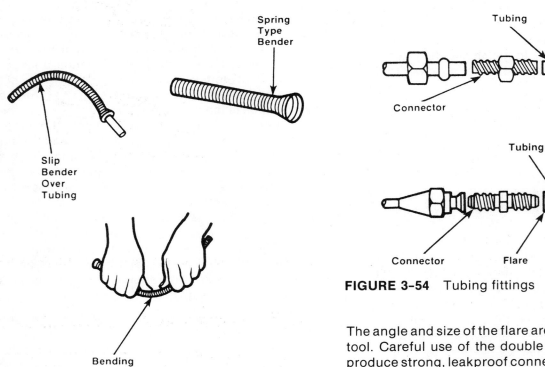

FIGURE 3-53 A bending tool avoids kinking

FIGURE 3-54 Tubing fittings

The angle and size of the flare are determined by the tool. Careful use of the double flaring will help to produce strong, leakproof connections. Figure 3-56 shows other metal fuel connections that are used by vehicle manufacturers.

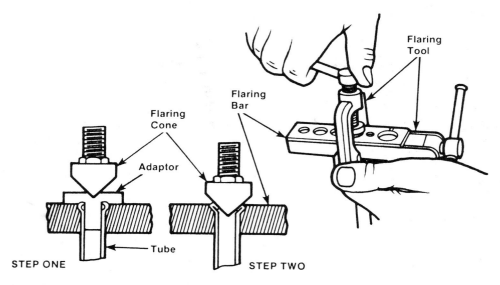

FIGURE 3–55   Double flare fit is made with a special tool

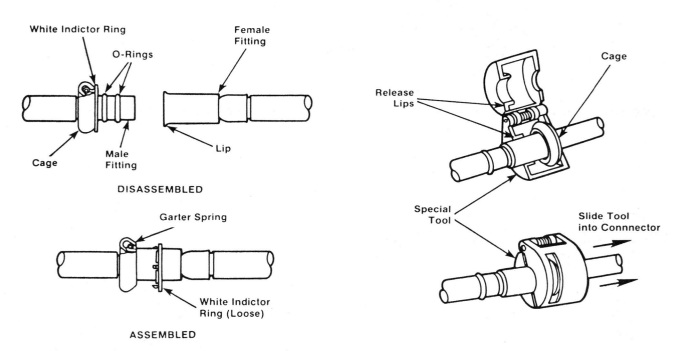

FIGURE 3–56   Metal fuel line connections

The flare tool can also be used to make sure that nylon and synthetic rubber hoses will stay in place. That is, to make sure the connection is secure, put a partial double-lip flare on the end of the tubing over which the hose is installed. This can be done quickly, with the proper flaring tool, by starting out as if it was going to be a double flare, but stopping halfway through the procedure (Figure 3–57). This provides an excellent sealing ridge that will not cut into the hose and holds tighter than a straight pipe, especial-ly if a clamp is placed directly behind the ridge on the hose.

 SHOP TALK _____

To insure a flexible fuel replacement hose is the right length, lay the old line alongside the new one. Then, with a sharp knife, cut the new line to the same length as the old one.

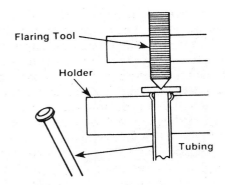

FIGURE 3-57 Cut hose connections must be secured in place.

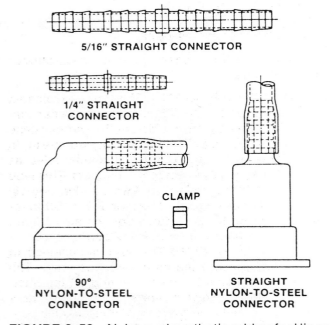

FIGURE 3-58 Nylon and synthetic rubber fuel lines

Nylon and synthetic rubber fuel line connectors are illustrated in Figure 3-58. Note that there are connectors available to combine rigid and flexible fuel lines. Of course, the methods of jointing nylon vacuum hose (Figure 3-11) covered earlier in this chapter apply to fuel lines. But, *never* use vacuum lines in a fuel system.

There are a variety of clamps used on fuel system lines, including the screw types (Figure 3-59). The crimp-type clamps shown in Figure 3-60 are used most for metal tubing, but they require a special tool to install.

FUEL FILTERS

There are two basic fuel delivery line filters:

- *Gasoline Filters* (Figure 3-61). These are designed to protect carburetion and fuel systems from contaminants that might be present in the gasoline. These contaminants, which include dirt, rust, scale, and water, can cause carburetor malfunction and engine wear.
- *Diesel Fuel Filters.* These are used with the diesel engines powering most heavy-duty over-the-road vehicles and off-the-road construction equipment.

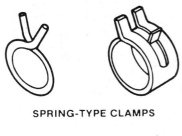

SPRING-TYPE CLAMPS

WORM DRIVE   ROLLED-EDGE

FIGURE 3-59 A variety of clamps used on fuel system hoses

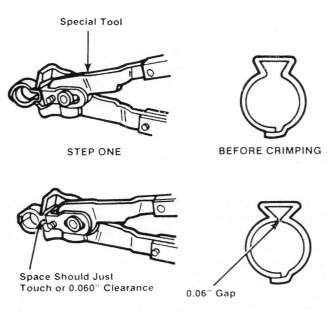

FIGURE 3-60 Crimp-type clamps

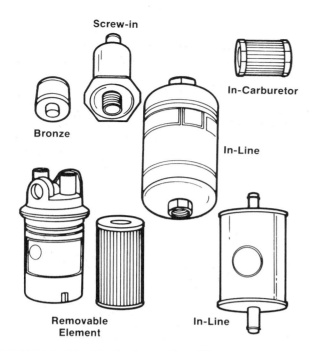

FIGURE 3-61   Typical gasoline filters

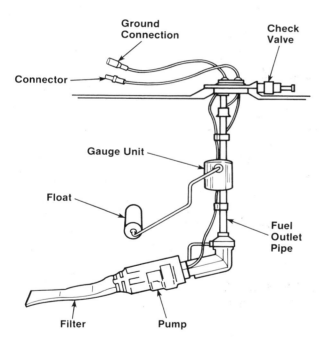

FIGURE 3-62   Cars and light trucks usually have an in-tank strainer and a gasoline filter.

Since gasoline filters and diesel fuel differ in application and servicing, each style is described separately. This chapter covers gasoline filters; diesel fuel filters are covered in Chapter 8.

Automobiles and light trucks usually have an in-tank strainer and a gasoline filter (Figure 3-62).

The strainer, located in the gasoline tank, is made of a finely woven fabric. The purpose of this strainer is to prevent large contaminant particles from entering the fuel system where they could cause excessive fuel pump wear. It also helps prevent passage of any water that might be present in the tank. Servicing of the fuel tank strainer is seldom required.

The second gasoline filter is usually located in the engine compartment and is the one this section will examine because it is replaceable and might require service on a regular basis. Most automobile applications are covered by the following types of gasoline filters:

- In-carburetor filters
- In-line carburetor filters
- Out-pump filters

In-Carburetor Filters

There are three basic types of in-carburetor gasoline filters:

1. *Pleated Paper Filter.* This type uses pleated paper as the filtering media. Paper elements are more efficient than screen-type elements, such as nylon or wire mesh, in removing and trapping small particles, as well as larger size contaminants. This type of filter is available in 1- and 2-inch lengths.
2. *Sintered Bronze Gasoline Filter.* Often referred to as a "stone" or "ceramic" filter, this filter is also an in-carburetor type.
3. *Screw-in Filter.* This type is designed to screw into the carburetor fuel inlet. The fuel line attaches to a fitting on the filter. One of newest filters of this type has a magnetic element to remove metallic contamination before it reaches the carburetor (Figure 3-63).

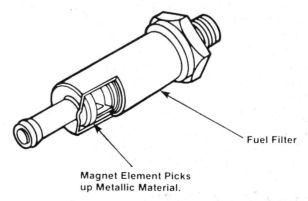

FIGURE 3-63   Filter with magnetic element to remove metallic contamination from the fuel

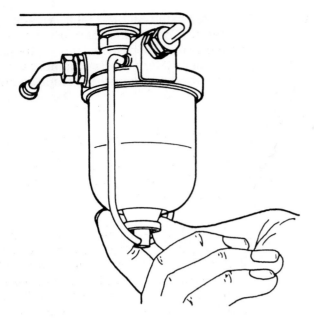

FIGURE 3-64 Sediment bowls can contain a paper, fiber, ceramic, or metal filter element.

Some imported vehicles as well as older domestic cars have a sediment bowl between the fuel pump and carburetor. The bowl contains a pleated paper and ceramic filter element. The bowl cover is held in place by a wire bail and clamp screw (Figure 3-64). It can be removed for filter cleaning or replacement. Ceramic can be cleaned and reused if necessary, but paper filter elements must be replaced when they are dirty.

In-Line Filters

This type of gasoline filter is installed in the fuel line (Figure 3-65). In carbureted engines, the in-line gasoline filter is usually installed between the fuel pump and the carburetor; in vehicles with a fuel injection system, the location of the fuel filter is determined by the manufacturer. Fitted with a pleated paper element, the in-line filter is sometimes installed as an extra protective measure. The optionally installed in-line filter will then work in conjunction with the in-tank and in-carburetor gasoline filters. Because of its large capacity, an in-line filter is often the most economical solution to a fuel system's contamination problems. The arrow on the filter shows the direction of fuel flow. Some in-line filters are called *cartridge-type filters*. They fit into a retainer housing (Figure 3-66).

Outlet Pump Filters

Some vehicles have fuel filters in the outlet side of the fuel pump (Figure 3-67).

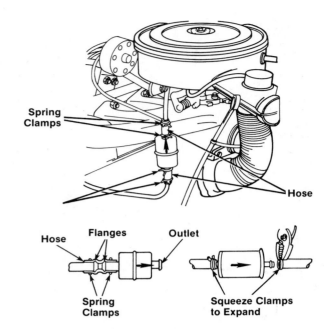

FIGURE 3-65 In-line fuel filter replacement

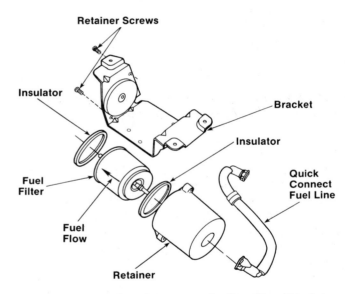

FIGURE 3-66 Cartridge-type in-line filter fits into a retainer housing

SERVICING FILTERS

In-line and in-carburetor filters and elements are serviced by replacement only. Replacing the gasoline filter or element at the intervals recommended by the vehicle or engine manufacturer is the most effective method of minimizing fuel starvation and other carburetor problems. On occasions when the fuel system has been subjected to excessive amounts of contaminants, more frequent filter changes may be required. Carburetor flooding or

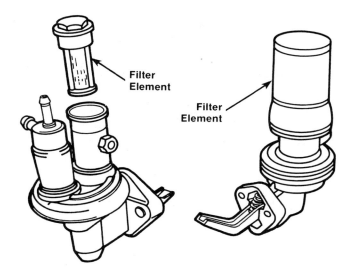

FIGURE 3-67   Some fuel filters are in the outlet side of the fuel pump

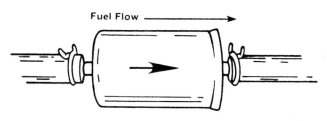

FIGURE 3-68   Inline fuel filters must be installed properly or fuel flow will be restricted.

power loss under high fuel demand can indicate possible filter plugging. The gasoline filter should be examined as a part of the diagnosis of these problems. If the problem is contaminants in the fuel, the type of contaminant held by the fuel filter can give the experienced technician an indication of the source and severity of the problem.

 SHOP TALK _____

Most inline filters are marked with arrows (Figure 3-68) and/or words (such as to carb).

Replacing an In-Carburetor Filter

To replace an in-carburetor gasoline filter, proceed as follows:

 1.   To remove an in-carburetor gasoline filter, remove the fuel line from the carburetor fuel inlet nut using two wrenches, one to hold the inlet nut, the other to turn the fuel line fitting (Figure 3-69).

2.   After disconnecting the fuel line, remove the inlet nut from the carburetor and discard the used filter. The inlet nut will have a gasket that is loose or permanently attached. This gasket can usually be reinstalled.

 SHOP TALK _____

Take care not to lose the spring located behind the filter. The spring is required to seal the filter to the inlet nut.

3.   The correct size and type of gasoline filter for the engine or vehicle being serviced is listed in the service manual. Installation of the filter is accomplished by simply reversing the steps used to remove the used gasoline filter.
4.   Check fuel line connections for leaks with the engine running.

Replacing a Sintered Bronze Filter

Replacement is similar to replacing the in-carburetor gasoline filter. Take care to reinstall the gasket at the fuel inlet to the filter.

Most in-line filters have tubes or threaded fittings. Tube fittings are used with rubber hoses and clamps. Threaded fittings may be inverted flare, or male or female pipe threads. When replacing an in-

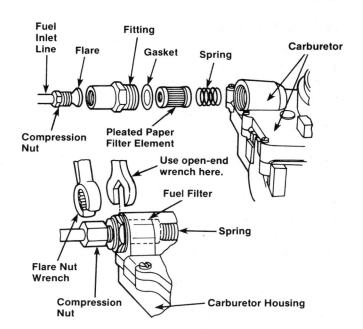

FIGURE 3-69   Removing parts of a carburetor fuel filter

line filter, take care to match the end of the gasoline line with the proper fitting on the filter housing. Most in-line gasoline filters have an arrow on the housing to show direction of fuel flow.

 SHOP TALK

Late-model automobiles have steel gasoline lines. Non-OE in-line gasoline filters can be installed using short rubber tubes and screw-type, tangential hose clamps. There should be no bends in the rubber tubes after installation. After installing an in-line gasoline filter, start the engine and check for leaks.

Replacing a Screw-in Filter

To replace a screw-in gasoline filter, proceed as follows:

1. Detach the fuel line from the gasoline filter.
2. Unscrew the filter from the carburetor using a box-end wrench. Be sure to discard the used filter.
3. Screw a new filter with gasket into the carburetor.
4. Reconnect the fuel line, start the engine, and check for leaks.

Most outlet pump or pump outlet filters are the screw-in type and are replaced/installed as described above.

Replacing an In-Line Filter

The in-line gasoline filter for some truck applications has a replaceable element. To replace, proceed as follows:

1. Unscrew the shell for access to the gasoline element.
2. Remove the used element by pulling it straight off the housing.
3. Install the new element specified in the vehicle's service manual.
4. Tighten the filter housing by hand.
5. Start the engine and check for leaks.

Cleaning Fuel Lines

If fuel filters clog frequently or the sediment bowl contains excessive amounts of contaminant, there is a good probability that the fuel lines and tank are dirty. Methods of cleaning fuel tanks were covered previously in this chapter.

To clean a fuel line, first remove the ground cable from the battery and the voltage supply wire.

Then disconnect the main fuel line at the fuel pump or the engine. Remove the fuel return line at the pump or liquid vapor separator. With both ends of the fuel line disconnected, blow out the contaminants with compressed air.

To complete the cleaning job, reconnect the fuel lines and fill the tank with about 5 gallons of fuel. Also reconnect the battery, but not the ignition system voltage supply wire. Connect one end of a hose from a supply line at the engine and place the end into a suitable approved gas container. Crank the engine with the starting motor until approximately 2 quarts of fuel are pumped into the container. This will purge any remaining contaminants from the line. Reconnect all components and test the fuel lines for any leaks.

FUEL PUMPS

The fuel pump is the device that draws the fuel in the fuel tank through the fuel lines to the engine's carburetor or injectors. Basically, there are two types of fuel pumps: mechanical and electrical. The latter is the most commonly used today.

Mechanical Fuel Pumps

Mechanical fuel pumps have a synthetic rubber diaphragm inside the unit that is actuated by an eccentric located on the engine's camshaft (Figure 3-70). As the camshaft rotates during engine opera-

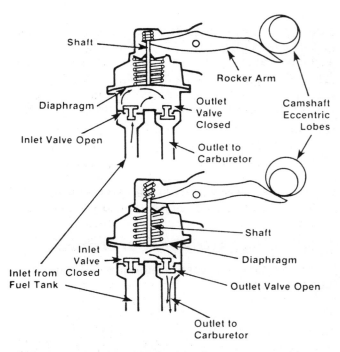

FIGURE 3-70 Mechanical fuel pump assembly

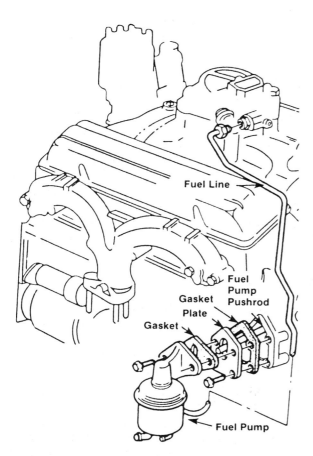

FIGURE 3-71 V-8 engines usually have pushrod between the camshaft eccentric and the fuel pump.

tion, a shaft or rocker arm in the pump is moved up and down or back and forth, depending on the fuel pump's position on the engine. This causes the diaphragm to move back and forth, drawing fuel from the fuel tank, through the fuel lines, and to the carburetor or injectors. On some V-8 engines, a pushrod, located between the camshaft eccentric and fuel pump, actuates the fuel pump rocker arm (Figure 3-71).

The mechanical fuel pump is located on the engine block near the front of the engine (Figure 3-72). This, of course, subjects it to engine heat. Further, during fuel intake, a low-pressure area is developed in the fuel line. High temperatures and low pressure can lead to a vapor lock condition. To overcome this condition, most modern mechanical fuel pump systems contain a fuel vapor separator and vapor return line to the fuel tank.

The modern fuel pump is a sealed unit that cannot be repaired. If the pump leaks from either the vent hole or from a seam, it must be replaced. If the engine performance indicates inadequate fuel, the

pump, while mounted on the engine, should be tested for pressure and volume. In fact, incorrect fuel pump pressure and low volume (capacity of flow rate) are the two most likely fuel pump troubles that will affect engine performance.

- Low pressure will cause a lean mixture and fuel starvation at high speeds.
- Excessive pressure will cause high fuel consumption and carburetor flooding.
- Low volume will cause fuel starvation at high speeds.

To determine that the fuel pump is in satisfactory operating condition, tests for both fuel pump pressure and fuel pump capacity (volume) should be performed. These tests are performed with the fuel pump installed on the engine and the engine at normal operating temperature and at idle speed.

 SHOP TALK _____

Before making any tests, be sure the replaceable fuel filter has been changed within the recommended maintenance mileage interval. When in doubt, install a new filter.

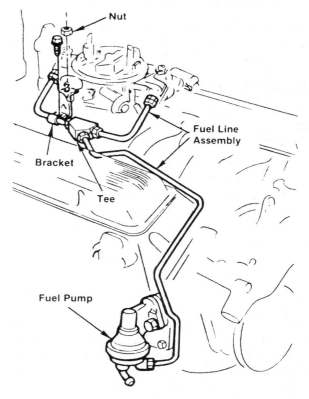

FIGURE 3-72 Typical fuel pump and line installation on a V-8 engine

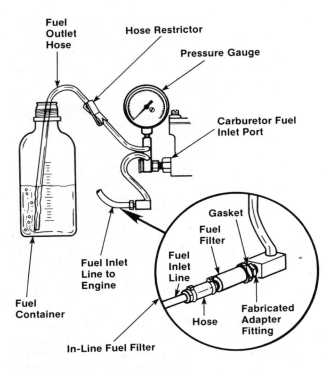

FIGURE 3-73   Fuel pump pressure test

To make a pressure test, proceed as follows:

1. Remove the air cleaner assembly. Disconnect the fuel inlet line or the fuel filter at the carburetor. **Use care to prevent combustion from fuel spillage.**
2. Connect a pressure gauge, a restrictor and a flexible hose (Figure 3-73) between the fuel filter and the carburetor.

WARNING: The fuel supply lines will remain pressurized for long periods of time after the engine is shut down. This pressure must be relieved before servicing of the fuel system is begun. A valve is provided on the throttle body for this purpose.

3. Position the flexible hose and the restrictor so the fuel can be discharged into a suitable, graduated container.
4. Before taking a pressure reading, operate the engine at the specified idle rpm and vent the system into the container by opening the hose restrictor momentarily.
5. Close the hose restrictor. Allow the pressure to stabilize and note the reading. If the pump pressure is not within specifications

Engine		Static Pressure (psi)[1]	Minimum Volume Flow[1] [3]
Liter	CID		
2.3		5.0–7.0[2]	1 pt in 25 sec
3.3	200	5.0–7.0	1 pt in 30 sec
4.1	250	5.0–7.0	1 pt in 20 sec
4.2	255	6.0–8.0	1 pt in 20 sec
5.0	302	6.0–8.0	1 pt in 20 sec
5.8	351W	6.0–8.0	1 pt in 20 sec
5.8	351M	6.0–8.0	1 pt in 20 sec
6.6	400	6.0–8.0	1 pt in 20 sec

[1]On engine with temperatures normalized and at normal curb idle speed, transmission in neutral
[2]With the pump-to-tank fuel return line pinched off and a new fuel filter installed in fuel line
[3]Inside diameter of smallest passage in test flow circuit must not be smaller than 0.220″

FIGURE 3-74   Typical fuel pump specifications

(Figure 3-74) and the fuel lines and filter are in satisfactory condition, the pump should be replaced.

 SHOP TALK _____

Fuel pump pressure specifications are given by the manufacturer in pounds per square inch (psi) at a certain engine speed.

If the fuel pump pressure is within specifications, test the capacity (volume) as follows:

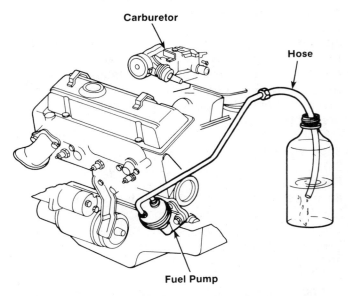

FIGURE 3-75   To perform a delivery flow test, open the clip and allow fuel to enter the container.

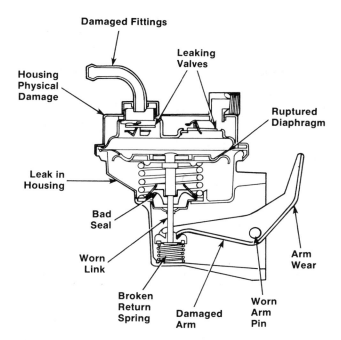

FIGURE 3-76   Common problems with a mechanical pump

 1.   Operate the engine at the specified idle rpm.
2.   Open the hose restrictor and allow the fuel to discharge into the graduated container (Figure 3-75) for the specified time. Then, close the restrictor. At least 1 pint of fuel should have been discharged within the specified time limit.
3.   If the pump volume is below specifications, repeat the test using an auxiliary fuel supply and a new fuel filter.
4.   If the pump volume meets specifications while using the auxiliary fuel supply, check for a restriction in the fuel supply from the tank and for improper tank ventilation.

 SHOP TALK _____

Manufacturer's specifications generally are given in terms of 1 pint of fuel delivered in a given number of seconds while the engine is at idle and also at a specified speed. Some manual specifications just state a good volume at cranking speed. This usually means a pint of fuel is delivered in 30 seconds or less at an engine speed of 500 rpm.

5.   If the pump fails the pressure and volume test, it generally requires replacement. But

before replacing the unit make sure that there are no restricted or leaking fuel lines or clogged filter hoses and that there is no dirt or sludge at the fuel pickup in the tank. Blow out any restriction with compressed air or replace any damaged parts. Figure 3-76 shows the common problems with a mechanical pump.

To replace the fuel pump (Figure 3-77), remove the inlet and outlet lines at the pump. Use a plug to stop the flow of fuel from the tank. With the correct size socket wrench, remove the mounting bolts, then remove the pump from the engine. Clean the old gasket material from the engine block. Apply gasket sealer to the mounting surface on the engine and to the threads of the mounting bolts. Install the new gasket, then position the pump by tilting it either toward or away from the block to correctly place the lever against the cam. If the pump is driven by a pushrod, the rod must be held up to permit the rocker arm to go under it. When the pump is properly positioned there should be an internal squeaking noise with each movement of the pump. Tighten the mounting bolts firmly. Attach the inlet and outlet lines, start the engine, and check for leaks.

Electric Fuel Pumps

Electric fuel pumps offer important advantages over mechanical fuel pumps (Figure 3-78). Because

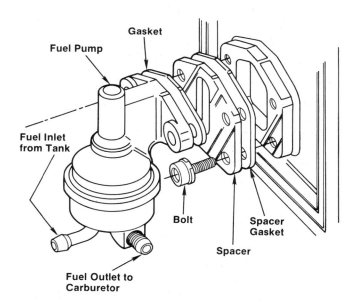

FIGURE 3-77   Before removing a fuel pump, disconnect the fuel lines at the pump. Using a proper-size wrench, remove the bolts holding the fuel pump to the engine block.

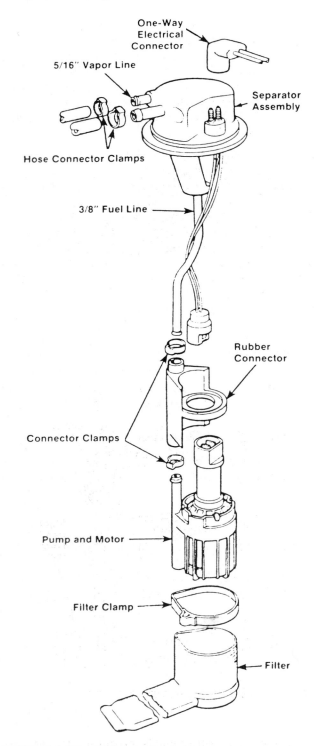

FIGURE 3-78 Electric fuel pump components

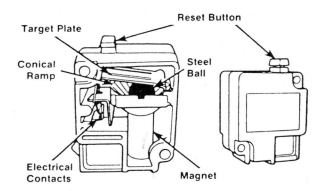

FIGURE 3-79 Some models use an inertia switch to turn off the electric fuel pump in an accident.

electric fuel pumps maintain constant fuel pressure, they aid starting and reduce vapor lock problems. Since they are not mechanically attached to the engine, fit problems are eliminated, making them especially useful when an exact replacement mechan-

ical fuel pump is unavailable. Unlike mechanical fuel pumps, the operation of electric pumps is not affected by worn cams. It is also easy to install a hidden on/off switch for the electric pump as an anti-theft precaution. An inertia switch (Figure 3-79) is sometimes installed in the electric fuel pump circuit to turn it off in case of an accident.

The electric fuel pump can be located inside (Figure 3-80) or outside (Figure 3-81) the fuel tank. As described in Chapter 1, there are four basic types of electric fuel pumps: the diaphragm, plunger, bellows, and impeller or rotary. The in-tank electric pump is usually a rotary type (Figure 3-82). The diaphragm, plunger, and bellows type (Figure 3-83) are usually the styles requested. That is, when the ignition is turned on, the pump begins operation. It shuts off automatically when the carburetor bowl is full and the fuel line is pressurized. When the carburetor demands more fuel, the electric pump pumps more. When demand is lower, it pumps less, so proper fuel flow and pressure are constantly maintained. In most installations, the rotary electric fuel pump is considered to be a continuous operated type. A typical wiring diagram for an electric pump is shown in Figure 3-84. Some major problem areas in an electric pump system are the following:

- Dead or inoperative pump
- Too much fuel pressure
- Too little fuel pressure

A high pressure reading usually indicates either a faulty pressure regulator or an obstructed return line. Disconnect the fuel return line at the tank. Use a length of hose to route the returning fuel into an appropriate container. Start the engine and note the pressure reading at the engine. If fuel pressure is now within specs, check for an obstruction in the in-tank return plumbing. The fuel reservoir check valve or aspirator jet might be clogged.

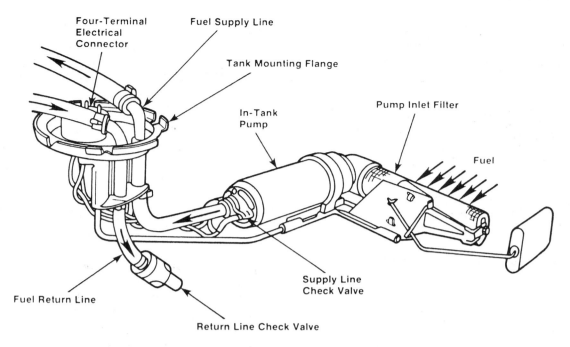

FIGURE 3-80   Electric fuel pump system located inside the fuel tank.

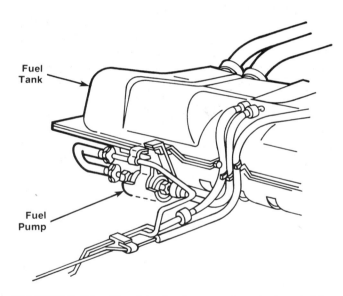

FIGURE 3-81   Electric fuel pump located outside the fuel tank.

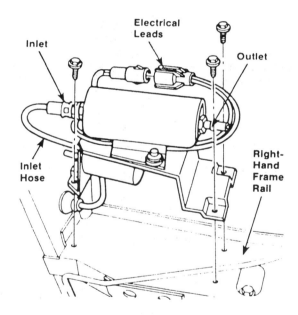

FIGURE 3-82   Rotary-type fuel pump

If the fuel pressure still reads high with the return line disconnected from the tank, note the volume of fuel flowing through the line (Figure 3-85). Little or no fuel flow can indicate a plugged return line. Shut off the engine and connect a length of hose directly to the fuel pressure regulator return port to bypass the return hose. Restart the engine and again check the pressure reading. If bypassing the return line brings the readings back within

specs, a plugged return line is the culprit. If pressure is still high, apply vacuum to the pressure regulator to see if that makes a difference (it should). If there is still no change, replace the faulty pressure regulator. If applying vacuum directly to the regulator lowers fuel pressure, the vacuum hose that controls the operation of the regulator might be plugged, leaking, or misrouted.

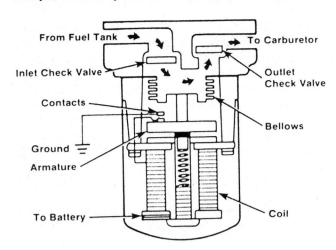

FIGURE 3-83   Bellows-type fuel pump

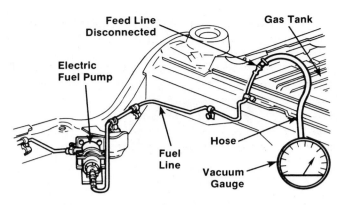

FIGURE 3-85   Vacuum test of electric pump

Low pressure, on the other hand, can be due to a clogged fuel filter, restricted fuel line, weak pump, leaky pump check valve, defective fuel pressure regulator, or dirty filter sock in the tank. It is possible to rule out filter and line restrictions as a cause of the problem by making a pressure check at the pump outlet. A higher reading at the pump outlet (at least 5 psi) means there is a restriction in the filter or line. If the reading at the pump outlet is unchanged, then either the pump is weak or is having trouble picking up fuel (clogged filter sock in the tank). Either way it will be necessary to get inside the fuel tank. If the filter sock is gummed up with dirt or debris, it is also wise to clean out the tank when the filter sock is cleaned or replaced.

Another source of trouble is the pump check valve. Some pumps have one, others have two (positive displacement roller vane pumps). The purpose of the check valve is to prevent fuel movement through the pump when the pump is off so residual pressure will remain in the injectors (which can be checked by watching the fuel pressure gauge after the engine is shut off).

Depending on the type of pump used, the check valve can also prevent the reverse flow of fuel and/or relieve internal pressure to regulate maximum pump output. Check valves can stick and leak, so if a pump runs but does not pump fuel, a bad check valve is to blame. Unfortunately, the check valve is usually an integral part of the pump assembly, which is sealed at the factory. Therefore, if the check valve is causing trouble, the entire pump has to be replaced.

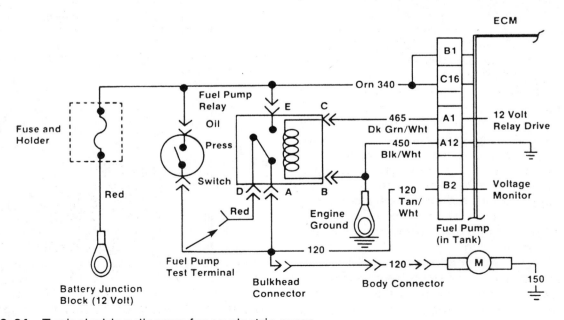

FIGURE 3-84   Typical wiring diagram for an electric pump

When an engine fails to start because there is no fuel delivery, the first step is to check the fuel gauge. A gauge that reads higher than a half tank probably means there is fuel in the tank, but not always. A defective sending unit or miscalibrated gauge might be giving a false indication. Sticking a wire or dowel rod down the fuel tank filler pipe will tell whether or not there is really fuel in the tank. If the gauge is faulty, repair or replace as discussed later in this chapter.

Listen for pump noise. When the key is turned on, the pump should buzz for a couple of seconds to build system pressure. The pump is usually energized through an oil pressure switch (the purpose of which is to shut off the flow of fuel in case of an accident that stalls the engine). But on most late-model cars with computerized engine controls, the computer energizes a pump relay when it receives a cranking signal from the distributor pickup or crankshaft sensor. An oil pressure switch might still be included in the circuitry for safety purposes and to serve as a backup in case the relay or computer signal fails. Failure of the pump relay or computer driver signal can cause slow starting because the fuel pump will not come on until the engine cranks long enough to build up sufficient oil pressure to trip the oil pressure switch.

Another element that might be included in the pump wiring circuit is an inertia safety switch. The switch usually located somewhere in the trunk, under the back seat, or behind a rear kick panel, is designed to turn off the fuel pump in a severe impact. A reset button should restore power to the pump. But if the switch is oversensitive so that ordinary bumps trip it, the switch might have to be replaced.

If a buzzing sound is not heard when the key is on or while the engine is being cranked, check for the presence of voltage at the pump electrical connectors (Figure 3-86). The pump might be good, but if it does not receive voltage and have a good ground, it will not run. To check the ground (Figure 3-87), connect a test light across the ground and feed wires at the pump to check for voltage, or use a voltmeter to read actual voltage and an ohmmeter to check ground resistance. The latter is the better test technique because a poor ground connection and/ or low voltage can reduce pump operating speed and output. If the electrical circuit checks out but the pump does not run, the pump is probably bad and should be replaced.

No voltage at the pump terminal when the key is on and the engine is cranking indicates a faulty oil pressure switch, pump relay, relay driver circuit in the computer, or a wiring problem. Check the pump fuse to see if it is blown. Replacing the fuse might

restore power to the pump, but until you have found out what caused the fuse to blow, the problem is not solved. The most likely cause of a blown fuse would be a short in the wiring between the relay and pump, or a hot short inside the oil pressure switch or relay.

A faulty oil pressure switch can be checked by bypassing it with a jumper wire. If that restores power to the pump and the engine starts, replace the switch. If an oil pressure switch or relay sticks in the *closed* position, the pump can run continuously whether the key is on or off depending on how the circuit is wired.

To check a pump relay, use a test light to check across the relay's hot and ground terminals. This will

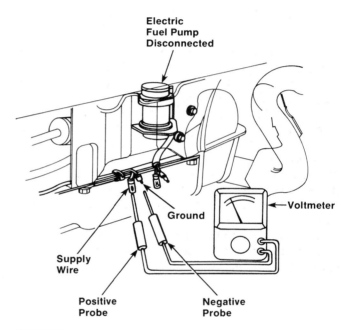

FIGURE 3-86   Checking for proper voltage to electric fuel pump

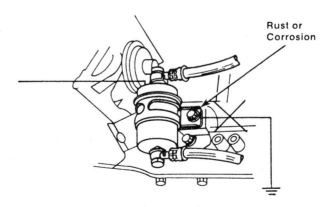

FIGURE 3-87   Rust or corrosion can prevent a good ground.

tell if the relay is getting battery voltage and ground. Next, turn off the ignition, wait about 10 seconds, then turn it on. The relay should click and you should see battery voltage at the relay's pump terminal. If nothing happens, repeat the test checking for voltage at the relay terminal that is wired to the computer. The presence of a voltage signal here means the computer is doing its job but the relay is failing to close and should be replaced. No voltage signal from the computer indicates an opening in that wiring circuit or a fault in the computer itself.

WARNING: When testing an electric fuel pump, do not let fuel contact any electrical wiring. The smallest spark (electric arc) could ignite the fuel.

When replacing an electric pump, be sure that the new or rebuilt replacement unit meets the minimum requirements of pressure and volume for that particular vehicle. This information can be found in the service manual.

To replace an electric fuel pump, proceed as follows:

1. Disconnect the ground terminal at the battery.
2. Disconnect the electrical connector(s) on the electric fuel pump. Label the wire(s) to aid in connecting it to the new pump. Reversing polarity on most pumps will destroy the unit.
3. Fuel and vapor lines should be removed from the pump. Label the lines to air in connecting it to the new pump.
4. The inside tank type pump can usually be taken out of the tank by removing the fuel sending unit retaining ring. Usually, however, it may be necessary to remove the fuel tank to get at the retaining ring. Removal of the fuel tank can be accomplished as described earlier in this chapter.
5. On a fuel pump that is outside of the tank remove the bolts holding it in place (Figure 3-88). On intake models, loosen the retaining ring by rotating it counterclockwise with a brass drift and hammer. Pull the pump and sending unit out of the tank (if they are combined in one unit) and discard the tank O-ring seal.
6. To remove the pump, twist off the filter sock, then push the pump up until the bottom is clear of the bracket. Swing the pump out to the side and pull it down to free it

from the rubber fuel line coupler. The rubber sound insulator between the bottom of the pump and bracket and the rubber coupler on the fuel line are normally discarded because new ones are included with the replacement pump. Some pumps have a rubber jacket around them to quiet the pump. If this is the case, slip off the jacket and put it on the new pump.

7. Compare the new or rebuilt electric fuel pump with the old one. If necessary, transfer any fuel line fittings from the old pump to the new one. Note the position of the filter sock on the pump so you can install a new one in the same relative position.
8. When inserting the new pump back into the sending unit bracket, be careful not to bend the bracket. Also make sure the rubber sound insulator under the bottom of the pump is in place. Install a new filter sock on the pump inlet and reconnect the pump wires. Be absolutely certain you have correct polarity. Replace the O-ring seal on the fuel tank opening, then put the pump and sender assembly back in the tank and tighten the locking ring by rotating it clockwise with a brass drift and hammer. Some pump/sender assembly units are secured by bolts.
9. If the fuel was removed from the vehicle, replace it.

FIGURE 3-88 Remove the bolts holding the fuel tank in place.

CAUTION: Avoid the temptation to test the new pump before replacing the fuel tank by energizing it with a couple of jumper wires. Running the pump dry can damage it because the pump relies on fuel for lubrication and cooling.

10. Reconnect the electrical connector(s).
11. Reconnect the ground terminal at the battery.
12. Start the engine and check all connections for fuel leaks.

FUEL GAUGES

A fuel gauge indicates to the driver the amount of fuel that is in the tank. Fuel gauges are generally located on the dashboard or console. As mentioned earlier in this chapter, the basic fuel gauge is operated in conjunction with the fuel sender unit (Figure 3–89). In addition to the fuel sender unit, the fuel gauge circuit (Figure 3–90) includes a power source (battery or charging system), an ignition switch, dash gauge assembly, connecting wire, and often a voltage regulator. The amount of current flowing through the tank rheostat unit controls the dash or console gauge needle movement. That is, when the tank is empty, the float and float arm swing downward toward the bottom of the float. This will change the resistance in the circuit so that the gasoline gauge pointer reads "E" (Empty). When the tank is half full, the float and float arm will move the slide contact to the center of the element, and the pointer deflects to indicate that the tank is half full. With a full tank of fuel, the resistance will move the pointer to "F" (Full).

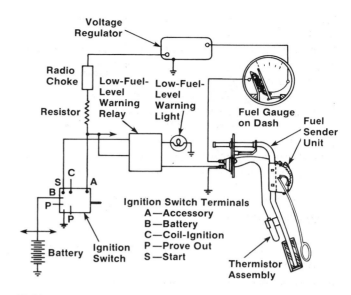

FIGURE 3–90   Fuel gauge circuit

In addition to the conventional variable rheostat fuel gauges, there are two other designs that are often used. They are:

1. *Balancing Coil (Magnetic) Fuel Gauge.* This dashboard gauge consists of two electric coils, an integral (combined) pointer and armature, connecting wires, and terminal connections. The coils are set at 90 degrees from each other to make them pull on the gauge needle armature. To prevent erratic movement of the gauge pointer, a dampening device is generally employed. The two gauge coils (one called the *empty coil,* the other the *full*) are capable of producing a magnetic field that will attract the integral armature and pointer to the coil with the strongest magnetic pull. When the fuel is low (Figure 3–91A), the tank sending unit resistance will be low. As a result, current shunts down through the empty coil and tank resistor; little current flows through the full coil. The magnetic field around the empty coil, working on the armature, moves the pointer to the left. On the other hand, when the tank is full, the sending unit resistance is high. This causes current to flow through both coils. Since the full coil is designed to have the strongest magnetic field, it will overpower the empty coil and move the gauge armature pointer to the *full* position (Figure 8–91B). When the ignition switch is off, the

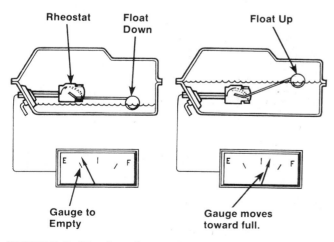

FIGURE 3–89   Sending unit of a fuel gauge

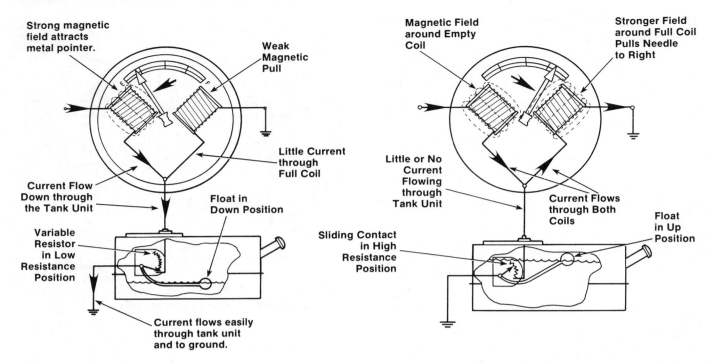

FIGURE 3-91   (A) When the fuel level is low, the tank unit has low resistance. (B) When the tank is full, the sending unit resistance is high.

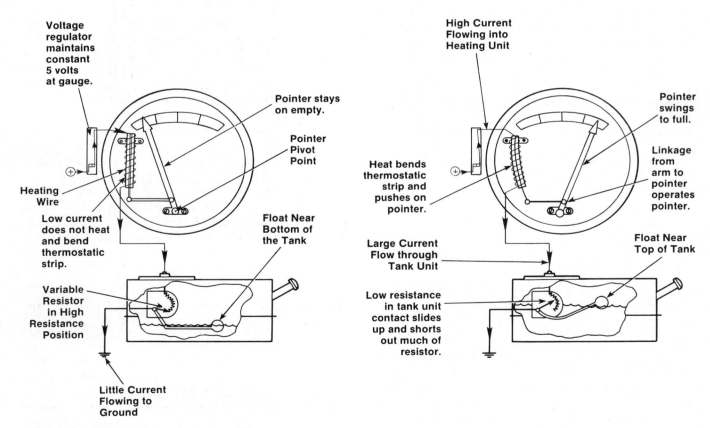

FIGURE 3-92   (A) With the fuel tank empty, the tank resistor has high resistance. (B) With the tank full, the thermostatic arm bends and pushes the gauge pointer toward full.

pointer might not completely return to the *empty* position.

2. *Thermostatic Fuel Gauge.* This design uses an electric bimetallic strip that can be bent or warped by heat. When the tank is empty, the fuel tank sender resistor has high resistance (Figure 3-92A). Very little current can flow through the circuit. The heating element in the gauge stays cool. The thermostatic arm and integral armature pointer will not move from *empty.* When the tank is filled, the float moves the tank sliding contact to the low resistance position. As a result, the current flows through the gauge. The current heats the element. The thermostatic arm bends and pushes the gauge pointer toward the *full* position (Figure 3-92B).

In addition to fuel gauges, there are several other fuel system devices that are in the control circuit that the technician should know about. For instance, some vehicles have a low fuel warning arrangement that is usually mounted on the face of the fuel gauge or in the dashboard instrument panel (Figure 3-93). Another similar device is the miles-to-empty indicator that generates a readout estimating the maximum number of miles a vehicle can be driven on the fuel remaining in the tank (Figure 3-94). Some vehicles have a dashboard indicator that informs the driver how much fuel the engine is consuming. The ultimate in dashboard indicators is the trip computer (Figure 3-95). It performs such calculations as:

1. Estimated time of arrival and estimated distance to destination
2. Time, day, month, and date
3. Trip distance and miles traveled
4. Average cruising speed
5. Elapsed driving time
6. Fuel consumed and amount remaining in tank
7. Miles traveled
8. Engine speed
9. Metric function comparison
10. Engine temperature

The number of calculations performed varies with the make and model of the trip computer. A keyboard is provided for requesting data and making some of the calculations.

Because both electronic and electrical fuel control devices—including gasoline gauges—vary greatly, it is necessary when servicing them to follow the information and troubleshooting procedures given in the vehicle's service manual.

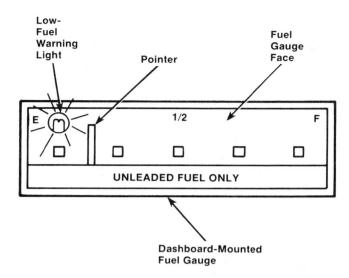

FIGURE 3-93 Light on fuel gauge comes on to warn the driver of low fuel level in the tank.

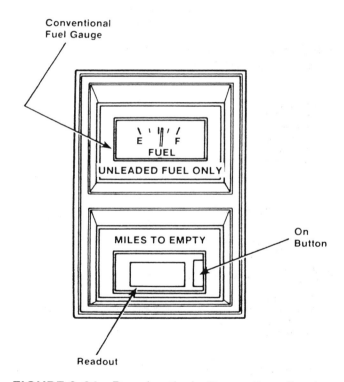

FIGURE 3-94 Pressing the button on the miles-to-empty gauge generates a readout for miles that can be driven on the fuel remaining in the tank.

EEC FUEL SYSTEM

The basic function of an electronic engine control fuel delivery system is the same as the function

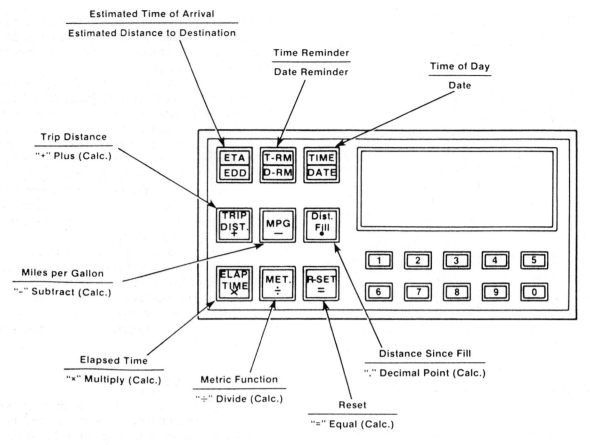

FIGURE 3-95 Keyboard of a trip computer

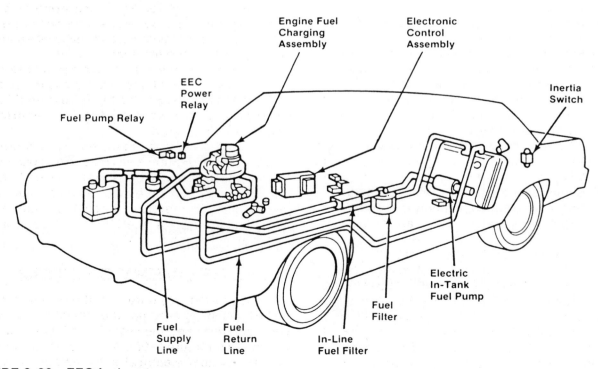

FIGURE 3-96 EEC fuel system

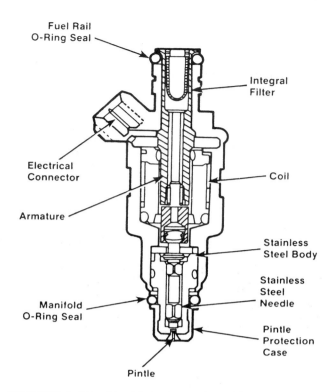

Fuel Rail
O-Ring Seal

Integral
Filter

Electrical
Connector

Armature

Coil

Stainless
Steel Body

Stainless
Steel
Needle

Manifold
O-Ring Seal

Pintle
Protection
Case

Pintle

FIGURE 3-97 Fuel injector solenoids

of a conventional system just described in this chapter. The main difference is that control of the system is obtained through the EEC power relay and electronic control assembly (ECA). Electrically the system consists of a fuel pump, an inertia switch, and a fuel pump relay (Figure 3-96).

When the ignition key is placed in the START or RUN position in a typical EEC fuel delivery system, the EEC power relay applies energizing voltage to the fuel pump relay. The ground path for the relay energizing circuit is through the ECA. This ground path is present only when the ignition key is in START or RUN. The fuel pump relay is usually located on a bracket above the ECA, the inertia switch is located in the left rear kick panel, and the fuel pump is mounted on a bracket at the fuel tank.

The inertia switch opens the power circuit to the fuel pump in the event of a collision and rollover of the vehicle. The switch must be reset by manually pushing the reset button on the switch.

The ECA controls operation of the fuel pump relay during the RUN mode by opening and closing the ground path to the relay coil. The ECA also has a time-out feature that shuts off the fuel pump during a key on/engine off situation. Under a condition of engine flooding, the ECA will shut off the injectors if the throttle position sensor (TPS) signals the wide-open throttle (WOT) valve during the crank mode.

This allows the driver to start a flooded engine, using the same technique used on engines with a carburetor, by simply holding the accelerator pedal down while cranking the engine.

Electronic fuel injection (EFI) systems spray fuel under pressure through injectors, or small nozzles, into the intake manifold. The amount of fuel sprayed is regulated by an electronic computer that receives signals from many sensors on the engine. The fuel injector solenoids (Figure 3-97) control the entry of fuel into the engine, according to commands received from the electronic control assembly (ECA). If the electrical connector to one or more of the fuel injector solenoids is loose, the injector(s) will not fire when commanded to do so by the electronic control assembly (ECA). This will cause the engine to miss under a load condition. During an idle condition, the ECA will try to compensate for a misfiring injector or injectors.

All electronic fuel injection systems are composed of two fuel subsystems: air intake and fuel delivery.

AIR INTAKE

Air intake for electronic fuel (EFI) systems is very much the same as on the carbureted systems. When the throttle plate is opened, the outside air under normal atmospheric pressure is forced through the air horn because the pressure in the intake manifold is lower. The primary difference between carburetion and fuel injection systems is that the EPI models use injectors, controlled by a computer, to spray fuel into the engine.

Careburetors rely on intake manifold pressure, which is lower than atmospheric pressure, to draw fuel into the engine. For the high-pressure EFI, two air horns are often used that are very similar to the air horns on a carburetor.

In most systems, air to the engine is controlled by a butterfly valve, or valves on some applications. The valve(s) is actuated by a linkage and pedal cable arrangement, operated by the driver, similar to the one used on carburetors. Full details on carburetors and injection system operations can be found in Chapters 4 through 8.

FUEL DELIVERY SYSTEM

The fuel delivery portion of the system consists of the fuel tank, pump, fuel filter, injectors, fuel pressure regulator, and the associated fuel supply and return lines (Figure 3-98). A typical electronic fuel delivery system electrical schematic is shown in Figure 3-99).

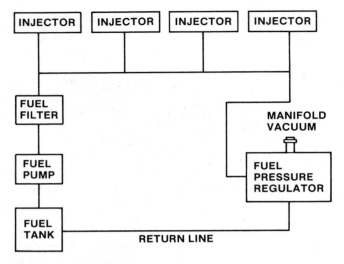

FIGURE 3-98 Fuel delivery system

Fuel Pump

Fuel is delivered to the injectors by an electric fuel pump or pumps, depending on vehicle application. While there are several systems, the three most common are:

1. High-pressure, in-tank system that has a high-pressure fuel pump located in the fuel tank.
2. Low-pressure, in-tank system that uses a low-pressure pump for the central fuel injection system.
3. Low-pressure, in-tank/high-pressure, in-line system that uses a low-pressure fuel pump located in the fuel tank to supply fuel to the externally mounted, high-pressure pump. The high-pressure pump supplies fuel to the injectors.

If the fuel pump malfunctions and does not provide sufficient fuel pressure, enough fuel might still be available to keep the engine running. However, the reduced pressure will not provide the fuel flow needed in high-demand (load) conditions such as acceleration or hill-climbing. Without the necessary fuel, the engine will lack power.

To test an EFI electric pump, proceed as follows:

1. Check for electrical continuity at the fuel pump(s) as described in the service manual.

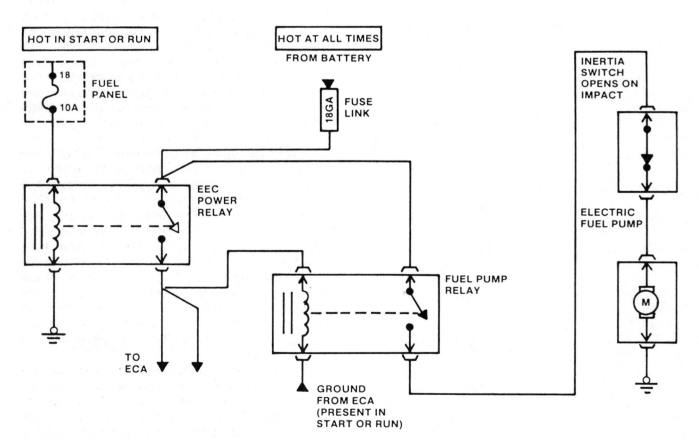

FIGURE 3-99 Fuel delivery system electrical schematic

SHOP TALK _____

The procedure given here contains the major steps required for all fuel-injected vehicles. Many models require additional steps. The fuel flow and pressure specifications vary from one model to the next. Refer to the shop manual for complete procedures and correct specifications.

2.   Disconnect the fuel line at the fuel rail or throttle body. Use care to avoid spillage.
3.   Connect the hose from the fuel return fitting to a calibrated container or at least a 1-liter (1 quart) container.
4.   Connect the pressure gauge tool to the fuel diagnostic valve (or Schrader valve) on the fuel rail or throttle body. A few models do not have diagnostic valves. Refer to the shop manual.
5.   Locate the fuel pump relay, then remove and replace it with the modified relay described in the shop manual.
6.   Energize the fuel pump(s) for 10 seconds by grounding the ground lead from the relay.
7.   Allow the fuel to drain from the hose into a container and observe the volume.
8.   The fuel pump(s) are operating properly if
      •   Fuel pressure reaches a specified amount.
      •   Fuel flow is at a specified amount.
      •   Fuel pressure remains at a specified amount.
9.   If all three conditions are met, the fuel pump is operating normally.
10.   If pressure condition is met but flow is not, check for blocked filter(s) and fuel supply lines. After correcting any blockage, recheck. If flow conditions are still not met, replace fuel pump.
11.   If both pressure and flow conditions are met but pressure will not maintain after de-energizing, check for leakage at the regulator or the injectors. If both check all right, replace fuel pump.
12.   If no flow or pressure is detected, the fuel system should be checked as in Step 1. If no trouble is found, replace the fuel pump, drop the fuel tank, and replace the fuel filter on the filler neck tube or on the low-pressure pump, depending on application.

Fuel Lines.   Two fuel lines are required. A suppy line connects the fuel pump(s) through the

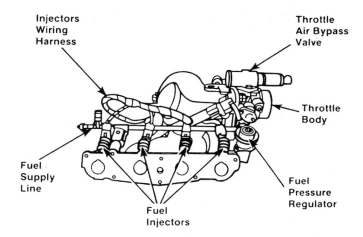

FIGURE 3-100   Fuel supply line with pressure regulator

filter to the fuel charging assembly (Figure 3–100) or fuel rails. A return line allows excessive fuel supplied by the pump(s), but not needed by the engine, to be returned to the fuel tank or to the reservoir (if so equipped).

Fuel Filter.   The fuel filter for a fuel injection system provides extremely fine filtration to protect the small metering orifices of the injector nozzles. The fuel filter is located in the fuel supply line, downstream from the fuel pump. It is mounted on the vehicle underbody or in the engine compartment.

If the fuel filter is completely plugged, no fuel will reach the injectors and the engine will not start. If the fuel filter is not totally plugged, it could allow just enough fuel flow to start the engine, but not enough to keep it running.

Fuel Pressure Regulator.   The fuel pressure regulator reduces the 60 psi pressure 39 to 40 psi for normal operating conditions (10 inches of manifold vacuum). At idle or other high manifold vacuum conditions, the regulator reduces this pressure to approximately 30 psi (20 inches of manifold vacuum). The fuel pressure regulator acts as a poppet valve and operates off manifold vacuum and spring tension. The regulator valve moves toward the *closed* position with high vacuum and toward the full *open* position with low vacuum. By referencing the pressure regulator to intake manifold vacuum, the constant equivalent of 40 psi can be maintained on the injectors.

The regulator (Figure 3–100) not only regulates the fuel pressure, but also traps fuel during engine shutdown, eliminating the possibility of vapor formation in the fuel line and providing for instant restarts and initial idle speed. It also provides for a regulated pressure drop across the injectors and excessive

TABLE 3-2: VACUUM/AIR/FUEL DELIVERY SYSTEMS DIAGNOSIS

Condition	Possible Cause	Remedy
Engine cranks normally, but will not start.	Improper starting procedure	Check with customer to determine if proper starting procedure is being used.
	No fuel in carburetor	Remove fuel line at carburetor. Connect hose to fuel line and run into metal container. Remove the wire from the BAT terminal of the distributor. Crank engine; if no fuel discharges from the fuel line, test fuel pump. If fuel supply is all right, check the following: • Inspect fuel inlet filter. If plugged, replace. • If fuel filter is all right, remove air horn and check for a bind in the float mechanism or a sticking inlet needle. If all right, adjust float as specified.
	Engine flooded	Remove the air horn. If foreign material is present, clean the fuel system and replace fuel filters as necessary. If excessive foreign material is found, completely disassemble and clean.
Poor gas mileage.	Customer driving habits.	Run mileage test with customer driving if possible. Make sure car has 2,000 to 3,000 miles for the break-in period.
	Loose, broken, or improperly routed vacuum hoses.	Check condition of all vacuum hose routings; correct as necessary.
	High fuel level in carburetor.	If excessive foreign material is present in the carburetor bowl, the carburetor should be cleaned.
Engine starts but will not keep running.	Fuel pump.	Check fuel pump pressure and volume; replace if necessary.
	Vacuum hose routing loose, broken, or incorrect	Check condition and routing of all vacuum hoses; correct as necessary. Check for free movement of fast-idle cam. Clean and/or realign as necessary.
	Choke vacuum break units not adjusted to specification or are defective.	Adjust both vacuum break assemblies to specification. If adjusted properly, check the vacuum break units for correct operation. Always check the fast-idle cam adjustment when adjusting vacuum break units.
	Insufficient fuel in carburetor	Check fuel pump pressure and volume. Check for partially plugged fuel inlet filter; replace if contaminated.
Engine starts hard (cranks normally).	Loose, broken, or incorrect vacuum hose routing	Check condition and routing of all vacuum hoses; correct as necessary.
	Improper starting procedures.	Check to be sure customer is using proper starting procedure.
	Insufficient fuel in bowl.	Check fuel pump pressure and volume. Check for partially plugged fuel inlet filter and replace if dirty. Check float mechanism and adjust as specified.

TABLE 3-2:   VACUUM/AIR/FUEL DELIVERY SYSTEMS DIAGNOSIS (CONTINUED)

Condition	Possible Cause	Remedy
	Fuel contaminated.	Check for contaminants in fuel. Clean system if necessary.
	Plugged fuel filter.	Check and replace as necessary.
	Float level too low.	Check and reset float level to specification.
	Malfunctioning float and/or needle and seat.	Check operation of system. Repair or replace as necessary.
Engine hesitates on acceleration.	Loose, broken, or incorrect vacuum hose routing.	Check condition and routing of all vacuum hoses; correct or replace.
	Malfunctioning air valve	Check operation of secondary air valve and spring tension adjustment.
	Power-enrichment system not operating correctly.	Check for binding or stuck power piston(s): correct as necessary.
	Inoperative air cleaner heated air control.	Check operation of thermostatic air cleaner system.
	Dirty or plugged fuel filter	Replace filter and clean fuel system as necessary.
	Malfunctioning distributor vacuum or mechanical advance	Check for proper operation.
	EGR valve stuck open.	Inspect and clean EGR valve.
Engine has less than normal power at normal accelerations.	Loose, broken, or incorrect vacuum hose routing.	Check condition and routing of all vacuum hoses.
	PCV system clogged or defective	Clean or replace as necessary.
	Sticking choke	Check complete choke system for sticking or binding. Clean and realign as necessary. Check adjustment of choke thermostatic coil. Check connections and operation of choke hot-air system. Check jets and channels for plugging; clean and blow out passages.
	Air cleaner temperature improperly regulated.	Check regulation and operation of air cleaner system.
Less than normal power on heavy acceleration or at high speed.	Plugged air cleaner element.	Replace element.
	Air valve malfunction (where applicable)	Check for free operation of air valve. Check spring tension adjustment. Make necessary adjustments and corrections.
	Fuel inlet filter plugged	Replace with a new filter element.
	Insufficient fuel to carburetor	Check fuel pump and system; run pressure and volume test.
	Fuel pump.	Check fuel pump pressure and volume; inspect lines for leaks and restrictions.

TABLE 3-2: VACUUM/AIR/FUEL DELIVERY SYSTEMS DIAGNOSIS (CONTINUED)

Condition	Possible Cause	Remedy
	Loose, broken, or incorrect vacuum hose routing.	Check condition and routing of all vacuum hoses; correct as necessary.
	Clogged or malfunctioning PCV system	Check PCV system; clean or replace as necessary.
	Loose carburetor, EGR, or intake manifold bolts and/or leaking gaskets.	Torque carburetor to manifold bolts. Spray light oil or kerosene around manifold to head mounting surface and carburetor base. If engine rpm changes, tighten or replace the carburetor or manifold gaskets as necessary. Check EGR mounting bolt torque.
	Fuel pump pressure low or erratic	Check fuel delivery and pressure.
	Fuel contaminated.	Check for contaminants in fuel. Clean system if necessary.
	Plugged fuel filter	Check and replace as necessary.
Engine starts but will not continue to run or engine runs but surges and backfires.	Faulty fuel pump.	Test fuel pump. Remove and replace as required.
Engine will not start.	Faulty fuel pump.	Perform diagnostic tests on the fuel pump. Remove and replace fuel pump as required.
Gasoline odor.	Overfilled tank	Do not "pack" tank. Fill to automatic shutoff.
	Leaking fuel feed or vapor return line	Correct as required.
	Leak in fuel tank.	Purge tank and repair or replace tank as required.
	Fuel tank vent lines or hoses disconnected	Connect lines or hoses as required.
	Purge lines not connected, improperly routed, plugged, or pinched.	Check, connect, and open lines as required.
	Faulty fill cap.	Install new cap.
Collapsed fuel tank.	Plugged or pinched vent lines or hoses, and defective cap.	Check all lines from tank to canister and replace cap.
Fuel tank rattles.	Loose mounting straps	Tighten straps to specifications.
	Loose baffle.	Replace fuel tank.
	Foreign material in tank.	Remove tank and clean.
	Improperly located felt strips	Install strips.
Fuel starvation.	Plugged tank gauge unit filter	Replace filter.
	Pinched, plugged, or misrouted fuel line	Check, open, or reroute as required.

TABLE 3-2:   VACUUM/AIR/FUEL DELIVERY SYSTEMS DIAGNOSIS (CONTINUED)

Condition	Possible Cause	Remedy
	Fuel pump not operating.	See fuel pump test.
	Air cleaner element plugged	Replace element.
Car feels like it is running out of gas; surging occurs in mid-speed range.	Plugged fuel filters.	Remove and replace filters.
	Faulty fuel pump.	Perform fuel pump test. Remove and replace fuel pump as required.
	Foreign material in fuel system or kinked fuel pipes or hoses.	Inspect pipes and hoses for kinks and bends; blow out to check for plugging. Remove and replace as required.

fuel not required by the engine is returned to the fuel tank by way of a fuel return line.

VACUUM/AIR/FUEL DELIVERY SYSTEM DIAGNOSIS

Driveability, performance, economy, and emission difficulties can result in vacuum/air/fuel delivery system problems. Table 3–2 presents the most common problems that can occur in these systems and lists remedies for these conditions.

REVIEW QUESTIONS

1. Vacuum system problems can produce or contribute to driveability symptoms such as
    _____ .
    a. backfire
    b. overheating
    c. poor fuel economy
    d. all of the above

2. Vacuum is used to control which of the following?
    a. speed control
    b. heater controls
    c. air conditioning controls
    d. all of the above

3. Technician A uses a vacuum gauge to check many of the vehicle's vacuum emission valves. Technician B uses a hand-held vacuum pump. Who is right?
    a. Technician A
    b. Technician B

c. Both A and B
d. Neither A nor B

4. What component prevents airborne contaminants and abrasives from entering the engine through the carburetor with the air/fuel mixture?
    a. air filter
    b. fuel filter
    c. vacuum pump
    d. air intake ductwork

5. What air filter uses a pleated paper element without chemicals to minimize dust entry to the engine?
    a. centrifugal air filter
    b. paper air filter
    c. oil-wet polyurethane
    d. oil bath air filter

6. Technician A uses compressed air to clean a dirty heavy-duty air filter element. Technician B says that dirty air filter elements must always be replaced. Who is right?
    a. Technician A
    b. Technician B
    c. Both A and B
    d. Neither A nor B

7. What percent of a modern fuel tank can ever be filled with fuel?
    a. 40 percent
    b. 60 percent
    c. 90 percent
    d. 100 percent

8. Technician A replaces any leaking metal fuel tank. Technician B steam cleans a damaged

fuel tank and then solders or brazes it. Who uses the best method?
a. Technician A
b. Technician B
c. Both A and B
d. Neither A nor B

9. When installing new flexible tubing, Technician A replaces damaged or worn clamps. Technician B replaces all old clamps. Who is right?
a. Technician A
b. Technician B
c. Both A and B
d. Neither A nor B

10. What component, located in the gasoline tank, prevents large contaminant particles from entering the fuel system?
a. fuel filter
b. diesel filter
c. air filter
d. strainer

11. A vehicle's engine performance indicates inadequate fuel, and the fuel pump is suspected. Technician A tests the fuel pump for pressure and volume while mounted on the engine. Technician B removes the fuel pump to inspect it. Who is right?
a. Technician A
b. Technician B
c. Both A and B
d. Neither A nor B

12. To unblock a clogged fuel line, Technician A disconnects the line and blows into it with a hand air pump in the direction of the normal fuel line flow. Technician B disconnects the line and uses a hand air pump to blow in the direction opposite of the normal fuel line flow. Who is right?
a. Technician A
b. Technician B
c. Both A and B
d. Neither A nor B

13. After a fuel pump is installed, an internal squeaking noise is heard with each movement of the pump. Technician A considers the noise to be normal. Technician B examines the pump for a defective or loose component Who is right?
a. Technician A
b. Technician B
c. Both A and B
d. Neither A nor B

14. Which of the following components exists in both a carbureted fuel delivery system and a fuel injector system?
a. inertia switch
b. electronic control assembly (ECA)
c. power relay
d. fuel pump

15. Which of the following is not a cause of poor gas mileage?
a. loose, broken, or improperly routed vacuum hoses
b. customer driving habits
c. insufficient fuel in bowl
d. high fuel level in the carburetor

CHAPTER FOUR

CARBURETOR SERVICING AND ADJUSTMENTS

Objectives

After reading this chapter, you should be able to:
- Explain the purpose and functions of a carburetor.
- Identify the main systems and components of a carburetor.
- Recognize carburetor-related performance problems.
- Troubleshoot and adjust a carburetor.

The carburetor is a device used to mix fuel with air and to meter the air/fuel mixture into the intake manifold in proportions that satisfy the energy demands of the engine in all operation modes. Although fuel injection systems have replaced carburetion in many passenger cars and light trucks, there are many carbureted engines still on the road, and all technicians must understand the principles of carburetion and how carburetors are constructed and operated before they can successfully tune carbureted engines.

HOW A CARBURETOR WORKS

Air contains approximately 20 percent oxygen. It is this oxygen content of air that combines with the fuel to support combustion in the engine. The number of oxygen molecules in air varies with sea level. At sea level, the pressure of the air is 14.7 psi. At 2000 feet above sea level, the air pressure is 13.6 psi. The higher the altitude, the lower the atmospheric pressure becomes.

Any pressure less than the earth's atmospheric pressure at any altitude is referred to as *vacuum.* A vacuum is a pressure that is lower than atmospheric pressure.

The pressure differential between atmospheric pressure and vacuum pressure can be used to move air or a liquid from one point to another. For example, consider the act of drinking a soda through a straw. The soda is not truly drawn up the straw. The pressure inside the straw is lowered by the drinker so that the atmospheric pressure pushing down on the liquid in the container forces the soda up the straw to the drinker's mouth. This same principle is used by a carburetor to move air and fuel into the combustion chambers of the engine. As an engine's cylinder moves downward on the intake stroke, it does not draw air in, but rather it creates an area of reduced pressure so that atmospheric pressure can force air into the cylinder.

Air enters the top portion of the carburetor through an air horn. The airflow passes the choke butterfly plates or valves, which are normally open and parallel to the airflow steam, and enters the carburetor barrel.

After the airflow stream passes into the barrel, it encounters a narrow passageway or restriction, often referred to as a *venturi* (Figure 4-1). The venturi forces the incoming airflow to speed up as it passes this restriction.

The increased speed of the air creates a vacuum or pressure drop in the venturi. Located inside the venturi is a fuel discharge nozzle. A mixture of fuel and air is forced out of the discharge tube by the atmospheric pressure acting on the fuel and air passageways in the carburetor (Figure 4-2).

The air/fuel mixture is then passed through a second butterfly valve called the *throttle valve* (Fig-

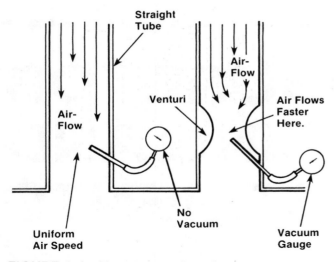

FIGURE 4-1 How a venturi works

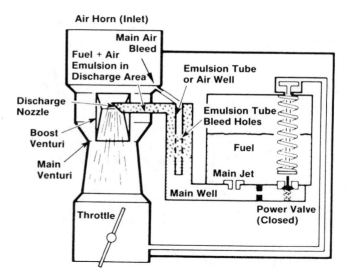

FIGURE 4-2 Air/fuel mixture routed to the booster venturi via the discharge nozzle.

ure 4-3). This valve is controlled by the linkage on the car's accelerator or "gas" pedal (Figure 4-4). The valve can be in a number of positions. When the throttle valve is parallel to the airflow stream, it offers little resistance to flow and is said to be wide open (Figure 4-5A). With the valve at a right angle to the airflow, it is fully closed, and the engine is said to be idling or decelerating (Figure 4-5B). Any position between these two indicates the engine is under varying degrees of power or deceleration. The air/fuel mixture then passes from the throttle valve into the engine's cylinders where it is ignited.

BASIC CARBURETOR CIRCUITS

Variations in engine speed and load demand different amounts of air and fuel (often in differing proportions) for optimum performance—and present complex problems to the carburetor. At engine idle speeds, for example, there is insufficient air velocity to cause fuel to be drawn from the discharge nozzle and into the airstream. Also, with a sudden change in engine speed, such as rapid acceleration, the venturi effect (pressure differential) is momentarily lost. Therefore, the carburetor must have special circuits or systems to cope with these situations. There are six basic circuits used on a typical carburetor.

1. Float
2. Idle _____ *Transistion*
3. Main metering
4. Full power (or power enrichment)
5. Accelerator pump
6. Choke

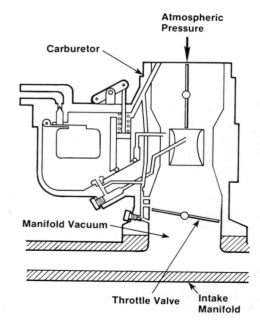

FIGURE 4-3 Manifold vacuum is created by downward movement of the engine's pistons.

FLOAT CIRCUIT

The float circuit (also called the *fuel inlet system*) (Figure 4-6) of a typical carburetor consists of the following:

- Fuel bowl
- Fuel inlet fitting
- Fuel inlet needle valve and seat
- Float

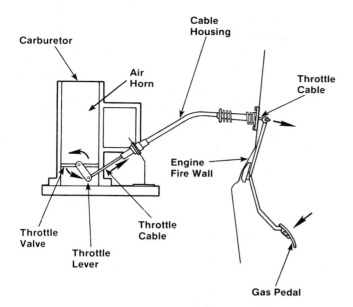

FIGURE 4-4   Throttle linkage connects gas pedal to throttle plate to control the engine's speed and power.

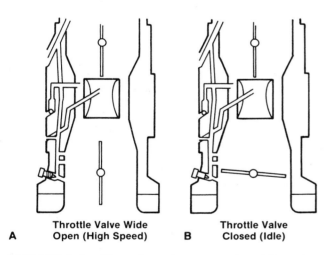

FIGURE 4-5   Throttle valve controls airflow into carburetor: (A) wide open; (B) closed

A fuel screen or filter is usually installed at the fuel inlet to prevent dirty fuel from mixing in the carburetor and causing a malfunction. Figure 4-7 illustrates the components of the float system.

The float system stores fuel and holds it at a precise level as a starting point for uniform fuel flow. Fuel enters the carburetor through the inlet line and passes through an inlet filter to the inlet needle valve and seat. The incoming fuel is captured and stored in the reservoir or fuel bowl. The level of the fuel in this bowl is maintained at a specified height by the rising and falling of the float in the fuel bowl. As fuel

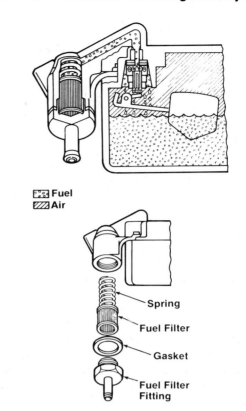

FIGURE 4-6   Fuel inlet system

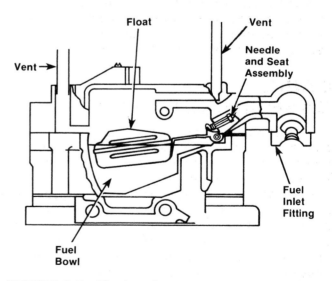

FIGURE 4-7   Float system

enters the bowl, the float rises and closes the inlet needle valve. With the needle valve closed, fuel is prevented from entering the carburetor. Fuel pressure against the inlet needle valve tends to force it open while the buoyancy of the float in the bowl tends to force it closed. This action establishes the precise fuel level for the carburetor. This basic fuel

111

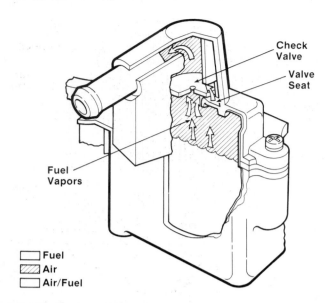

Check Valve

Valve Seat

Fuel Vapors

☐ Fuel
▨ Air
☐ Air/Fuel

FIGURE 4-8 External vent system

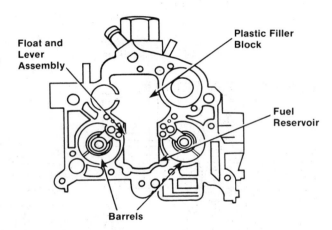

Float and Lever Assembly

Plastic Filler Block

Fuel Reservoir

Barrels

A   FUEL   RESERVOIR

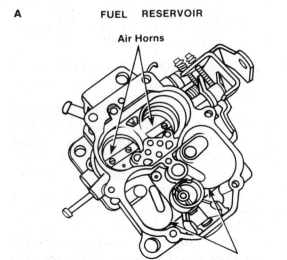

Air Horns

Fuel Bowl

B

FIGURE 4-9 Fuel bowl locations: (A) Front of bores; (B) center of carburetor

metering system is calibrated to deliver the proper mixture only when the fuel is adjusted to the correct level by either an external or internal fuel inlet valve. To stabilize float movement, the carburetor utilizes a bumper spring installed under the float level.

The fuel bowl is vented internally to the air horn by a vent tube in the carburetor body. Prior to the introduction of emission controls most primary fuel bowls were also vented to the atmosphere when the engine was at idle or turned off. Since the introduction of evaporative control systems almost all fuel bowls are vented internally. Most applications are vented by a valve to a charcoal canister (Figure 4-8) and the vapors are returned to the engine when it is restarted. Further details on the vapor venting system can be found in chapters 10 and 11.

Two basic types of fuel bowls are used:

1. *Integral Fuel Bowl.* In this carburetor design, the bowl is cast as an integral part of the air horn body. It can be located in front of the bores (Figure 4-9A), to one side, or even in the center of the carburetor throat (Figure 4-9B). The front-mounted bowl always has an externally adjustable inlet valve. A side-mounted bowl can have either an external or internal adjustment technique. The only difference between the two types is their application. In applications such as circle-track or road racing, where side-to-side fuel movement is predominant, the front-mounted bowl is ideal. The side-mounted bowl is best suited for drag racing applications where fore-and-aft fuel movement is predominant.

2. *Removable Fuel Bowl.* In this carburetor design, the fuel bowl can be removed from the main body (Figure 4-10). A gasket prevents fuel flow between the body and bowl.

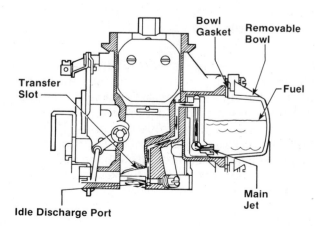

Bowl Gasket

Removable Bowl

Transfer Slot

Fuel

Idle Discharge Port

Main Jet

FIGURE 4-10 Cutaway shows removable fuel bowl

Floats are constructed of synthetic material (generally nitrophyl) or brass. Brass floats are prone to leaks and should never be used in turbo- or super-charged applications because the float will collapse. Synthetic floats can survive the pressure of a blow through application but can only be used with gasoline fuel.

The inlet needle valve and seat are available in various sizes and designs. Adjustment of externally adjustable needle valves can be accomplished without removing the carburetor or any of its components (Figure 4-11). On the other hand, the carburetor must be removed from the engine to adjust an internally adjustable needle valve (Figure 4-12). In-

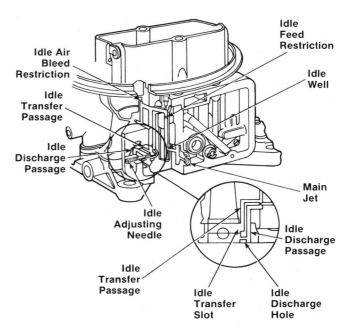

FIGURE 4-13   Idle system

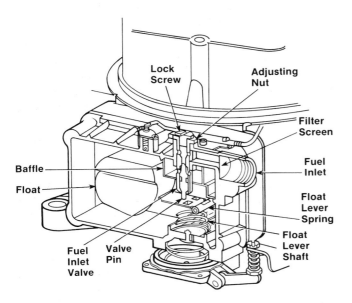

FIGURE 4-11   Externally adjustable fuel inlet valve

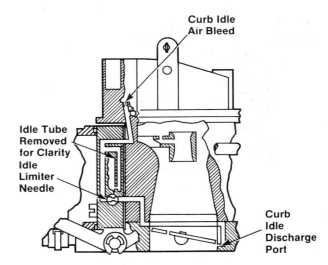

FIGURE 4-14   Primary idle transfer system

creasing the needle valve seat size produces two effects: an increase in fuel flow rate and a compromised or sacrificed fuel control capability. A larger needle valve seat should not be installed in an engine unless it really needs the added fuel flow capacity.

IDLE CIRCUIT

The idle circuit (Figure 4-13) supplies the air/fuel mixture to operate the engine at idle and low speeds. Fuel is drawn into the idle system from the main well. Some applications have an idle tube in the main well to meter the fuel (Figure 4-14). Other ap-

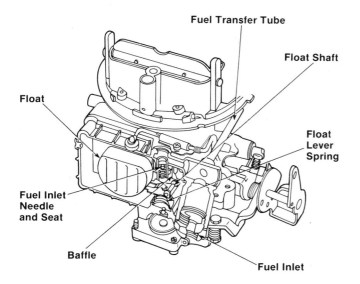

FIGURE 4-12   Internally adjustable fuel inlet valve

plications use a horizontal passage from the main well to the idle well and an idle channel restriction for metering purposes.

In either type, air enters the idle air bleed and mixes with the fuel after the fuel flows through the idle tube or restriction. At curb idle the throttle valves are almost closed. This creates a high vacuum below the throttle valve with near atmospheric pressure above the valve. Fuel is drawn out of the curb idle port below the throttle plate. As the throttle valves are opened a transfer slot located above the throttle plate is progressively exposed to vacuum and the air/fuel emulsion is also discharged from the transfer slot.

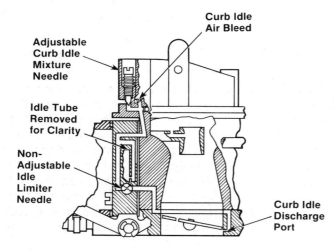

FIGURE 4-15   Central adjustment

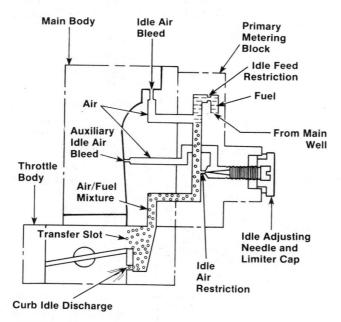

FIGURE 4-16   Variable air bleed idle system

The increased air/fuel mixture flow provides a smooth transition between idle and cruising modes of operation. Some carburetors have a series of holes called off-idle air passages, instead of transfer slots. Like the transfer slot, the holes permit increased fuel delivery as the throttle opens (Figure 4-15).

To improve idle quality when meeting emission standards, a variable air bleed idle system is used on some carburetors (Figure 4-16). In this system there is the normal fixed idle fuel restriction. However, there are two idle air bleeds. One idle air bleed is installed normally in the air horn. An auxiliary idle air bleed is drilled into the lower skirt of the venturi. The air entering through the auxiliary passage is adjusted by an idle air adjusting screw. The screw is turned clockwise to enrich the idle mixture and counterclockwise to lean out the idle air/fuel mixture. An aluminum decal on the metering body has an arrow that indicates the lean direction (Figure 4-17).

 SHOP TALK

On most modern carburetors, the idle mixture is factory set and the adjustment screw is covered with a plate or plug (Figure 4-18). This ensures that the CO percentage in the exhaust meets emissions standards and prevents anyone from tampering with the factory setting. Earlier carburetors have an idle limiter cap attached to the idle mixture screw. The limiter cap also prevents the factory setting from being tampered with, but allows the idle mixture to be adjusted within a narrow range (about 3/4 of a turn).

As the throttle valve is opened somewhat wider the speed and quantity of the air flowing through the venturi is increased. A low pressure or partial vacuum is created in the booster venturi, and the main metering system starts to flow. This moderate opening (transition) of the throttle is commonly called *tip in.* When the throttle valves are opened, the accelerator pump system mechanically squirts raw fuel into the venturi to prevent a stumble until the mixture from the main metering system catches up with the increased airflow.

Another variation of this air bleed design is the so-called off-idle system (Figure 4-19) that acts as additional air bleeds for the idle circuit. It is actually an extension of the circuit intended to provide additional fuel during the transition from idle operation to higher operating speeds. The system consists of a

FIGURE 4-17　An arrow on a decal indicates lean direction.

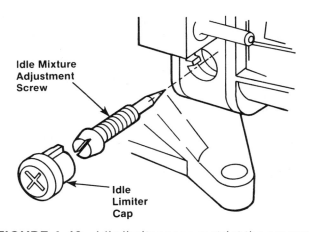

FIGURE 4-18　Idle limiter caps restrict the amount of adjustment allowed for the idle mixture.

slot or series of holes (called *idle transfer passages*) in the air just above the throttle plate. When the throttle plate opens, these extra passages are exposed to intake manifold vacuum. This action permits an increased amount of air to rush past the opened butterfly plate, which provides extra fuel to keep the proper air/fuel ratio during the initial acceleration from idle.

MAIN METERING CIRCUIT

Just as the idle circuit handles lower engine speeds, the main metering circuit (Figure 4-20)

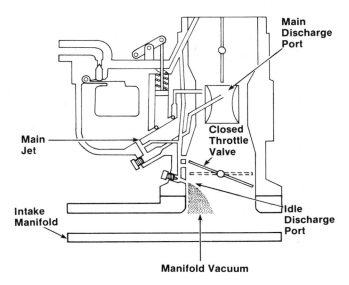

FIGURE 4-19　Off-idle system

comes into operation when the engine speed is about 20 mph or higher. Increasing the speed of the engine causes the throttle plate in the carburetor to open past the idle position. This increases the air flowing through the carburetor venturi and creates enough vacuum to allow atmospheric pressure to force fuel through the main metering system and out the main fuel discharge nozzle, located in the center of the booster venturi (Figure 4-21). As engine speed is increased, the vacuum at the discharge nozzle increases.

This vacuum or pressure differential causes fuel to flow out of the fuel bowl, through the main metering jet and into the main well. On most carburetors

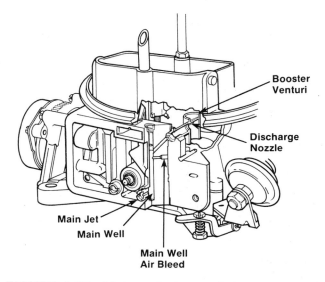

FIGURE 4-20　Main metering system

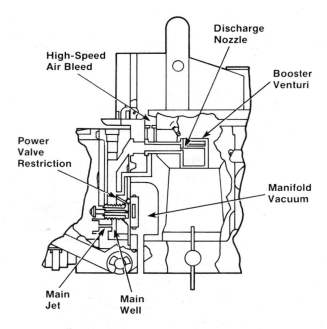

FIGURE 4-21 Main metering system—side view

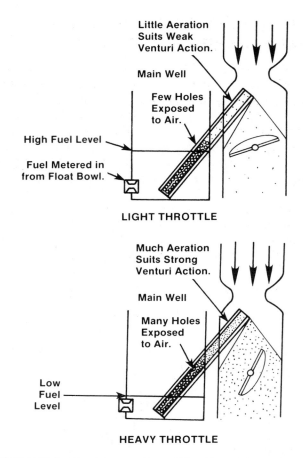

FIGURE 4-22 How main well fuel level draws down to aerate fuel

the main well is vented through a precisely sized opening called the *main well air bleed.* The main well air bleed allows air to enter at the top of the main well. Fuel then flows from the main well up the main well tube to the discharge nozzle. When this happens, the pressure in the main well drops off. Air entering the calibrated main air bleed prevents a vacuum from developing in the main well. The air also allows for aeration of the fuel as it leaves the main well and travels up the well tube. Air enters the main well tube through small holes in it. This will allow the fuel to be partially atomized as it travels toward the discharge nozzle (Figure 4-22).

The more air that flows through the carburetor, the more fuel it draws from the main well. This will lower the fuel level in the main well and expose more air bleed openings in the main well tube. This will cause extra air to enter the well tube, mix with the fuel, and, in essence, dilute the fuel. This action circumvents the richening effect caused by the increased carburetor airflow. If the fuel were not diluted as such, the fuel mixture would richen at high speed as the venturi vacuum increased faster than the engine's need for additional fuel.

The main air bleeds draw air from the high-pressure areas in the carburetor barrel above the venturi. Air flows through the bleeds because the high-pressure areas are at atmospheric pressure levels while at the same time, the main metering discharge nozzle is in the low-pressure venturi level. Air and fuel quantities, as well as bleed airflow, are controlled by the throttle opening.

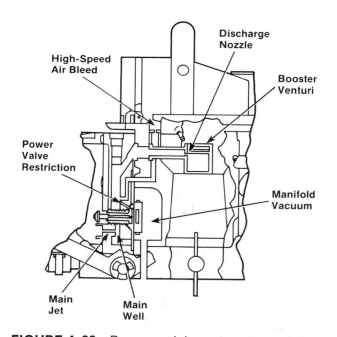

FIGURE 4-23 Power enrichment system

FULL-POWER OR POWER ENRICHMENT CIRCUITS

At wide open throttle, the engine needs a richer than normal air/fuel mixture. This mixture cannot be supplied by the main metering system, so an additional fuel enrichment or full-power system is provided on most carburetors (Figure 4-23). The power enrichment system meters additional fuel into the mixture. This can be accomplished in several ways. But in most, the key component of the fuel enrichment circuit is the metering rod which moves within the main jet to allow more fuel to be drawn through the main metering circuit. Most metering rods are tapered or stepped to increase the extra fuel flow gradually (Figure 4-24). The metering rod can be operated by mechanical means, vacuum, or electrical solenoid.

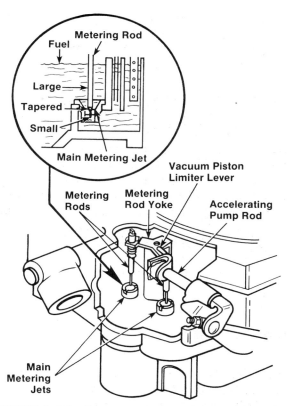

FIGURE 4-25   Some power systems consist of metering rods placed in the main jets.

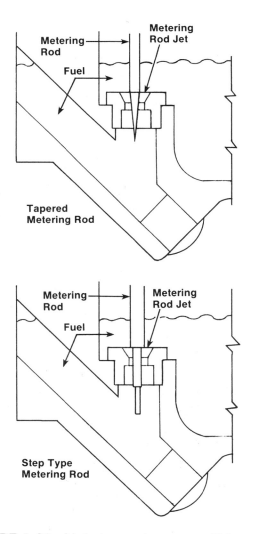

FIGURE 4-24   Metering rods move within a carburetor to vary fuel flow.

Mechanical Metering Rod

In this design, the throttle butterfly plate and metering rod operate as a unit. That is, when the throttle plate is not completely open, the throttle linkage keeps the rod in the jet. As acceleration increases, the mechanical linkage lifts the metering rod out of the main jet as the throttle opens wide (Figure 4-25).

Vacuum Metering Rod

A vacuum passage in the throttle body of the carburetor is used to sense engine load or speed. At high speeds, manifold vacuum is low; at low speeds, it is just the opposite. The manifold vacuum passageway transmits the sensed vacuum level to a power valve chamber in the main body of the carburetor (Figure 4-26). A power valve in the chamber is usually activated by a diaphragm or piston. Manifold vacuum is applied to the vacuum side of the diaphragm to hold it closed at idle or low engine speeds and hold it open at high engine speeds.

When manifold vacuum drops, the power valve spring opens the valve to admit additional fuel into the main well. Fuel is metered into the main well by a power valve channel restriction in the metering body. Fuel flows through the restriction, into the main well, and is added to the fuel already flowing from the main metering system.

Electric Metering Rod

Some carburetors have an electric gas solenoid in the engine's exhaust system. An onboard computer moderates the signals from the exhaust sensor and then energizes the metering rod solenoid (Figure 4-27). That is, the electrical impulses to the solenoid control the metering rod. The solenoid pulls the metering rod in and out of its jets to match the fuel mixture to the engine's needs. This system reduces fuel waste and keeps exhaust emissions low.

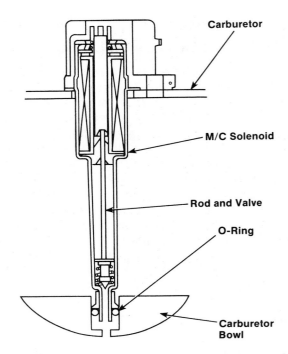

FIGURE 4-27 Electrical impulses to the solenoid control an electric metering rod.

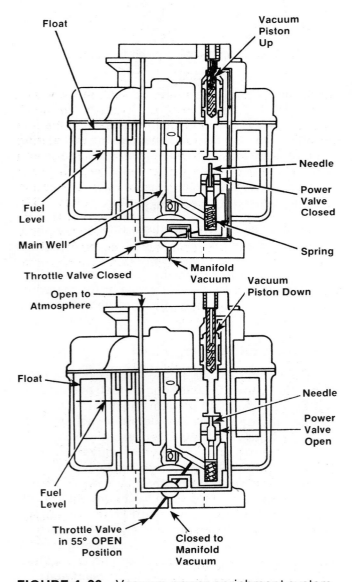

FIGURE 4-26 Vacuum power enrichment system

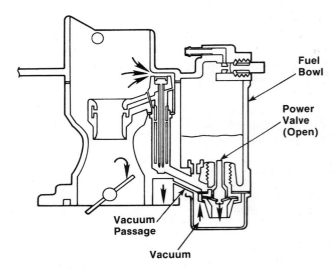

FIGURE 4-28 The power valve enriches the high-speed fuel mixture.

Power Valve

The power valve (Figure 4-28) is basically a switch that serves the same purpose as a metering rod. It consists of a flexible diaphragm attached to a spring valve that opens at a predetermined vacuum level. A vacuum passage goes to the bottom of the carburetor air horn. On one side of the power valve or jet diaphragm is a supply of fuel; on the other side is the engine vacuum.

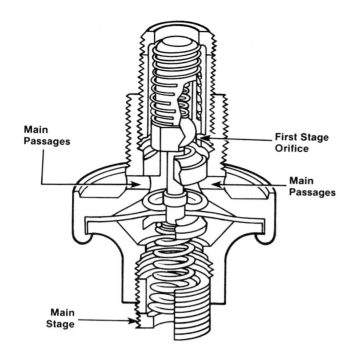

FIGURE 4-29 Two-stage power valve

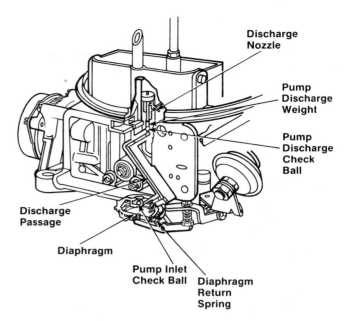

FIGURE 4-30 Accelerator pump system

Another type of power or jet valve is the two-stage valve that reacts to different vacuum levels to meter fuel more precisely into the engine. The valve itself (Figure 4-29) is threaded into a cavity at the bottom of the fuel bowl. It consists of a seat, stemmed valve, spring, first stage orifice, and main passageways.

ACCELERATOR PUMP CIRCUIT

During acceleration the air flowing through the carburetor barrel reacts almost immediately to each change in the throttle opening. Since fuel is heavier than air, it has a slower response time. Fuel in the main metering system and/or idle system takes a fraction of a second to respond to the throttle opening. This lag in time creates a hesitation of fuel flow whenever the accelerator pedal is depressed. The accelerator pump system solves this problem by mechanically supplying fuel until the other fuel metering systems are able to supply the proper mixture.

The typical accelerator pump (Figure 4-30) is usually the diaphragm type and is located in the bottom of the fuel bowl. Locating the pump in the bottom of the fuel bowl assures a more solid charge of fuel (fewer bubbles).

When the throttle is opened, the pump linkage, activated by a cam on the throttle lever (Figure 4-31), forces the pump diaphragm up. As the diaphragm moves up, the pressure forces the pump

inlet check ball or valve onto its seat, thereby preventing the fuel from flowing back into the fuel bowl. The fuel passes through a short passage in the fuel bowl into the long diagonal passage in the metering body. It next goes into the main body passage and then the pump discharge chamber. The pressure of the fuel causes the discharge jet or valve to raise and fuel is then discharged into the venturi (Figure 4-32).

The pump duration or override spring (Figure 4-33) is an important part of all accelerator pump systems. When the accelerator is moved rapidly to the wide open position, the duration spring is com-

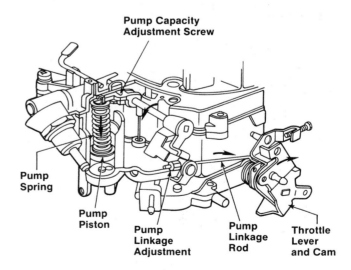

FIGURE 4-31 The mechanical link to the throttle lever activates most accelerator pump systems.

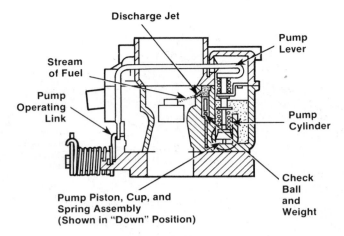

FIGURE 4-32 The accelerator pump forces a stream of fuel into the carburetor throat every time the gas pedal is pressed.

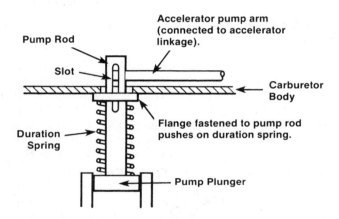

FIGURE 4-33 Pump duration spring

pressed and allows the full pump travel. The spring applies pressure to maintain the pump discharge. Without the spring the pump linkage would be bent or broken due to the resistance of the fuel which is not compressible.

As the throttle moves toward the *closed* position, the linkage returns to its original position and the diaphragm return spring forces the diaphragm down. The pump inlet check valve is moved off its seat and the diaphragm chamber is refilled with fuel from the fuel bowl.

CHOKE CIRCUIT

A cold engine needs a very rich air/fuel mixture during cranking and start-up. Providing the rich mixture is the job of the choke circuit.

During a cold start-up, the choke should be closed. This creates a very high vacuum level in the air horn below the choke plate. As the air pressure

outside the carburetor forces its way into the low-pressure areas, it draws with it a rich air/fuel mixture. When the throttle plate is closed, the mixture is forced out through the idle port or ports below the throttle valve. If the throttle valve is opened to assist in starting the engine, additional ports are exposed to the low-pressure manifold pressure and additional fuel is forced into the air horn. After the engine starts, a leaner mixture can be used to keep the engine running. Therefore, the choke should be opened some to allow increased airflow. After the engine has warmed to normal operating temperatures, the choke should be opened completely to allow the throttle to control airflow and fuel metering.

Before the introduction of automatic chokes, the opening and closing of the choke plate was manually controlled by the driver. A choke cable was connected to a knob inside the passenger compartment on the dash. To close the choke, the driver simply pulled the knob out. As the engine warmed, the choke knob would be gradually pushed in to open the choke.

Most modern vehicles have an automatic choke (Figure 4-34) that operates without any driver assistance. Being more sensitive to engine temperature, an automatic choke is more efficient. Although several designs are used, the basic automatic types that are the most common are:

- Integral choke
- Divorced or remote choke
- Electric choke
- Hot air choke

Integral Choke

When the accelerator pedal is depressed and released, the choke thermostatic spring pushes the choke valve to a fully or partially closed position and the fast idle cam to a corresponding fast idle position (Figure 4-35). A cold engine at low temperatures allows the choke valve to close completely and the fast idle cam to move to its top step.

A manifold vacuum passage (Figure 4-36) through the carburetor body is connected to the bottom end of the choke piston cavity. When the engine starts, the vacuum acting on the bottom of the choke piston opens the valve to a predetermined position established by an adjustable stop. This initial vacuum opening is called the choke qualifying dimension, vacuum break, vacuum kick, or vacuum pull-down position, depending upon the terminology used by the manufacturer. The choke plate might open further under increasing airflow conditions.

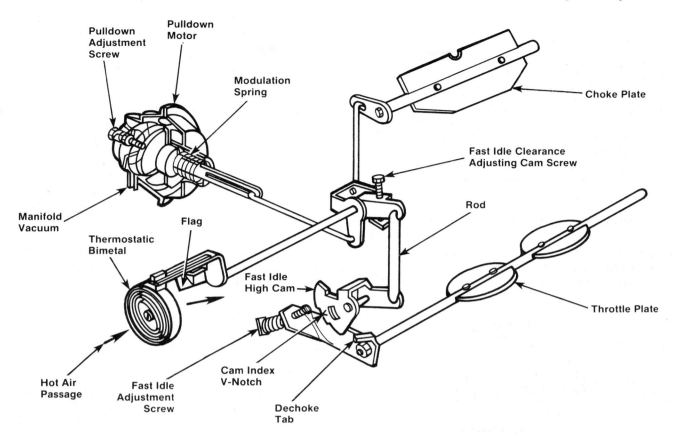

FIGURE 4-34   Typical automatic choke system

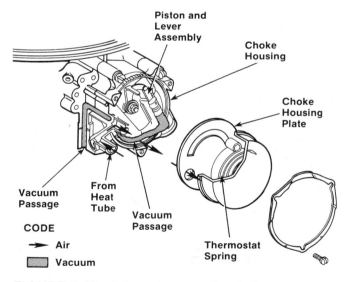

FIGURE 4-35   Integral automatic choke

The choke heat tube connects the heat source with the heat tube connection of the choke housing. After the piston is pulled down to its stop, a vacuum bypass slot or hole in the center of the piston is opened and increased hot air is circulated in the housing to further heat the bimetallic spring. As the

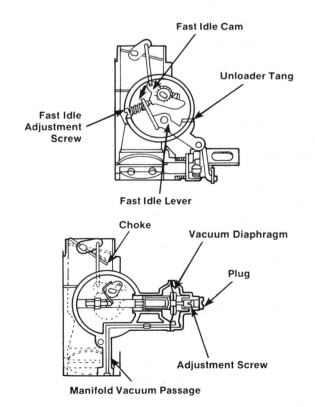

FIGURE 4-36   Choke system

spring warms up, its tension on the choke shaft is relaxed and allows the choke valve to open gradually to the wide-open position (Figure 4–37). The choke bimetallic cover usually has index marks and an arrow indicating the lean direction (Figure 4–38). Different applications require different settings. When the engine first starts, the fast idle cam can be kicked down to a lower step and speed by tapping the accelerator pedal lightly.

Divorced or Remote Choke

In the remote or divorced automatic choke applications the thermostatic spring assembly is mounted in a heat well (Figure 4–39A) or on a pad on the intake manifold and connected to the choke lever by a rod (Figure 4–39B). A vacuum diaphragm is mounted on the carburetor body and connected to a vacuum passage in the throttle body.

To close the choke, the accelerator pedal is depressed to the floor and released. When the engine

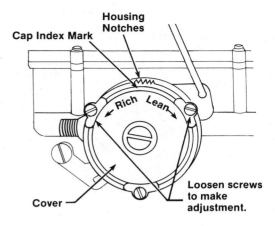

FIGURE 4–38 Arrows on the choke cover indicate the direction of adjustment.

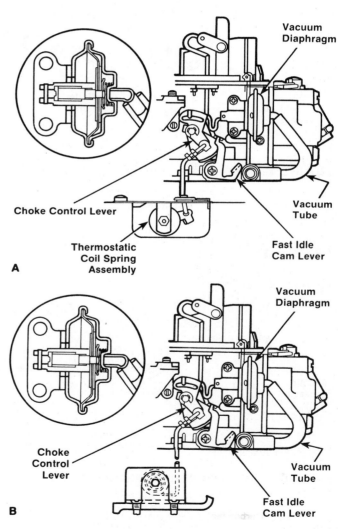

FIGURE 4–39 (A) Divorced choke thermostatic element mounted in heat well; (B) divorced choke thermostatic element mounted on intake manifold pad

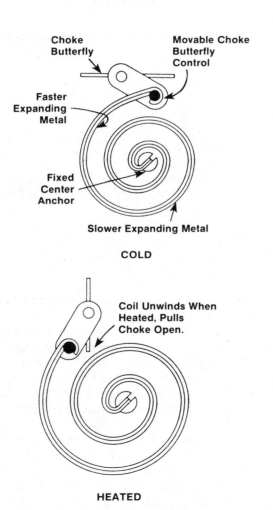

COLD

HEATED

FIGURE 4–37 Principle of a bimetallic spring

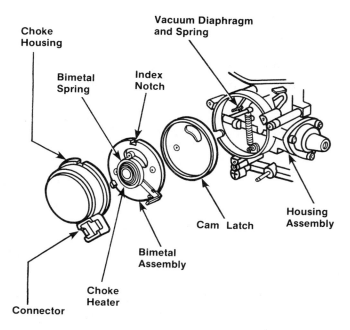

FIGURE 4-40 Constantly operating electric assisted choke

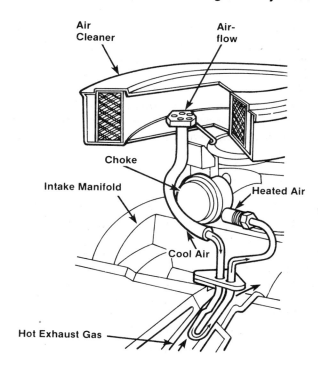

FIGURE 4-41 Late-model hot air choke system

starts, the manifold vacuum acting on the choke vacuum diaphragm opens the choke to the choke qualifying position. Incoming air acting on the offset choke plate assists in the opening and holds the linkage snugly against the action of the choke bimetallic spring. As the bimetallic spring warms up, it permits the choke valve to travel to the wide-open position. Early bimetal assemblies were adjustable. Most late units are not adjustable.

Electric Choke

Electric assisted chokes function somewhat similarly and require nearly the same hardware as integral systems, except for the bimetallic cover assembly. The electric choke differs in bimetallic cover assembly design. A positive temperature coefficient (PTC) ceramic resistor heater is usually built into the cap assembly and provides a direct heat supply for the bimetal spring (Figure 4-40). This type of choke system can be adjusted for the choking duration on the vehicle by readjusting or repositioning the index setting of the bimetallic cap. There are several methods of controlling the electric heating. Check the service manual to determine the type used.

Hot Air Modulator Chokes

Many engines today that do not have electric chokes have some type of hot air modulator system. Most hot air choke setups use heated air flowing

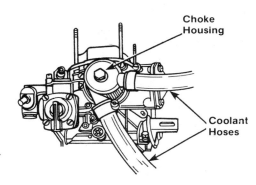

FIGURE 4-42 Hot water heated choke

from the exhaust manifold to the choke housing through a heat tube.

A hot air modulator choke system that has increased in use in recent years is shown in Figure 4-41. In this system, air is drawn through a fresh air intake in the air cleaner housing to a tube in the intake manifold. Passages in the manifold allow hot exhaust gases to circulate around the tube and heat the housing.

Some choke systems use hot coolant (Figure 4-42) instead of hot air to heat the integral choke spring housing. When the engine coolant is cold, the choke spring keeps the choke valve closed. As the engine warms up, the heated coolant causes the thermostatic spring to open the choke valve.

ADDITIONAL CARBURETOR CONTROL

To meet complex fuel economy and emission requirements, modern carburetors have some extra controls for better performance. But remember that no carburetors have all of the following idle speed controls. Check the vehicle's service manual to verify the carburetor's control.

DASHPOT

The dashpot (Figure 4-43) is used on some carburetors to retard the closing of the throttle in the last few degrees. This allows a smooth transition from the main metering system to the idle system

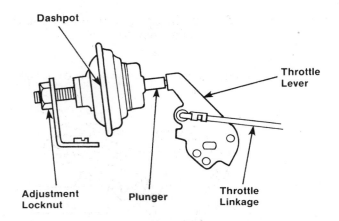

FIGURE 4-43 Typical carburetor dashpot installation

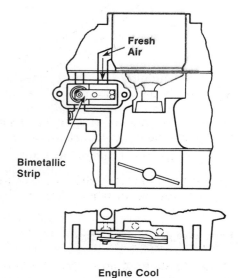

Engine Cool

FIGURE 4-44 Hot-idle compensator operation

and prevents stalling due to an overly rich air/fuel mixture. It also controls the level of HC in the exhaust during deceleration.

The dashpot consists of a small chamber with a spring-loaded diaphragm and a plunger. A link from the throttle comes in contact with the dashpot plunger as the throttle closes. As the throttle linkage exerts force on the plunger, air (or hydraulic fluid) slowly bleeds out of the diaphragm chamber through a small hole. This slows the closing action of the throttle plate.

HOT-IDLE COMPENSATOR (HIC) VALVE

When the engine is overheated, a hot-idle compensator opens an air passage to lean the mixture slightly (Figure 4-44). This increases idle speed to help cool the engine and to prevent excess fuel vaporization within the carburetor. A bimetal thermostatically controlled air bleed valve admits additional air into the idle system; this relieves an overrich idle condition and the resulting high engine temperatures during prolonged hot idle.

IDLE-STOP SOLENOID

The idle-stop solenoid (Figure 4-45), which was adopted early in the emission control program, prevents engine dieseling. Dieseling or run-on is a result of the high idle speed required to compensate for the leaner idle fuel mixtures of recently produced engines. The idle-stop solenoid allows the carburetor throttle valves to close as soon as the ignition is switched off. This helps prevent additional mixture

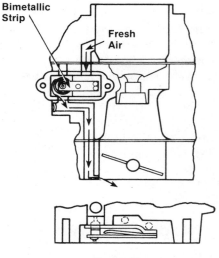

Engine Hot

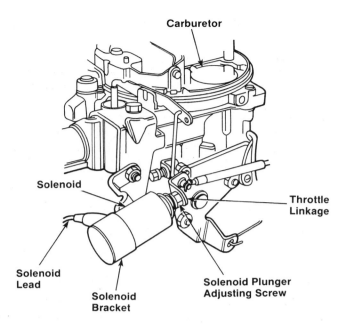

FIGURE 4-45   Typical throttle stop solenoid installation

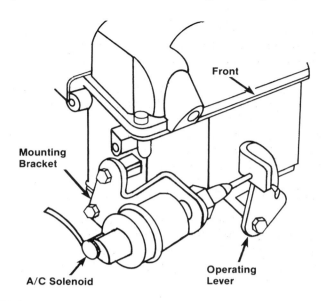

FIGURE 4-46 Air conditioning throttle solenoid slightly opens the throttle when the A/C is on.

from being drawn into the engine and becoming ignited by chamber heat, thereby eliminating engine dieseling. On applications where this part is used, it is usually located next to the carburetor's throttle valve lever.

AIR CONDITIONING SPEED CONTROL

The purpose of a solenoid on an engine with air-conditioning is to open the throttle valves slightly when the air conditioning is on (Figure 4-46). This compensates for the additional load placed on the engine by the compressor at idle speed and helps prevent rough idling, hesitation, and stalling.

DUAL VACUUM BREAK

Most modern carburetors are equipped with a dual vacuum break system (Figure 4-47)—a primary and a secondary. The primary vacuum diaphragm opens the choke valve slightly as soon as the engine starts to keep the engine from overchoking and stalling. The secondary vacuum diaphragm, which is also attached to the choke lever, opens the choke valve slightly wider in warm weather or when a warm engine is being started. Vacuum to the secondary diaphragm is controlled by a thermal vacuum switch or valve. The TVV releases vacuum to the secondary vacuum break when the engine reaches a certain temperature. This prevents a rich fuel mixture and

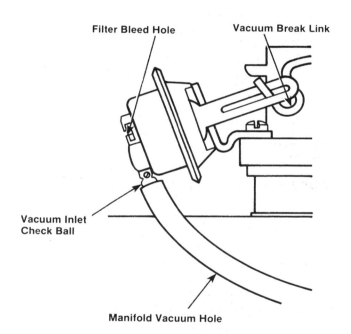

FIGURE 4-47 The vacuum break opens the choke slightly as soon as the engine starts.

the high emissions that result when a cold engine is started in warm weather or when a warm engine is started.

CHOKE UNLOADER

To be able to start a cold engine that has been flooded with gasoline, a choke unloader is required. The choke unloader (Figure 4-48) is throttle linkage

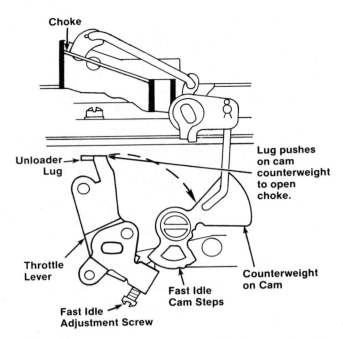

FIGURE 4-48 The choke unloader opens the choke any time the gas pedal is "floored."

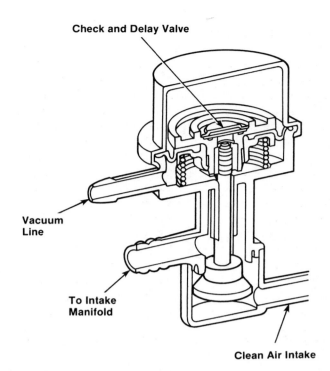

FIGURE 4-50 Deceleration valve

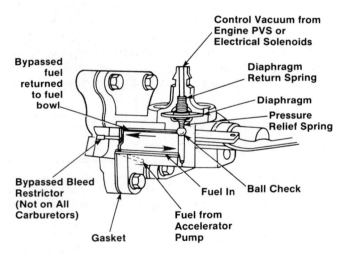

FIGURE 4-49 Components of a temperature-compensated accelerator pump

actuated and opens the choke whenever the accelerator paddle is "floored." At wide-open throttle the partially opened choke allows additional air to lean out the mixture and reduce fuel flow.

TEMPERATURE-COMPENSATED ACCELERATOR PUMP

This temperature-sensitive device (Figure 4-49) on the accelerator pump allows more fuel to be pumped when the engine is cold. This helps the engine accelerate well when cold. The device also causes less fuel to be pumped when the engine is hot.

DECELERATION VALVE

This valve (Figure 4-50) is designed to prevent backfire during deceleration as the fuel mixture becomes richer. The valve, which operates only during deceleration, is usually located between the intake manifold and the air cleaner. A typical valve has a cam-shaped diaphragm housing on one end. A control manifold-vacuum line is attached to a port under the diaphragm housing. The other end of the valve is connected by hoses to the intake manifold and air cleaner. When deceleration causes an increase in manifold vacuum, the diaphragm opens the deceleration valve and allows air to pass from the air cleaner into the intake manifold, leaning the fuel mixture and preventing exhaust system backfire.

THROTTLE POSITION SOLENOID

The throttle position solenoid (TPS) is an electrical device used to control the position of the throttle plate (Figure 4-51). It can have several functions, depending on its application. When the basic function is to prevent dieseling, the solenoid is called a throttle stop solenoid or an idle stop solenoid. When the engine is started, the solenoid is energized and the plunger extends, pushing against the throttle linkage. This forces the throttle plate(s) open slightly to the curb idle position. When the ignition switch is turned off, the solenoid is de-energized and the plunger retracts, allowing the throttle plate to close completely and shutting off the air/fuel supply.

The throttle position solenoid is also used to increase the curb idle speed to compensate for extra loads on the engine. When this is its primary function, the solenoid might also be called an idle speed-up solenoid or a throttle kicker. This feature is most often used on cars with air conditioners; when the A/C is turned on, a relay energizes the solenoid so that the plunger extends farther, raising the idle rpm. This keeps the engine running at the high rpm's necessary to ensure adequate emission control.

The throttle position solenoid is also used to control idle speed when the transmission is engaged. A relay in the PARK/NEUTRAL switch signals the solenoid to extend when the transmission is shifted into gear. This opens the throttle slightly to compensate for the increased load on the engine.

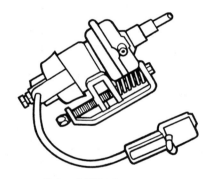

Solenoid Diaphragm

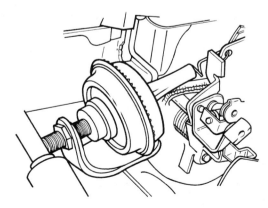

Vacuum Operated Throttle Modulator (VOTM)

FIGURE 4-51 Throttle positioners

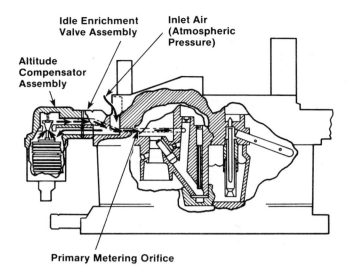

Idle Enrichment Valve Assembly
Inlet Air (Atmospheric Pressure)
Altitude Compensator Assembly
Primary Metering Orifice

FIGURE 4-52 Pressure-sensitive bellows operates an air valve

ALTITUDE COMPENSATION VALVE

At higher altitudes, leaner fuel mixtures are required. On most carburetors, a pressure-sensitive (aneroid) bellows (Figure 4-52) operates an air valve to increase the amount of airflow at high-altitude operation. Without the altitude compensator, the air/fuel ratio would be too rich for high-altitude operation.

CARBURETOR DESIGN FEATURES

While all carburetors contain the basic circuit designs, the following features are important to the service technician:

- Airflow directions
- Number of barrels
- Number and type of venturi
- Number of stages
- Size

AIRFLOW DIRECTION

With most standard carburetor designs there are three airflow patterns through the air horn (Figure 4-53).

- *Downdraft Carburetor.* Located on top of the engine's intake manifold, this type of carburetor is the most used today. Fuel is drawn up through the air and falls as it is atomized into fuel droplets ready for combustion.

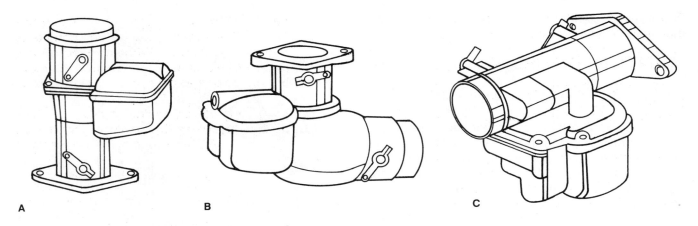

FIGURE 4-53 Carburetor classifications for airflow direction: (A) downdraft; (B) updraft; (C) sidedraft

- *Updraft Carburetor.* This type is seldom found in modern passenger vehicles because of the high velocity required to lift air into the air horn.
- *Sidedraft (Crossdraft) Carburetor.* This type of carburetor is usually found in small or compact vehicles because it requires less clearance between the top of the engine and the vehicle's hood than the other two designs. The carburetor and air cleaner are located on one side of the engine; airflow is horizontal through the air horn.

CARBURETOR BARRELS

A carburetor can have one, two, or four air horns or barrels (throats). The single-barrel carburetor (Figure 4-54A) has only one air horn or throat. Although this carburetor is good for fuel economy, its airflow capacity is small, thus its power is low. For this reason, a single carburetor is usually found on four-cylinder and small six-cylinder vehicles.

The two-barrel carburetor (Figure 4-54B) is really two single barrels built into one body or flange. That is, each barrel is equipped with its own idle circuit, choke plates, and other carburetor components. The front air horn is called the *primary throttle* and is often smaller in diameter than a rear or secondary throttle. Because the two-barrel carburetor represents a good compromise between fuel economy and horsepower, it is used extensively on six-cylinder cars.

The four-barrel carburetor has four air horns in a single body (Figure 4-54C). It has two primary and two secondary throttles. When the secondary throttle plates are open more than the front primary plates, the carburetor is classified as a spread bore.

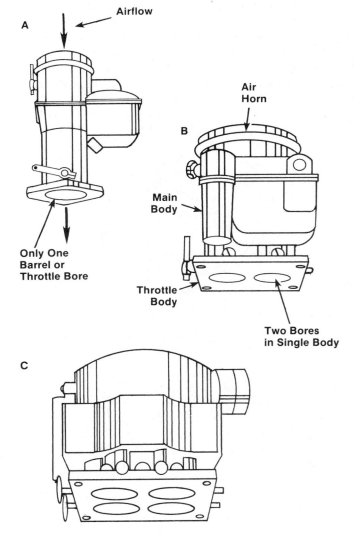

FIGURE 4-54 (A) Single-barrel carburetor; (B) two-barrel carburetor; (C) four-barrel carburetor

Four-barrel carburetors are used primarily on larger displacement V-8 engines.

VENTURI DESIGNS

Several venturi designs have been used in automotive carburetors over the years. Currently, the most commonly used designs are the following:

- *Double-Stage Venturi (2V).* This venturi uses both a primary venturi in the bore wall and a secondary venturi suspended in the center of the air horn. This configuration of venturis is usually found in the primary barrel of the carburetor. The double-stage venturi provides good performance at both low and high engine speeds.
- *Triple-Stage Venturi (3V).* This design has a primary venturi as well as two secondary venturis located in the middle of the carburetor bore. The third venturi improves atomization and vacuum at low engine speeds.
- *Vaned Venturi.* The vaned venturi carburetor was designed to increase fuel economy by improving the atomization at low engine speeds. Vanes or fins in the secondary venturi that run radially from the center to the outer edge create a turbulence in the airflow

passing through the carburetor. This turbulence aids in breaking up the fuel droplets as they are discharged. While a vaned venturi provides excellent fuel economy, the extra restriction of the vanes reduces engine power slightly.

- *Variable Venturi.* Considered by many automotive experts as the most efficient design (Figure 4-55), the variable venturi tends to

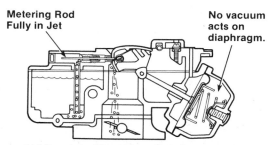

Metering Rod Fully in Jet — No vacuum acts on diaphragm.

At idle, venturi is very small and metering rod reduces fuel flow to air horn.

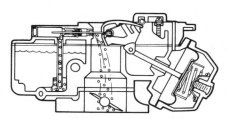

At idle, air horn is slightly larger. More fuel is drawn into it.

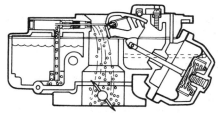

At immediate speed, still more fuel and air are drawn into throat.

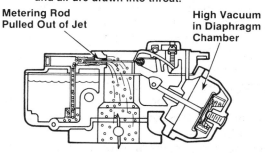

Metering Rod Pulled Out of Jet — High Vacuum in Diaphragm Chamber

With wide-open throttle, variable venturi is fully withdrawn, as is metering rod.

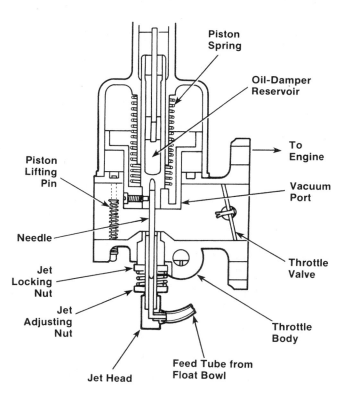

Piston Spring
Oil-Damper Reservoir
To Engine
Vacuum Port
Throttle Valve
Throttle Body
Feed Tube from Float Bowl
Jet Head
Jet Adjusting Nut
Jet Locking Nut
Needle
Piston Lifting Pin

FIGURE 4-55 Variable venturi carburetor assembly

FIGURE 4-56 Diaphragm controls variable venturis in this carburetor design

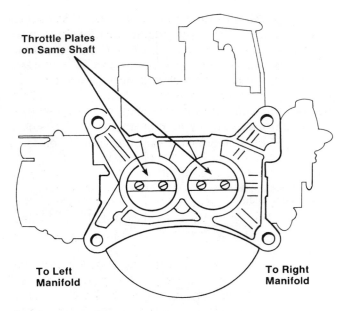

Throttle Plates on Same Shaft

To Left Manifold

To Right Manifold

FIGURE 4-57 Single-stage carburetor

maintain constant air velocity independent of engine speed. It can vary its area to match engine speed, vacuum, and load. Figure 4-56 shows the operation of a varible venturi carburetor. Most assist systems are not necessary when a variable venturi is used. A variable venturi increases in size as engine demands increase. In this way, airflow speed through the venturi and the resulting pressure differential remains fairly constant. Thus, a variable venturi carburetor is also known as a *constant velocity* carburetor or a *constant depression* (vacuum) carburetor.

 SHOP TALK _____

The single venturi had only one restriction formed into the wall of the carburetor bore. Although this design was used extensively at one time, such venturis are not found on modern production vehicles.

SECONDARY FUEL METERING SYSTEMS

Carburetor stage refers to the way the throttle butterflies are opened. When both throttle plates open at the same time (Figure 4-57), the carburetor is called a *single stage*. When the primary throttles open independently of the secondary throttles, the carburetor is classified as a two-stage (Figure 4-58). The secondary throttle opens when more power is

needed. That is, the secondary system operates only when the primary throttle reaches about 40 degrees of travel, contacts a free-floating lever on the primary-throttle shaft, and opens the secondary throttle (Figure 4-59). A staged carburetor has a small primary venturi that creates high venturi velocities capable of providing good fuel mixing, atomization,

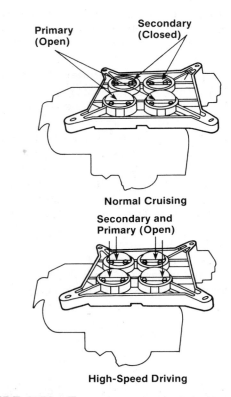

Primary (Open) **Secondary (Closed)**

Normal Cruising

Secondary and Primary (Open)

High-Speed Driving

FIGURE 4-58 Two-stage carburetor; secondary throttle opens when more power is needed.

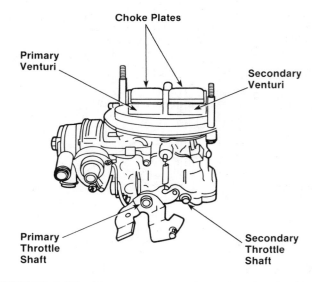

Choke Plates

Primary Venturi

Secondary Venturi

Primary Throttle Shaft

Secondary Throttle Shaft

FIGURE 4-59 Two-stage, two-barrel carburetor

and vaporization. The secondary venturi provides the added flow capacity necessary to raise the engine output power (Figure 4-60).

At low speeds the secondary throttle valves remain closed, allowing the engine to maintain proper air/fuel velocities and distribution for lower speed and light load operation. When engine demand increases to a point where additional breathing capacity is needed, the vacuum-controlled secondary throttle valves begin to open automatically (Figure 4-61).

Vacuum from one of the primary venturi and one of the secondary venturi is channeled to the top of the secondary diaphragm (Figure 4-62). The bottom of the diaphragm is open to atmospheric pressure. At higher speeds and higher primary venturi vacuum, the diaphragm, operating through a rod and secondary throttle lever, begins to open the secondary throttle valves. This action starts to compress the secondary diaphragm spring.

As the secondary throttle valves open further a vacuum signal is created in the secondary venturi. This additional vacuum assists in opening the secondary throttle valves to the maximum designed opening. The secondary opening rate is controlled by the diaphragm spring and size of the vacuum restrictions in the venturi.

When engine speed is reduced, venturi vacuum decreases and the diaphragm spring starts to push the diaphragm down to start the closing of the secondaries. Closing the primary throttle valves moves the secondary throttle connecting link. Some carburetor designs feature a secondary air valve that is placed over the top of the secondary plate (Figure 4-63). This plate prevents secondary barrel operation until sufficient airflow is present in the primary.

Most production applications have a ball check and bypass bleed installed in the diaphragm passage. The ball permits a smooth, even opening of the secondaries, but lifts off the inlet bleed to cause their

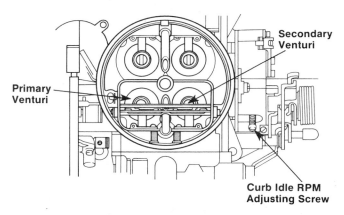

FIGURE 4-60 A two-stage venturi uses both a primary venturi in the bore wall and a secondary venturi suspended in the center of the air horn.

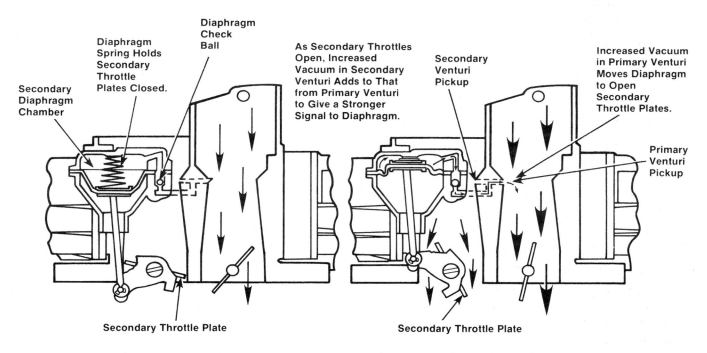

FIGURE 4-61 Vacuum-controlled secondary system

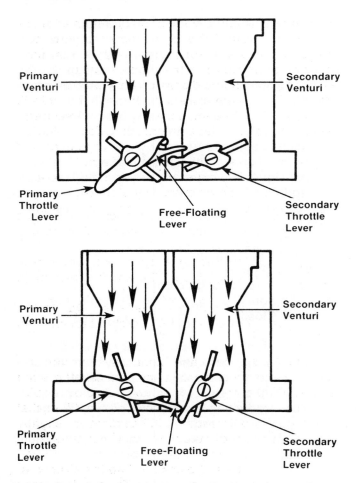

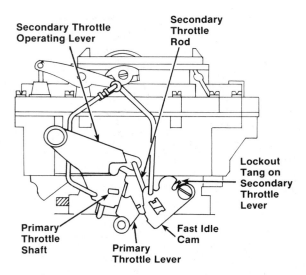

FIGURE 4-64 Mechanical secondary throttle linkage on a four-barrel carburetor

FIGURE 4-62 Secondary diaphragm responds to vacuum from ports within the carburetor barrels.

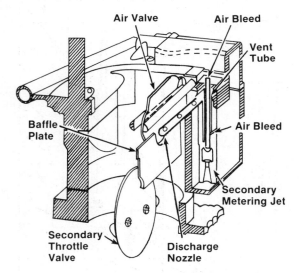

FIGURE 4-63 To prevent secondary barrel operation until sufficient airflow is present in the primary, a secondary air valve is sometimes placed over the top of the secondary throttle plates.

rapid closing when the primary throttle valves are closed.

Mechanically operated secondaries function simply. Secondary throttles are opened by a direct link from the primary throttle lever. Usually the secondary throttle linkage (Figure 4-64) is progressive so that the secondary opening is delayed until the primary throttle valves have opened approximately 40 to 45 degrees. When the primary valves are closed, the secondary link pulls the secondaries closed. Air pressure on the throttle valves aids the closing.

Secondary Float System

Secondary systems have a separate fuel bowl. Fuel is usually supplied to the secondary bowl by a transfer tube from the primary fuel inlet fitting. High performance models or bowls with center hinged floats use exterior plumbing.

The secondary fuel bowl is equipped with a fuel inlet valve and float assembly similar to the primary side.

The specified fuel level on the primary side is usually slightly lower than the secondary side. Many applications include a balance tube, which vents vapors and excess fuel from backing into the primary bowl. The secondary fuel inlet system is calibrated to deliver the proper mixture to the other systems only when the fuel is at the specified level.

Secondary Idle System

If the secondary system were to remain inoperative over long periods, portions of the system would

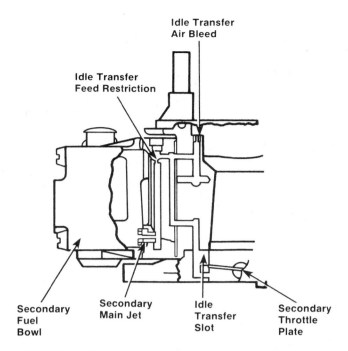

Idle Transfer Air Bleed

Idle Transfer Feed Restriction

Secondary Fuel Bowl

Secondary Main Jet

Idle Transfer Slot

Secondary Throttle Plate

FIGURE 4-65   Secondary idle and transfer system

become blocked with gum and varnish. To prevent this condition and maintain stable fuel level control, an idle system is incorporated on the secondary side to insure the use of fuel from the secondary side under various driving conditions and also maintain stable fuel levels and balanced idle mixtures.

Fuel flows from the secondary fuel bowl through the secondary main metering jet, through a passage into the idle well, then up a vertical passage and through the idle feed restriction (Figure 4-65). Here the fuel blends with a small amount of air entering through the secondary idle air bleed. This air/fuel emulsion flows down a vertical passage to the secondary idle air bleed. As the secondary throttle valve opens, exposing the transfer slot to vacuum, the air/fuel mixture flows from both the idle passage and the idle transfer slot. A carburetor with a secondary idle and transfer is considered very efficient.

CARBURETOR SIZE

The carburetor's size is rated in cubic feet per minute (cfm), which is the amount of air that can flow through a wide-open throttle. Carburetor sizes range from approximately 250 cfm on small, one-barrel carburetors to more than 1000 cfm on high-performance four barrels.

To determine the correct carburetor size for any given engine, volumetric efficiency (VE) must be known. Volumetic efficiency is an indicator of how

well an engine can breathe. The better an engine's "breathing ability," the higher its volumetric efficiency. It is expressed as the ratio of the actual mass (weight) of air taken into the engine compared to the mass that the engine displacement would theoretically take in if there were no losses. The ratio is expressed as a percentage. It is quite low at idle and low speeds and varies with engine speed. Volumetric efficiency should be computed at the expected operating rpm of the engine application.

Use the following examples as a guide to estimating the volumetric efficiency of an engine:

- An ordinary low-performance engine has a volumetric efficiency of about 80 percent at maximum torque.
- A high-performance engine has a volumetric efficiency of about 85 percent at maximum torque.
- An all-out racing engine has a volumetric efficiency of about 95 percent at maximum torque.

A highly tuned intake and exhaust system with efficient cylinder head porting and a camshaft ground to take full advantage of the engine's other equipment can provide such complete cylinder filling that a volumetric efficiency of 100 percent, or slightly higher, is obtained at the speed for which the system is tuned.

Figure 4-66 can be used to find the airflow requirements for any given engine. The graph is based

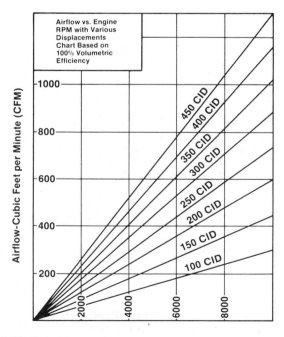

FIGURE 4-66   Airflow requirements for engines

on 100 percent volumetric efficiency, so any indicated airflow must be multiplied by the volumetric efficiency of your particular engine. Use a carburetor with an airflow rating equal to or slightly smaller than the air requirement for the engine. Take, for example, a 300 cubic inch displacement (cid) V–8 that has a maximum rpm limit of 8000 rpm. It has been determined that this particular engine has a volumetric efficiency of 85 percent. According to the graph the airflow requirement of this engine is 700 cfm at 100 percent volumetric efficiency. At 85 percent, however, the cfm requirement is 595 cfm. This engine would, therefore, require a 600 cfm carburetor.

When selecting a carburetor remember that it is not wise to overcarbureate. This means using a carburetor with an excessively high cfm rating. Overcarburetion will not only cut gas economy, it will also decrease engine power. Fortunately for the service technician, the carburetor installed by the auto manufacturer will be the best size. When installing a new or rebuilt carburetor, identification including make, model, and year, is important information required for servicing carburetors and ordering parts. Some methods used for carburetor identification are shown in Figure 4-67. Manufacturer's catalogs and vehicle service manuals are good sources of carburetor size information.

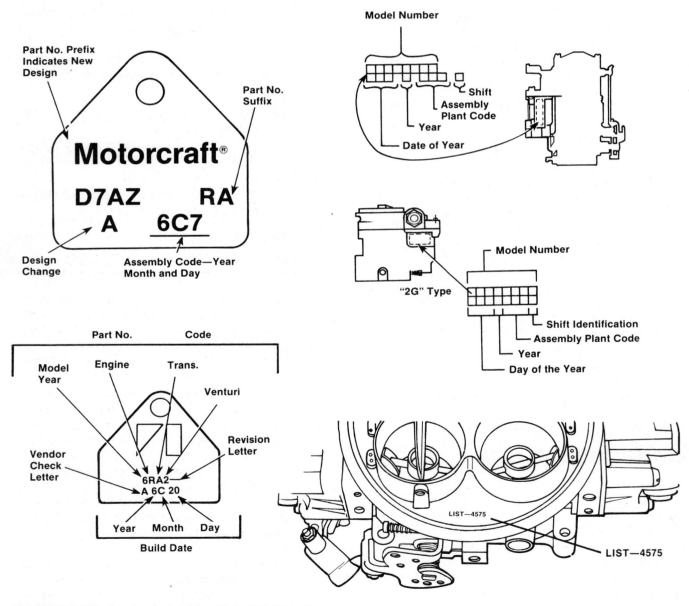

FIGURE 4-67 Methods for identifying carburetors

SHOP TALK _____

To understand carburetor manufacturers' catalog terms, the service technician should know that:

- *Carburetors listed under the heading "Emission Design/Street Performance" are the result of extensive engineering programs to design and produce performance carburetors with emissions capability. They are recommended for use in areas where local emission legislation does not allow the use of non-emission (competition/off-road) performance carburetors.*
- *Carburetors listed under this heading may or may not be legal for street use in California. Those carburetors that are not legal for street use will have the following statement below the carburetor description: "Not legal for street use in California."*
- *Carburetors listed under the heading Competition/Off-road are maximum-performance designs; however, these carburetors can be used on performance street vehicles where local emission legislation permits. These carburetors are calibrated with richer air/fuel ratios to allow for higher compression ratios, increased spark advance, and changes in camshaft timing. All other carburetors are universal designs that will fit a variety of applications and are not necessarily true bolt-ons. Some modifications to the throttle and choke linkage might be necessary to complete installation. Carburetors listed under this heading are not legal for street use in California.*

Carburetors listed under the heading Race Only are calibrated for fully modified engines used solely for racing. These carburetors offer no compromise for street use and therefore are not recommended for street-driven vehicles.

For high-performance engines, a carburetor with mechanical secondaries has an inherent advantage over a carburetor with a controlled secondary system (air valve or vacuum diaphragm). This is possible because a controlled secondary carburetor, until it reaches wide-open throttle, will not have as great a pressure drop below the throttle plates as would a mechanical secondary unit. The greater the pressure drop below the throttle plates the more dense will be the air/fuel charge to the engine and, hence, the more output. Greater care, however, must be taken in selecting the correct size mechanical secondary carburetor for an application. Double pump, mechanical secondary carburetors initially depend only on the accelerator pumps to provide adequate fuel until enough airflow can be established to begin pulling in the main system. The larger the carburetor the higher the airflow required to accomplish this. If the carburetor is too large, the pump shot will be consumed before the main system starts. The result is a bog or sag.

CARBURETOR CONSTRUCTION MATERIALS

Carburetor components are made from various materials:

1. Steel (choke, shafts, throttle plates, and fast idle cam)
2. Cast iron (throttle body)
3. Aluminum (main body, air horn body, venturi, and throttle)
4. Brass (jets, valve seats, floats, and needle valves)
5. Plastic (baffles and floats)
6. Synthetic rubber (seals, diaphragms, needle tips, and pump pistons)

FEEDBACK CARBURETOR SYSTEMS

The latest type of carburetor system is the electronic feedback controlled design (Figure 4–68), which provides better combustion by control of the air/fuel mixture. Though these carburetors provide the required performance and economy, the better combustion reduces exhaust emissions.

In the late 1970s, the feedback carburetor was introduced. Two other developments made the use of a feedback carburetor feasible. One was the exhaust gas oxygen sensor and the other was the three-way catalytic converter. Prior to 1978, two-way catalytic converters controlled the output of hydrocarbons, (HC) and carbon monoxide (CO) in the exhaust. In 1978, both Ford and General Motors began using a three-way converter that not only oxidized HC and CO but also chemically reduced oxides of nitrogen.

However, for the three-way catalyst to work efficiently, the air/fuel ratio must be maintained very close to a 14.7 to 1 ratio. If the air/fuel mixture is too lean, NO_x will not be converted efficiently. If the

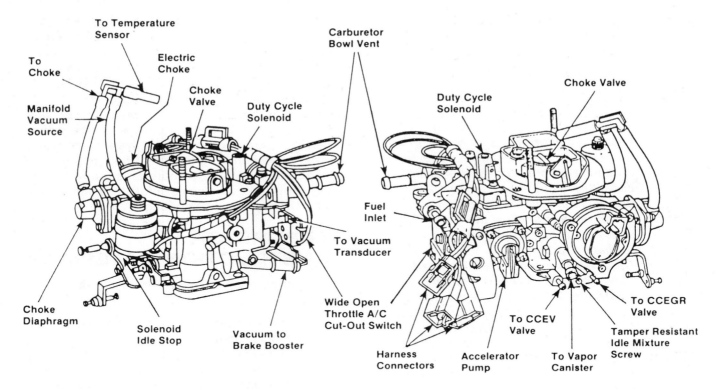

FIGURE 4-68 A 2-V two-stage electronic feedback carburetor

mixture is too rich, HC and CO will not oxidize efficiently. Monitoring the air/fuel ratio is the job of the exhaust gas oxygen sensor (Figure 4-69).

An oxygen sensor (as its name implies) senses the amount of oxygen present in the exhaust stream. (For a full explanation of how an oxygen sensor is constructed and operates, refer to Chapter 12). A lean mixture will produce a high level of unburned oxygen in the exhaust. A rich mixture will produce little oxygen in the exhaust. The oxygen sensor, placed in the exhaust upstream from the catalytic converter, produces a voltage signal that varies in intensity in direct proportion to the amount of oxygen the sensor detects in the exhaust. If the oxygen level is high (a lean mixture), the voltage output is low. If the oxygen level is low (a rich mixture), the voltage output is high.

The electrical output of the oxygen sensor is monitored by an electronic control module (ECM). This microprocessor is programmed to interpret the input signals from the sensor and in turn generate output signals to a control device that will meter more or less fuel into the air charge as it is needed to maintain the 14.7 to 1 ratio (Figure 4-70).

Whenever these components are working to control the air/fuel ratio, the carburetor is said to be operating in closed loop. Closed loop is illustrated in

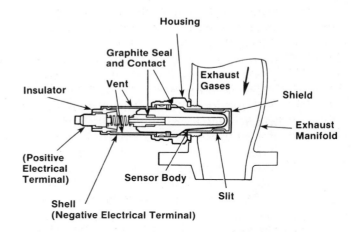

FIGURE 4-69 Exhaust gas oxygen sensor

the schematic shown in Figure 4-71. The oxygen is constantly monitoring the residual oxygen the exhaust, and the control module is constantly making adjustments to the air/fuel mixture based on the fluctuations in the sensor's voltage output. However, there are certain conditions under which the control unit ignores the signals from the oxygen and does not regulate the ratio of fuel to air. During these times, the carburetor is functioning in a convention-

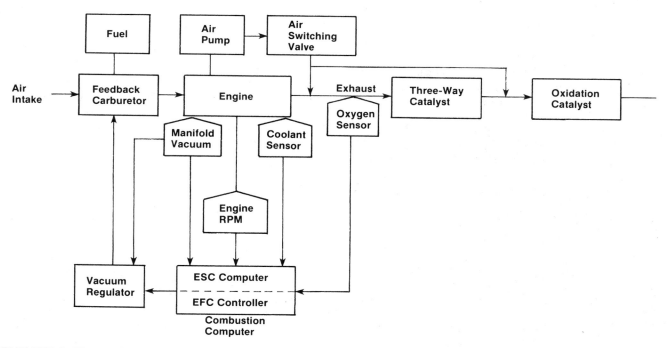

FIGURE 4-70   EFC system block diagram

al manner and is said to be operating in open loop (the control cycle has been broken).

The carburetor must operate in open loop until the oxygen sensor reaches a certain temperature (approximately 600 degrees Fahrenheit). The carburetor also goes into open loop when a richer than normal air/fuel mixture is required—such as during warmup and heavy throttle application. Several other sensors are needed to alert the electronic control unit of these conditions. A coolant sensor provides input relating to engine temperature. A vacuum sen-

sor and a throttle position sensor indicate wide open throttle.

Early feedback systems used a vacuum switch to control metering devices on the carburetor. Figure 4-72 shows a feedback carburetor introduced on Ford vehicles in 1978. Closed loop signals from the

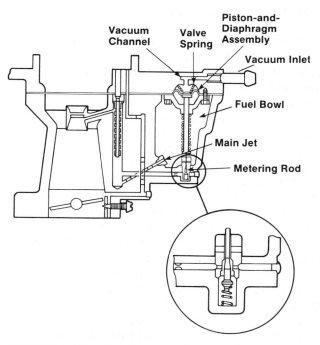

FIGURE 4-72   Vacuum feedback carburetor

Lean Mixture
O₂ in Exhaust Gas

Carburetor Control "Leans" Mixture

Low Sensor Voltage

Electronic Control Energizes Carburetor Solenoid

Electronic Control De-energizes Carburetor Solenoid

High Sensor Voltage

Carburetor Control Enriches Mixture

Less O₂ in Exhaust Gas

FIGURE 4-71   Closed loop operation

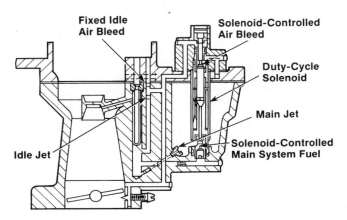

FIGURE 4-73 Electronic feedback carburetor

electronic control module are sent to a vacuum solenoid regulator, which in turn controls vacuum to a piston and diaphragm assembly in the carburetor. The vacuum diaphragm and a spring above the diaphragm work together to lift and lower a tapered fuel metering rod that moves a tapered fuel metering rod in and out of an auxiliary fuel jet in the bottom of the fuel bowl. The position of the metering rod in the jet controls the amount of fuel allowed to flow into the main fuel well.

The more advanced feedback systems use electrical solenoids on the carburetor to control the metering rods (Figure 4-73). These solenoids are generally referred to as duty-cycle solenoids. The solenoid is normally wired through the ignition switch and grounded through the electronic control unit. The solenoid is energized when the ECU completes the ground. The ECU is programmed to cycle

(turn on and off) the solenoid ten times per second. Each cycle lasts 100 ms (milliseconds). The amount of fuel metered into the main fuel well is determined by how many milliseconds the solenoid is on during each cycle. The solenoid can be on almost 100 percent of the cycle or it can be off nearly 100 percent of the time. The M/C solenoid can control a fuel metering rod, an air bleed, or both. Figure 4-74 illustrates two of the most common types of mixture control solenoids; Figure 4-75 shows the action of an M/C solenoid.

CARBURETOR SERVICE

A complete and thorough road test and check of minor carburetor adjustments should precede major carburetor service. Specifications for some adjustments are listed on the vehicle emission control information label found in each engine compartment.

Many poor performance complaints are incorrectly blamed on the carburetor. Therefore, the following items should be carefully checked, repaired, or adjusted before removing the carburetor from the vehicle.

- Ignition wires
- Spark plugs
- Distributor
- Ignition timing
- Vacuum lines and hoses
- EGR valve, PVC system, manifold heat valve, and thermal vacuum switch
- Compression

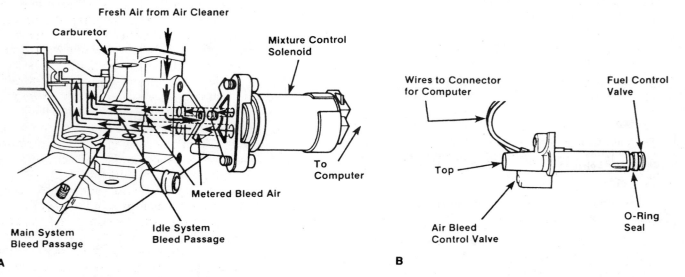

FIGURE 4-74 Mixture control solenoids: (A) opens and closes air passages to control mixture ratio; (B) controls both air and fuel flow.

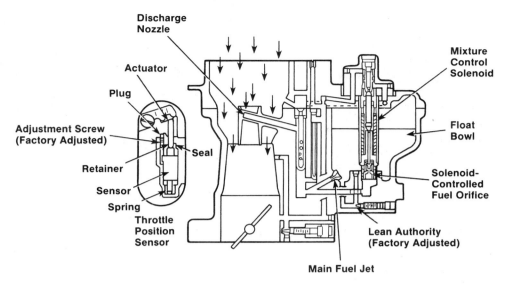

FIGURE 4-75 Cutaway of mixture control solenoid

- Fuel filter
- Float bowl
- Air filter
- Fuel pump

If the problem remains after making the above checks and taking the proper action, then proceed with carburetor removal and repair or service as directed in the service manual.

CARBURETOR REMOVAL

Carburetor removal and replacement procedures vary from vehicle to vehicle as well as between models of the same manufacturer. Therefore, procedures presented are general, intended to cover most carburetor installations.

 SHOP TALK ─────────

The first and most important step before removing and replacing a carburetor is to read the manufacturer's service manual or the instructions enclosed with the replacement carburetor. Doing this can prevent costly mistakes and will familiarize the technician with replacement procedures and various parts of the carburetor.

───────────────────────

WARNING: Be sure that the engine is cold before attempting replacement. Personal injury or a car fire can result when attempting to remove a carburetor from a hot engine.

───────────────────────

To remove and replace the carburetor do the following:

1. Disconnect the battery ground cable.
2. Tag all fuel lines, vacuum lines, and electrical connections for identification before disconnecting, or sketch or photograph the routing arrangements, if more convenient.
3. Disconnect all vacuum lines from the carburetor.
4. Disconnect any electrical connections to components such as the choke, etc.
5. Remove air cleaner.
6. Wrap the fuel line connection to the carburetor with a shop rag or towel to ensure that fuel does not spray when the line is disconnected. If the carburetor uses a brass fuel line and flare nuts, use a flare nut wrench to disconnect the line.
7. Disengage the throttle (Figure 4-76) and choke linkage by removing the appropriate retaining clips.
8. Most carburetors have two or three rods and springs that connect to the throttle lever. These should also be tagged and identified before disconnecting.
9. If the vehicle uses a hot-water choke system, do the following:
   - Loosen the radiator cap to release any pressure.
   - Place a shop rag or towel around the hot-water choke to soak up escaping coolant.
   - Remove the hot-water choke.

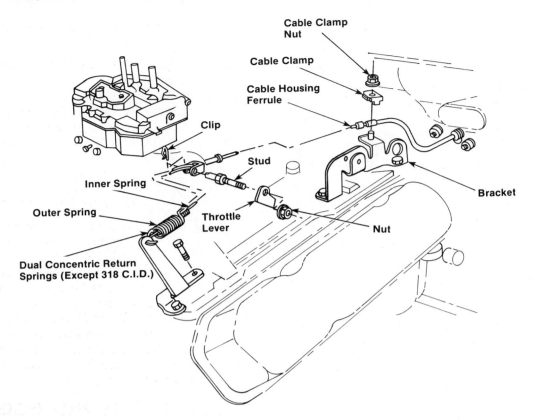

FIGURE 4-76 Typical carburetor throttle linkage connects the throttle shaft to the accelerator pedal.

10. Once the linkage, fuel lines, hoses, and electrical connections are disconnected, loosen and remove the carburetor mounting nuts.

11. Ensure all connections are disconnected from the carburetor before it is lifted off the manifold. Hold the carburetor level to avoid spilling fuel from the fuel bowl. Drain fuel from the carburetor; dispose of the fuel in recommended manner.

12. Ensure that no dirt or parts get into the manifold once the carburetor is removed. If they do, use a vacuum cleaner or magnet to remove the foreign matter.

13. Once the carburetor is removed, inspect all fuel lines, hoses, and clamps. Replace any that are old, cracked, or brittle.

14. Replace the fuel filter.

15. Remove the old gasket on the manifold. If a scraper must be used to remove the old gasket, first plug up the openings in the intake manifold with a hand cloth. This will prevent any scrapings from falling into the manifold.

16. Disassemble the carburetor as recommended in the vehicle's service manual.

Mount the carburetor on a carburetor repair stand (Figure 4-77). This prevents accidental damage to the throttle plates.

CAUTION: When disassembling a carburetor, be careful not to drop any components, such as check balls or weights. Note which holes these ports fit into in the carburetor body. Use special carburetor service tools for this procedure (Figure 4-78).

Cleaning Carburetor Parts

Although there are several carburetor or decarbonizing cleaners available, they must be used with some caution. That is, the choke diaphragm, choke heater, and plastic or rubber parts can be damaged by those solvents. Also, carburetors that use special coating on the linkage and shafts should not be subjected to carburetor cleaners. Clean the external surfaces of these parts with a clean cloth or soft brush. Shake dirt or other foreign material from the stem (plunger) side of the diaphragm. Compressed air can be used to remove loose dirt but should not be connected to the vacuum diaphragm fitting.

WARNING: Carburetor cleaner can cause serious chemical burns to the eyes and skin. Always wear eye protection and rubber gloves when working with carburetor type cleaning solvents.

If the commercial solvent or cleaner recommends the use of water as a rinse, hot water will produce better results. After rinsing, all water must be blown from the passages with air pressure.

CAUTION: Do not use wire, drill, or other mechanical means to clean out carburetor jets. These devices could enlarge or scratch the passages, upsetting carburetor operation.

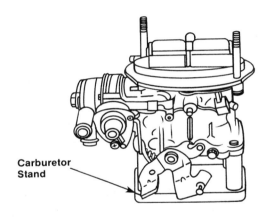

Carburetor Stand

FIGURE 4-77 A carburetor stand facilitates carburetor overhaul

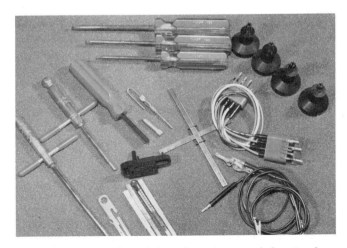

FIGURE 4-78 Special carburetor servicing tools

After the parts are cleaned, check them for the following:

- Warpage, cracks, or other damage to carburetor bodies
- Bent float hinge arm
- Wear between the throttle body and shaft
- Weakened or damaged springs
- Burrs, nicks, and dirt on gasket mating surfaces
- Damage to the tips of idle mixture screws
- Stripped fasteners
- Binding of the choke linkage and plate

When inspecting parts, it is recommended that new ones should be installed whenever the old are questionable. Carburetor kits are available that contain such items as gaskets, seals, retaining clips, check balls or valves, needle and seat, accelerator pump, and the necessary tamper proofing plugs and rivets (Figure 4-79). Some kits contain parts for several different models of the same carburetor; in this case, some of the kit parts may not be needed. Careful examination is required to ensure that all the correct kit parts are used.

CARBURETOR REASSEMBLY AND REINSTALLATION

Follow the vehicle's service manual for instructions to reassemble a carburetor. Some car manufacturers have as many as fifteen to twenty-five different carburetors for a single model year. There can be hundreds of different procedures and adjust-

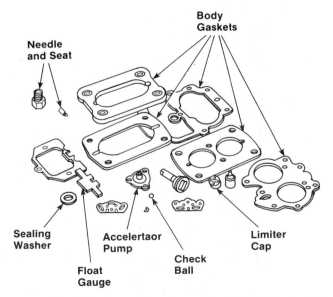

Needle and Seat

Body Gaskets

Sealing Washer

Accelertaor Pump

Float Gauge

Check Ball

Limiter Cap

FIGURE 4-79 A carburetor rebuild kit includes most of the parts needed to restore a carburetor to good condition.

TABLE 4-1: CARBURETOR TROUBLESHOOTING GUIDELINES

Condition	Possible Cause	Remedy
Engine cranks but will not start or starts hard when cold.	Choke valve not closing sufficiently when cold.	Adjust the index of the choke thermostatic (bimetal) coil.
	Choke valve or linkage binding or sticking.	Realign the choke valve or linkage as necessary. If caused by dirt and gum, clean with automatic choke cleaner. DO NOT OIL CHOKE LINKAGE. If parts are replaced, readjust to specifications.
	No fuel in carburetor.	Remove fuel line at carburetor. Connect hose to fuel line and run into metal container. Remove the high tension coil wire from center tower on distributor cap and ground. Crank over engine—if there is no fuel discharge from the fuel line, check for kinked or bent lines. Disconnect fuel line at tank and blow out with air hose, reconnect line and check again for fuel discharge. If none, replace fuel pump. Check pump for adequate flow, as outlined in service manual. If fuel supply is all right, check the following: Inspect fuel filter(s). If plugged replace. If filters are all right, remove air horn or fuel bowl and check for a bind in the float mechanism or a sticking float needle. If all right, adjust float to specifications.
	Engine flooded. To check for flooding remove the air cleaner. With the engine off look into the carburetor bores. Fuel will be dripping off nozzles and/or the carburetor will be very wet.	Check to determine if customer is using proper carburetor unloading procedure. Depress the accelerator to the floor and check the carburetor to determine if the choke valve is opening. If not, adjust the throttle linkage and unloaded to specifications.
	Carburetor flooding	Note: Before removing the carburetor air horn, use the following procedure which may eliminate the flooding. Remove the fuel line at the carburetor and plug. Crank and run the engine until the fuel bowl runs dry. Turn off the engine and connect fuel line. Then restart and run engine. This will often flush dirt past the carburetor float needle and seat. If dirt is in fuel system, clean the system and replace fuel filter(s) as necessary. If excessive dirt is found, remove the carburetor unit. Disassemble and clean. Check float needle and seat for proper seal. If the needle is defective, replace with a matched set. Check float for being loaded with fuel, bent float hanger or binding of the float arm. Adjust float to specifications.
Engine starts hard when hot.	Choke valve not opening completely.	Check for binding choke valve and/or linkage. Clean and free up or replace parts as necessary. DO NOT OIL CHOKE LINKAGE. Check and adjust choke thermostatic coil. Check for choke thermostatic coil binding in well or housing. Check for vacuum leak with integral choke system.
	Engine flooded; carburetor flooding.	See procedure under "Engine cranks, will not start."

TABLE 4-1:   CARBURETOR TROUBLESHOOTING GUIDELINES (CONTINUED)

Condition	Possible Cause	Remedy
	No fuel in carburetor.	Check fuel pump. Run pressure and volume test. Check float needle for sticking in seat, or binding float.
	Leaking float bowl.	Fill bowl with fuel and check for leaks.
	Fuel percolation.	Open throttle wide and operate starter to relieve over rich condition.
Engine idles rough and stalls.	Idle mixture adjustment.	Adjust idle mixture screws to lean best idle. Repeat the operation on 2- and 4-volt carburetors. Turn mixture screws in until idle speed drops 25 rpm on tachometer.
	Idle speed setting.	Reset idle speed per instructions on decal in engine compartment. Check solenoid operation.
	Manifold vacuum hoses disconnected or improperly installed.	Check all vacuum hoses leading to the manifold or carburetor base for leaks, being disconnected or connected improperly. Install or replace as necessary.
	Carburetor loose on intake manifold.	Torque carburetor to manifold bolts (100 in-lb.).
	Intake manifold is loose or gaskets are defective.	Using a pressure oil can, spray light oil or kerosene around manifold legs and carburetor base. If engine rpm changes, tighten or replace the manifold gaskets or carburetor base gaskets as necessary.
	Hot idle compensator not operating (where used).	Normally the hot idle compensator should be closed when engine is running cold and open when engine is hot (approx. 140° F); replace if defective.
	Carburetor flooding.	Correct by using procedure outlined under "Engine cranks, will not start."
Engine starts and stalls.	Engine does not have enough fast idle speed when cold.	Check and re-set the fast idle setting and fast idle cam.
	Choke vacuum diaphragm unit is not adjusted to specification or unit is defective.	Adjust vacuum break to specification. If adjusted correctly, check the vacuum opening for proper operation as follows. On externally mounted vacuum diaphragm unit, connect a piece of hose to fitting on the vacuum diaphragm unit and apply suction preferably by hand vacuum pump or another vehicle. Plunger should move inward and hold vacuum. If not, replace the unit. On the integral vacuum piston unit, remove cover and visually check piston and vacuum channel. If piston is corroded or sticking replace assembly. Note: Always check the fast idle cam adjustment before adjusting vacuum unit.
	Choke coil rod out of adjustment.	Adjust choke coil rod.

TABLE 4-1: CARBURETOR TROUBLESHOOTING GUIDELINES (CONTINUED)

Condition	Possible Cause	Remedy
	Choke valve and/or linkage sticking or binding.	Clean and align choke valve and linkage. Replace if necessary. Readjust if part replacement is necessary.
	Idle speed setting	Adjust idle speed to specifications on decal in engine compartment.
	Not enough fuel in carburetor.	Check fuel pump pressure and volume. Check for partially plugged fuel inlet filter. Replace if dirty. Remove air horn or fuel bowl and check float adjustments.
	Carburetor flooding	Correct by using procedure outlined under "Engine cranks, will not start."
Engine runs uneven or surges.	Fuel restriction	Check all hoses and fuel lines for bends, kinks, or leaks. Check fuel filter. If plugged or dirty, replace.
	Dirt or water in fuel system.	Clean fuel tank and lines. Remove and clean carburetor.
	Fuel level	Adjust float. Check for free float and float needle valve operation.
	Main metering jet defective, loose, or incorrect part.	Replace as necessary. See service manual.
	Power system in carburetor not functioning properly.	Power valve or piston sticking in DOWN position. Free up or replace as necessary. Power valve loose, incorrect gasket or leaking around threads: Tighten or replace as necessary. Leaking diaphragm: Test with hand vacuum pump; replace as necessary.
	Vacuum leaks	It is absolutely necessary that all vacuum hoses and gaskets are properly installed, with no air leaks. The carburetor and manifold should be evenly tightened to specified torque.
Engine hesitates on acceleration.	Defective accelerator pump system. Note: A quick check of the pump system can be made as follows: With the engine off, remove air cleaner and look into the carburetor bores and observe pump stream, while briskly opening throttle valve. A full stream of fuel should emit from pump jet and strike near the center of the venturi area.	Piston type: Remove air horn and check pump cup. If cracked, scored, or distorted, replace the pump plunger. Piston and diaphragm types: Check the pump discharge ball for properly seating and location. The pump discharge ball is located in a cavity next to the pump well. To check for proper seating, remove air horn and gasket and fill cavity with fuel. No leak down should occur. Restake and replace check ball if leaking. Make sure discharge ball, spring, and retainer are properly installed. Diaphragm type: Check pump discharge as above. Inspect diaphragm; replace if defective. Check pump inlet ball valve clearance. Adjust pump operating lever clearance.

TABLE 4-1: CARBURETOR TROUBLESHOOTING GUIDELINES (CONTINUED)

Condition	Possible Cause	Remedy
	Dirt in pump passages or pump jet.	Clean and blow out with compressed air.
	Fuel level.	Check for sticking float needle or binding float. Free up or replace parts as necessary. Check and reset float level to specification.
	Leaking air horn to float bowl gasket.	Torque air horn to float bowl using proper tightening procedure.
	Carburetor loose on manifold.	Torque carburetor to manifold bolts. (100 in.-lb.)
No power on heavy acceleration or at high speed.	Carburetor throttle valve(s) not going wide open. (Check by pushing accelerator pedal to floor)	Adjust throttle linkage to obtain wide open throttle in carburetor.
	Dirty or plugged fuel filter(s).	Replace with a new filter element.
	Power system not operating.	Piston type: Check power piston for free up and down movement. If power piston is sticking check power piston and cavity for dirt, or scores. Check power piston spring for distortion. Clean or replace as necessary. Piston and diaphragm types: Check power valve channel restrictions. Clean if necessary.
	Float level too low.	Check and reset float level to specification.
	Float not dropping far enough into float bowl.	Check for binding float hanger and for proper float alignment in float bowl.
	Main metering jet(s) dirty, plugged or incorrect part.	If main metering jets are plugged or dirty and excessive dirt is in fuel bowl, carburetor should be completely disassembled and cleaned.
Engine backfires.	Choke valve, fully or partially open, binding or sticking.	Free up with choke solvent. Realign or replace if bent.
	Accelerator pump not operating properly.	Remove air cleaner and observe pump discharge. Replace pump cup or diaphragm. Readjust pump to specifications. Restake or replace pump intake or discharge valve
	Old or dirty (fouled) spark plugs.	Clean or replace spark plugs.
	Old or cracked spark plug wires.	Test with a scope if possible or observe wires on dark night with engine running. Replace wires.
	Partially clogged fuel filter.	Replace filter on regular maintenance schedule.
	Backfire on deceleration. Defective air pump diverter valve.	Check hoses and fittings for tightness and leakage. Disconnect valve signal line. With engine running a vacuum must be felt. With engine idling, hold hand at exhaust port. No air should be felt. If valve or hoses defective then must be replaced.
Secondary system inoperative	Sticking throttle valves.	Readjust secondary throttle valve stop screw. Throttle valves nicked or throttle shaft binding.

TABLE 4-1: CARBURETOR TROUBLESHOOTING GUIDELINES (CONTINUED)

Condition	Possible Cause	Remedy
		Repair or replace throttle valve. Check throttle body for warpage. Torque throttle body screws evenly.
	Ruptured or leaking secondary diaphragm.	Inspect diaphragm. Replace or install properly.
	Venturi vacuum ports plugged.	Try cleaning ports with choke solvent or lacquer thinner. It may be necessary to remove the diaphragm assembly and back blow into the venturi.

ments. Each is critical to the carburetor performance. Of course, the exposed views of a carburetor in a service manual are the best guides for reassembling a unit.

Once it has been decided to install a rebuilt unit or new carburetor, the following is a typical reinstallation procedure:

1. Install a new base gasket.
2. Install the carburetor mounting bolts, but do not tighten.
3. Connect the fuel line while the carburetor can still be moved.
4. Once the fuel line is connected, tighten the carburetor mounting bolts in a criss-cross pattern and do not exceed a torque of 115 foot-pounds. Overtightening will strip the threads.
5. Install the remaining vacuum lines and electrical connections.
6. If the vehicle used a hot-water choke, install the choke mechanism at this time.
7. Engage the linkage with the appropriate retaining clip(s).
8. Connect the throttle lever springs and rods. Make sure the throttle can be closed and opened.

CAUTION: Never pour gasoline into the throat of the carburetor (commonly known as priming) to start the engine. Before cranking the engine, attach the air cleaner assembly. Then allow the fuel pump to pump fuel to the carburetor by simply cranking the engine with the starter. Remember to stop after 30 to 40 seconds of continuous cranking. Permit the starter to cool off for 3 or 4 minutes and try again.

9. Install a new oil filter and the filter top.
10. Install a ground back or battery.
11. Start the engine. It will crank longer than usual because the fuel bowl is empty and needs time to fill.
12. Set the idle according to manual specifications as described later in this chapter.

CARBURETOR DIAGNOSIS AND ADJUSTMENT

The diagnosis and troubleshooting chart (Table 4-1) focuses only on specific carburetor problems. The charts provide a problem area, possible causes, and remedies for the cause. Use of the charts assumes that the engine is in good mechanical condition and properly tuned. As previously stated, the technician should be aware that many ignition and carburetor problems have the same symptoms. The technician should not assume that the fault is with the carburetor.

As mentioned, late-model vehicles have many assist devices that help the engine perform better and improve emission control. It is important that each part of the carburetor system is properly adjusted and in good working order. A typical test inspection and adjustment procedure for a carburetor system is as follows:

- Choke
- Float
- Idle mixture and speed
- Accelerate pump system
- Other carburetor assist adjustments

Many of the parts and assist devices mentioned in this portion of the chapter have been illustrated earlier. When reading the adjustment procedure, refer to these illustrations.

CHOKE SYSTEMS

The choke plate must close for easy cold starts and initial cold driving. It must open as the engine warms, or there will be excessive enrichment. Because the choke should operate only when the engine is cold, most problems in this system occur during or right after cranking. Problems are caused by the following:

- *Choke Sticks Shut.* This causes an extra-rich air/fuel mixture, which, in turn, causes the engine to operate extremely roughly. Also, the exhaust will be black. A partially closed choke can even cause a warm engine to slightly miss or roll. The engine will lack power and stall.
- *Choke Sticks Open.* This causes a lean air/ fuel mixture that causes hard starting in cold weather. No problem may appear in warm weather. This condition can also cause a cold engine to stall during acceleration.

Exact choke servicing and adjustment procedures vary with choke type and carburetor model. Before making any choke adjustments, move the link by hand (Figure 4-80) to be certain there is no binding or sticking in the linkage, shaft, or choke plate. Use an approved carburetor cleaner to remove any gum deposits that interfere with free operation.

Integral Automatic Choke

With an integral choke, manufacturers usually recommend specific alignment of the choke index marks. A mark on the plastic choke cover must be lined up with a mark on the metal choke housing or body. This initial choke setting will usually provide adequate choke operation. That is, check the thermostat housing "index" setting (Figure 4-81) against

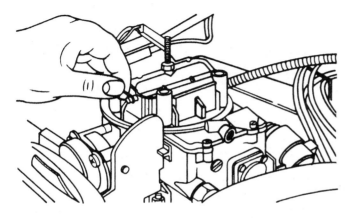

FIGURE 4-80   Checking choke linkage

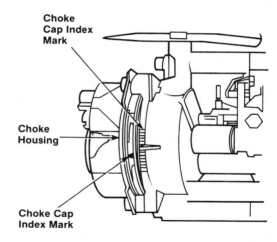

FIGURE 4-81   Choke index adjustment

the specification on the engine decal or the vehicle's service manual. Turn the cap in the proper direction and line up the cap index mark with the specified housing mark.

CAUTION: Most late-model carburetors have sealed idle mixture screws preset at the factory. Unless major carburetor repairs have been made, do not remove and tamper with these screws. To avoid a violation of federal law, follow the manufacturer's prescribed procedures when making carburetor adjustments.

Most tamper-resistant adjustment carburetors use rivets rather than screws to hold the cap in place to prevent misadjustment. The procedure for choke cap rivet removal and installation is as follows:

 1. Remove and drain the carburetor, then check each rivet mandrel. The mandrel should be below the rivet head. If the mandrel is high, drive it down with a 1/16-inch diameter tip punch (Figure 4-82).
   2. Drill out the rivet head using a 1/8 inch drill or #30 drill (0.128 inch). Drill gently into the head until the rivet head comes loose from the rivet body (Figure 4-83).
   3. After the rivet head is removed, drive the rivet body out of the hole with a 1/8-inch diameter punch.
   4. Repeat the procedure for each rivet.

To install the new rivet, use a pop-rivet gun or equivalent and proceed as follows:

 1. Position the choke cap gasket and hold the cap against the gasket after the bimetal

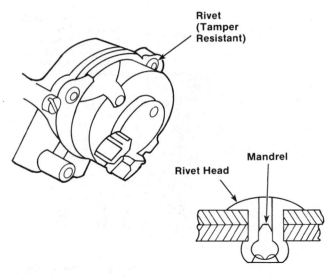

FIGURE 4-82 Rivet inspection

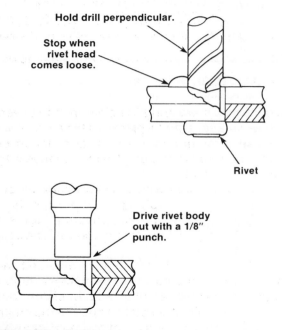

FIGURE 4-83 Rivet removal

outer tang is inserted in the slot of the bimetal lever.

2. Position the cap retainer on the cap and be sure to align all holes.

3. Place the rivet in the gun and trigger lightly just to retain the rivet. Rotate the cap to the proper index mark, then press the rivet fully into position. Install the other rivets.

4. Install the carburetor.

The integral choke can be tested quickly and easily with an automatic choke tester. The tester

connects to the shop air and supplies either hot or cold air to the choke housing. It attaches to the housing at the heated air connection from the manifold. When cold air is at the choke, the choke plate should close (Figure 4-84A). Then, when fed hot air, the plate should open smoothly (Figure 4-84B). If the choke does not test properly, replace the thermostatic coil or repair the choke as required.

Remote or Divorced Choke

To adjust the remote or divorced choke, proceed in the following manner (Figure 4-85):

1. Disconnect the choke rod from the choke plate and close the plate.

2. Move the rod up or down as directed in the service manual. Check whether or not the

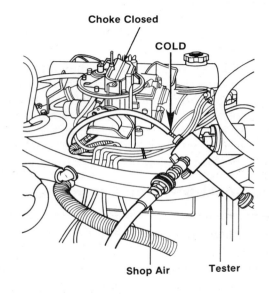

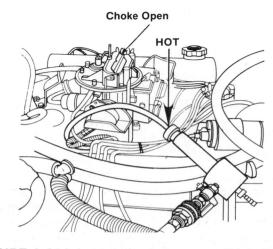

FIGURE 4-84 Automatic choke tester (A) Cold air feed; (B) hot air

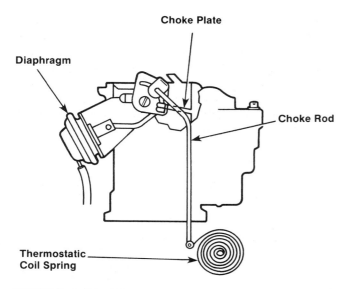

FIGURE 4-85 Divorced choke adjustment

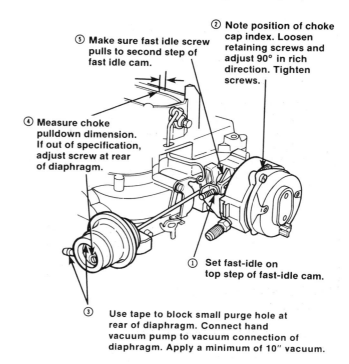

② Note position of choke cap index. Loosen retaining screws and adjust 90° in rich direction. Tighten screws.

⑤ Make sure fast idle screw pulls to second step of fast idle cam.

④ Measure choke pulldown dimension. If out of specification, adjust screw at rear of diaphragm.

① Set fast-idle on top step of fast-idle cam.

③ Use tape to block small purge hole at rear of diaphragm. Connect hand vacuum pump to vacuum connection of diaphragm. Apply a minimum of 10" vacuum.

FIGURE 4-86 Ford delayed choke pulldown adjustment

rod is above or below the hole in the check plate. The bottom of the rod should align with the top edge of the hole. With choke plates that use a sliding fit, the rod is usually set at the middle of the slot. However, always check the manufacturer's instructions.

3. Bend the choke rod to lengthen or shorten it as required.

CAUTION: Improper bending will cause the rod to bind.

---

4. Test for free movement between open and closed choke positions. There must not be any binding or interference.

CAUTION: Never attempt an adjustment on the thermostatic coil spring.

Choke Pulldown or Break Adjustments

Choke pulldown (often referred to as to *vacuum break* or *vacuum kick*) means partially opening the choke after the engine starts to prevent overrichness. It is accomplished by applying engine manifold vacuum to a pulldown piston or diaphragm.

The adjustment is checked by mechanically forcing the piston or diaphragm to the *pulldown*

position and measuring the clearance between the lower edge of the choke plate and the air horn wall. If a pulldown diaphragm is used, a hand vacuum pump can be attached to the diaphragm to move it to the *pulldown* position.

On carburetors with a diaphragm pulldown, a ruptured or leaky diaphragm will prevent the pulldown and cause cold choke loading. Figure 4-86 shows a typical adjustment procedure for a delayed choke pulldown.

All carburetors with remote or divorced chokes and some with integral chokes employ a diaphragm vacuum break instead of a vacuum piston to open the choke to a position that will permit the engine to operate without stalling or unloading (Figure 4-87). The vacuum break adjustment for a piston-type choke is shown in Figure 4-88.

Since the various heated chokes—air heated, electric, and electric assist, as well as the hot coolant system—vary by model in their adjustment procedure, check the manufacturer's instructions for details.

CHOKE CIRCUIT PROBLEMS

The choke cannot affect engine performance until it is running. All electrical and related tune-up work should be done prior to carburetor and/or choke diagnosis:

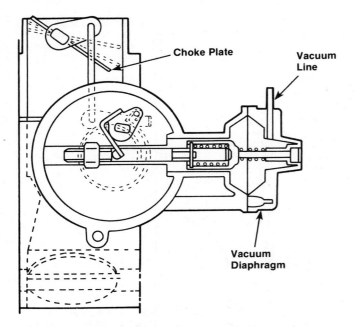

FIGURE 4-87 Diaphragm vacuum break opens the choke

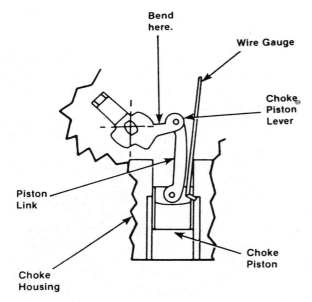

FIGURE 4-88 Vacuum break adjustment for a piston-type choke

 1. Check all linkage for any binding or sticking. Check choke valve to make certain it is free and not binding. Overtightening the air cleaner can cause a warpage or binding.

2. For an integral choke, check for air leaks between the choke cover and gasket. A leak in this area will slow the opening of the choke valve.

3. The choke heat tube should be free of any carbon; this, too, will delay choke opening. Check the heat tube in the exhaust manifold for leaks. If it is burned out, it will allow exhaust gases to enter and damage the choke.

4. Choke setting, fast idle, and unloader adjustment should be to specifications.

5. Check the diaphragm choke pulloff for leaks and proper adjustment.

6. When the crossover type choke is used, the carburetor mounting gasket is most important. If it is not to specified thickness, it upsets choke calibration due to the length of the choke rod. Most crossover chokes are mounted in a pocket in the intake manifold and are nonadjustable.

Carburetors with an integral choke having a binding or stuck choke valve should be checked for a burned out heat tube located in the exhaust manifold. This is not visible from the outside but can be checked by removing the climatic choke cover. Whenever carbon deposits are found inside the choke housing, a burned out heat tube in the manifold is indicated. The most common choke circuit problem areas are shown in Figure 4-89.

Fast Idle Adjustment

As mentioned earlier in this chapter, it is necessary during the warm-up to provide a fast idle speed to prevent engine stalling. This is accomplished by a fast idle cam connected to the choke linkage (Figure 4-90). The fast idle adjusting screw on the throttle

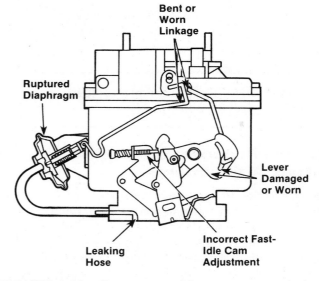

FIGURE 4-89 Common problem areas on a choke

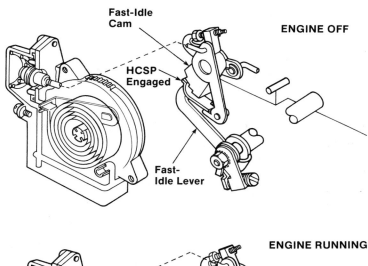

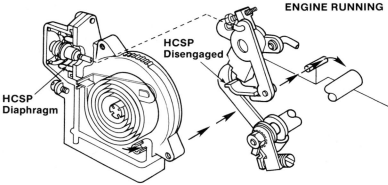

FIGURE 4-90 Typical fast idle cam operation

lever contacts the fast idle cam and prevents the throttle valves from returning to a normal warm engine idle position while the choke is in operation.

Fast idle speed adjustment is made as follows:

1. With the engine idling at normal operating temperature, place the fast idle lever on the specified step of the fast idle cam (Figure 4-91). In many cases, the EGR is disconnected and the vacuum line plugged.
2. Make sure the high cam speed positioner lever is disengaged.
3. Turn the fast idle adjusting screw clockwise to increase speed and counterclockwise to decrease speed.

The fast idle cam set is adjusted in the following way:

1. Remove the choke cap.
2. Place the fast idle lever in the corner of the specified step of the fast idle cam (counting the highest step as the first) with the high cam speed positioner retracted.
3. Hold the throttle lightly closed with a rubber band to maintain cam position.

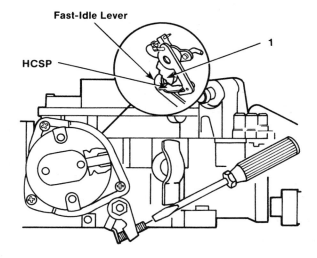

FIGURE 4-91 Typical fast idle adjustment points

This step is not required if the adjustment is done on the vehicle.

4. Install the stator cap and rotate it clockwise until the lever contracts the adjusting screw.

5. Turn the fast idle cam adjusting screw until the index mark on the stator cap lines up with the specified notch on the choke casting.

6. Remove the stator cap and reinstall the choke cap.

Unloader

If during the starting period the engine becomes flooded, the choke valve can be opened manually to clean out excessive fuel in the intake manifold. This is accomplished by depressing the accelerator pedal to the floor mat and engaging the starter. The unloader projection on the throttle lever contacts the unloader lug on the fast idle cam and partially opens the choke valve (Figure 4–92).

The unloader adjustment can be made as follows:

1. With the throttle valves wide open, close the choke valve as far as possible without forcing.

2. Bend the unloader tang on the throttle lever to obtain the dimension as listed in the specifications between the upper edge of the choke valve and the inner wall of the air horn.

FLOAT, INLET VALVE, AND FUEL BOWL

The float system provides fuel for all of the other carburetor circuits. An improperly operating float

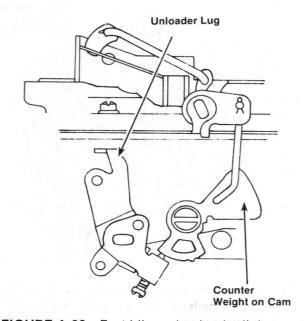

FIGURE 4–92 Fast idle and unloader linkage

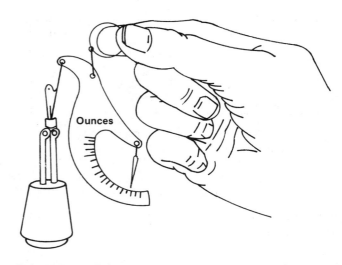

FIGURE 4–93 Float scale

system can result in flooding, rich fuel mixture, lean fuel mixture, fuel starvation (no fuel), stalling, low-speed engine miss, and high-speed engine miss. Thus a faulty float system can affect engine operations at all speeds.

To inspect the float, remove the air horn and check that the float is undamaged and that it operates freely. If the float or any related parts are damaged, make the needed repairs and perform a float adjustment.

CAUTION: It has been determined that some carburetor cleaners can affect the composition of the float material and cause a slightly spongy condition. The float, therefore, should not be immersed in or exposed to solvents.

If a rich engine performance condition occurs shortly after a carburetor has been serviced, the float should be checked for sponginess. Use a float scale (Figure 4–93) and read the float weight on the scale. Compare it with the specifications from a service manual or with a known good float. If the old float is heavier than the specifications or the good float, install a new float and adjust it to specifications.

Float Adjustments

An error in float adjustment as small as 1/32 of an inch can change the air/fuel ratio sufficiently to make other carburetor adjustments difficult, if not impossible. Adjustments include the float level, float drop, float toe, and float adjustment. Since most do not require all four adjustments, check the vehicle's

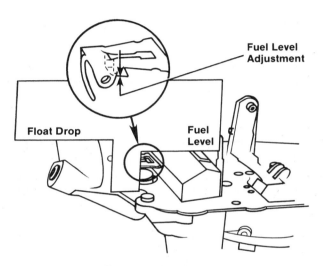

FIGURE 4-94 Check the service manual for exact dimensions

FIGURE 4-95 Fuel level adjustment

service manual or the rebuilding kit instruction sheet for the adjustment that must be performed.

Float Level Adjustment. A typical float level adjustment is as follows:

1. Remove the upper body assembly and the upper body gasket; install a new gasket before making the adjustment.
2. From cardboard, fabricate a gauge to the specified dimension following the specifications given in the service manual or rebuilding kit. Figure 4-94 shows a typical float gauge. Check the service manual for the exact dimensions.
3. With the upper body inverted, place the fuel level gauge on the cast surface of the upper body (not on the gasket) and measure the vertical distance from the cast surface or the upper body and the bottom of the float (Figure 4-95).

4. To adjust, bend the float operating lever away from the fuel inlet needle to decrease the setting and toward the needle to increase the setting.
5. Check and/or adjust the float drop.

Although most float level checks are performed with the carburetor removed from the vehicle, a wet test can be performed with fuel in the unit. With the bowl filled with fuel, remove the air horn and measure the fuel by either method shown in Figure 4-96. The float should be adjusted to the manufacturer's specifications in the same manner as for the dry test (without fuel).

Some carburetors have a sight glass in the fuel bowl which makes it possible to check the float level while the engine is operating. Floats on these carburetors can be adjusted by an external screw. Checking the fuel lever on a carburetor with a sight

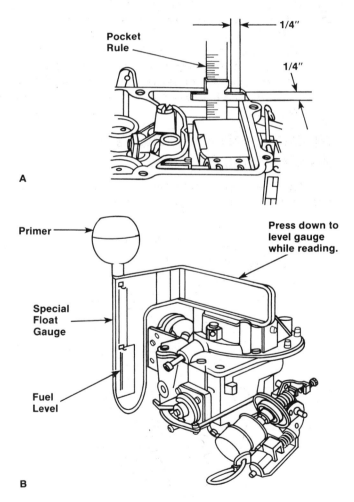

FIGURE 4-96 Wet float level adjustment: (A) using a ruler to check the level; (B) using a special gauge to check the float level

glass not only aids in adjusting the float level but also in locating float problems.

Float Drop Adjustment. On floats hung on the air horn, a float drop adjustment such as this is required:

1. Fabricate a gauge to the specified dimension as in the float level test (Figure 4-97).
2. With the upper body assembly held upright, place the gauge against the cast surface of the upper body (not on the gasket) and measure the vertical distance between the cast surface of the upper body and the bottom of the float.
3. To adjust, bend the stop tab on the float lever away from the hinge pin to increase the setting and toward the hinge pin to decrease the setting.
4. Reinstall the upper body assembly.

Float Toe Adjustment. In addition to the float level and drop adjustments, some carburetors also have a float toe adjustment. To make this adjustment, the air horn is turned over (no gasket installed) and checked to make sure the float toes are flush with the air horn casting (Figure 4-98A). If the float has dimples (Figure 4-98B), measure from the dimples to the top of the gasket. In either case the float arm is bent to where it meets the float.

After making a toe adjustment, be sure to recheck the float level and drop to see if they have changed.

Float Alignment Adjustment. The float pontoon must be parallel in all cases to the edges of the float bowl or air horn. Badly aligned floats can rub against the float bowl walls and stick with the inlet needle valves open. Straighten by bending the float arms.

Float Inlet Valve (Needle and Seat)

While checking float operation, see if the needle valve moves freely. With the float removed, check the valve and seat for wear or damage. Replace the float valve assembly if it is damaged. Remember that carburetor flooding can be caused by dirt inside the needle valve, by an incorrect float drop setting, or by a ruptured hollow brass float. If light tapping on the bowl temporarily corrects the flooding, check the float drop adjustment and action of the needle valve. Generally, a carburetor rebuild is required to permanently correct a carburetor flooding problem.

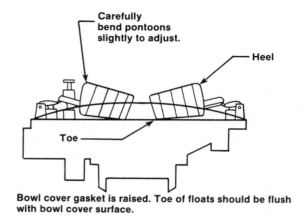

Bowl cover gasket is raised. Toe of floats should be flush with bowl cover surface.

A   FLOAT TOE ADJUSTMENT WITHOUT DIMPLE

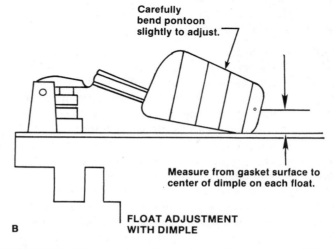

B   FLOAT ADJUSTMENT WITH DIMPLE

FIGURE 4-98 Typical float toe adjustments: (A) without dimple; (B) with dimple

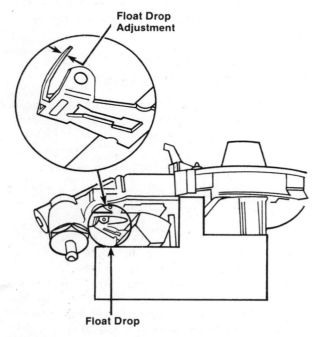

FIGURE 4-97 Float drop adjustment

Gum can cause the float needle to stick shut, which can cause fuel starvation. The engine might not start or will not continue to run after initial starting. Part cleaning and usually a rebuild are required.

Fuel Bowl

Inspect the fuel bowl interior for dirt. Any excess foreign material in the bowl can indicate a need to clean out the tank and fuel lines.

Venting of the fuel bowl is an important part of the carburetor. Air must enter or leave the bowl every time the level of fuel changes in the bowl. During acceleration the liquid fuel level in the bowl is lowered. This allows the float to drop, allowing the needle to come off its seat so that fuel can enter the bowl. The increased space (above the liquid) caused by the lowering of the fuel level must be filled with air to maintain a constant pressure in the bowl. This is the purpose of the bowl vent. During constant throttle operation, the amount of fuel entering the bowl is the same as the amount being discharged from the low-speed or high-speed circuit.

Many carburetors use combination inside/outside vents. Because pressure is higher at the outside vent, air movements from the outside vent to the inside vent are used to keep all fuel vapors swept from the fuel bowl. Later models use a mechanically operated bowl vent that opens at *closed throttle* position. When the engine is turned off, underhood temperatures increase, causing vapors to rise from the fuel in the bowl. The outside vent improves starting characteristics as it prevents vapors from entering the bore of the carburetor by way of the inside vent. A dirty air cleaner would cause a rich mixture or a loss of volumetric efficiency depending on the type of carburetor vents used.

All off-idle engine operation is with the inside vent only. With this type of vent, the pressure in the fuel bowl and carburetor air horn is the same. In other words, a balance is effected between the pressure in the air horn and the pressure in the bowl. This prevents a rich condition should the air cleaner become restricted. A restricted air cleaner with this type of venting will not change the air/fuel ratio. However, it will affect volumetric efficiency.

Bowl vapor vent adjustment must be to specifications given in the vehicle's service manual. If the valve does not open to specifications with the throttle valves closed, bowl vapors cannot escape freely; this can cause hard hot starting. If it opens too far, or hangs open, it will allow an external vent to the bowl, resulting in poor mileage. Further information on the topic of bowl vapor vent adjustment can be found in Chapter 11.

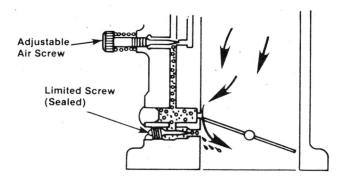

FIGURE 4-99   Sealed limiter screw acts much like idle mixture

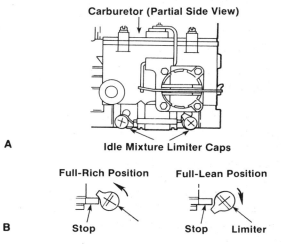

FIGURE 4-100   (A) Use of limiter caps on the idle mixture adjusting screws; (B) installation of new limiter caps on the idle mixture adjustment screws

IDLE MIXTURE AND IDLE SPEED ADJUSTMENTS

Various automobile manufacturers provide different, but generally similar procedures for idle mixture and idle speed adjustments. The more common directions are given here. For exact details on the adjustment procedure for a given vehicle, check the service manual.

Idle Mixture Adjustment

Probably the most critical adjustment on the carburetor is the idle mixture. Since 1966, however, federal and state emission laws require some means for limiting the adjustability of the idle mixture screws (Figure 4-99). This allows for proper idle adjustment while assuring that the emission limits will not be exceeded. Figure 4-100 illustrates two of the more common limiting arrangements.

Idle Adjustment Using Propane Enrichment Method

Because of the combination of the catalytic converters and the very lean mixtures used on today's engines, it is difficult to properly perform curb idle speed and mixture adjustments. These adjustments must be properly made to insure that emission control devices limit CO, HC, and NO_x to specified levels. The solution adapted by Ford, Chrysler, and General Motors is to use propane gas to assist in achieving correct idle settings.

CAUTION: The carburetor idle mixture has been adjusted and limiter(s) installed at the factory under controlled conditions. These limiters should not be removed during normal service operations. Idle mixture should be adjusted only during carburetor repair when specified or when necessary as a result of government inspection laws. Adjusting idle mixture by a method other than the following propane enrichment procedure might not be in compliance with federal, state, or provincial laws.

Although the propane injection method varies slightly from one carmaker to another (check the service manual), the following procedure can be considered basic:

1. Apply the parking brake and block the wheels. Disconnect the automatic brake release and plug the vacuum connection.
2. Connect a tachometer to the engine.
3. Disconnect the fuel evaporative purge return hose at the engine and plug the connection.
4. Disconnect the fuel evaporative purge hose at the air cleaner and plug the nipple.
5. Disconnect the flexible fresh air tube from the air cleaner duct or adapter. Using a propane enrichment tool, insert the tool hose approximately 3/4 inch of the way into the duct or fresh air tube (Figure 4-101). If necessary, secure the hose with tape and hold the bottle upright to even the propane flow.
6. For vehicles equipped with an air injector system, revise the dump valve vacuum hoses as follows:
   - For dump valves with two vacuum fittings, disconnect and plug the hoses(s).
   - For dump valves with one fitting, remove the hose at the dump valve and plug it. Connect a slave hose from the

dump valve vacuum fitting to an intake manifold vacuum fitting.

7. Verify that the idle mixture limiter(s) is set to the maximum *rich* position (counterclockwise against the stop); correct if required.
8. Check the engine curb idle rpm (or A/C–OFF rpm). If necessary, reset to specification.
   - With the engine off, turn the mixture screws clockwise until lightly seated.
   - Turn the mixture screws counterclockwise two turns. Start the engine and proceed with Step 9.
9. With the transmission in neutral, run the engine at approximately 2500 rpm for 15 seconds before each mixture check.
10. With the engine idling at normal operating temperature, place the transmission selector in the position specified for the fuel mixture check. Gradually open the propane tool valve and watch for engine speed gain, if any, on the tachometer. When the engine speed reaches a maximum and then begins to drop off, note the amount of speed gain. If the engine speed will not drop off, check the bottle gas supply. If necessary, repeat Step 10 with a new bottle gas supply.
11. Compare the measured speed gain to the specified speed gain on the engine decal or specification sheet. If it is determined that an idle fuel mixture adjustment will be necessary, adjust the mixture according to the reset rpm specification.

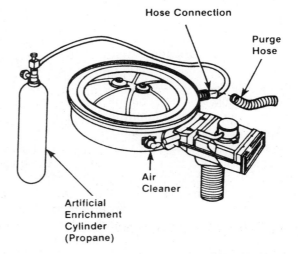

FIGURE 4-101 Typical artificial enrichment test setup

- If the measured speed gain is within specification, except as noted here, proceed to Step 12.
- If the measured speed gain is higher than the speed gain specification: Turn the mixture screw(s)/limiter(s) counterclockwise (rich) in equal amounts and simultaneously repeat Steps 8 through 10 until the measured speed rise meets the reset rpm specification. After the final adjustment, proceed to Step 12.
- If the measured speed gain is lower than the speed gain specification: Turn the mixture screw(s)/limiter(s) clockwise (lean) in equal amounts and simultaneously repeat Steps 8 through 10 until the measured speed rise meets the reset rpm specification. After the final adjustment, proceed to Step 12.
- Speed drop test information is specified with the artificial enrichment requirements whenever the speed gain specification is zero rpm on engine decal or specification sheet. If there is no rpm rise and the minimum speed gain specification is zero rpm, perform the following speed drop test:
  —While watching the tachometer, adjust the mixture screw(s)/limiter(s) clockwise (lean) by the number of turns specified for the speed drop test. Note the drop engine speed.
  —If the measured speed drop is equal to or drops off by more than (including stall) the speed drop specification, return the mixture limiter(s) to the maximum *rich* position or the mixture screw(s) to the position before adjustment. Proceed to Step 12.
  —If the measured speed drop is less than the specified minimum speed drop specification, leave the mixture limiter(s) in the adjusted position and repeat Steps 8 through 11.
12. If the vehicle has less than 100 miles of operation, reset the curb idle speed to the green engine specification.
13. Remove all test equipment. Reconnect all system components and the reinstall air cleaner; torque the air cleaner wing nut(s) to specification.

Hot (Curb) Idle Speed

Curb idle rpm settings are higher today because of emission controls and because engine accessories, such as air conditioner compressors. Also, an automatic transmission requires a higher hot idle setting to prevent stalling when the vehicle stops in gear.

To prepare for an idle rpm check, proceed as follows:

1. To make a curb idle check and/or adjustment the engine must be hot.
2. Connect a tachometer.
3. Remove the air cleaner assembly and plug the vacuum line(s) to the air cleaner. Also, on light truck models with V-8 engines and equipped with air injector systems, revise the dump valve vacuum lines as follows:
   - For dump valves with one or two vacuum lines at the side, disconnect the plug and line(s).
   - For dump valves with one vacuum line at the top, check the line to see if it is connected to the intake manifold. If not, remove the line at the dump valve and plug it. Connect a slave line from the dump valve vacuum fitting to an intake manifold vacuum fitting.
4. Check throttle and choke linkage for freedom of movement. Correct as required.
5. If so equipped, remove the spark delay valve and route the distributor vacuum advance line directly to the advance side of the distributor.
6. If so equipped, disconnect the fuel deceleration valve hose at the carburetor and plug the hose.
7. Remove the EGR vacuum line at the EGR valve and plug the line.
8. Turn off all accessories including the air conditioner.

 SHOP TALK ——————

Refer to the emissions decal for adjustment conditions (air conditioner on, air conditioner off, and so on) and for idle specifications.

9. Start the engine.
10. If the rpm is high, adjust the vacuum-operated throttle modulator bracket adjusting screw counterclockwise to the specified curb idle rpm and recheck.
11. If the rpm is low, shut off the engine. Turn the vacuum-operated throttle modulator adjusting screw one full turn clockwise.
12. Restart the engine and run at 2000 rpm for 10 seconds. Let the engine idle stabilize for

60 seconds (not to exceed 120 seconds) before checking or adjusting the rpm.

13. Stop the engine. Remove all test equipment, reconnect the automatic parking brake release (if disconnected), and tighten the air cleaner wing nut(s) to specification.

14. Verify that the air cleaner heat riser tube and fresh air pickup connections are correct.

Curb Idle Speed (with Solenoid) Adjustment

Curb idle speed must be adjusted to manufacturer's specifications. Before making the adjustment, be sure the engine is at normal operating temperature, the parking brake applied, air cleaner or filter in place, transmission in neutral, and air conditioning off. Also be sure that the throttle stop solenoid is energized and extended. Use the curb idle speed screw (Figure 4–102), which contacts the throttle stop solenoid to adjust the idle speed.

When adjusting the idle, identify and disconnect the vapor storage canister, distributor vacuum, and EGR hoses and plug the ends of these hoses. After adjustments are completed, unplug and reconnect the hoses to their proper fittings.

To prevent variations of fuel and temperature when setting idle mixture, observe the following precautions:

1. Do not idle the engine for more than 3 minutes at one time.
2. After 3 minutes of idling, increase the engine speed to 2000 rpm for 1 minute.
3. Continue with the adjustment.
4. Do not idle the engine for more than 3 minutes without repeating Step 2.

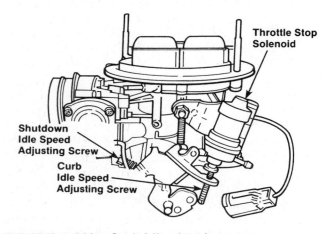

FIGURE 4–102 Curb idle speed screw

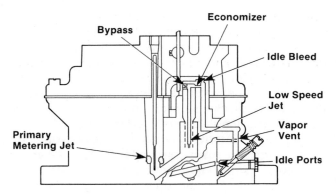

FIGURE 4–103 Low-speed circuit

Many applications do not use an idle solenoid. To make a curb idle speed adjustment when no solenoid is used, follow the above procedures. The carburetor throttle lever tang should then be contacting the curb idle adjusting screw instead of the solenoid plunger. Adjust the idle adjusting screw to specifications.

The curb idle speed can also be set with an infrared analyzer (see Chapter 11).

Low-Speed Circuit

Fuel for idle and early part-throttle operation is metered through the low-speed circuit. The low-speed circuit is located on the primary side only. Fuel enters the idle wells through the main metering jets. The low-speed jets measure the amount of fuel for idle and early part-throttle operation (Figure 4–103).

The quality of the mixture is determined by the size of the idle jet, the bypass and the idle bleed. The quantity of the mixture to the idle port is adjusted by the idle adjusting screws. The idle ports, located directly above the idle adjusting screws ports, are slot shaped. As the throttle valves are opened, more of the idle ports are uncovered, allowing a greater quantity of the air/fuel mixture to enter the carburetor bores. Although the idle port is located below the fuel level in the bowl, no siphoning action takes place because the bypass and the idle bleed serve as vents to prevent siphoning of fuel from the bowl. Opening the throttle further causes the throttle valve to move away from the idle port. This increases the pressure at the port, diminishing fuel delivery from the low-speed circuit. The increase in air velocity with the throttle opening causes fuel flow from the high-speed nozzle. This is known as the *transfer point*.

A low float setting can cause a problem at the transfer point because the high-speed nozzle will not begin fuel delivery at the correct time. Float setting plays its part in timing the nozzle.

An idle adjusting screw is used for trimming the idle mixture to individual engine requirements. Turning the idle adjustment screw toward its seat reduces the quantity of air/fuel mixture supplied by the idle circuit. This is an overrich mixture that emerges from the port but is leaned to a proper combustible mixture by the air that enters the engine manifold around the cracked primary throttle valve. Consequently, rotating the idle adjusting screw inward leans the idle mixture; rotating it outward enriches the idle mixture.

Dirt or foreign material in the economizer will cause a lean idle condition; any restriction in the bypass or idle bleed will result in a rich condition.

The low-speed jet, idle bleed, economizer, and bypass bushings are pressed in place. Do not remove in servicing. To insure proper alignment of the low-speed mixture passage, the primary venturi assemblies were designed with interlocking bosses so they can be installed only in the proper locations. (When the primary venturi assemblies are placed in the wrong side of the carburetor, they will not fit all the way into the casting.)

The bypass, economizer, idle bleed, idle port, idle adjustment screw port, and the bore of the carburetor flanges must be clean and free of carbon. Obstructions will cause poor low-speed engine operation.

Air leakage at the gasketed surface surrounding the low speed (idle) mixture passages or between the flange and manifold may cause poor idle and low speed operation. Always use new gaskets when servicing the carburetor.

Port relation is the position of the throttle valve relative to the idle port at curb idle. Port relation could be out of specifications due to carbon in the bore of the carburetor, throttle shaft or throttle body wear. In either case, it can upset the idle and affect the transfer point. Improper idle adjustment can also upset port relation.

The idle adjusting screws can be used to quickly test the low-speed circuit on the car. The idle adjusting screws should be sensitive when turning in or out. If turning the mixture screws does not appreciably affect idle, there could be a problem in the low-speed circuit.

If the best idle is attained with the idle mixture screws seated or near the seat, it indicates a rich idle mixture. This could be traced to dirt in the bypass or idle bleed or both of them partially plugged.

If the best idle is attained with the idle mixture screws near the outward (counterclockwise) position, it indicates a lean idle mixture. This could be dirt or a restriction in the economizer.

HIGH-SPEED CIRCUIT

Fuel for part-throttle and full-throttle operation is supplied through the high-speed circuit. (Figure 4-104). The position of the step-up rod in the main metering jet controls the amount of fuel admitted to the nozzles. The position of the step-up rod is controlled by manifold vacuum applied to the vacuum piston.

During part-throttle operation, manifold vacuum pulls the step-up piston and rod assembly down, holding the large diameter of the step-up rod in the main metering jet. This is true when the vacuum under the piston is strong enough to overcome the tension of the step-up piston spring. Fuel is then metered around the large diameter of the step-up rod in the jet.

Under any operating condition, when the tension of the spring overcomes the pull of vacuum under the piston, the step-up rod will move up so its smaller diameter, or power step, is in the jet. This allows additional fuel to be metered through the jet. The step-up rod does not require adjustment. A vent tube aerates the fuel as it leaves the high speed well. Both downhill and uphill nozzles have been used.

Many carburetor designs use staged step-up rods to better control air/fuel mixtures throughout the entire high-speed range. When the manifold vacuum is high (above 12 inches of mercury), both springs are compressed allowing the step-up piston to bottom and the top step of the rod to remain in the jet. When the manifold vacuum drops to 10 inches to 12 inches of mercury, the upper spring lifts the step-up piston and rod so that the second step of the rod is in the orifice of the jet. When the vacuum drops to

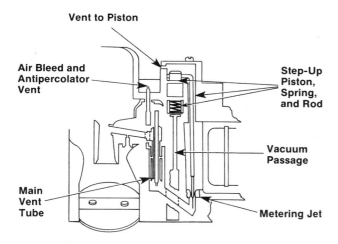

Vent to Piston

Air Bleed and Antipercolator Vent

Step-Up Piston, Spring, and Rod

Vacuum Passage

Main Vent Tube

Metering Jet

FIGURE 4-104 High-speed circuit—primary side

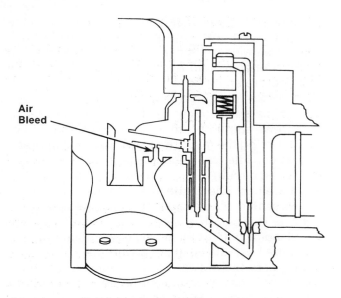

FIGURE 4-105   Understrut primary air bleed

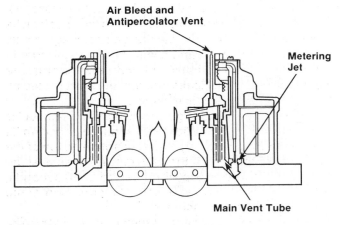

FIGURE 4-106   High-speed bleeds act as antipercolator vents

4 inches to 6 inches of mercury, the lower spring lifts the step-up piston and rod to the wide-open throttle or power step of the rod. The metering jets used with the staged step-up are 1/8 inch higher to allow greater travel of the step-up rod to make the staging less sensitive. The step-up piston uses a raised or extruded cover plate to allow for the greater travel of the metering rod.

Understrut primary air bleeds are used in some models to give a smoother transfer from the low-speed circuit to the high-speed circuit and increase part-throttle economy (Figure 4-105). This bleed is subject to venturi pressure changes that follow engine load conditions. This is a variable type bleed and, under some conditions, could act as a discharge port.

High-Speed Circuit Problems

The three major problems with the high-speed circuit are

1. Use of the wrong step-up piston spring, or one that has been stretched or cut off, can seriously upset carburetor calibration.
2. A clogged air bleed or main vent tube can cause excessively rich mixtures. The high-speed bleed and main vent tubes are permanently installed.
3. An incorrect float setting can also adversely affect high-speed circuit operation.

High-Speed Circuit— Secondary Side

Fuel for the high-speed circuit of the secondary side is metered at the main metering jets. No step-up rods are used. On some applications the secondary metering jets are located under the secondary venturi cluster. Due to the use of step-up rods on the primary side, the primary main metering jets are larger than the secondary jets.

The high-speed bleeds also act as antipercolator vents when a hot engine is stopped or at idling speed (Figure 4-106). They vent fuel vapor pressure in the high-speed wells before it is sufficient to push fuel out of the nozzles and into the intake manifold.

Erratic Engine Idle Checks

If the engine idle is erratic or rough after correct idle adjustments have been made, do the following:

 1. Recheck spark plugs and spark plug wires. Be sure all cylinders are firing. A missing cylinder or occasional misfire will turn an engine that is a minor emitter into a gross emitter.
2. Check the EGR (if used) to be sure the valve is closed at idle and not held open by dirt or engine deposits.
3. Check for vacuum leaks (vacuum lines or manifold).
4. Check the hot idle compensator (Figure 4-107) valve.

Many late carburetor applications include a bimetal hot idle compensator (HIC) valve. If the valve is leaking, the engine idle will be erratic. To test the valve use the following procedure:

 1. With the engine running, remove the air cleaner.
2. Cover the hot idle compensator air inlet port in the carburetor air horn. If the idle

quality improves, the hot compensator valve is not functioning properly.

3. Remove the compensator cover and bimetal valve assembly and check that the valve seat gasket is in good condition and in place in the counterbore.

4. If the replacement of a faulty gasket does not improve idle quality, the bimetal valve is bad and should be replaced.

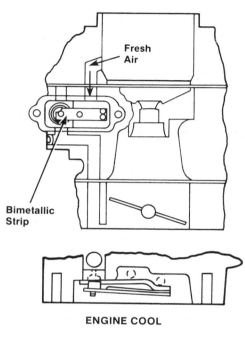

ENGINE COOL

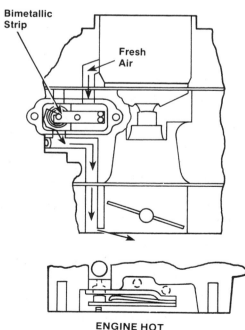

ENGINE HOT

FIGURE 4-107   Hot-idle compensator operation

PUMP CIRCUIT

The accelerating pump circuit, located in the primary side, provides a measured amount of fuel necessary to insure smooth engine operation on acceleration at low car speeds.

Pump Problems

Troubleshooting the pump circuit should include a thorough check of the intake and discharge checks, pump plunger, duration spring, and all linkage.

If the discharge check is not seating, air will be drawn into the pump circuit during deceleration. On acceleration, air will be discharged before solid fuel is delivered to the pump jet resulting in a stumble or hesitation.

If the intake check is not seating, some of the fuel during acceleration will be returned to the fuel bowl, resulting in lack of fuel for acceleration.

To determine if the checks are seating, proceed as follows: With fuel in the bowl and pump passages, use a suitable tool to hold the discharge check on its seat. While holding the discharge check against its seat, manually raise and lower the pump plunger in its cylinder several times. This will create a pressure in the circuit. If fuel leaks into the fuel bowl (noted by bubbles), the intake is leaking.

To test the discharge check, make certain fuel is in the entire circuit. With the discharge check on its seat (by its own weight), squirt gasoline on top of the check and observe. If the fuel does not leak past the check (remains on top) for a period of 30 seconds, then the check is good. Check the condition of the pump plunger and cylinder.

The linkage connecting the pump plunger to the throttle requires an adjustment to obtain the correct pump stroke for proper pump delivery. This adjustment must be made according to specifications listed by the manufacturer.

Accelerator Pump System Inoperative or Improperly Adjusted

The best tool for checking an accelerator pump is an infrared analyzer. Full use of this tool is given in Chapter 11. To adjust an accelerator pump with an infrared analyzer, proceed as follows:

1. Disable the air injector reactor pump (if used) by disconnecting and plugging the air supply tube(s) leading to the exhaust system.

2. Connect an infrared analyzer to the vehicle (if not already connected).

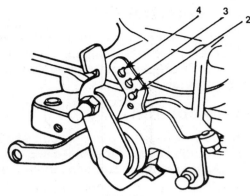

**AUTOMATIC OVERDRIVE
TRANSMISSION**

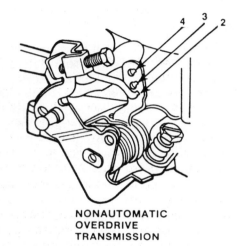

**NONAUTOMATIC
OVERDRIVE
TRANSMISSION**

FIGURE 4-108 Accelerator pump stroke adjustment

3. With the engine idling and at normal operating temperature, usually 158 degrees to 212 degrees Fahrenheit engine oil temperature, note the percent CO meter reading.
4. Quickly open and close the throttle. Again note the percent CO meter reading.
   - If the accelerator pump action is good, it will show less than a .5 percent lean out before the percent CO increase and 1 percent or more increase in CO after opening the throttle.
   - If the accelerator pump action is poor, adjust the accelerator pump linkage or service the pump system.

Accelerator Pump Lever and Rod

To replace the pump link and rod assembly or to reset the pump position to specification, perform the following (Figure 4-108):

1. Using a blunt-tipped punch, remove (and retain) the roll pin from the accelerator pump cover.
2. Rotate the pump link and rod assembly until the keyed end of the rod is aligned with the keyed hole in the pump overtravel lever.
3. Reposition the rod in the specified hole and reassemble the pump link and rod assembly.

Accelerator Pump Stroke

The accelerator pump stroke is preset at the factory and should not be adjusted to improve drive performance.

CAUTION: Take care when disassembling the carburetor so that the accelerator pump operating link is not bent, and make sure it is returned to the proper slot in the throttle return spring arm.

The inner pump rod slot is identified as number one; the outer slot is number two (Figure 4-109). If the pump link is damaged or bent, check the length and readjust by bending the loop provided to achieve the dimension specified in the service manual.

Checking Accelerator Pump Lever Clearance

To check a typical accelerator pump lever clearance, perform the following procedure:

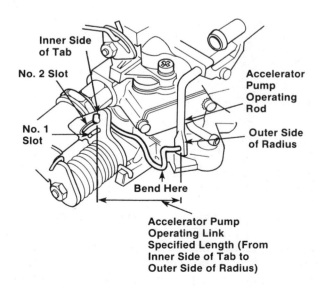

FIGURE 4-109 Inner and outer pump rod slots

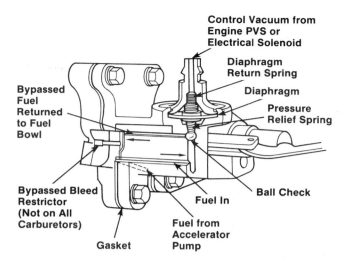

FIGURE 4-110   Typical temperature compensated accelerator

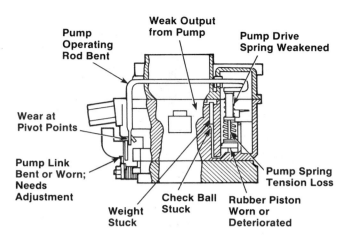

FIGURE 4-111   Check linkage adjustment on an accelerator pump

 1.   With the throttle valves held wide open and the pump lever held down, it should be possible to insert a 0.015-inch (minimum) to a 0.062-inch (maximum) gauge between the adjusting nut lever.

2.   If necessary, adjust the pump override screw to obtain the correct clearance.

3.   There must be no free movement of the pump lever when the throttle lever is at curb idle.

4.   Secondary system pump linkage should be adjusted in a similar manner on carburetors so equipped (a two-barrel carb).

Many modern accelerator pumps have a temperature-sensitive device that permits more fuel to be pumped when the engine is cold (Figure 4-110). This aids the engine better in cold weather. It also causes less fuel to be pumped when the engine is hot.

A bad accelerator pump can cause an engine to hesitate or stumble on acceleration. After checking the adjustment of the linkage, check the pump parts shown in Figure 4-111.

COMPUTER-CONTROLLED CARBURETORS

Although computer-controlled carburetors require specialized service techniques, fortunately, on most vehicles, a built-in diagnostic system indicates a computer problem. On some systems, when the "check engine" light goes on, a trouble code number appears. On some computerized engines it is neces-

sary to read the self-diagnosis code with an analog (needle type) voltmeter. When activated by a jumper wire, the on-board computer sends voltage pulses to the test socket and the voltmeter. Each deflection (sweep) of the voltmeter needle indicates a code number. For example, as shown in Figure 4-112, two needle deflections, a short pause, and three more needle deflections represent a trouble code of 23.

With a known trouble code, the technician can then pinpoint the trouble by checking the trouble code in the vehicle's service manual. The manual will give complete test and repair procedures for the vehicle. Remember that system codes and test procedures vary from car to car.

CAUTION: An analog (needle type) meter should *not* be used to check many electronic parts in a computer system because it draws too much current through the system and damages the components.

OTHER COMPUTER SYSTEM CHECKS

Once the problem area is discovered by using the trouble code and proper service manual, there are other checks that should be taken. Figure 4-113 shows some of the more common troubles that can occur with a computer-controlled carburetor. Actually, this type of computer is subject to the same problems as a conventional carburetor. But, before condemning the carburetor, check all of the other possible problem areas.

Unlike the conventional carburetor, sensors provide the input to a preprogrammed on-board

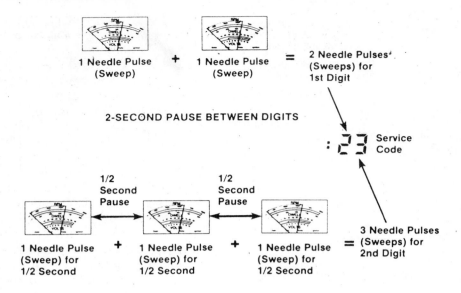

2-SECOND PAUSE BETWEEN DIGITS

4-SECOND PAUSE BETWEEN SERVICE CODES
WHEN MORE THAN ONE CODE IS INDICATED.

FIGURE 4-112 Finding the trouble code with an analog voltmeter in a typical computer circuit

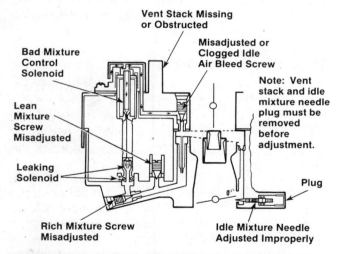

FIGURE 4-113 Some of the troubles that can occur with a computer-controlled carburetor

computer. The three common sensors that affect a computerized carburetor circuit are:

1. Mixture control solenoid
2. Throttle position sensor
3. Oxygen sensor

Mixture Control Solenoid

To keep the proper air/fuel ratio, a mixture control solenoid (Figure 4-114) is used on most computerized carburetor systems. It contains a needle—either tapered or stepped—in the carburetion jets

that operates for mixture enrichment as described earlier in this chapter. The computer provides an electrical output signal to operate the mixture control solenoid on the carburetor. The solenoid pulsates or cycles ten times a second to control the position of the metering valve in the fuel bowl of the carburetor. This cycle rate is fast enough to provide almost instantaneous response to computer commands. The valve cycles between *open* and *closed* positions. Enriching or leaning out the mixture oc-

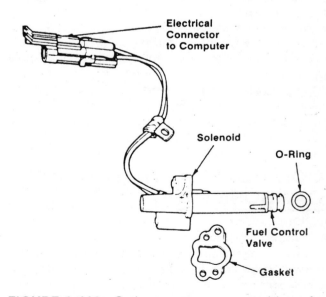

FIGURE 4-114 O-rings are common problems for mixture control solenoids

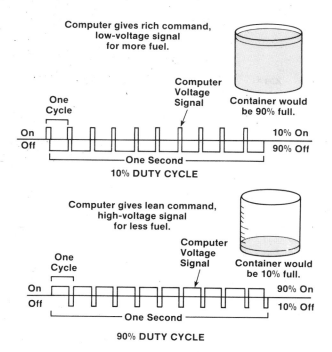

FIGURE 4-115   The computer rapidly pulses mixture control solenoid off and on to control the air/fuel ratio.

FIGURE 4-116   Dwell meter/tachometer

curs when the solenoid and valve are in the rich or lean mode more than 50 percent of the time during each second. The relative amount of time that the mixture control solenoid closes the fuel passage is called the *duty cycle* (Figure 4-115).

To check the computer output system to the mixture control, a dwell meter/tachometer (Figure 4-116) is needed. Attach the test clips of the dwell meter between the mixture control solenoid test lead and ground as directed in the service manual. Re-

gardless of the actual number of cylinders, use the six-cylinder meter setting. Hook up the tachometer in the normal manner in the ignition system. Then start the engine and read the dwell. Figure 4-117 shows the relationship between dwell and mixture control solenoid operation.

A defective mixture control solenoid can create carburetor problems at both idle and cruising speeds. Compare the dwell meter readings with those in the service manual. The manual generally gives a step-by-step servicing procedure using a digital VOM or special test equipment.

To check the M/C solenoid, remove the O-ring, retainer, and spacer from the solenoid stem, then attach the hose of a manual vacuum pump as shown in Figure 4-118. Put 12 volts across the solenoid terminals, apply 25 inches mercury, then time the leakdown from 20 to 15 inches mercury. If it is less than 5 seconds, a new solenoid is required. To find out if the plunger is sticking in the *down* position, pump up a vacuum, then de-energize the solenoid. The reading should go to zero in less than 1 second. Always use a new O-ring and retainer (leave a little space for the rubber to expand), and coat the O-ring with silicone grease or engine oil before reinstalling the M/C solenoid.

Throttle Position Sensor

This sensor (Figure 4-119) sends electrical signals to the computer, telling it how far the throttle plates in the carburetor are opened. If the sensor becomes faulty, the computer will be given false information that could affect the operation of the carburetor.

To check the throttle position sensor, place a digital VOM across the terminals as directed in the service manual. If the reading is off specifications, driveability and performance problems can result. A reading that varies erratically as the throttle is opened rather than going up smoothly indicates a bad sensor. Compare maximum throttle reading when flooring the acclerator to see if throttle linkage can be opened wide by hand. If there is a discrepancy, the problem is generally a mechanical one between the foot and carb linkage (look for binding or maladjustment).

Oxygen Sensor

As mentioned earlier in this chapter, the oxygen sensor informs the computer how much oxygen is present in the exhaust gases. Since this amount is proportional to rich and lean mixtures, the computer will adjust the duty cycle lower if exhaust gas is too lean or higher if too rich (see Chapter 12).

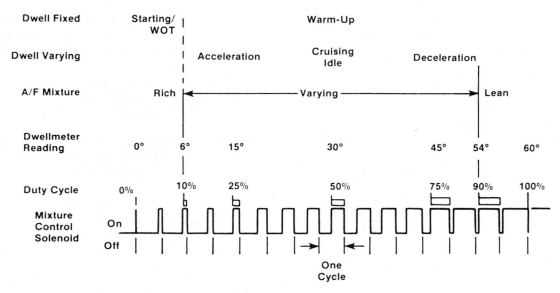

	Starting/ WOT				Warm-Up			
Dwell Fixed								
Dwell Varying		Acceleration		Cruising Idle		Deceleration		
A/F Mixture	Rich ◄──────────── Varying ──────────► Lean							
Dwellmeter Reading	0° 6°	15°		30°		45° 54°		60°
Duty Cycle	0% 10%	25%		50%		75% 90%		100%
Mixture Control Solenoid On Off				One Cycle				

FIGURE 4-117   Relationship between dwell and mixture control solenoid operation. Note that as the dwell increases the mixture becomes leaner and vice versa.

FIGURE 4-118   Manual vacuum pump

REVIEW QUESTIONS

1. A carburetor _____ .
    a. controls engine power output
    b. mixes fuel and air in correct proportions
    c. atomizes and vaporizes the air/fuel mixture
    d. all of the above

2. Which carburetor system is most important on a cold morning?
    a. float system
    b. choke system
    c. main metering system
    d. accelerator pump system

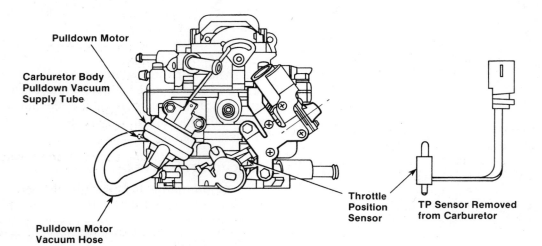

Pulldown Motor

Carburetor Body
Pulldown Vacuum
Supply Tube

Pulldown Motor
Vacuum Hose

Throttle
Position
Sensor

TP Sensor Removed
from Carburetor

FIGURE 4-119   Throttle position sensor assembly

3. Which carburetor system is most important when the accelerator is floored to speed up?
   a. main metering system
   b. full power system
   c. accelerator pump system
   d. none of the above

4. Which carburetor system is most important when cruising at a high rate of speed?
   a. float system
   b. choke system
   c. main metering system
   d. full power system

5. The vacuum break on an automatic choke _____ .
   a. closes the choke
   b. opens the choke
   c. sets the choke
   d. has nothing to do with the choke

6. Engine idle can be set to specifications by using a _____ .
   a. tachometer
   b. vacuum gauge
   c. either a tachometer or a vacuum gauge
   d. none of the above

7. A car idles roughly. Technician A says the roughness is caused by the engine getting more air than it needs. Technician B says the rough idle is caused by a vacuum leak. Who is correct?
   a. Technician A
   b. Technician B
   c. Both A and B
   d. Neither A nor B

8. The venturi in a carburetor barrel _____ .
   a. creates a pressure drop or differential
   b. forces the incoming airflow to slow down
   c. atomizes the fuel
   d. vaporizes the fuel

9. The primary system of a carburetor _____ .
   a. provided a high venturi velocity to the airflow
   b. provides good fuel metering
   c. increases fuel atomization and vaporization
   d. none of the above
   e. all of the above

10. What carburetor system provides added flow capacity that allows increased power at higher engine speeds?
    a. primary system
    b. main metering system
    c. secondary system
    d. full power system

11. A car in the shop has an erratic idle problem. Technician A says that the EGR is being held open by dirt or foreign matter. Technician B says that the hot idle compensator is the culprit. Who is correct?
    a. Technician A
    b. Technician B
    c. Both A and B
    d. Neither A nor B

12. A passenger car carburetor is experiencing hesitation or "say" at tip in (moderate opening of the throttle). Technician A says the problem lies in the accelerator pump system. Technician B says that there is manifold leak or the possibility of gummed-up air bleeds. Who is correct?
    a. Technician A
    b. Technician B
    c. Both A and B
    d. Neither A nor B

13. Which of the following is most likely to be an adjustment on feedback carburetors?
    a. idle speed
    b. idle mixture
    c. fast idle speed
    d. all of the above

14. Which of the following prevents stalling during deceleration?
    a. dashpot
    b. hot-idle compensator valve
    c. choke unloader
    d. all of the above

15. Which of the following is not a function of a throttle position solenoid?
    a. Prevents dieseling.
    b. Increases idle speed to compensate for a load.
    c. Compensates for high altitudes.
    d. All of the above

16. Which of the following could cause a rich fuel mixture?
    a. vacuum leak

b. inoperative accelerator pump
c. heavy float
d. choke that will not close

17. Dirt in the _____ will cause poor low-speed engine operation.
    a. idle port
    b. main metering jet
    c. accelerator pump circuit
    d. booster venturi

18. A car is experiencing a low-speed stumble. Technician A thinks the problem is a defective choke pulloff diaphragm but checks the ignition system first. Technician B thinks the problem is a heavy float but checks for a vacuum leak first. Who is correct?
    a. Technician A
    b. Technician B
    c. Both A and B
    d. Neither A nor B

CHAPTER FIVE

BASICS OF FUEL INJECTION

Objectives

After reading this chapter, you should be able to:
* Compare the operation of a fuel injected system to a carbureted system.
* Explain the two methods of fuel injection: electronic fuel injection (EFI) and continuous injection system (CIS).
* Explain the basic operation of a fuel injection system.
* Name the advantages and limitations of a fuel injection system.
* List the components of a basic fuel injection system.
* Describe the functions of the components of a basic fuel injection system.
* Differentiate between closed and open loop operations.

A few years ago, fuel injection was something one occasionally read about, rarely saw, and hardly ever tinkered with. To shop technicians as well as home mechanics, fuel injection was as foreign as the cars it came on. This status prevailed because the American automotive industry built its reputation on big cars with big engines that came factory equipped with carburetors.

Today, the idea of squirting fuel directly into the intake manifold is in vogue. America now manufactures cars that are downsized, fuel efficient, have front-wheel drive, and come factory equipped with fuel injection systems.

The carburetor, although it has served well, is on its way out.

Learning the operation of fuel injection systems and how to service them takes effort, time, and some specialized tools, but the rewards will be apparent. Fuel injection is here to stay and new service opportunities will be for the taking by those having the knowledge and expertise of this fuel delivery concept.

This chapter presents the basics of fuel injection systems and looks at the components that make up these systems. As the chapter progresses, it concentrates on gasoline engine fuel injection systems and touches on the subject of diesel fuel injection systems as well.

There is one thing in the technician's favor when it comes to understanding fuel injection—similarity of design. The majority of injection systems in use today derived their technology from one company—Robert Bosch. It was Bosch who actually perfected and mass-produced diesel fuel injection systems back in 1927. Once one or two systems are learned, with minor exceptions, all other systems will be very similar in operation, layout, arrangement, and servicing.

FUEL INJECTION VERSUS CARBURETION

Fuel injection has often been hailed as a superior principle, and the carburetor has too often been coined a compromise. Modern carburetors have been designed to a high state of perfection, incorporating so many design compromises that they cannot perform their functions optimally. On the other hand, when one considers general principles of operation, both systems have similar characteristics. Fuel and air are mixed by an injection system to form a combustible mixture. A carburetor does the same thing. Fuel is sucked into the airflow by a carburetor; it is squirted into the airflow by an injection system. The distinction between an injection system and a carburetor requires closer scrutiny to recognize the differences.

It might be helpful to list the comparable functions of carburetor parts with parts of the injection system. The following list shows the parts and their functions of these two systems.

Fuel Injection System	Carburetor
Throttle switch	Accelerator pump
Fuel pressure regulator	Float
Inlet manifold pressure sensor or airflow sensor	Metering rods
Thermo timer (switch)	Fast idle cam
Injector valves and electronic control unit	Metering jets and idle fuel system

If you do not have trouble comprehending or relating the functions of the above parts between the two systems, you will have no trouble understanding the fuel injection system.

On the other hand it might be helpful to contrast the characteristics of the injection system against the carburetor to highlight their unique differences.

PRINCIPLES OF OPERATION

In an injection system, air and fuel are measured separately. The metering of fuel takes into account variable parameters such as airflow, engine temperature, engine speed, throttle position, and other engine/control factors. For any given set of engine conditions, uniform quantities of fuel are always delivered individually to each cylinder.

Fuel and air are not measured in a carburetor system. A form of fuel metering is maintained at a nearly constant proportion by the manifold vacuum. The fuel mixture (air and fuel) is prepared upstream of the intake manifold and is prepared together for all cylinders. In other words, all cylinders get the same fuel mixture.

AIR/FUEL MIXTURE

In an injection system, fuel droplets are squirted into the inrushing airstream under pressure. The injectors are mounted close to the intake valve preventing the manifold walls from getting wet. All fuel goes into the cylinder. Precise control of the air/fuel ratio is constantly maintained (Figure 5-1).

In a carburetor system, fuel is drawn into the airstream by the vacuum created in the manifold. Because fuel and air are mixed upstream of the mani-

fold, wetting of the manifold walls takes place, thereby upsetting the air/fuel ratio balance (Figure 5-2).

IDLE

When no power is needed, fuel delivery is cut off completely in an injection system. When engine speed comes or is close to an idle speed, fuel delivery is turned on to prevent stalling. This is a big fuel energy saving.

On a carburetor system, fuel is continually mixed into the airstream in predetermined proportions when no power is needed.

ALTITUDE COMPENSATION

Compensation for differing altitudes can be built into the basic controls of an injection system.

At higher altitudes, with standard carburetor settings, the engine wastes fuel because of a too-rich mixture; that is, less oxygen intake, but the same quantity of fuel to be mixed.

FUEL CONSUMPTION

An injection system has lower fuel consumption and exhibits a clear advantage at all reasonable

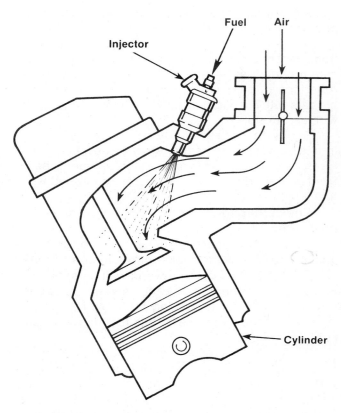

FIGURE 5-1 Mixing fuel and air in an injection system

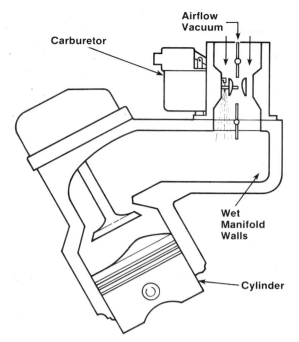

FIGURE 5-2   Mixing fuel and air in a carburetion system

speeds. This claim cannot be applied to a carburetor system.

COMPLEXITY

Carburetors can be as complex if not more complex than an injection system. Some carburetors can be a maze of mechanical, pneumatic, and hydraulic (fuel) systems. Complexity of an injection system is contained within an electronic control unit of which the technician needs to have little knowledge of its internal circuitry. An injection system seems complex because of the number of sensors it employs. On par, an injection system is cheaper than a carburetor when it comes to servicing.

HISTORY OF FUEL INJECTION

Fuel injection has been with us since the turn of the century. As early as 1912, the Robert Bosch Company of Stuttgart began experimentation with gasoline injection, but it did not lead to practical results. Twenty years later, this work resulted in the first fuel-injected German aircraft engines. However, it was not until 1952 that Bosch installed a fuel injection system in an automobile.

American development of fuel injection systems goes back as far as the Wright Brothers, who actually used an engine that employed continuous injection of gasoline into the intake manifold. Development of an injection system actually began in the 1940s. At that time, both Enderle and Hilborn were working on fuel injection systems to use in racing cars. However, it was not until 1957 that the first injection system was used on a domestic vehicle. It was that year that Chevrolet introduced a Rochester produced fuel injection system on its Corvette line and on some other models. At about the same time, Pontiac produced a limited number of cars equipped with fuel injection. All of these systems were mechanical.

The first electronic fuel injection system was offered as an option by Chrysler in 1958. The system was manufactured by Bendix and coined the "Electrojecter" system. A major factor in its lack of popularity was its high cost ($400 to $500). Bendix decided to sell the manufacturing rights to this system when the popularity of the system regressed. The Bosch Company bought those rights and continued with development of the system believing that there might be a future for electronic fuel injection. In 1968 the first Bosch electronic fuel injection system was offered on a Volkswagen.

The U.S. government became quite concerned about the state of the environment and increasing dependence on foreign oil in the early 1970s. Legislation was enacted to establish minimum fuel efficiency standards and maximum pollution levels for vehicles sold in the United States. Automobile manufacturers began to realize that their present fuel systems would be hard pressed to meet any future standards imposed by legislation. It was at this time that the complexity and cost of carburetion systems began to rise.

In the meantime, advances in the electronics industry were resulting in the availability of reliable and inexpensive solid state components. Eventually these advances were applied to computer control of fuel injection systems. As a result fuel injection became more dependable and cost competitive. In 1975 the first electronic fuel injection system reappeared in a domestic vehicle—the Cadillac Seville. Today, every domestic car manufactured as well as most import cars are equipped with electronic fuel injection.

FUEL INJECTION—MANUAL VERSUS ELECTRONIC

There are two methods of fuel injection in use— electronic fuel injection (EFI) and the mechanical or continuous injection system (CIS). The most widely used system is electronic fuel injection. These systems are often referred to as intermittent or pulsed because the fuel flows in short spurts as the injectors turn on and off. An electronic control unit (ECU) determines how long an injector is held open. As the name implies, this system relies on electronic circuitry for its operation. The ECU, with memory capability, also outputs electrical signals to control certain vehicle systems and inputs electrical signals to monitor or sense vehicle operating conditions. The ECU operates in conjunction with several engine and vehicle sensors that are fed a voltage reference signal. Each sensor operates on the resistance principle—a changing temperature or a changing pressure. Each sensor in turn sends a voltage signal back to the ECU where it is compared and computed with a preprogrammed memory system. Based on a reference mode of operation, the ECU then sends out a voltage signal to adjust or change the air/fuel ratio for various speeds, loads, and general operating parameters. If a sensor fails or an abnormal condition exists, the ECU will sense the problem and illuminate or flash a check light on the instrument panel. The ECU has several other names usually tied to the automotive manufacturer. General Motors refers to it as an electronic control module (ECM), Ford uses the term electronic engine control microprocessor control unit (EEC/MCU), and Chrysler refers to their system as computer-controlled combustion (CCC) or an ECU.

The continuous injection system utilizes limited electronics and is basically mechanical in nature. This type of system is often termed CIS because it supplies fuel to the injectors in a continuous, constant flow. Fuel flow and/or injector timing is not computer or electronically controlled in this type of system. Fuel flow is controlled by a distributor that feeds fuel to the engine proportional to airflow. The fuel distributor can be thought of as an ECU that is mechanical in design.

BASIC OPERATION OF A FUEL INJECTION SYSTEM

Most electronic fuel injection systems only inject fuel during part of an engine combustion cycle. Fuel requirements are measured either by airflow across a sensor or by intake manifold pressure (vacuum). These sensors convert airflow or vacuum parameters into electrical signals that are fed to a microprocessor in the ECU (Figure 5–3). The ECU in turn processes these signals to determine the engine fuel requirements. Once these requirements are determined, the ECU generates an electrical

signal to operate the fuel injectors. Basically, the ECU sends a signal that controls the length of time each injector stays open. This open time interval is known as the injector pulse width. This pulsing or intermittent opening and closing of the injectors is a typical characteristic of an electronic fuel injection system. Other sensors mounted within the engine compartment monitor other parameters such as air temperature, coolant temperature, engine speed, and throttle position. These sensors send electrical signals to the ECU that are also used to determine the overall fuel requirements for the engine. The fuel delivery system, consisting of fuel tank, fuel pump, and filter, are typical of any gasoline engine. A fuel pressure regulator is used, however, to control the amount of fuel pressure that is delivered to each injector.

Two modes of operation are characteristic of electronic injection systems. The open loop operational mode uses memory information within the ECU to determine air/fuel ratio, injection timing, and so on, as opposed to using "real time" sensor inputs. This mode usually occurs during cold engine operation or whenever a sensor or sensors malfunction. The cold start mode uses the thermo time switch, the auxiliary air, and the cold start injector. The auxiliary air valve provides additional air into the intake manifold. The thermo time switch operates on coolant temperature and the time it takes for a bimetallic element in the switch to heat up and open the electrical circuit to close the cold start injector.

The other operational mode is the closed loop system. This type of operation utilizes an oxygen sensor to determine air/fuel ratio. This type of operation is also referred to as the feedback system. Basically, this mode of operation limits the amounts of harmful substances in the vehicle's exhaust gases by controlling the metering of the air/fuel mixture very closely. By contantly measuring the oxygen content of the exhaust gases, the fuel delivered by the injection system can be adjusted as needed to obtain the proper oxygen output.

There is one other type of fuel injection system that can be covered here—the continuous injection system (Figure 5–4). Technically speaking, this can be considered a mechanical fuel injection system because the only electrical component within the system is the fuel pump or, on some systems, the in-tank fuel pump. Computer control of fuel flow and injector timing is not used in this system; therefore, the ECU is not used. In fact, fuel is supplied to the injectors in a constant, continuous flow. The volume of fuel is controlled by flow rate and not by injector pulse width. An airflow sensor moves valves that alter the fuel flow. Newer models use an electronic system that controls fuel pressure and affects the

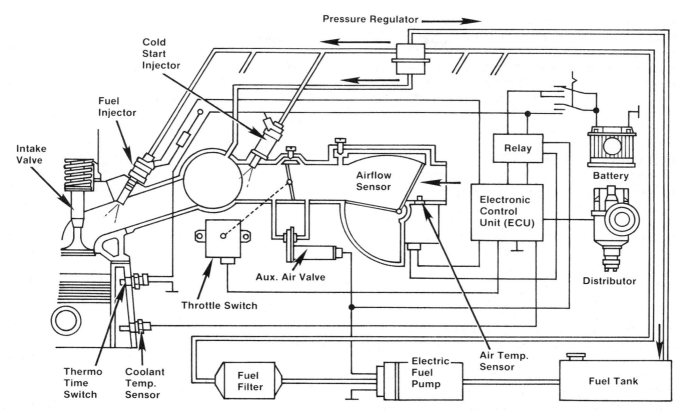

FIGURE 5-3   Schematic view of an electronically controlled fuel injection system

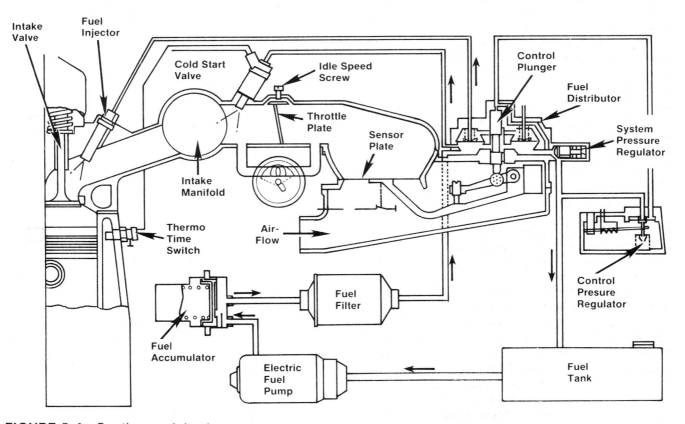

FIGURE 5-4   Continuous injection system

flow rates. This latter system is one of the earliest fuel injection designs developed by the Bosch Company. The main attraction of the CIS is the fuel distributor. The distributor is supplied with fuel from the fuel tank and leaves the distributor via "one" fuel line for each injector at a constant predetermined pressure.

ADVANTAGES OF FUEL INJECTION SYSTEMS

Fuel injection offers the following advantages over its carburetor counterpart:

1. Improved fuel distribution to each cylinder, thereby providing a more even load distribution between cylinders with less tendency toward detonation because of the leaner air/fuel ratios.
2. Engine power increases on an average by 10 percent. This is due to better volumetric efficiency and is achieved by using larger air inlet passages. The efficiency is further assisted by the fact that cooler fuel is delivered to the injectors and also accounts for better fuel vaporization.
3. A wider range of fuel can be used because of the mechanical atomization of the fuel.
4. Faster acceleration is possible because the atomized fuel is delivered directly to the cylinder.
5. Leaner air/fuel ratios are more easily attained.
6. Provisions are included for fuel shutoff when a vehicle decelerates.
7. Icing is minimized because the fuel is atomized in the cylinder.
8. Higher engine torque, quicker starts, and engine warm-up
9. Fewer exhaust emissions
10. Backfiring in the inlet manifold or throttle body is reduced or practically eliminated.
11. Many pollution control devices normally required by a carburetion system are eliminated.
12. Better fuel economy
13. No choke requirements and no mixture screw adjustments as predominantly used on carburetion systems
14. Accurate control of air/fuel mixture ratios
15. Maintain stoichiometric conditions over a wide range of operating conditions

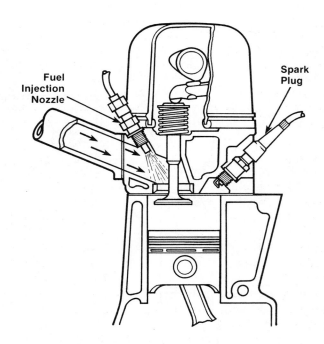

FIGURE 5–5 Port injection systems inject fuel into an intake port instead of into a manifold or cylinder.

LIMITATIONS OF FUEL INJECTION SYSTEMS

Although fuel injection has its many advantages, there are a few limitations of the systems in use that can be addressed. These limitations can be classified as cost, environmental operating conditions, and complexity.

For a while fuel injection systems were usually more expensive than carburetion systems. At the present, however, the cost differences are being reduced. Carburetors are becoming more expensive because of the additional feedback systems and emission controls they require to meet certain environmental regulations.

Because fuel injection systems must meter extremely precise quantities of fuel, they must have clean fuel as well as clean air to operate properly and efficiently. Dirt is the number one enemy of fuel injection systems.

Although fuel injected systems have been considered difficult to service, they are not all that complex, especially when compared to the modern carburetion systems. Like any new concept or technique, once the principles of operation are understood and mastered, the servicing comes easy.

TYPES OF FUEL INJECTION SYSTEMS

Injector location determines the type of fuel injection system that is used. There are basically two types in use:

1. Port or multi-point injection
2. Throttle or single-point injection

Most systems deliver fuel in a fine, atomized mist at the intake port of a cylinder head as shown in Figure 5-5. These systems are the port type, with one injector for each cylinder. Fuel lines run to each cylinder from a fuel manifold, usually referred to as a fuel rail. With port injection, only air passes through the intake manifold. With little or no fuel to wet the manifold walls, there is no need for manifold heat or an early fuel evaporation system. Fuel will not collect in puddles at the base of the manifold. This means that the intake manifold passages can be tuned or designed for better low-speed power availability (Figure 5-6). The port type systems provide a more accurate and efficient delivery of fuel.

Other fuel injection systems deliver fuel at a throttle body (Figure 5-7). When fuel is delivered at a central point, the systems are called single-point, or throttle body, even though more than one injector is used. The throttle-body systems are lower in cost,

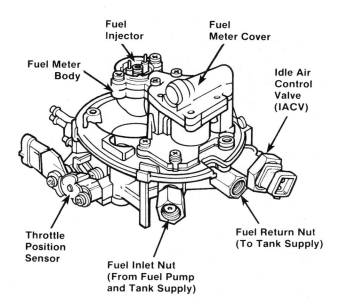

FIGURE 5-7 Throttle body injection system provides more precise fuel control than a carburetor without the precision of port injection.

requiring only one injector, or two in a V-type engine. Under the control of a computer (or ECU), these systems can produce results approaching port-type injection.

The throttle body type of injection system can be thought of as a compromise between a carburetor and a complete electronic fuel injection system. Throttle body injection provides a more correct air/fuel ratio than a carburetor, is simpler and less expensive than a port type injection system, but is less efficient. Like a carburetion system, fuel is not distributed equally to all cylinders.

DESIGN VARIATIONS

Design variations of the basic two types of fuel injection systems can be seen throughout the automotive industry. The variations occur in the firing sequence, fuel fed points, and injector timing.

Sequential Fuel Injection

On this type of system, fuel is sprayed just before the opening of each individual intake valve. Therefore, each injector sprays once every two crankshaft revolutions. The fuel mixture is drawn into the cylinder immediately. Sequential fuel injection, or SFI, uses an ECU that can be programmed to provide a more precise air/fuel mixture than systems that pulse alternate injector groups.

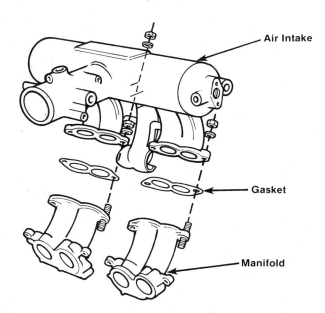

FIGURE 5-6 Port injection systems usually have tuned intake manifold runners to optimize low-speed power.

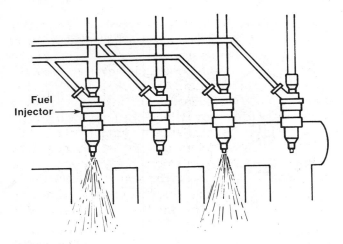

FIGURE 5-8 Injector groupings

Injection Grouping

On most electronic fuel injection systems, the injectors spray fuel in groups. Half of the injectors sprays once each crankshaft revolution, and the other half sprays on the next crankshaft revolution (Figure 5-8).

Injection Timing

Because the four-stroke cycle requires two complete revolutions, the air/fuel mixture is not drawn into the cylinders immediately. Consequently, some of the mixture remains in the manifold until an individual intake valve opens.

Top-Feed and Bottom-Feed Injectors

Two types of injectors are currently in use (Figure 5-9). In a top-feed injector, fuel is delivered

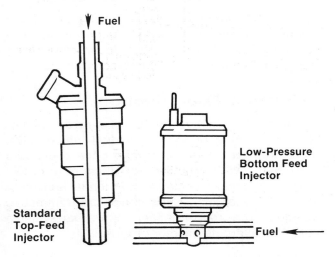

FIGURE 5-9 Top-feed and bottom-feed injectors

under higher pressure (up to 79 psi) to help prevent vaporization. The higher pressure requires a more expensive fuel pump. One disadvantage of this type of injector is that fuel vapors tend to rise into the incoming fuel stream and can prevent or block fuel delivery. Bottom-fed injectors are able to use fuel pressures as low as 10 psi. Fuel vapors flow upward through the injector to a return line that feeds the fuel tank. Using bottom-feed injectors requires a less expensive pump and fuel vapors do not tend to block fuel delivery.

GASOLINE AND DIESEL FUEL INJECTION SYSTEMS

There are two major classifications of injection systems that are based on the type of fuel—gasoline and diesel. Gasoline fuel injection systems deliver fuel into the intake port or the throttle body. Fuel is then injected into the air before the mixture reaches the cylinder. This type of system requires operational fuel pressures that are higher than those required of a carburetion system but much less than in diesel fuel injection.

In contrast, diesel fuel injection systems deliver fuel directly into the cylinder or to a precombustion chamber. In operation, diesel injection fuel pressures are relatively high to overcome cylinder compression pressures that might be as high as several hundred psi. The process of removing and replacing injectors provides a good contrast between the two systems. Gasoline injectors are pushed into place, sealing themselves with an O-ring. Diesel fuel injectors must be torqued against gaskets and shields because of the higher operating pressures.

Considering the fact that gasoline injection pressures are quite lower than diesel injection pressures, they still are several times greater than those utilized in a carburetion system. Care must be exercised when loosening a pressurized line in a fuel injection system. A jet of high-pressure fuel spray can be unleashed onto a hot manifold causing personal injury or a fire.

GASOLINE ELECTRONIC FUEL INJECTION SYSTEMS

Many European cars imported to the U.S. as well as many domestic cars are now factory equipped with electronic fuel injection systems. These intermittent (or pulse) systems use electronics to completely control the metering of fuel. These systems measure the intake airflow, electronically modify the signal according to engine speed, temperature and other conditions, compute the fuel required for the airflow, and signal the injectors to open for the required time pulse width.

The fuel injection system must provide the correct air/fuel ratio for all engine loads, speeds, and temperature conditions. Unlike a carburetor, a fuel injection system uses the same basic single system to provide these difficult air/fuel ratios. This basic system includes:

- Fuel delivery system
- System sensors
- Electronic control unit (ECU)
- Fuel injectors

Fuel Delivery System

The fuel delivery system is similar to that used for a carburetion system. This system typically includes such components as an electric fuel pump, fuel filter, pressure regulator, fuel tank, fuel rail assembly, and individual connecting lines to each injector (Figure 5-10). It also includes an injector for cold-start operation.

Fuel Tank and Connecting Lines. The fuel tank and connecting lines are like those of a carburetion system; the main difference is that a return line is used to return excess fuel to the tank. This is accomplished by the pressure regulator.

Fuel Pumps. The fuel pump, as shown in Figure 5-11, is a roller-cell pump that is mounted

directly on the shaft of the electric motor. The motor runs surrounded by gasoline to cool and lubricate it. An explosive atmosphere is avoided because the mixture in the pump housing is never an ignitable one (restricted air ratio). The pump pressurizes the fuel and thereby prevents air bubbles in the lines and vapor lock. The rotary roller pump consists of rollers on a centrally mounted eccentric that rotates within a housing. The rollers are pushed outboard by springs against the pump housing allowing fuel to flow in behind the rollers. The eccentric continues to rotate thereby compressing the rollers and creating pressure on the fuel behind them. As the eccentric

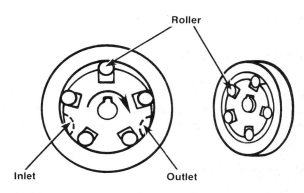

FIGURE 5-11   As the rotor spins, centrifugal force keeps the rollers pressed against the pump walls, forcing fuel through the pump.

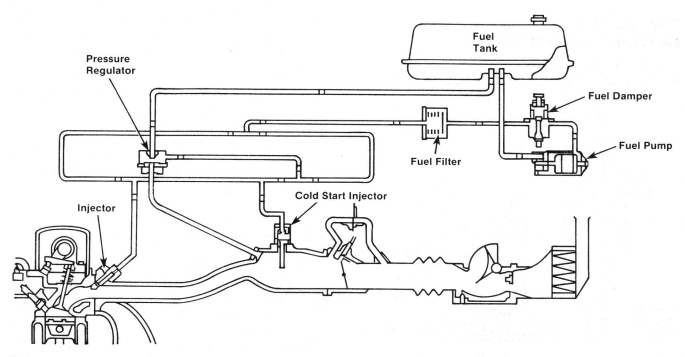

FIGURE 5-10   Fuel circuit of a fuel injection system

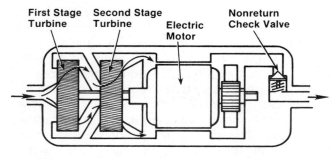

First Stage Turbine   **Second Stage Turbine**   **Electric Motor**   **Nonreturn Check Valve**

FIGURE 5-12   Turbine fuel pump

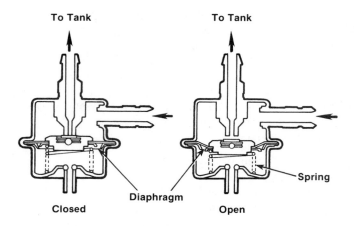

To Tank   To Tank

Diaphragm   Spring

Closed   Open

FIGURE 5-13   Pressure regulator provides constant pressure differential between fuel pressure and manifold pressure.

rotates past an outlet port, the fuel is squeezed out under pressure and discharged into the connecting line. Two other types of pumps are also in use: the diaphragm type and the turbo type. The diaphragm pump is identical to the type of mechanical pump used on carburetion systems. This type of pump is used only as a transfer pump to deliver fuel to a high-pressure pump. The turbo type (Figure 5-12) is similar to a water pump or turbocharger compressor wheel. Fuel is drawn inward at the center of the wheel and forced outward through centrifugal force. The centrifugal force creates the pressure.

Power to the fuel pump is controlled by a relay in the fuel injection electrical system. One circuit delivers fuel pressure during cranking and another circuit delivers fuel when the engine is running and cuts off fuel pressure if the engine stops (even if the ignition is in the run mode).

Fuel Filter

Small contaminant particles can lodge in an injection nozzle and block it partially open or closed. Water can also seep into the system and corrode closely machined parts of the injectors. To eliminate dirt and water contaminants, one or more fuel filters are used in the system and are very similar to those used on carburetion systems. The filter can be located under the hood or near the fuel tank and pump at the rear of the car. The filter is made especially of fine construction and functions at fuel pressures many times greater than carburetion systems.

Fuel Pressure Regulator

Fuel pressure must be regulated according to manifold absolute pressure (MAP). The pulse width, which is the time the injector is open, and the pressure determine the amount of fuel delivered. Since the difference in pressure between the fuel in the injector and the air pressure in the manifold must remain constant, the fuel pressure must change as the MAP changes. The manifold pressure will be

higher at full throttle than at a light load cruising speed. In a turbocharged engine, the injector will sometimes have to inject against higher MAP. There is an air hose connection between the bottom of the fuel pressure regulator and the intake manifold, as shown in Figure 5-10. Figure 5-13 shows how MAP is applied to the bottom side of the diaphragm to regulate the fuel pressure. Thus, the diaphragm controls the amount of fuel entering the connection from the pump. Excess fuel returns through the return line to the tank. The fuel pressure regulator is adjusted by the manufacturer; it cannot be field adjusted. The fuel pressure in a typical electronic fuel injection system is regulated at a value higher than MAP. It is expressed as 2.55 bar, 255 kilopascals, or 37 pounds per square inch.

Fuel Rail Assembly

The fuel rail assembly (Figure 5-14) is a mechanical assembly that contains or houses the fuel injectors (usually on a port fuel injection system). On some automotive models the fuel pressure regulator can also be mounted on this assembly.

Airflow Sensor

The airflow sensor is in the intake system between the air cleaner and throttle (Figure 5-15). The airflow sensor is not usually attached rigidly to the engine. Once the air passes the airflow sensor, it moves through a flexible duct to the engine, either to the intake manifold or to the turbocharger.

The air sensor is not a throttle that controls airflow but a device used to measure it. The flap deflection is proportional to the amount of airflow. The flap is deflected against a spiral spring. Flap

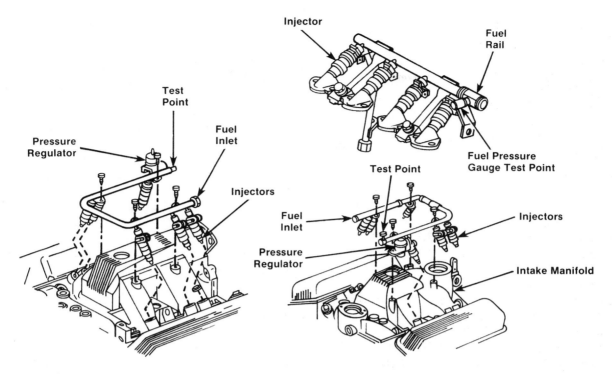

FIGURE 5-14   Typical fuel rail assemblies

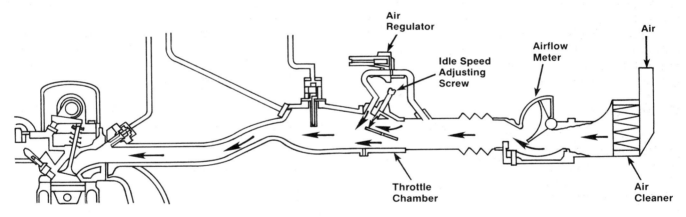

FIGURE 5-15   The airflow sensor measures the flow of incoming air.

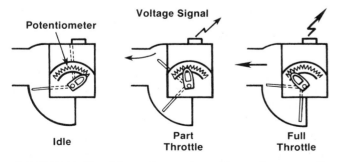

FIGURE 5-16   The voltage signal from the potentiometer varies in response to the amount the airflow sensor flap is deflected.

deflection moves a potentiometer, which is on the same shaft as the flap. The potentiometer, or voltage control, adjusts the electrical signal proportional to the intake air (Figure 5–16). An air temperature sensor adjusts the potentiometer signal. This signal adjustment compensates for the higher density of the cold air. The denser air requires more fuel for efficient combustion.

Air Leaks.   The engine cannot run properly if air leaks allow extra air into the engine without being measured. Since the extra air (also called *false air*) does not pass the airflow sensor, the electronic control unit does not provide fuel for the extra air. As a result, the air/fuel mixture is too lean. When servicing fuel injection systems that depend on airflow movement, checking for leaks is especially important.

Damping Chamber. As shown in Figure 5-17, the damping chamber is that part of the airflow sensor that is curved in shape. The damping flap (also called the *compensating flap*) is on the same shaft as the airflow sensing flap. It smooths out any pulsations caused by the opening and closing of the intake valves. This prevents the airflow measurement from becoming a meaningless, erratic signal. Instead, the signal remains steady and closely related to airflow, which is throttle controlled.

Backfire Protection. A spring-loaded valve on the airflow sensor flap provides for backfire protection (Figure 5-17). If a backfire causes the intake manifold pressure to increase suddenly, this valve releases the pressure and the system is protected from damage.

Idle Bypass. As shown in Figure 5-17, the airflow sensor assembly has a channel near the top that allows some air to bypass the airflow sensor plate. The purpose of the idle bypass is to smooth the flow of idle intake air, which insures steady signals to the electronic control unit for regular injections to the engine. The idle bypass is needed because the opening and closing of the intake valves can cause pulsations in the intake manifold when the throttle is closed at idle. Without the idle bypass, these pulsations could cause the flap to shudder, which would result in an uneven air/fuel mixture.

Adjustments. Unless the system is sealed, both idle speed and mixture are adjustable. The idle speed screw is located near the throttle. It adjusts the idle speed by changing the air already measured by the airflow sensor. The mixture screw changes the air before it is measured and also adjusts the level of carbon monoxide in the exhaust. Mixture is adjusted at all speeds.

The idle speed must be adjusted before the mixture is adjusted. The idle speed adjusting screw is in the air passage that bypasses the throttle plate (Figure 5-15). When the throttle plate is closed, as it is during idle, the amount of air flowing through this passage will control the engine's rpm. The rpm is lowered by turning in the screw, which reduces this air.

The mixture can also be adjusted. Turning the mixture adjustment screw clockwise decreases the bypass air (Figure 5-17). This enriches the mixture and increases the CO level in the exhaust. Turning the screw counterclockwise increases the bypass air, which makes the mixture leaner and reduces the CO level. Some electronic fuel injection engines run

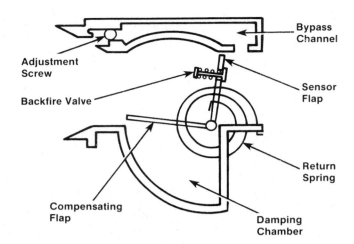

FIGURE 5-17 The airflow sensor flap is equipped with a backfire valve to prevent damage to the sensor.

very lean. For this reason, special enrichment procedures might be required for the CO meter to react to the mixture adjustment screw. Instructions are usually given in the engine manual for the particular application.

Other Sensors

Signals from the airflow sensor are used by the electronic control unit to establish the minimum injector pulse width (open time). However, the electronic control unit modifies this base pulse according to engine operating conditions. Other sensors are used to provide the electronic control unit with information about these conditions.

Coolant Temperature. When the engine needs cold enrichment, as it does during warm-up, the coolant temperature sensor signals the electronic control unit. The control unit increases the pulse width. Then as the engine warms up, the pulse width is reduced to the base pulse again.

Air Temperature. When colder air is coming in from outside, the coolant temperature sensor signals the electronic control unit. Since the colder air is heavier and more fuel is needed for a proper mixture, the control unit sends a longer pulse width signal to the fuel injectors.

Throttle Position. When the throttle is closed, the switches on the throttle shaft send a signal to the electronic control unit for idle enrichment (Figure 5-18). These switches also signal the electronic control unit when the throttle is near the wide-open position so that full load enrichment is provided.

Engine Speed. A signal corresponding to engine speed is sent to the electronic control unit

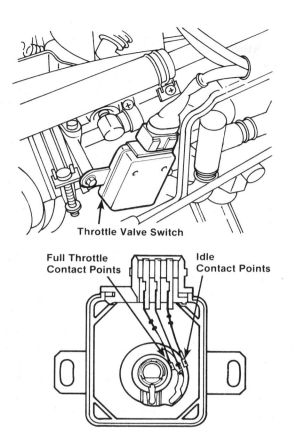

Throttle Valve Switch

Full Throttle Contact Points   **Idle Contact Points**

FIGURE 5-18 The throttle position switch signals the ECU when the throttle is wide open or at idle.

either from the coil or from the distributor of the ignition system. This signal causes the control unit to modify the pulse width of the injectors for engine speed. The signal also synchronizes the start of the injection with the intake stroke cycle.

Cranking Enrichment. During cranking operations, the starter circuit signals for fuel enrichment even when the engine is warm. This signal for enrichment is independent of any cold-start enrichment demands.

Additional Input Information Sensors. Additional sensors are also used to provide the following information about engine conditions:

- Crankcase position
- Camshaft position
- Timing of ignition spark
- Air conditioner operation
- Gearshift lever position
- Battery voltage
- Amount of oxygen in exhaust gases
- Emission control device operation

Sequential Injection paper ASE Test

Figure 5-19 shows the inputs and outputs of an onboard computer (ECU) on a typical vehicle.

ELECTRONIC CONTROL UNIT

The heart of the fuel injection system is the electronic control unit (ECU). The ECU is a small computer that is usually mounted within the passenger compartment to keep it away from the heat and vibration of the engine. The ECU includes solid state devices, including integrated circuits and a microprocessor.

The ECU receives signals from all the system sensors, processes them, and transmits programmed electrical pulses to the fuel injectors. Both incoming and outgoing signals are sent through a wiring harness and a multiple-pin connector.

Electronic feedback in the ECU means the unit is self-regulating, controlling the injectors by the ECU on the basis of operating performance or parameters rather than on preprogrammed instructions. An ECU with a feedback loop, for example, reads signals from the oxygen sensor, varies the pulse width of the injectors, and again reads the signals from the oxygen sensor. This is repeated until the injectors are properly pushed to give the required oxygen output in the exhaust gases from the engine.

Injectors

The injector is similar in size and shape to a typical spark plug (Figure 5-20). The ECU signal energizes the solenoid winding of the injector. As a result, the winding pulls back the armature and the nozzle valve is opened. When the ECU signal shuts off, a spring pushes the nozzle back, closing the valve. The valve might be open as long as 10 milliseconds at full load and as short as 1 millisecond at idle.

Some injectors are all connected in parallel so that they open and close at the same time. Since the ECU has to develop only one signal for all injectors, it simplifies the electronics, yet injection efficiency is not compromised. Half of the total amount of fuel required for each four-stroke cycle is delivered every 360 degrees of crankshaft rotation, as shown in Figure 5-21. Thus, there are two injections in a four-stroke cycle. The firing order of cylinders is 1-5-3-6-2-4. The first injection to all cylinders starts during the intake stroke of cylinder 3. The intake valve of cylinder 3 is open when the injection begins; however, the other cylinder intake valves are closed. Thus, the injected fuel forms a vapor in the manifold that is ready to be drawn in when the other intake valves are opened. The same thing happens during the second injection when the fourth cylinder intake valve is opened. There is little difference in perfor-

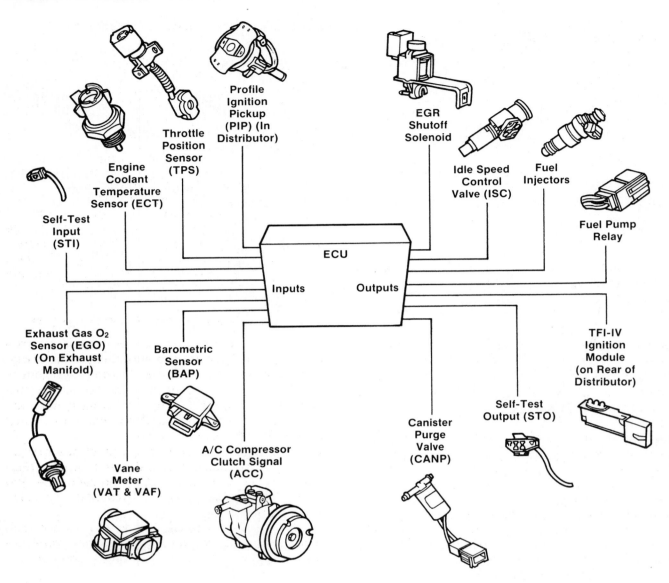

FIGURE 5-19 Onboard computer system

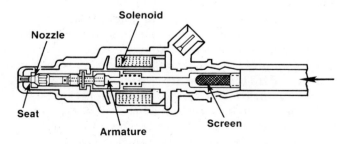

FIGURE 5-20 Solenoid-operated injector

mance between this simultaneous injection and sequential (individually timed) injections.

The injectors in some engines operate at a lower voltage than that of the battery. A ballast resistor (also called a *dropping resistor*) is included in each injection circuit. A portion of the battery voltage is dropped across this resistor. These resistors protect the injectors from voltage surges in the electrical system.

Cold Start Operation. The units already described are for normal warm engine operation. Additional units provided for cold starting and warm-up include the cold start injector (valve) with its thermo-time switch and the auxiliary air device.

The cold start injector is supplied with fuel under pressure whenever the fuel pump is running. The extra injector delivers the additional fuel when the cold engine is cranking. The coldness of the engine determines the amount of extra fuel injected

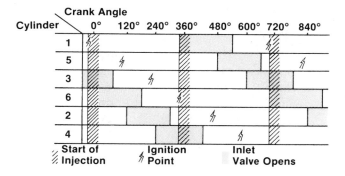

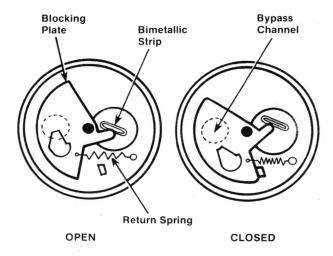

FIGURE 5-21   Simultaneous injection system. Fuel is injected once per crankshaft revolution.

during cranking. When the engine is warm, extra fuel is not injected.

The cold start injector is powered through the starting circuit and grounded through the thermo-time switch. The switch operates on both coolant temperature and time. The time involved is changed by an electric heating element in a bimetal switch. This bimetal switch is also powered by the starting circuit. When the bimetal switch opens the circuit after several seconds of cranking, the cold start fuel injection stops. Thus, when the engine is cold, and the ignition key is turned to START, cold start fuel injection is supplied. When the engine is cold and the ignition key is released to RUN, cold start fuel injection shuts off. However, continued starting or cranking of the engine could cause flooding so the cold start fuel injection shuts off after running for approximately 20 seconds at 0 degrees Fahrenheit. The switch might allow cold start fuel delivery for about 5 seconds when the engine is warmer, about 32 degrees Fahrenheit. When the engine is about room temperature and the ignition is turned to START, there is no cold start fuel injection.

The auxiliary air device provides extra air during start and warm-up. This device is an idle bypass of the throttle that provides extra measured air. As a result, more injected fuel is provided and the idling engine has more power to overcome the increased friction of the cold engine. Figure 5-22 shows how the auxiliary air device opens up and closes off the bypass hose. When the engine is cold, the blocking plate opens the channel and extra air bypasses the throttle. A bimetal strip regulates the plate opening. As it warms up, it bends and the blocking plate is able to rotate as it is pulled by the return spring. As the temperature increases, the plate gradually blocks the opening until there is no auxiliary air.

Some engines, such as air-cooled, use lubricating oil to warm the auxiliary air device. The auxiliary air device is usually located on the block or cylinder head, however, so that it is warm when the engine is warm.

Usually the auxiliary air must be shut off sooner from a cold start than it would be by rising engine temperature above. An electric heating element powered from the circuit of the ignition switch heats the bimetal strip that moves the blocking plate directly. It is not controlled by the ECU but is powered continuously when the ignition key is set to the *run* position. The auxiliary air device also works independently of the cold start injector.

The passage opens for extra air when the engine is started cold. Once the engine is running and still cold, the heater begins operating to gradually close the open passage. It requires 8 minutes at the most, if the engine started at 0 degrees Fahrenheit. When a warm engine is started, the passage is closed and the normal amount of air is supplied for idle.

Closed Loop Feedback System

Manufacturers are required to control the metering of the air/fuel mixture closely because of regulations limiting the amounts of harmful substances in exhaust gases. The feedback or closed loop fuel control system is the method used by most manufacturers, whether the vehicle is equipped with a fuel injection system or carburetor.

The system includes an exhaust gas oxygen sensor in the exhaust manifold near the engine (Figure 5-23), additions to the ECU, and the connecting wires. The exhaust gas oxygen sensor is also called a *Lambda sensor.*

The system is called *feedback* or *closed loop* because the exhaust conditions provide the information needed to adjust the exhaust conditions.

FIGURE 5-22   The auxiliary air device controls an air bypass passage to increase idle speed during cold operation.

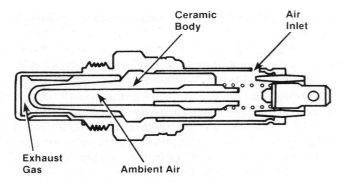

FIGURE 5-23 The Lambda or oxygen sensor tells the ECU if the fuel mixture is lean or rich.

The Lambda sensor measures the concentration of oxygen in the exhaust fumes. This information is passed to the ECU in the form of a variable voltage signal. When the incoming mixture is too rich, the oxygen content of the exhaust will be low, and the sensor will put out a high voltage signal. When the mixture is too lean, the sensor will put out a low voltage signal. The ECU uses this information along with the input from other sensors to determine and regulate the air/fuel mixture. Even when a mixture is only slightly lean or slightly rich, the change in oxygen sensor voltage is so quick and pronounced that it provides the ECU with a responsive and accurate method for monitoring the air/fuel ratio. Thus, the feedback system offers tight control over the air/fuel mixture so that the three-way converter is able to function properly reducing NO_x and oxidizing HC and CO.

Closed Loop Operation. The signals from the Lambda sensor are received by the ECU for processing and adjustment of the fuel injection pulse width. This adjustment changes the engine exhaust, which, in turn, changes the Lambda sensor signal. This closed loop control continually adjusts the fuel injection according to changing conditions so that the right air/fuel ratio is maintained.

There are a number of conditions that must be fulfilled before the closed loop is operated. For instance, if the closed loop is operating before the engine is warm, the ECU will take away the extra fuel that is added for the warm-up. In order for the oxygen sensor to send reliable signals, it must also be hot (above 570 degrees Fahrenheit). The engine must be between idle and full throttle, or else the ECU will remove the extra fuel. The loop is open, operating on preprogrammed instructions in the ECU, until all of these conditions are met. Signals are sent to the ECU, which decides when to switch from open loop to closed loop and back.

Open Loop Operation. The ECU ignores signals from the Lambda sensor during open loop operation and sends programmed signals for warm-up, idle, and full load. In this way, the ECU acts like it does on engines without Lambda control until the conditions for closed loop operation are met.

During engine warm-up, the engine might operate under an open loop condition for up to 10 minutes. The open loop condition might last only 10 seconds for a hot start after a brief stop. A loose connection to the coolant temperature sensor or slow-moving traffic can also cause an open loop condition.

Advantages and Disadvantages. When a feedback control unit is functioning properly, it offers accurate control of fuel injection to satisfy varying operating conditions. As well as making possible the use of the three-way converter for better emission control, it also improves power, economy, and driveability. However, the system tends to make certain engine malfunctions, such as a failed EGR or misfiring spark plug, difficult to defect. Not all malfunctions are detected by normal maintenance testing. Poor driveability, black smoke exhaust, and reduced fuel economy are warning signals that both operators and service technicians should be aware of. It should be noted, however, that a closed loop system will not cause hard starting, poor idle, or lack of full power.

Other Options

Coasting shutoff can be found on a number of engines. It can improve fuel economy as well as reduce emissions of hydrocarbons and carbon monoxide. Fuel shutoff is controlled in different ways depending on the type of transmission (manual or automatic) the car is equipped with. The ECU makes a coasting shutoff decision based on two input signals:

1. A closed throttle as indicated by the idle switch
2. Engine speed indicated by the signal from the ignition coil

Fuel injection and turbochargers are well suited for each other because the manifold system conveys only air as opposed to the typical air/fuel mixture. This offers flexibility in the installation of the turbocharger and the airflow control sensor unit.

As the car operates at higher altitudes, the thinner air needs less fuel. Altitude compensation in a fuel injection system is accomplished by installing a sensor to monitor barometric pressure. Although the oxygen sensor can compensate for altitude to a slight degree, it can reach the limit of its correction scale, usually about 20 percent on either side of its setting. Signals from the barometric pressure sensor are sent to the ECU to reduce the injector pulse width (or reduce the amount of fuel injected).

GASOLINE CONTINUOUS FUEL INJECTION SYSTEM

A continuous fuel injection system (CIS) is sometimes contrasted with an electronic fuel injection system and described as "mechanical" because its basic operation is not electronically controlled. The CIS uses fuel under pressure to modulate or change the fuel injection rate. Because of this, it is sometimes referred to as a "hydraulic" system. In later systems, some electronics were used to modify the basic mechanical fuel-metering function, but not enough to call them electronic fuel injection systems.

The CIS delivers fuel continually to the intake manifold and maintains a constant relative fuel pressure. The fuel delivery rate is not varied by injector pulse width, as in an electronic system, but by varying the amount of fuel coming from the injector.

Figure 5-24 shows the operation of the CIS. Air entering the system through the air filter is measured by an airflow sensor. The driver controls the airflow using a regular throttle valve. The air travels through the intake tubes into the combustion chamber.

An electric fuel pump delivers fuel through a fuel accumulator and fuel filter. The movement of the airflow sensor in the mixture control unit controls the amount of fuel delivered through the injectors. When the intake valve is open, the injected fuel in the intake manifold is delivered to the combustion chamber by intake air. The fuel vapor is stored in the intake manifold if the intake valve is closed.

Figure 5-24 shows the air entering beneath the airflow sensor plate. As the air moves toward the throttle and intake manifold, it lifts the sensor plate. This is called an *updraft system.* In a down draft system, however, the movement of intake air presses down on the airflow sensor plate.

Fuel System

The fuel system for CIS is similar to that for an electronic system. It includes the fuel tank, fuel

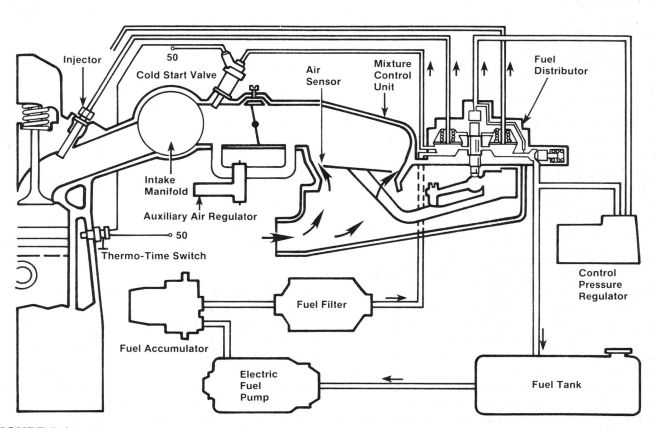

FIGURE 5-24   A continuous flow system that can be considered mechanical or hydraulic in operation.

pump, accumulator, and filter. The principal difference between the two fuel systems is the control pressure regulator.

The regulator includes a push valve, as shown in Figure 5-25. The regulator with its push valve controls system pressure with the engine running. However, it also retains fuel pressure in the connecting lines when the engine is turned off. This will ensure quick restarting. It also means that the fuel pressure must be relieved before opening any of the fuel lines. Failure to relieve this pressure can cause fuel to be sprayed around the engine compartment, possibly causing a fire or personal injury.

The fuel system also includes a return line to the fuel tank and associated fuel relays to control the fuel pump during starting and running. The relays also shut off the pump, for safety reasons, if the engine should stop with the ignition on.

Airflow Movement

The cone shape of the air funnel in the mixture regulator causes the airflow sensor plate to rise farther when more air flows into the engine. The movement of the sensor plate is used to measure how much air is entering the system.

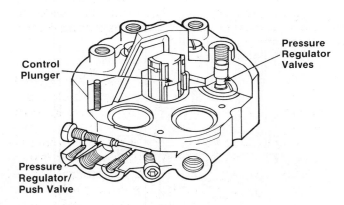

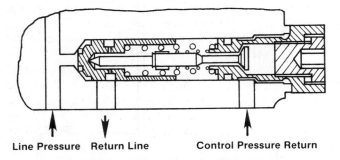

LINE PRESSURE REGULATOR, FUEL PUMP IDLE

FIGURE 5-25 The push valve maintains line pressure with the engine off to ensure a fast start.

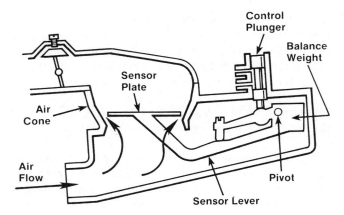

FIGURE 5-26 The CIS airflow sensor is coupled directly to the control plunger in the mixture regulator.

If any air enters the intake system without passing the sensor plate, the proper air/fuel mixture will not be obtained and the engine will run lean. Just as in the electronic system, the CIS must be free of any vacuum leaks for proper operation.

As shown in Figure 5-26, the airflow sensor is part of the mixture control unit. The sensor plate, moved by incoming air, raises the control plunger in the fuel distributor. This action controls how much fuel is injected. Although the fuel distributor resembles an ignition distributor, it does not rotate, distributing fuel to the cylinders in firing order. The unit sends continuously the same amount of fuel to all cylinders at the same time.

The sensor lever moves the control plunger. Hydraulic counterpressure is applied to the top of the plunger by pressurized fuel. Since 1983, a small return spring has been used to help the plunger follow the rapid movements of the sensor plate.

Fuel Metering

As shown in Figure 5-27, the control plunger on the mixture regulator moves up and down in a barrel with vertical metering slits. There is one slit in the barrel for each cylinder in the engine. Each slit is approximately 0.2 mm wide.

Figure 5-27 also shows how the plunger movement is used for metering the fuel. In the first position, the plate is raised only slightly by the small airflow at idle. Thus, the control plunger is raised slightly and the metering slit is opened to allow only enough fuel for idle. More of the metering slit is exposed as the airflow increases until, at full throttle, the fuel flow is at maximum.

While the fuel control plunger must fit snugly in the barrel to minimize leaks, it must still move freely enough to respond to the smallest movements of the sensor plate. Even a small particle of dirt can jam the control plunger. When this occurs, the sensor plate is locked in place, and the engine cannot respond to the variations in airflow.

Each injector has its own differential pressure valve (Figure 5–28) within the mixture control unit. By maintaining a constant fuel pressure difference across the metering port, these valves maintain equivalent fuel flows at each injector.

A steel diaphragm separates the two chambers of each differential pressure valve. Inside the valve, the spring presses down on the diaphragm. At the same time, the fuel system pressure is pressing up on the diaphragm from underneath. The diaphragm must be deflected slightly against fuel system pressure for the fuel to flow to the injector.

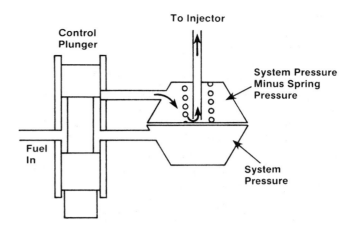

FIGURE 5–28   A pressure regulator valve for each cylinder maintains a constant pressure difference on either side of the slit.

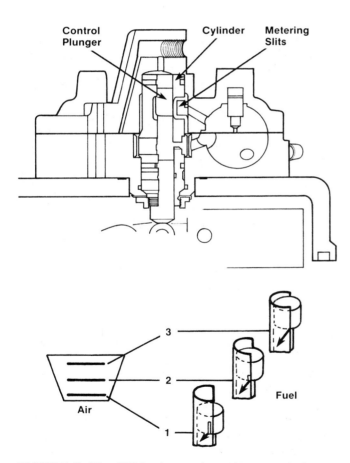

FIGURE 5–27   CIS fuel metering: (A) control plunger is raised and lowered in a barrel that has one metering slit for each cylinder of the engine; (B) as the plunger is raised in response to increased airflow, it uncovers more of the metering slit.

These valves are not adjustable. Their operation allows the control plunger to control fuel volume precisely without being affected by pressure changes.

In a CIS without Lambda control or feedback, the system pressure regulator maintains constant pressure in the lower part of each differential pressure valve. In systems with auxiliary electronics such as Lambda control, this pressure is adjusted to modify the amount of fuel that flows to the injector.

Fuel Pressures

The CIS operates on fuel pressure at several different levels. The system pressure, which is determined by the fuel pressure regulator, is 65 to 75 psi.

When the engine is warm, control pressure is reduced from system pressure by 50 to 55 psi. When the engine is cold, control pressure is reduced by 19 to 24 psi. When under full load, or under turbo boost, control pressure decreases to enrich the mixture. Injection pressure is 51 to 60 psi and nonadjustable.

Control Pressure Regulator.   In a CIS, the control pressure regulator helps to regulate the air/fuel mixture by determining the fuel pressure on top of the control plunger. The fuel pressure applied to the top of the control plunger balances the force from the airflow sensor. This control pressure holds the control plunger down, which, in turn, limits how much fuel passes through each metering slit to its injector.

In order to change the amount of fuel delivered to the injectors, the control pressure on top of the control plunger must be changed. By lowering the control pressure, the control plunger can rise farther

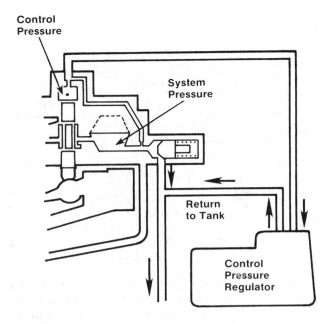

Control Pressure

System Pressure

Return to Tank

Control Pressure Regulator

FIGURE 5-29 Altering the control pressure will affect regulation of the air/fuel mixture.

under pressure from the airflow sensor. As a result, more fuel is able to pass to the injectors, and the mixture is enriched.

Cold Start Provisions. Mixture adaption in the CIS, as for cold start operations, is similar to the electronic fuel injection system. Refer to the electronic fuel injection system description for cold start operation (includes the cold start valve, thermotime switch, coolant temperature swtich, and auxiliary air device).

Warm-Up Enrichment. The control pressure regulator, the device used to change control pressure, is usually located on the engine block. Thus, it is sensitive to engine temperature. It is sometimes called a *warm-up regulator* because its only use in some engines is to enrich the mixture

during warm-up. In order to reduce control pressure, the control pressure regulator increases the return flow to the tank. The control pressure regulator and the mixture control unit are connected by hoses. The valve spring and bimetal strip resist the opening of the valve diaphragm. The bimetal strip is electrically heated by a resistance coil (Figure 5-30). In this manner, the warm-up regulation may be accurately related to specific engine characteristics.

When the engine is cold, the bimetal strip exerts downward pressure against a spring-loaded valve (Figure 5-31). This causes the valve to open a wider fuel return passage in the control pressure regulator, allowing more fuel to return to the fuel tank. This reduces the fuel pressure above the control plunger in the fuel distributor. As a result, the plunger is allowed to rise higher and admit more fuel, which makes a richer mixture. The colder the engine, the lower the control pressure and the higher the plunger is allowed to rise, enriching the mixture.

As the engine warms up, the bimetal strip is heated by an electric heating coil. As the bimetal strip rises, it gradually releases pressure on the spring, so the spring raises the valve diaphragm and thereby reduces the return flow (Figure 5-32). As a result, the control pressure on top of the plunger gradually increases, and the airflow sensor plate cannot lift the control plunger as high. The amount of fuel delivered to the injectors is reduced, which causes the mixture to become more lean as the engine warms up.

The control pressure regulator cannot be adjusted. If there is a low reading when warm, a break might have occurred in the ground circuit or current supply. A restricted return line could result in a high reading. The control pressure regulator should be replaced when any other incorrect readings occur.

Full Load Enrichment. While some control pressure regulators have the capacity only to enrich the mixture for the engine during warm-up,

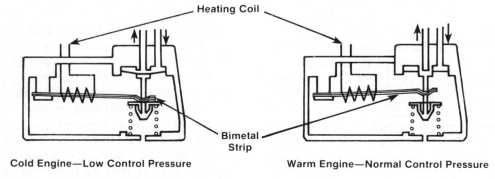

Heating Coil

Bimetal Strip

Cold Engine—Low Control Pressure   **Warm Engine—Normal Control Pressure**

FIGURE 5-30 The cold pressure regulator reduces control pressure during cold engine operation to create a richer mixture.

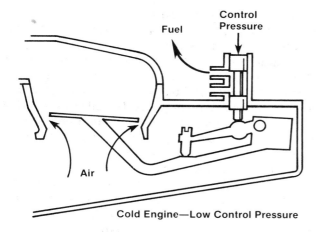

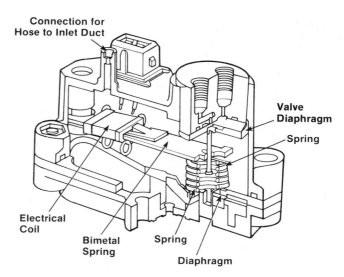

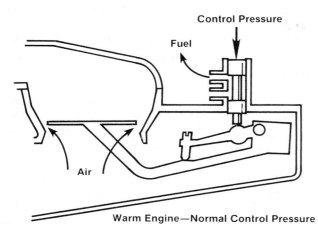

FIGURE 5-31 Lower control pressure allows the control plunger to rise further, exposing more of the metering slit and allowing more fuel flow.

FIGURE 5-32 Some control pressure regulators are connected to the intake manifold vacuum to allow mixture enrichment during full throttle operation.

others are able to provide full load enrichment as well.

Vacuum from the intake manifold is applied through a port, and air is drawn from the control pressure regulator during normal part load cruise. The atmospheric pressure applied to the bottom of the diaphragm presses it up, causing the valve diaphragm to shut. Thus, the control pressure increases, and the mixture is made lean. Opening the throttle to full load increases the manifold absolute pressure, which presses the diaphragm down. As a result, the valve diaphragm is lowered. This reduces the control pressure and enriches the fuel mixture.

The control pressure regulator can also enrich the mixture when manifold absolute pressure increases due to boost from a supercharger or turbocharger. It also has the ability to compensate for thinner air at high altitude.

Fuel Injectors. The fuel injectors (Figure 5-33) are pressed into their manifold openings and

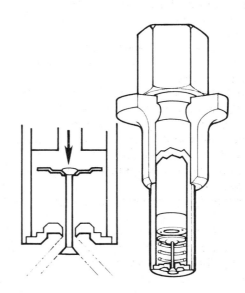

FIGURE 5-33 Typical CIS fuel injector

sealed with O-rings. Each injector has an internal fuel filter to insure that clean fuel reaches the injector opening.

Whenever the engine is running, these injectors are continuously spraying finely atomized fuel into the intake port. The vibrating valve needle makes a chattering noise when the valve is operating properly. The vibrations of the valve needle atomize fuel. When the engine is shut off, spring pressure tightly closes the pin to trap the fuel under pressure in the line. This prevents vapor lock and makes a quick start possible the next time.

Lambda Control Feedback System

Continuous injection systems can be fitted with a Lambda sensor (also called an *oxygen sensor*) for feedback control (Figure 5–23). The Lambda sensor is located in the exhaust manifold so that it heats up quickly when the engine is started.

The Lambda control unit receives signals from the Lambda sensor and adjusts the fuel flow in the mixture control unit. This is the manner in which the proper air/fuel ratio for engine operation is maintained. Figure 5–34 shows this closed loop operation. The Lambda sensor, affected by changes in the exhaust gas, sends a signal in a loop through the mixture control unit to the engine.

Figure 5–35 shows the mixture control unit for Lambda operation. A bleed passage to the Lambda control valve is connected to the lower chamber of each differential pressure valve.

Operating on signals from the Lambda control unit, the Lambda control valve opens and closes, controlling the amount of fuel returned to the tank. Lengthening the time the Lambda control valve is open reduces the amount of pressure in the lower part of the differential pressure valve. This increases the fuel flow to the injector and enriches the mixture. When the open time of the Lambda control valve decreases, the fuel flow decreases, and the mixture is made leaner.

The amount of time the Lambda control valve is open is called the *dwell time.* The Lambda control valve is usually open about 50 percent of the time. Based on a series of signals from the sensor from oxygen-rich to oxygen-lean, the control valve continually cycles from an open time of about 40 percent to 60 percent.

When the Lambda sensor is cold or when the engine is cold, the Lambda control unit switches to the open loop mode. During open-loop operation, the Lambda control valve is held open for a fixed amount of time, usually about 60 percent. The change in sound caused by the change in open time can be heard while testing the operation of the Lambda control unit and its control valve.

On those models equipped with Lambda control, the CO output level can be adjusted in one of two ways. First, the Lambda sensor wire can be disconnected and the exhaust sample taken at the pipe provided on the exhaust manifold. This requires drill-

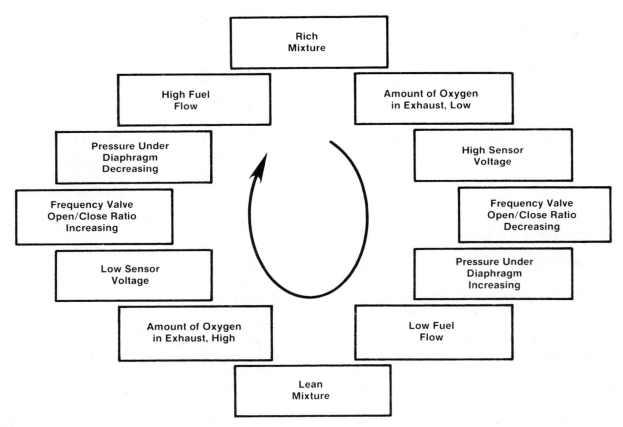

FIGURE 5–34 Closed loop operation refers to the way various components respond to each other.

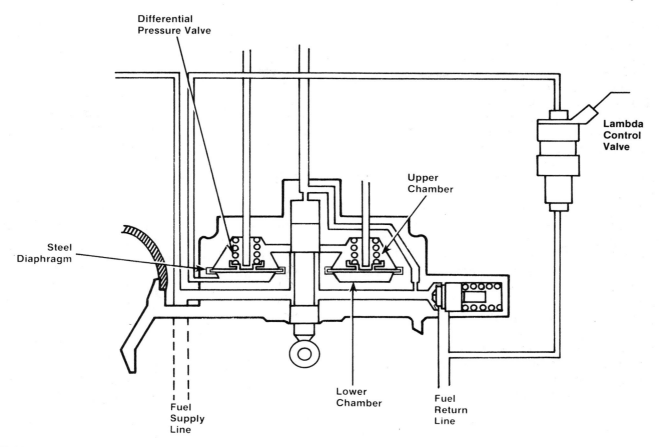

FIGURE 5–35   The frequency valve lowers the pressure in the lower portion of the pressure regulator valves to alter the fuel quantity to the injectors in response to the oxygen sensor.

ing and removing the tamper-proof plug (if there is one) from the mixture control unit. The alternative is to use a dwell meter and adjust the mixture in closed-loop operation with the Lambda sensor still connected. Additional instructions can be found on the underhood decal.

REVIEW QUESTIONS

1. Which of the following fuel injection system components is comparable in function to the carburetor's accelerator pump? *pg. 170*
   a. injector valve
   b. throttle switch
   c. inlet manifold pressure sensor
   d. electronic control unit

2. In which system are air and fuel measured separately? *pg. 170*
   a. injection system
   b. carburetor system

   c. exhaust system
   d. all of the above

3. Which of the following is an advantage of an injection system over a carburetor system?
   a. Saves fuel.
   b. Compensates for differing altitudes.
   c. Less complex than a carburetor system.
   d. All of the above   *pg. 170*

4. Which of the following is a mechanical type of fuel injection?
   a. EFI
   b. CIS   *pg. 171*
   c. ECU
   d. none of the above

5. Which component of the continuous injection system is comparable in function to the electronic control unit in an electronic fuel injection system?
   a. sensors
   b. monitor

c. control pressure regulator
d. voltage signal

6. The interval of time that each injector stays open is known as _____ .
   a. injector pulse width
   b. vacuum parameters
   c. intake manifold pressure
   d. open loop

   pg-176

7. A vehicle has a problem with the amount of fuel pressure being delivered to each injector. Technician A checks the intake manifold. Technician B checks the fuel pressure regulator. Who is right?
   a. Technician A
   b. Technician B
   c. Both A and B
   d. Neither A nor B

   pg 189

8. What is the number one enemy of fuel injection systems?
   a. cold weather
   b. unleaded fuel
   c. dirt
   d. long-distance driving

   pg174

9. Fuel injection systems that deliver fuel at a central point are called _____ systems.

a. port-type
b. multi-point
c. throttle body
d. all of the above

pg. 175

10. What component is a mechanical assembly that contains or houses the fuel injectors?
    a. port injector
    b. fuel regulator
    c. fuel distributor
    d. fuel rail assembly

    pg 178

11. What sensor in a closed loop feedback system measures the oxygen concentration in the exhaust gases and passes this information to the ECU via a voltage signal?
    a. oxygen sensor
    b. lambda sensor
    c. both A and B
    d. neither A nor B

    pg 190

12. The control pressure in a vehicle with a continuous injection system is not changing with temperature changes. Technician A claims the problem lies with the control pressure regulator. Technician B says the problem is the warm-up regulator. Who is right.
    a. Technician A
    b. Technician B
    c. Both A and B
    d. Neither A nor B

CHAPTER SIX

THROTTLE BODY AND PORT FUEL INJECTION SYSTEMS

Objectives

After reading this chapter, you should be able to:
- Explain the operation of a throttle body fuel injection system.
- Identify the major parts of a throttle body fuel injection system and explain the function of each part.
- Explain the operation of a port fuel injection system.
- Identify the major parts that are unique to port fuel injection and those that are common to both throttle body and port injection systems.
- Identify and describe the physical and operating differences among the fuel injection systems marketed by major domestic car manufacturers.
- Explain the advantages and disadvantages of both throttle body and port injection.

This chapter covers the operation of basic throttle body and port fuel injection. Specific injection systems are also described to show how designs vary among domestic car manufacturers.

THROTTLE BODY INJECTION SYSTEM

The throttle body injection system distributes fuel in such a manner that exhaust emissions from vehicles are controlled within legislated limits. The system accomplishes this by precisely controlling the air/fuel mixture for all engine operating conditions and by providing for complete combustion of the fuel. The heart of the system is an electronic control unit (ECU) that receives electrical signals from strategically located engine sensors and outputs electrical signals to control fuel delivery and combustion. The ECU is basically an on-board microcomputer (Figure 6–1). It uses information from the sensors to vary the amount of fuel delivered to the injector thereby attempting to achieve the ideal air/fuel ratio of 14.7:1. This ratio allows the three-way catalytic converter to efficiently reduce exhaust emissions and increase fuel economy at the same time.

This system uses a throttle body assembly that is mounted on the intake manifold. The assembly usually contains one or two fuel injectors. The ECU signals the injector in the throttle body to provide the correct fuel quantity for the given range of operating parameters.

The throttle body injection (TBI) assembly is located on the intake manifold at a central point where fuel and air are distributed through a single bore in the throttle body. Air for combustion is metered or controlled by a single throttle valve, which is connected directly to the accelerator pedal linkage by a throttle shaft and lever assembly. A plate, located beneath the throttle valve, helps to distribute the mixture. The fuel needed for combustion is supplied by a single fuel injector, which is mounted on the throttle body assembly with its metering tip lo-

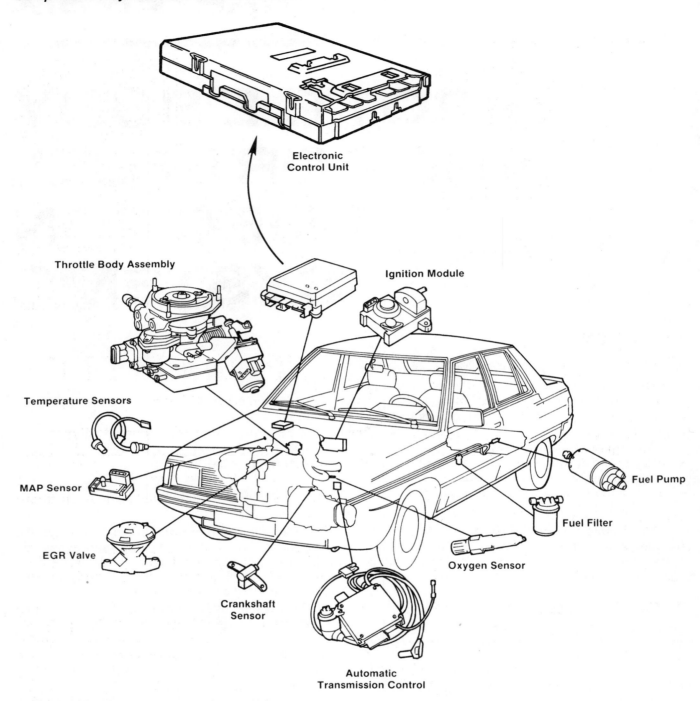

FIGURE 6-1 Electronic control unit (ECU)

cated directly above the throttle valve (Figure 6-2). The ECU pulses the injector on or off (open or closed) for the appropriate operating condition—cold starting, cranking, acceleration, deceleration, and so on.

When the ignition key is turned on, the ECU energizes the fuel pump to pressurize the fuel system. The distributor provides reference signals or

pulses to the ECU when the engine is started and turning. These reference pulses represent engine speed. If the ECU does not receive the pulses within a specified time, it de-energizes the fuel pump. If the reference pulses are received at a later time, the ECU will turn on the fuel pump again.

The pulsing signals for the fuel injectors can be sent by the ECU in two modes: synchronized and

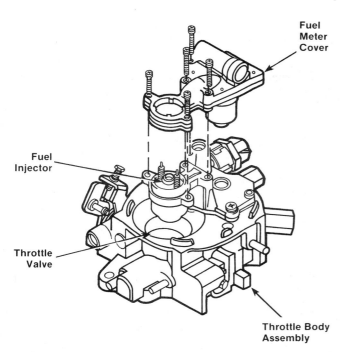

FIGURE 6-2 Throttle body assembly

nonsynchronized. When the ECU is in the synchronized mode, the fuel injector is pulsed (or opened) once for each received distributor reference pulse. If a dual throttle body system is employed, then the injectors are alternately pulsed on and off (figures 6-3 and 6-4).

In the nonsynchronized mode of operation, the fuel injector is pulsed once every 6.25 to 12.5 milliseconds. This pulse time is totally independent of the incoming distributor reference pulses to the ECU (figures 6-5 and 6-6). The nonsynchronized

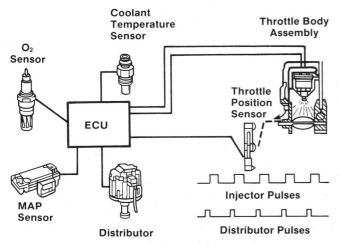

FIGURE 6-3 Synchronized mode single throttle body system

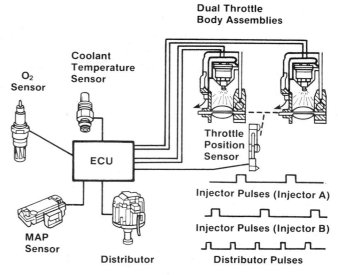

FIGURE 6-4 Synchronized mode dual throttle body system

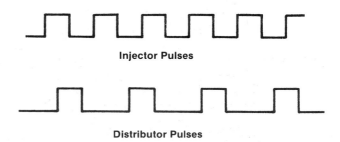

FIGURE 6-5 Nonsynchronized mode single throttle body system

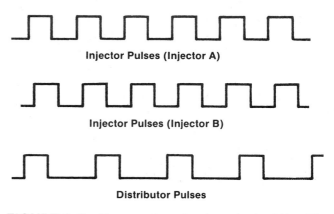

FIGURE 6-6 Nonsynchronized mode dual throttle body system

mode of operation is implemented by the ECU only when the following conditions exist:

• Fuel pulse width is too small to be delivered accurately by the injector.

- During deceleration lean-out
- During acceleration enrichment

Under all other engine operating conditions, the ECU implements the synchronized mode. Such engine operating conditions include cranking, clearing a flooded condition, and normal running mode.

RUNNING MODES OF OPERATION

There are two types of running modes: open loop and closed loop. When the engine is first started and the rpm is well above 600, the TBI system goes into open loop operation. In this mode, the ECU does the following:

- Ignores the receipt of any signals from the oxygen sensor
- Uses input from the coolant temperature sensor and the MAP sensor to calculate the injector-pulsed on-time (amount of fuel enrichment)

The ECU determines when the TBI system is ready to go to closed loop operation by monitoring the following:

- Coolant temperature is above a specified level
- The output signal from the oxygen sensor varies
- A specific amount of time has elapsed after the engine has been started

When these conditions have been attained, the TBI system switches to the closed loop operating mode. In this mode, only signal data from the oxygen sensor is used by the ECU to vary the injector-pulsed on-time. The on-time will be decreased if the air/fuel ratio is too rich and the on-time will be increased if the air/fuel ratio is too lean.

The injector-pulsed on-time is determined by changes in the coolant temperature, manifold pressure, and throttle angle. The wider the opening of the throttle and the higher the manifold pressure, the longer the pulsed on-time.

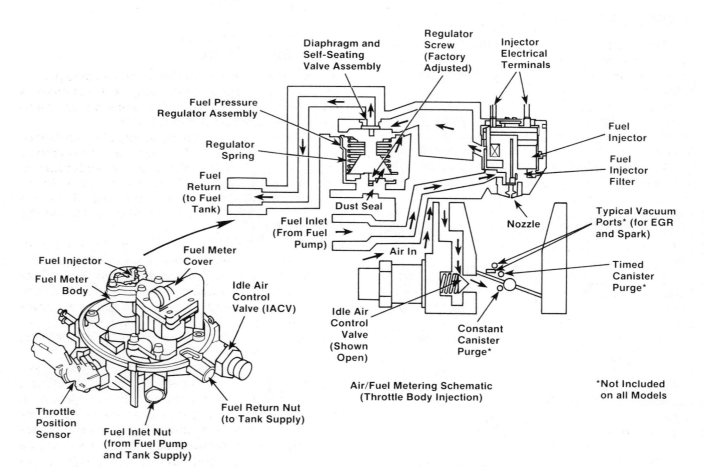

FIGURE 6-7 Throttle body assembly

Accelerating the engine opens the throttle valve and causes a rapid increase in manifold absolute pressure (MAP). The rapid increase in MAP causes fuel to condense on the walls of the manifold. The ECU senses the increase in MAP and throttle angle and supplies additional fuel for a short time to prevent the engine from stalling. Any reduction in throttle angle will cancel the enrichment pulsing of the injector. This prevents overenriching the mixture if the accelerator is subjected to rapid movement.

Upon deceleration, a leaner fuel mixture is mandatory to reduce exhaust emissions. The ECU adjusts the injector on-time during deceleration by using the decrease in MAP and throttle angle to calculate a decrease in injector-pulse on-time.

Sometimes a deceleration fuel cutoff feature is used to remove fuel from the engine when extreme deceleration conditions exist. This feature momentarily cuts off fuel delivery to the TBI system and overrides the deceleration lean-out mode. The ECU determines deceleration fuel cutoff by evaluating manifold pressure, throttle position, and engine rpm.

THROTTLE BODY ASSEMBLY

The basic throttle body assembly consists of two major castings: a throttle body with a valve to control airflow and a fuel body to supply the required fuel. A pressure regulator and fuel injector are integral parts of the fuel body (Figure 6–7). Also included as part of the assembly is a device to control idle speed and one to provide throttle valve positioning data.

The throttle body casting has ports that can be located above, below, or at the throttle valve depending on the manufacturer's design. These ports generate vacuum signals for the manifold absolute pressure sensor and for devices in the emission control system, such as the EGR valve, the canister purge system, and so on.

The pressure regulator used on the throttle body assembly is similar to a diaphragm-operated relief valve. The fuel injector pressure bearing is on one side of the diaphragm and the air cleaner pressure is on the other side. The regulator is designed to provide a constant pressure on the fuel injector throughout the range of engine loads and speeds (Figure 6–8). If regulator pressure is too high, a strong fuel odor is emitted, and there is a chance that detonation could take place. On the other hand, regulator pressure that is too low results in poor engine performance.

The fuel injector is solenoid operated and pulsed on and off by the ECU. Surrounding the injector inlet is a fine screen filter where the incoming fuel is directed (see Figure 6–7). When the injector's solenoid is energized, a normally closed ball valve is lifted. Fuel under pressure is then injected at the walls of the throttle body bore just above the throttle valve. Excess fuel is returned to the fuel tank via the pressure regulator.

IDLE AIR CONTROL SYSTEM

The idle air control (IAC) system controls engine idle speeds and prevents changing engine loads from stalling the engine. Mounted on the throttle body assembly, the IAC controls the volume of bypass air around the throttle plate. The amount of air that moves around the throttle plate is regulated by extending or retracting a conical valve (figures 6–9 and 6–10). If rpm is too low, the valve allows more air to be diverted around the throttle plate, which increases engine revolutions.

When the engine is at idle, the correct position of the IAC valve is determined by the ECU. The valve positioning signal from the ECU is based on the following:
- Engine load
- Engine rpm
- Coolant temperature
- Battery voltage

When a condition arises in which the throttle plate is closed and the speed of the engine drops below a predetermined rate, the ECU will immediately calculate a new position for the IAC valve.

The IAC motor has a capability of extending or retracting the valve in 255 positions. At the zero or reference position, the valve is fully extended and no air is permitted to bypass the throttle plate. At the 255th position, the valve is fully retracted and the maximum amount of air is permitted to bypass the throttle plate.

The ECU calculates or "knows" the exact position of the IAC motor because it always monitors the number of positions the valve has been extended or retracted. Once the engine reaches a specified rpm, the ECU fully extends the IAC valve from whatever position it was in. The ECU then refers to this position as "0" and thereby keeps its zero reference updated.

The IAC system affects only engine idle performance. If the IAC valve malfunctions when closed, the idle speed will be too low (insufficient airflow). If it malfunctions when open, engine idle speed will be too high (excessive airflow). If the valve malfunctions somewhere between open and closed, engine idle will be rough and, consequently, will not respond to idle load changes.

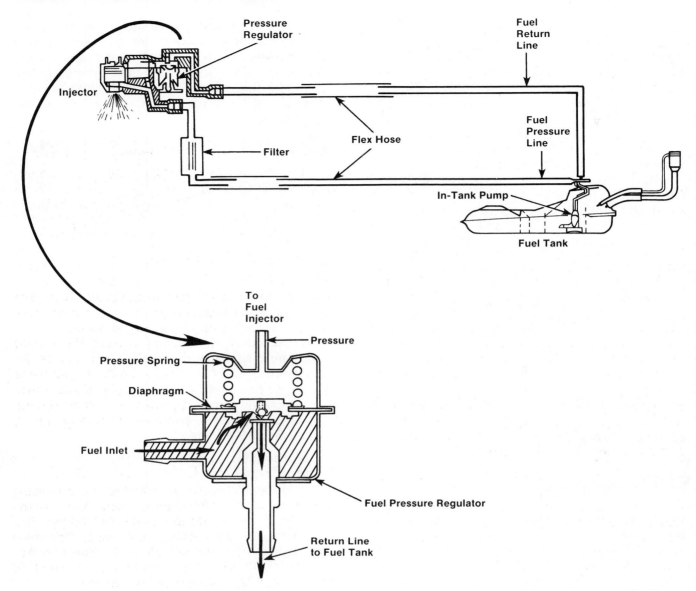

FIGURE 6-8 TBI fuel pressure regulator and circuit

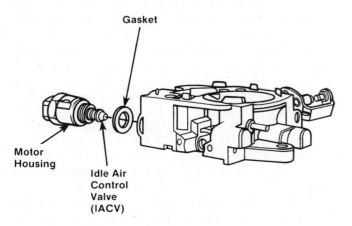

FIGURE 6-9 Idle air control assembly

IDLE SPEED CONTROL

Incorrect diagnosis and/or misunderstanding of the idle speed control systems used on EFI engines sometimes lead to unnecessary replacement of the IAC valve.

Engine idle speed is controlled by the ECU, which changes idle speed by moving the IAC valve. The ECU adjusts idle speed in response to fluctuations in engine load (air conditioning, power steering, electrical load, and so on) to maintain acceptable idle quality and proper exhaust emission performance. For example, as engine load increases and idle speed decreases, the ECU moves the IAC valve to maintain proper idle speed. After engine

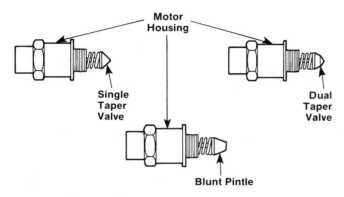

FIGURE 6-10 IAC valve designs

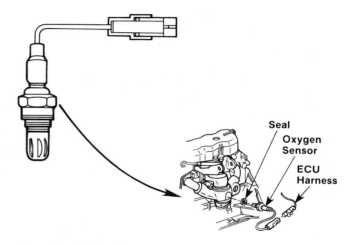

FIGURE 6-12 Oxygen sensor

load is reduced, the ECU returns the idle speed to the proper level. The ECU also increases idle speed whenever the air conditioner compressor is cycling or high-power steering loads are encountered.

During cold starts idle speed can be as high as 2100 rpm to quickly raise the temperature of the catalytic converter for proper exhaust emissions. Idle speed that is attained after a cold start is controlled by the ECU. The ECU will maintain idle speed for approximately 40 to 50 seconds even if the driver attempts to alter it by "kicking" the accelerator. After this preprogrammed time interval, depressing the accelerator pedal rotates the throttle position sensor (TPS) and signals the ECU to reduce idle speed.

DATA SENSORS

Numerous on-board sensors provide information to the ECU that relate to the operating characteristics of the engine. These sensors include engine coolant, oxygen, manifold absolute pressure, and throttle position.

Engine Coolant Sensor

The coolant sensor (Figure 6-11) is a thermister (a resistor that changes value based on temperature) mounted in the engine coolant stream. As the temperature of the engine coolant changes, the re-

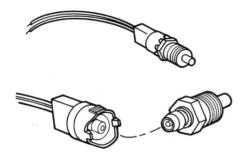

FIGURE 6-11 Typical coolant sensor

sistance of the coolant sensor changes. Low coolant temperature produces a high resistance; high coolant temperature produces a low resistance.

The ECU supplies a voltage signal to the coolant sensor and measures the voltage that returns. By measuring the voltage change, the ECU determines the engine coolant temperature. This information is used to control fuel management, IAC, spark timing, EGR, canister purge, and other engine operating conditions.

Oxygen Sensor

The oxygen sensor is mounted in the exhaust system where it monitors the oxygen content of the exhaust gas stream (Figure 6-12). The oxygen content in the exhaust reacts with the sensor to produce a voltage output. This voltage ranges from approximately 100 millivolts (high oxygen, lean mixture) to 900 millivolts (low oxygen, rich mixture).

By monitoring the voltage output of the oxygen sensor, the ECU determines what fuel mixture command to give to the injector (lean mixture: low voltage, rich command; rich mixture: high voltage, lean command).

Remember that the oxygen sensor indicates to the ECU what is happening in the exhaust. It does not cause things to happen. It is a type of gauge: a high oxygen content equals a lean mixture; a low oxygen content equals a rich mixture. The ECU adjusts fuel to keep the system working.

Manifold Absolute Pressure Sensor

The MAP sensor measures changes in the intake manifold pressure that result from changes in engine load and speed (Figure 6-13).

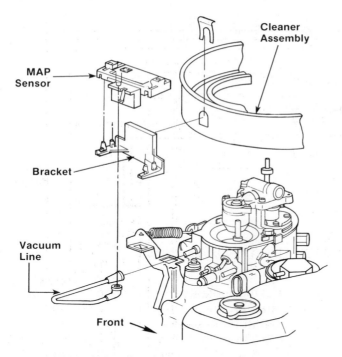

FIGURE 6-13 MAP sensor

The pressure measured by the MAP sensor is the difference between barometric pressure (outside air) and manifold pressure (vacuum). A closed throttle engine produces a low MAP value; a wide-open throttle produces a high value. This high value is produced when the pressure inside the manifold is the same as pressure outside the manifold, and 100 percent of the outside air is being measured. This MAP output is the opposite of what is measured on a vacuum gauge. The use of this sensor also allows the ECU to adjust automatically for different altitudes.

The ECU sends a voltage reference signal to the MAP sensor. As the MAP changes, the electrical resistance of the sensor also changes. The ECU can determine the manifold pressure by monitoring the sensor output voltage. A high pressure, low vacuum (high voltage) requires more fuel; a low pressure, high vacuum (low voltage) requires less fuel.

Throttle Position Sensor

The throttle position sensor (TPS) is connected to the throttle shaft and is controlled by the throttle mechanism (Figure 6-14). A voltage reference signal is sent to the TPS by the ECU. As the throttle valve angle is changed (accelerator pedal moved), the resistance of the TPS also changes. At a closed throttle position, the resistance of the TPS is high, so the output voltage to the ECU will be low. As the throttle plate opens, the resistance decreases so that at wide-open throttle, the output voltage should be approximately 5 volts.

By monitoring the output voltage from the TPS, the ECU can determine fuel delivery based on throttle valve angle (driver demand).

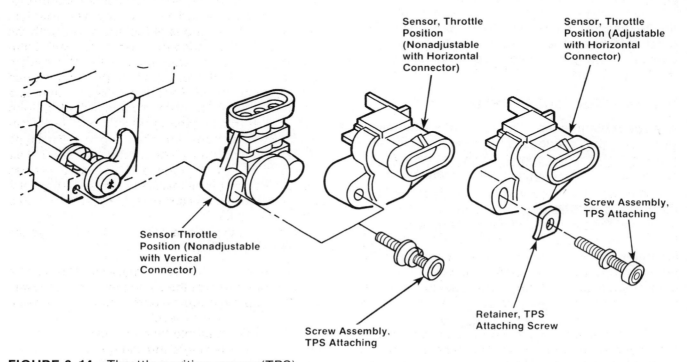

FIGURE 6-14 Throttle position sensor (TPS)

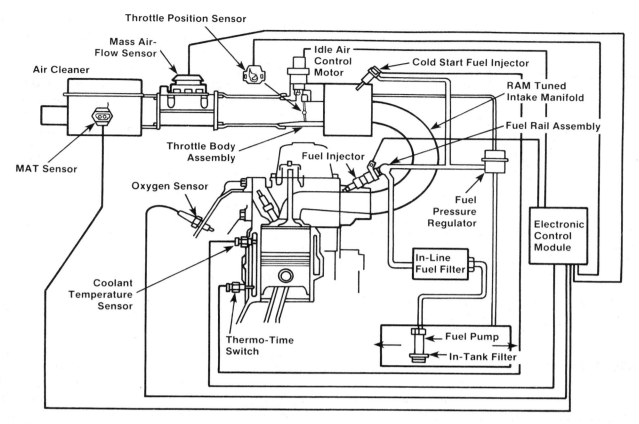

FIGURE 6-15 Typical port fuel injection system

The following symptoms can indicate a misadjusted, open, shorted, or loose TPS:

- Poor idle or wide-open throttle (misadjusted TPS)
- Constantly closed throttle (open TPS)
- Constantly wide-open throttle (shorted TPS)
- Intermittent bursts of fuel from the injector and unstable idle speed (loose TPS)

PORT FUEL INJECTION SYSTEM

The port fuel injection (PFI) system differs from the throttle body injection system in that it uses individual injectors, one at each intake port, rather than one or two injectors located in a throttle body assembly. Some PFI systems use an additional injector to supply an enriched fuel mixture to all cylinders simultaneously during cold start-ups. This additional injector is referred to as the cold start injector.

A schematic of a typical port fuel injection system is shown in Figure 6-15. This type of injection system is also controlled by an on-board computer,

the ECU. The main attraction of a PFI system is its simultaneous double-firing fuel injection feature. All fuel injectors are pulsed once per engine revolution. Therefore, two injections of fuel are mixed with the incoming air to produce the mixture for each combustion cycle. Other variations of the PFI system include sequential fuel injection (SFI) and tuned port injection. The SFI variation pulses the injectors in a sequential order, normally in spark plug firing order. Tuned port injection takes a qualitative approach to the composition of the mixture. This type of injection makes the air/fuel ratio a function of air mass in place of air volume. This way a precise measurement of the air mass is achieved through the use of a hot wire, mass airflow (MAF) sensor located in the intake duct.

The advantages of a port fuel injection system include the following:

1. Improved fuel economy: Axle ratios can be optimized at the same performance level.
2. Improved vehicle performance at the same fuel economy level
   - No manifold fill required.
   - Large single-throttle bore.
   - Increased torque output.

3. Improved driveability—high technology
4. Improved emission performance in air/fuel distribution
   - No choke cycle
   - Leaner operation during warm-up
   - Improved airflow sensing
   - Improved transient fuel control
     —Three-way catalytic converter
   - Primary emission control system
5. Increased torque output
   - Ram tuning provides a denser cylinder charge.
   - Lower mixture temperatures increase cylinder charge density.

Some of the devices and components used in the typical PFI system are identical to those used in throttle body injection. Devices such as the pressure regulator, ECU, and the multitude of sensors are common to both system types. Only those devices or principles of operation that are unique to the PFI system will be covered here.

FUEL INJECTORS

Although common to both types of injection systems, the PFI fuel injector is installed in the intake manifold at each cylinder (Figure 6–16). The injector is mounted so that it has a nozzle spray pattern that is on a 25-degree angle. Two O-ring seals are used on the injector; the lower ring seals the intake manifold injector interface and the upper ring seals the injector fuel line interface (Figure 6–17). The O-rings also provide thermal insulation and prevent the formation of vapor bubbles and excessive injector vibration. Any air leakage at the injector intake manifold interface creates a lean cylinder condition, which causes driveability problems.

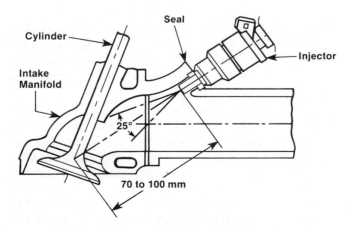

1-Pulse at Crankshaft Revolution

FIGURE 6–16 Typical fuel injector installation

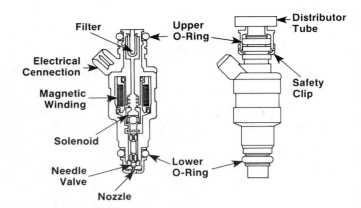

FIGURE 6–17 Typical fuel injector mechanical arrangement

The injector is a solenoid-operated device that consists of a valve body and nozzle body. The nozzle body has a specially ground pintle. At the rear of the valve body is the solenoid winding. In the front of the valve body is the guide for the nozzle valve. A movable armature is attached to the nozzle valve that is compressed against the nozzle body sealing seat by a helical spring.

FUEL RAIL ASSEMBLY

The fuel rail assembly on a PFI system usually consists of a left- and right-hand rail assembly. The two rails can be connected either by crossover and return fuel tubes or by a mechanical bracket arrangement. Typical fuel rail arrangements are shown in figures 6–18 and 6–19.

The fuel rail assembly shown in Figure 6–18 has the pressure regulator straddling both rails. Inlet and outlet fuel connections are also provided at this point. Each rail has dual flow passageways that provide an even fuel distribution to all the injectors. The assembly shown in Figure 6–19 is basically the same except that fuel tubes crisscross between the two rails and the pressure regulator is attached to the back of one of the rails. A cutaway view of a typical fuel rail assembly is shown in Figure 6–20.

COLD START OPERATION

When starting a cold engine, additional fuel is required to provide sufficient fuel vapors and atomization for combustion. A cold start injector is used to provide this additional fuel during the cold start mode. This device is important when engine coolant temperature is low because the main injectors are not pulsed on long enough to provide the fuel necessary to start the engine.

During engine cranking, fuel is injected into the cylinder ports through the individual fuel injectors

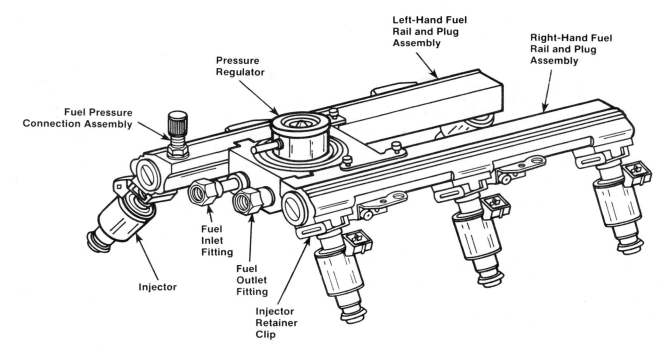

FIGURE 6-18   Fuel rail assembly, center-base type

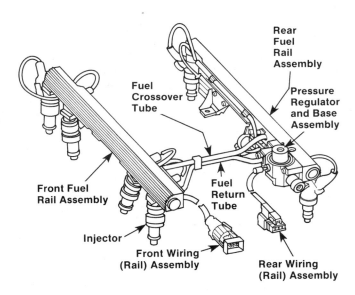

FIGURE 6-19   Fuel rail assembly, crossover type

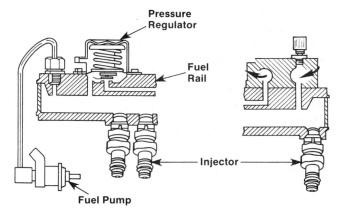

FIGURE 6-20   Fuel rail cutaway

and the required additional fuel is injected into a separate passage within the inlet manifold by the cold start injector. The cold start injector is controlled by the thermal time switch and operates at an engine temperature below 95 degrees Fahrenheit (Figure 6-21).

PFI THROTTLE BODY

The throttle body in a PFI system controls the amount of air that enters the engine as well as the

amount of vacuum in the throttle body manifold. It also houses and controls the throttle position sensor (TPS). The TPS enables the ECU to know where the throttle is positioned at all times. A typical PFI throttle body is shown in Figure 6-22.

The throttle body is a single cast aluminum housing with a single throttle blade attached to the throttle shaft. The TPS and the IAC valve/motor are also attached to the housing. The throttle shaft is controlled by a cable running from the accelerator pedal. The throttle shaft extends the full length of the housing. The throttle bore controls the amount of incoming air that enters the air induction system. A small amount of coolant is also routed through a

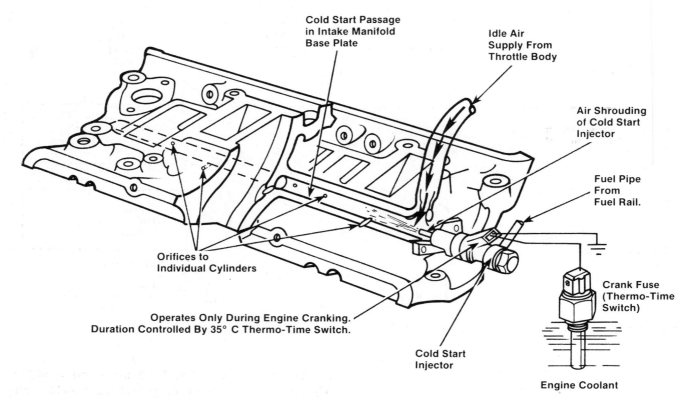

FIGURE 6-21   Cold start operational mode

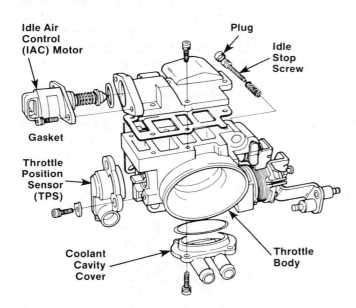

FIGURE 6-22   Typical PFI throttle body

passage in the throttle body to prevent throttle icing during cold months.

AIRFLOW SENSING

The PFI system uses two types of air measurement techniques: speed density and mass airflow.

The main reason for air sensing measurements is to monitor more accurately the conditions and amount of incoming air to further refine the air/fuel mixture.

The speed density technique uses MAP and temperature information to calculate the mass airflow rate (MAF) in the ECU. This technique also utilizes the MAP's estimate of engine variables such as volumetric efficiency and EGR parameters to determine the calculation. The use of this engine data makes this technique quite sensitive to variations in EGR parameters and engine operating conditions (Figure 6-23).

The MAF technique combines the airflow measurement into a single sensor (Figure 6-24). The mass airflow rate is calculated by processing the MAF sensor signal in the ECU. The ECU utilizes a software look-up table during the calculation process.

The MAF sensor (Figure 6-25) can be classified as a thermal measurement device. The typical MAF sensor consists of a screen to break up the incoming airflow, a resistor to measure the incoming air temperature, a heated film, and an electronic module to provide the ECU with an input signal. The electronic module is environmentally protected with a silicone gel sealant.

In operation, the mass of air entering the induction system is measured by the MAF sensor, which is

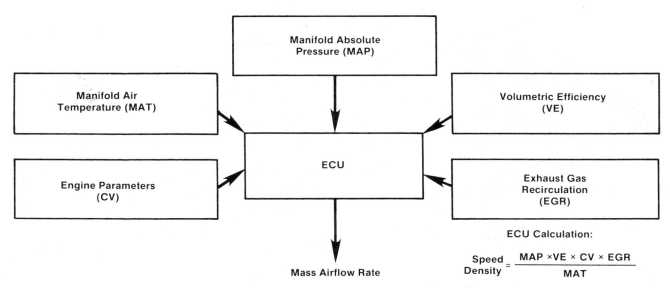

FIGURE 6-23   Speed density technique of air measurement

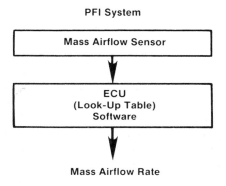

FIGURE 6-24   Mass airflow technique of air measurement

capable of compensating for altitude and humidity. By measuring the electrical power required to maintain the MAF sensor at 167 degrees Fahrenheit above the incoming air temperature, air mass can be determined. As air enters the air induction system, it passes over and cools the sensing element. This requires additional electrical power to maintain the sensor at 167 degrees Fahrenheit above the incoming air temperature.

The electrical power requirement is a measure of MAF and is converted to a digital signal (30 to 150 Hz) that is supplied to the ECU. It is used to calculate engine load. An increase in air mass creates a pro-

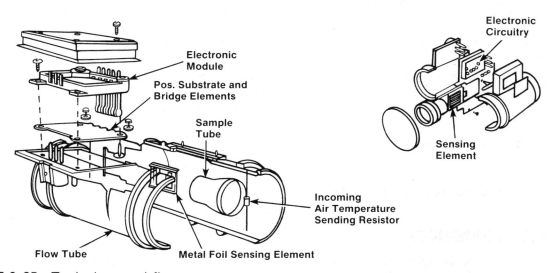

FIGURE 6-25   Typical mass airflow sensor

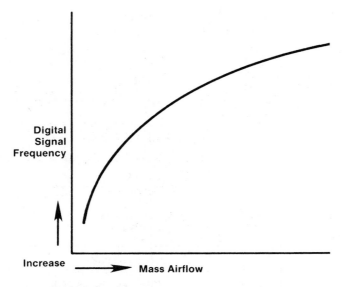

FIGURE 6-26 Mass airflow sensor transfer function

portional increase in the 30 to 150 Hz frequency (Figure 6–26).

Using calculations of MAF engine, temperature, and rpm, the ECU can calculate the exact amount of fuel required to provide an ideal air/fuel ratio (14.7 parts air to 1 part fuel). The ECU also determines the amount of spark advance required using input signals provided by the MAF sensor, manifold air temperature (MAT) sensor (if equipped), engine coolant temperature sensor, and engine rpm signals from the distributor.

TYPICAL FUEL INJECTION SYSTEMS

The remaining sections of this chapter cover the operation of fuel injection systems in vehicles produced by the major domestic manufacturers in the United States. Each manufacturer has a unique way

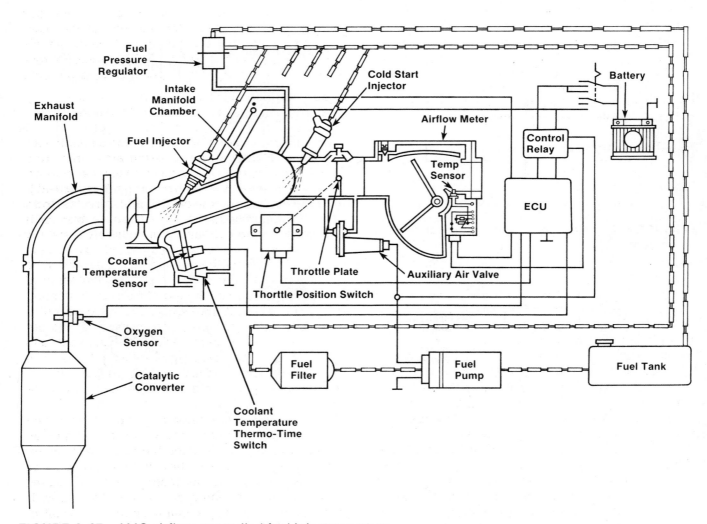

FIGURE 6-27 AMC airflow-controlled fuel injector system

of implementing electronic fuel injection into engine designs. For example, General Motors (GM) incorporates the throttle body injection system and its variations, whereas American Motors (AMC) employs both throttle body and port fuel injection systems.

AMERICAN MOTORS AIRFLOW-CONTROLLED FUEL INJECTION

AMC models (such as the Alliance/Encore) that are manufactured for sale in California use an airflow-controlled electronic fuel injection system. This system, manufactured by Bosch, measures intake airflow to determine fuel requirements. It is basically a port fuel injection system. This input information combined with data from the various engine sensors determines the specific fuel requirements for any given engine operating condition. The system consists of a fuel subsystem and a control subsystem. The fuel subsystem is typical of systems found in most domestic vehicles. The control subsystem consists of an ECU, control relay, airflow meter, auxiliary air valve, TPS, coolant temperature sensor, coolant temperature thermo time switch, and oxygen sensor. Figure 6-27 illustrates the AMC airflow-controlled fuel injection system.

Fuel Subsystem

An in-tank electric fuel pump delivers fuel to the fuel rail assembly (Figure 6-28). The pressure regulator maintains a constant pressure of approximately 36 psi. The regulator contains a spring-controlled diaphragm that is subjected to intake manifold pressure on one side and fuel pressure on the other. Excess fuel is bypassed by the regulator and routed back to the fuel tank.

The fuel air assembly supplies fuel to the intake manifold via the injectors, which are electromagnetic solenoid valves. Each injector has a needle valve that is held against a seat by a coiled spring. An electrical armature at the rear of the valve responds to electrical signals from the ECU by pulling and pushing the needle valve off and onto the seat. Pulling the needle valve off the seat allows fuel to be injected into the intake manifold. All injectors are fired simultaneously (twice per engine revolution).

During a cold start, the cold start injector supplies the additional fuel required for this mode. Power for the injector solenoid is routed through the coolant temperature thermo time switch. This switch is usually mounted within the cylinder head water jacket. When the engine is cold, the switch is closed

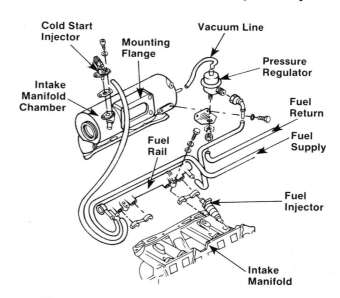

FIGURE 6-28 Fuel delivery system

and electrical power is supplied to the cold start injector solenoid. The switch will pulse on the injector for a maximum of 8 seconds at a coolant temperature of 4 degrees Fahrenheit. At 95 degrees Fahrenheit or higher, the switch opens, closing the cold start injector.

Airflow Meter

The airflow meter, through which all engine air is drawn, is located between the air cleaner and intake manifold. The meter has a passageway with a measuring flap and a dampering flap (Figure 6-29). The two flaps are on the casting and are offset from each other by 90 degrees. The measuring flap oscillates within the air stream against the pressure of a calibrated spring. The ECU is supplied an electrical signal by a potentiometer connected directly to the flap. Also within the airflow meter is a temperature sensor to measure the temperature of the incoming air stream. An electrical signal from this sensor plus the electrical signal from the measuring flap are combined into one electrical signal that is routed to the ECU. This composite signal indicates air density as well as volume and is used by the ECU to determine fuel requirements under various load conditions.

At idle speed, the measuring flap is almost completely closed. Built into the airflow meter is an air bypass that accommodates the idle speed air requirements. The idle airflow rate can be adjusted with the bypass adjustment screw but is not required under normal operating conditions. The screw is only adjusted during a major engine overhaul or when an airflow meter is removed and replaced.

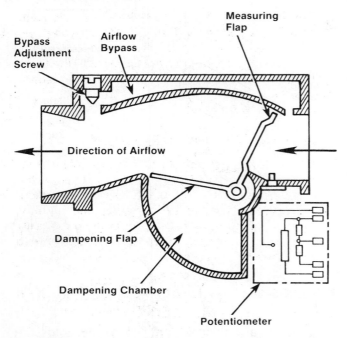

FIGURE 6-29 Cutaway of airflow meter assembly

Adjustment of the screw will vary the carbon monoxide (CO) content of the exhaust gases.

Throttle Plate Assembly

The throttle plate assembly, located on the intake manifold chamber (Figure 6-30), consists of two throttle plates that are connected to the car's accelerator by a throttle linkage. A throttle position switch is also built into the assembly. The switch

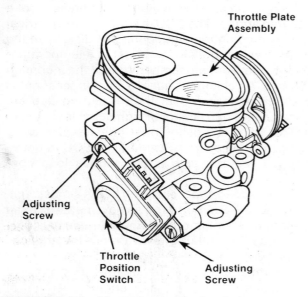

FIGURE 6-30 Throttle position switch

gives the ECU an electrical signal that indicates a wide-open throttle to an idle condition. This positioning signal is used by the ECU to adjust the engine's air/fuel mixture. Some assemblies have a built-in fast idle valve that compensates for decreasing idle speed during air conditioning operation, an extreme turn, or engine start-up. An auxiliary air circuit in the throttle plate assembly provides the additional idle speed air requirements.

Auxiliary Air Valve

To increase idle speed during cold engine start-up, extra air is supplied to the engine by the auxiliary air valve. The air valve is mounted to the cylinder head and receives air from the front of the throttle valve assembly. An air passage in the valve opens and closes in response to changes in engine temperature. Air from the valve is routed to the intake manifold chamber during cold engine start-up, bypassing the throttle valve assembly.

Oxygen Sensor

The oxygen sensor is located in the exhaust manifold. The sensor's outer surface is always in contact with the exhaust gases and its inner surface is in contact with outside air. The difference in oxygen contents between the two surfaces creates an electrical signal, which is routed to the ECU and represents a measurement of the unburned oxygen in the vehicle's exhaust. This measurement is directly related to the air/fuel mixture in the intake manifold. The ECU, based on the input electrical signal, is kept current as to the operating air/fuel ratio and can make the necessary adjustments to the mixture for the proper output of oxygen content.

THROTTLE BODY FUEL INJECTION SYSTEM

The AMC throttle body fuel injection system injects a metered fuel spray above the throttle blade in the throttle body assembly (Figure 6-31). This injection system also consists of a fuel subsystem and a control subsystem. Major components of the control subsystem include:

- Manifold air/fuel mixture temperature sensor
- Manifold absolute pressure sensor
- Coolant temperature sensor
- Oxygen sensor
- Throttle position sensor
- Wide-open throttle switch
- Closed throttle switch
- Idle speed control motor
- ECU

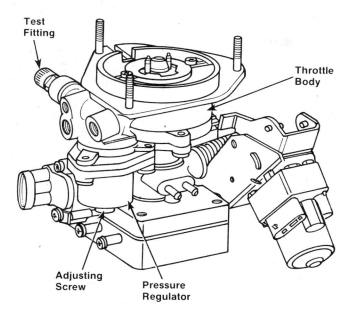

FIGURE 6-31   Throttle body assembly

Figure 6-32 illustrates the AMC throttle body fuel injection system. In the control subsystem the throttle position sensor, idle speed control motor, and fuel pressure regulator are located on the throttle switch. The discussion of this AMC fuel injection system will focus on those components within the throttle body assembly.

Idle Speed Control Motor

The idle speed control (ISC) motor is an actuator that is driven electrically and can change the throttle stop angle by acting as a movable idle stop. It controls the idle speed of the engine and maintains a smooth idle speed when the engine is suddenly decelerated. The throttle stop angle is determined by the following:

- Transaxle in PARK or NEUTRAL
- Position of the throttle (wide open or closed)
- Air conditioner compressor on or off

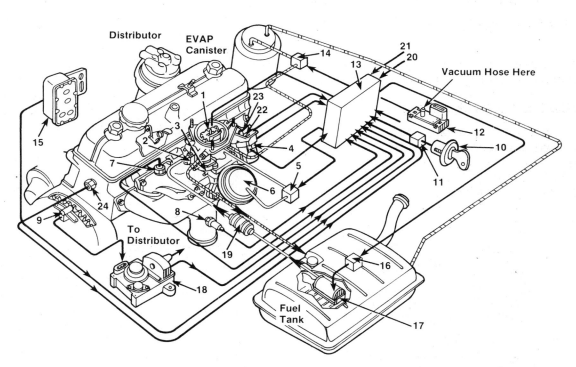

BASIC COMPONENT LOCATIONS

1. Injector
2. Throttle Position Sensor
3. Pressure Regulator
4. Idle Speed Control Motor
5. Solenoid-to-EGR Valve
6. EGR Valve
7. Manifold Air/Fuel Temperature Sensor
8. O₂ Sensor

9. Speed Sensor
10. Ignition Switch
11. Power Relay
12. MAP Sensor
13. Electronic Control Unit (ECU)
14. Solenoid-to-EVAP Canister Control
15. Starter Motor Relay
16. Fuel Pump Relay

17. Fuel Pump
18. Ignition Control Module
19. In-Line Fuel Filter
20. Air Conditioner On
21. Transaxle Neutral/Park Switch
22. Closed-Throttle (Idle) Switch
23. Wide-Open Throttle (WOT) Switch
24. Temperature Sensor (Coolant)

FIGURE 6-32   AMC throttle body fuel injection system

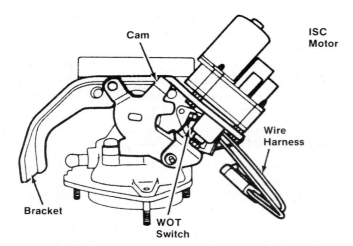

FIGURE 6-33   Typical ISC arrangement

During cold engine start-ups, the throttle will be held open for a longer period of time to provide an adequate engine warm-up period. When starting a hot engine, the throttle open time period is drastically shortened.

Normally, idle speed is maintained at a preprogrammed rpm, which might vary slightly because of engine operating conditions. When the vehicle is decelerated, the throttle might be held open slightly. Figure 6-33 illustrates a typical ISC motor.

Wide-Open Throttle Switch

The wide-open throttle (WDT) switch is integral with the throttle body assembly. It provides a signal to the ECU when the throttle is wide open. The ECU response to this signal is to increase the amount of fuel delivered to the injector.

Closed Throttle Switch

Often referred to as the idle switch, the closed throttle switch is located on the ISC motor. When the throttle is closed, this switch sends a signal to the ECU, which, in turn, will increase or decrease the throttle stop angle to respond to the engine operating condition.

Sensors

The manifold air/fuel temperature (MAT) sensor provides a signal to the ECU that is representative of the air/fuel mixture temperature in the intake manifold.

Absolute pressure in the intake manifold as well as the ambient atmospheric pressure are monitored by the MAP sensor. The ECU sees the signal from this sensor as representative of engine loading. This

sensor is mounted under the dash in the passenger compartment, away from the extreme environments under the hood. Manifold pressure data is supplied to the sensor via a vacuum line from the throttle body assembly.

The coolant temperature sensor enriches the fuel mixture when it senses that the engine is cold (cold engine start-ups).

The oxygen sensor generates a low level signal that is sent to the ECU indicating that the amount of oxygen in the exhaust system is high (lean mixture). A high level signal indicates that the amount of oxygen is low (rich mixture).

Throttle Position Sensor

The TPS is a variable resistor located on the throttle body assembly and connected to the throttle shaft. Manipulation of the throttle cable rotates the throttle shaft, which causes the throttle to open and close. The TPS senses this movement and signals the ECU, which in turn uses the signal to determine operating conditions for the automatic transmission system. This sensor is only found on vehicles equipped with automatic transaxles.

CHRYSLER CORPORATION MULTI-POINT FUEL INJECTION SYSTEM

One form of electronic fuel injection employed by the Chrysler Corporation is marketed as a multi-point fuel injection system. The system combines the techniques of electronic fuel injection and electronic spark advance and is typically a PFI system. The main subsystems consist of the air induction, fuel delivery, fuel control, emission control, logic module, power module, and appropriate data sensors. The logic module and power module on 1988 models have been combined into a single electronic control unit that Chrysler calls the single module engine controller (SMEC). Figure 6-34 illustrates a typical Chrysler multi-point electronic fuel injection system.

The air induction system consists of the air cleaner, throttle body, TPS, automatic idle speed motor, and turbocharger assembly. An exploded view of an air induction system used by Chrysler is illustrated in Figure 6-35.

The fuel delivery system provides fuel from the fuel pump to the fuel control system. It also returns excess fuel to the fuel tank. The system is composed of an in-tank electric fuel pump, fuel filter, two check valves, and a return line. Power is provided to operate the fuel pump through an automatic shutdown

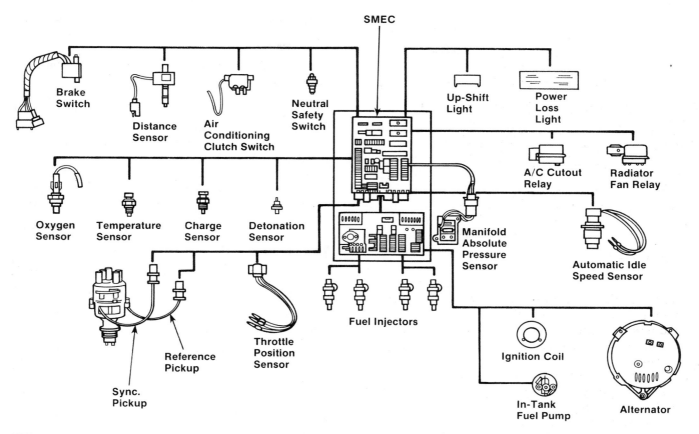

FIGURE 6-34 Typical Chrysler multi-point electronic fuel injection system

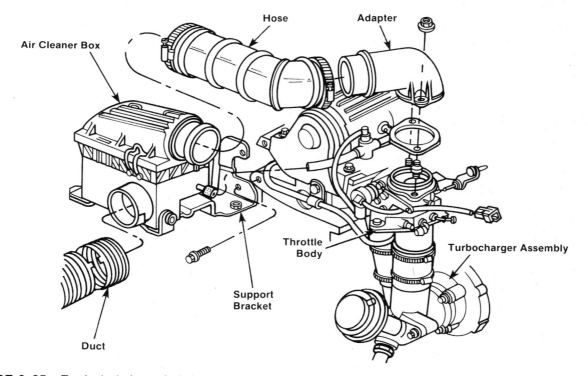

FIGURE 6-35 Exploded view of air induction system

(ASD) relay inside the power module. The ASD relay also controls the ignition coil, fuel injectors, and portions of the power module.

The fuel control system handles the actual delivery of fuel into the engine. The fuel pressure regulator maintains constant fuel pressure. In addition to the regulator, the system consists of the fuel rail and four or five fuel injectors typical of a port fuel injection system.

The emission system, although directly operated by the logic module in the SMEC, is not unique to the multi-point EFI engine. Emission system controls include the EGR, aspirated air system, evaporative emission control, and crankcase ventilation.

The logic module in the SMEC is a microprocessor computer. This module receives input signals from various switches and sensors. It then computes the fuel injector pulse width, spark advance, ignition coil dwell, idle speed, canister purge cycles, alternator field control, feedback control, and boost level from this information.

The power module in the SMEC is a small computer that contains necessary circuitry to power the ignition coil and fuel injectors. This module supplies most of the operating current for the entire system.

Data sensors provide the logic module with engine operating information. The computer analyzes this information and corrects air/fuel ratio, ignition timing, and emission control as needed to maintain efficient engine operation.

The SMEC, used on 1988 models, contains the circuits necessary to operate the ignition coil, fuel injector, and alternator field. It computes the fuel injector pulse width, spark advance, ignition coil dwell, idle speed, purge, cooling fan operation, and alternator charge rate. An externally mounted ASD relay is turned on and off by the SMEC.

THROTTLE BODY INJECTION SYSTEM

The Chrysler throttle body fuel injection system is computer controlled and utilizes a throttle body assembly with a single fuel injector.

Fuel is supplied to the engine through an electronically controlled injector located in the throttle

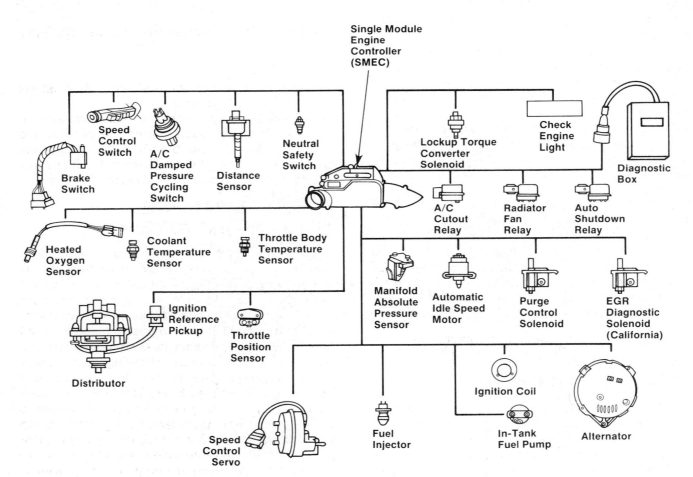

FIGURE 6-36 Typical TBI system for a 4-cylinder vehicle

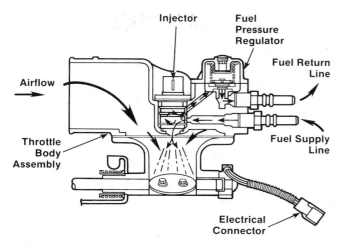

FIGURE 6-37 Low-pressure CFI fuel charging assembly

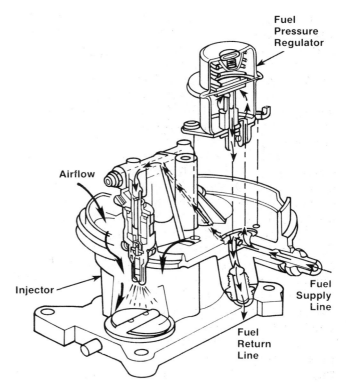

FIGURE 6-38 High-pressure CFI fuel charging system

body assembly on top of the intake manifold. On 1987 and earlier models, power to the injector is supplied by the power module, and the length of time the injector is left open is determined by the logic module. On 1988 models, the single module engine controller replaces both the power module and logic module.

The logic module is a preprogrammed microcomputer that controls ignition timing, emission control devices, and idle speed in addition to the air/fuel ratio. The amount of fuel to be metered through the injector is based on engine operating information as supplied by various engine sensors and switches. These include the MAP sensor, TPS, oxygen sensor, coolant temperature sensor, throttle body temperature sensor, and vehicle distance/speed sensor.

The logic module converts the input signals from these sensors/switches into control signals. Control signals are sent to the power module, which responds by issuing electrical signals to specific components to alter the air/fuel ratio and/or ignition timing to meet the indicated conditions.

A typical TBI system for a 1988 four-cylinder vehicle is shown in Figure 6-36.

FORD MOTOR COMPANY CENTRAL FUEL INJECTION SYSTEM

The CFI system is a single point, pulse time modulated injection system. Engine demands determine how much fuel is metered into the air intake stream. The CFI system is basically a throttle body fuel injection system. It can use either a high-pressure system (about 39 psi) or a low-pressure

system (14.5 psi). Figures 6-37 and 6-38 illustrate both types of CFI systems.

A basic CFI system consists of four subsystems: fuel delivery, air induction, engine data sensors, and electronic control assembly (ECA). Other than the fuel pressures used, the CFI system functions in a manner that is identical to a TBI system.

MULTI-POINT FUEL INJECTION SYSTEM

The Ford Motor Company uses a multi-point, pulse time injection system that uses a mass airflow sensor on its smaller engine models (1.6 to 2.3 L) or a barometric and manifold atmospheric pressure (BMAP) sensor and a separate air temperature sensor on all its other larger engines. Fuel is metered into the intake air stream in accordance with the demands of the engine. Metering takes place through injectors that are mounted onto a tuned intake manifold. The on-board ECA uses input sensor data to compute the required fuel flow rate to maintain an optimum air/fuel ratio. Figures 6-39 and 6-40 illustrate Ford multi-point injection systems for one of its smaller engine models. A multi-point injection system that is used on larger Ford engine models is illustrated in Figure 6-41.

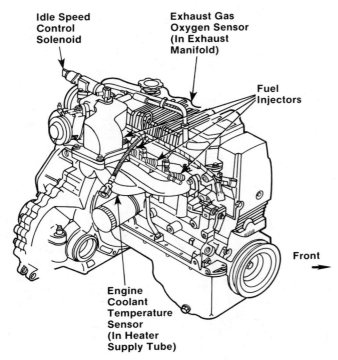

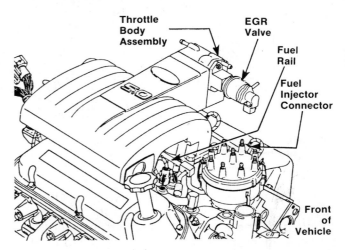

GENERAL MOTORS DIGITAL FUEL INJECTION SYSTEM

General Motors digital fuel injection (DFI) system consists of six subsystems: air induction, fuel

FIGURE 6-39 Ford engine with multi-point injection system

FIGURE 6-41 Ford 5-liter engine with multi-point injection system

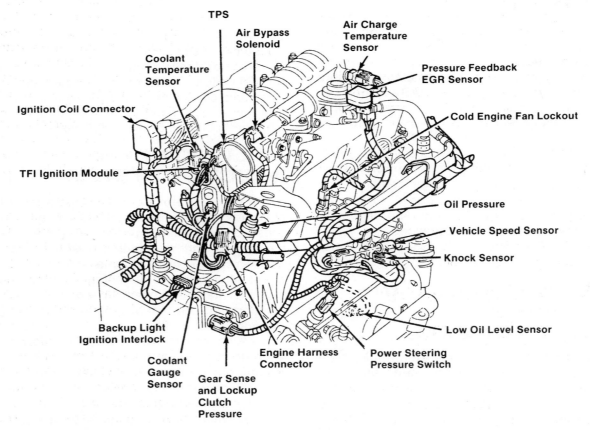

FIGURE 6-40 Ford 3-liter engine with multi-point injection system

DIGITAL FUEL INJECTION/CLOSED LOOP OPERATION

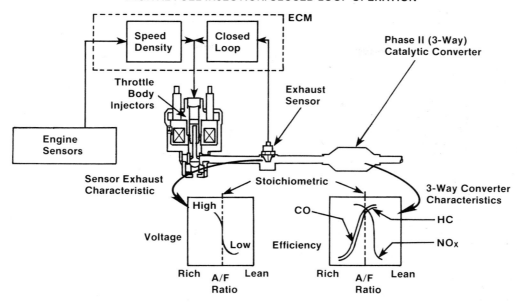

FIGURE 6-42   Closed loop operation of a DFI system

delivery, data sensors, electronic control module (ECM), idle speed controller, and system diagnostics. The system supplies fuel to the engine through two injectors located in the throttle body on top of the intake manifold. The ECM is an electronic microcomputer that controls or meters the fuel through the injectors based on the demands of the engine.

The DFI system is a speed density fuel system that can accurately control the air/fuel mixture entering the engine. The density amount of air entering the engine is measured by monitoring the following:

• Manifold absolute pressure
• Intake air temperature
• Barometric pressure

The system also utilizes a high-energy ignition (HEI) distributor to provide engine speed data for the ECM when it calculates the optimum fuel flow rate. Figure 6-42 illustrates the operation of a typical DFI system.

DUAL UNIT FUEL INJECTION SYSTEM

This system uses dual throttle bodies or a crossfire injection technique as shown in Figure 6-43. The system consists of the following major subsystems: fuel delivery, two throttle body injector (TBI) assemblies, idle air control (IAC), ECM, electronic spark timing and control, emission controls, and applicable data sensors.

An electric fuel pump is located inside the fuel tank and is an integral part of the fuel gauge sending unit. The pump supplies fuel under pressure to the fuel pressure compensator on the front TBI unit. Pump operation is controlled by a fuel pump relay located on the left side of the engine compartment (behind the driver's seat on 1982 Corvettes). When the ignition switch is turned on, the fuel pump relay activates the fuel pump for 1-1/2 to 2 seconds to prime the injectors. If the ECM does not receive reference pulses from the distributor after this period, the ECM deactivates the fuel pump circuit. The fuel pump circuit will be activated again through the relay when ECM receives distributor reference pulses.

Each TBI unit is composed of two casting assemblies. One assembly contains a throttle body with an idle air control valve to control airflow and a TPS. The second assembly is a fuel body meter cover with a built-in pressure regulator (rear unit) or pressure compensator (front unit) and fuel injector. The throttle body casting can also contain vacuum ports for EGR valve, MAP sensor, and canister purge system.

The pressure regulator is a diaphragm-operated relief valve. Injector pressure is present on one side of the diaphragm and air cleaner vacuum on the other. By controlling the return of excess fuel to the fuel tank, the pressure regulator maintains a constant pressure drop of about 10 psi across both injectors throughout all engine operating conditions.

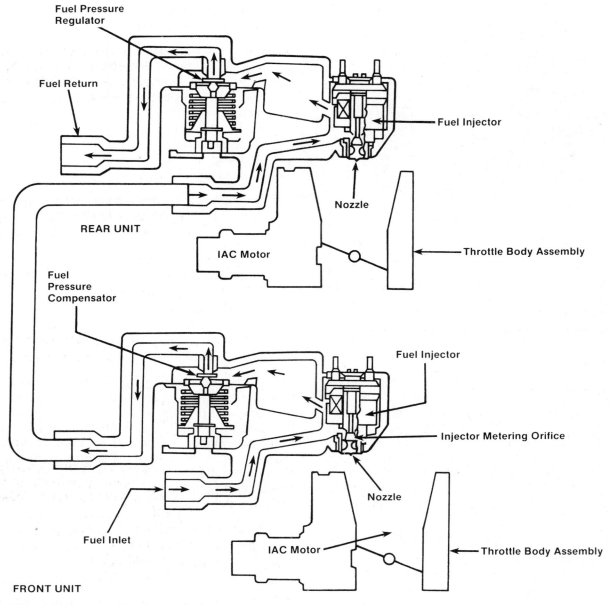

FIGURE 6-43 Sectional view of dual throttle body assemblies

The pressure compensator is similar in design to the pressure regulator. The compensator makes up for any momentary drop in fuel pressure between the front and rear TBI units, thereby maintaining consistent operating pressures in the fuel delivery lines.

FUTURE OF FUEL INJECTION VERSUS CARBURETION

Most experts agree that carburetors are doomed for several reasons: tougher government mileage standards, strict emissions regulations (which are becoming stricter in other areas of the world), and increased emphasis on performance.

What the experts do not agree on is what kind of fuel injection system will predominate in the future. Regardless of which injection system becomes the predominant choice of automakers in the future, certainly some combination of throttle body injection and multi-port injection will virtually eliminate carburetors at the OE level within the next few years.

In fact, the aftermarket manufacturers are looking at the fuel injection/carburetor situation. As shown in Figure 6-44, conversion kits are available that permit the carburetor to be replaced by a throt-

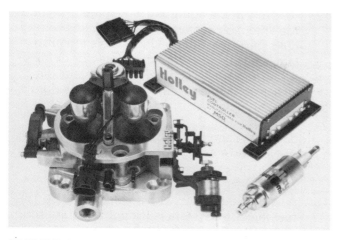

FIGURE 6–44 Conversion kit to replace carburetor

tle body injection unit. At the nucleus of this typical V-8 engine TBI conversion kit are two high-low fuel injectors. Through the use of a pressure regulator, these low pressure injectors have an adjustable fuel pressure range of 10 to 15 psi. Fuel pressure is easily adjustable by turning a screw on the throttle body. Both injectors are factory-set at 13 psi but each can flow 80 pounds of fuel per hour at 15 psi, capable of sustaining 320 horsepower.

The in-line electric fuel pump ensures a continuous supply of pressurized fuel. When activated, the pump's anti-flood mode uses a signal to shut off the fuel supply in the event the engine stalls. (Fabrication of a return fuel line is necessary if not so originally equipped. Also, use of a 4-bbl. intake manifold and mechanical fuel pump block off plate are required.)

The system includes its own throttle linkage, computer wiring harness and fuel fittings. Provisions for PCV, timed distributor spark, manifold vacuum, and EGR are also included. The electronic control module (ECM) attaches in a convenient location such as under the dash. A high capapcity Holley electric fuel pump connects into the fuel line outside the tank. Sensors are installed, and electrical, fuel line, and vacuum hose connections are made.

REVIEW QUESTIONS

1. The reference pulses from the distributor in a throttle body injection system represent _____ .
   a. a fuel injector cycle
   b. throttle position
   c. engine speed
   d. none of the above

2. If fuel injector pulses from the ECU are sent at a fixed rate but do not coincide with the distributor references pulses, the TBI system is said to be operating in the _____ mode.
   a. closed loop
   b. open loop
   c. speed density
   d. nonsynchronized

3. During cranking or cold start-ups, acceleration enrichment or deceleration leaning, the TBI system operates in the _____ mode.
   a. nonsynchronized
   b. synchronized
   c. Lambda control feedback
   d. mass airflow

4. The fuel injector pulse-on time for a TBI system operating in a closed loop mode is determined by _____ .
   a. coolant temperature
   b. manifold pressure
   c. throttle angle
   d. all of the above

5. On a TBI system operating in the closed loop mode, the wider the opening of the throttle and the higher the manifold absolute pressure, the longer the _____ .
   a. distributor reference pulses
   b. injector pulse-on time
   c. ECU must calculate airflow
   d. nonsynchronized mode

6. Closed loop operation in a fuel injection system is based on _____ .
   a. oxygen content of exhaust gases
   b. volume of intake air
   c. fuel pressure
   d. none of the above

7. In the _____ running mode of operation, the ECU in a TBI system disregards signals from the oxygen sensor and sends preprogrammed signals for warm-up, idle, and full-load enrichment.
   a. close loop
   b. open loop
   c. speed density mode
   d. feedback mode

8. The fuel rail assembly is characteristic of a(n) _____ injection system.
   a. throttle body
   b. electronic fuel
   c. port injection system
   d. all of the above

9. The pressure regulator used on a TBI system is similar to a diaphragm-operated relief valve in that it has _____ .
   a. mass airflow pressure on one side and oil pressure on the other
   b. fuel pressure bearing on one side and air cleaner pressure on the other
   c. fuel pressure bearing on both sides
   d. air pressure bearing on both sides

10. The central fuel injection (CFI) system is used on vehicles manufactured by
    _____
    a. GM
    b. Chrysler
    c. AMC
    d. Ford

11. The PFI system uses two types of air measurement techniques. One of these is the _____ technique.
    a. speed density
    b. synchronized
    c. nonsynchronized
    d. none of the above

12. A Ford vehicle equipped with a multi-point fuel injection system is roughly idling during cold start-ups. Technician A says the problem is a leaking or defective injector or injector O-ring seal. Technician B says the idle speed adjustment is incorrect. Who is right?
    a. Technician A
    b. Technician B
    c. Both A and B
    d. Neither A nor B

13. A vehicle equipped with an electronic port fuel injection system is brought into the shop with a rough, high idling problem. Technician A says a vacuum or air leak exists between the airflow meter and the throttle plates. Technician B says the vehicle has a leaking valve cover gasket, which is allowing excess air to enter the PCV system. Who is right?
    a. Technician A
    b. Technician B
    c. Both A and B
    d. Neither A nor B

14. The throttle body in a PFI system controls the _____ .
    a. amount of air that enters the engine
    b. amount of vacuum in the throttle body manifold
    c. throttle position sensor
    d. all of the above

CHAPTER SEVEN

FUEL INJECTION SERVICE

Objectives

After reading this chapter, you should be able to:
* Relieve fuel pressure in the system prior to performing maintenance or repairs.
* Perform preventive maintenance checks.
* Use troubleshooting guides to isolate a faulty component.
* Check for fuel delivery on a fuel injection system.
* Describe the diagnostic systems of the three major domestic car manufacturers.
* Know and identify the various trouble codes generated by diagnostic systems.
* Use testers with the applicable diagnostic systems.

Fuel injection systems are extremely complicated and vary greatly among manufacturers, models, and similar systems installed on different engines. Therefore, it is important that the technician use and refer to the applicable manufacturer's service manual before attempting any fuel injection maintenance, repairs, or adjustments.

This chapter covers fuel injection servicing techniques as well as safety (Figure 7–1), troubleshooting, and preventive and corrective maintenance. Preventive maintenance for fuel injection systems consists of preliminary checks and adjustments where possible. Corrective maintenance includes performing diagnostic test procedures, isolating the faulty component, and replacing it. Diagnostic systems used by domestic car manufacturers are also covered.

SAFETY PRECAUTIONS

The standard industry use of fuel injection on today's automobiles has brought about some new safety precautions and rules for the shop and the technician. The most prominent safety concern is a fuel fire. The two basic types of fuel injection systems currently in production are the multi-point fuel injection (MFI) system and the throttle body injection (TBI) system. The basic difference between the systems lies in where the fuel injectors are located. What both systems have in common, however, is

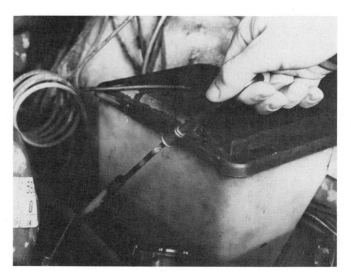

FIGURE 7–1 Observe safety precautions in the shop even when doing something as simple as checking the oil.

that each system operates with fuel under pressure (Figure 7–2). Both systems have fuel under pressure that is at least twice the psi rating (if not higher) of a typical carbureted engine. Normal fuel pressure in a multi-point injection system runs about 65 to 85 psi. Even after the engine is shut down the fuel lines can remain in a highly pressurized state for a long period of time. Because of these operating and design conditions, the fuel pressure in a fuel injection system

FIGURE 7-2 Fuel injection under pressure

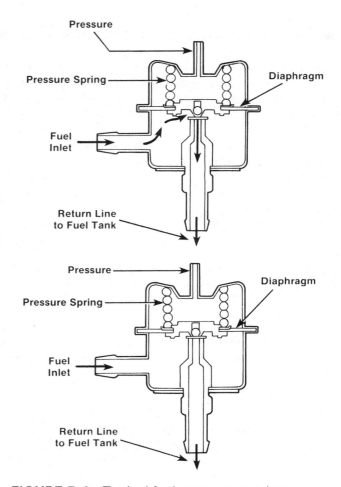

FIGURE 7-4 Typical fuel pressure regulator

must be relieved before the technician attempts to repair, test, or service the system. A hot manifold, a hot exhaust, or a spark mixed with spraying fuel from a disconnected high-pressure fuel line all are *definite* safety hazards to the technician as well as the shop. Methods of relieving the fuel pressure will vary from one car manufacturer to another. It therefore is crucial for the technician to know the various techniques and procedures for bleeding the fuel pressure on the fuel injection systems in use today. He or she should also be aware of the safety precautions involved with each procedure.

One step the technician can take to ensure a safe work environment when working on a fuel injection system is to understand how each particular system works. Basically, a fuel injection system has a fuel pump that delivers fuel under pressure to a fuel rail and injectors and then to a pressure regulator. Figure 7-3 illustrates a typical fuel delivery system

for a TBI system. The fuel pump capacity exceeds the engine's consumption rating in most cases so the system incorporates a fuel return line to return the excess fuel to the tank. A pressure regulator always maintains a constant pressure to the fuel injectors so that they are ready to deliver a precise,

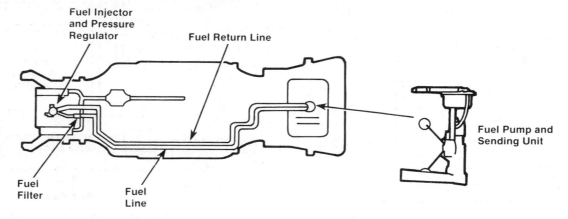

FIGURE 7-3 Typical TBI fuel delivery system

metered amount of fuel when the on-board computer initiates the injection command. The fuel injectors are solenoid-operated and readily respond to the electrical command signal from the computer. The majority of fuel injection systems use an electric fuel pump that has a maximum pump pressure ranging between 60 to 90 psi.

The typical fuel pressure regulator has a pressure chamber and a vacuum chamber. Both chambers are separated by a diaphragm relief valve with a calibrated spring (Figure 7-4). The diaphragm-spring assembly is usually located within the vacuum chamber. Regulation of the system fuel pressure is accomplished when the fuel pump pressure acting on the bottom side of the diaphragm overcomes the force of the spring acting on the top side of the diaphragm. Basically, the vacuum and spring pressure acting on the top side of the diaphragm controls

the system fuel pressure by opening a fuel return valve. It stands to reason, therefore, that if the vacuum should decrease, the fuel pressure will increase, and vice-versa.

A system used on most import cars is the Bosch constant injection system (CIS) (Figure 7-5). This system differs from domestic systems in that the CIS controls fuel injection by varying the fuel flow rate through the injectors as opposed to the on/off time or pulsing of the fuel injector solenoids on domestic systems. Until 1983 the CIS was often referred to as a mechanical system. A fuel distributor is the heart of the system. The distributor controls the amount of fuel passing through each pressure regulator to the injectors. A fuel accumulator, not found on domestic systems, maintains the system pressure or prevents it from dropping off when the engine is off. This feature is used to prevent vapor lock of the system

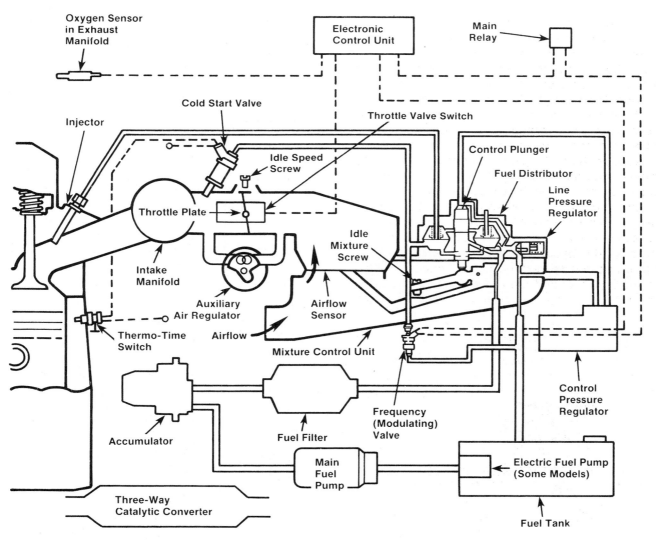

FIGURE 7-5 Bosch CIS fuel injection system

during a hot restart or when the ambient temperature is high.

FUEL INJECTION SERVICING SAFETY PRECAUTIONS

The technician can never be too careful when servicing a fuel injection system. The main concerns should be preventing fire and an explosive atmosphere. The following safety precautions should be taken:

- When servicing an injection system, always disconnect the negative battery cable to prevent an electrical spark or short and to keep the fuel system components from operating.
- Make sure the work area has adequate ventilation to remove fuel vapors.
- Do not use an electrical drop light to inspect the fuel system components. Use a good flashlight (Figure 7-6).
- Always use clean shop rags to catch or clean up residual fuel when fuel lines are disconnected.
- Always wear proper eye protection devices when working with high-pressure fuel components.
- Be sure that all fuel line connections are clean and tight.
- Never substitute a normal fuel line for any fuel injection supply or return line. Use reinforced, high-pressure fuel lines.

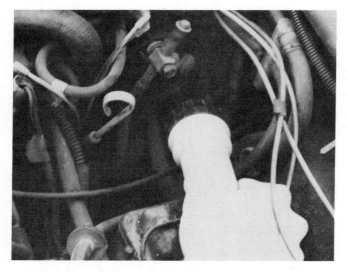

FIGURE 7-6 To inspect fuel system components or connections use a flashlight instead of a drop light.

FIGURE 7-7 Fuel pressure venting

- Do not use or install paper or plastic fuel filters. Use the correct fuel filter for the system.
- Keep all open flames, sparks, or welding equipment away from your work area.
- Do not work on the car if the manifold or exhaust is still hot. Allow the engine to cool down before servicing the fuel system components.
- Do not smoke when servicing the fuel system.
- Always have in close proximity to your work area a dry chemical (Class B) fire extinguisher.
- Always dispose of fuel-soaked rags properly.
- Maintain a clean work area—dirt is the number one enemy of a fuel injection system.

RELIEVING FUEL PRESSURE

Procedures for relieving fuel pressure vary among car manufacturers. The procedures here can be followed when working on most domestic and imported cars equipped with a fuel injection system.

Imports

The fuel pressure in a CIS can be relieved by energizing the cold start valve for about 10 seconds. An alternate procedure is to wrap the fuel line connection on the pressure regulator with a clean shop rag and then carefully crack loose the connection (Figure 7-7). The rag will catch any fuel spray from the connection. To test fuel pressure on the CIS, tee a pressure gauge into the line running between the fuel distributor and warm-up compensator. Always test the cold control pressure first followed by the hot control pressure next, and finally, the rest pressure.

The Bosch fuel injection systems used on 1980 through 1983 European and Japanese imports have an operating fuel pressure of about 36 psi. Pressure on these systems can be relieved by using a hand, vacuum pump. Connect the pump to the signal port on the pressure regulator and apply a vacuum of 20 inches of mercury for about 10 seconds (see Figure 7-7). This effectively energizes the cold start valve for 10 seconds. An alternate procedure is to start the engine and then remove the fuel pump fuse or relay until the engine stalls. To use a fuel pressure gauge to test this type of system, tee the gauge between the cold start injector and the fuel rail. Figure 7-8 illustrates another method of connecting a fuel pressure gauge to test the pressure on this type of system. When taking fuel pressure readings with the gauge, look for an average pressure reading (Figure 7-9) instead of a specific pressure reading. Normal pulsa-

FIGURE 7-10   Fuel supply and return lines

tion of the injector will cause the gauge to fluctuate, especially on four-cylinder engines.

American Motors Models

Jeep/Eagle models can be equipped with a TBI system or an MFI system. Normal fuel pressure on a TBI system is 14.5 psi, and the MFI system has a fuel pressure of 31 psi. If the fuel pressure is not to specification, the technician should check the fuel supply and return lines for kinks and restrictions (Figure 7-10). This procedure applies to all vehicle types and not necessarily to American Motors models. The fuel pressure on Jeep/Eagle models is relieved by connecting a fuel pressure gauge to the pressure test port on the fuel rail. The gauge should be equipped with a bleed valve. After the gauge has been properly connected, carefully and slowly open the bleed valve to release the system fuel pressure. Once the system pressure has been relieved, separate the quick-connect fuel lines by squeezing the two retaining tabs against the fuel line and then pulling the tube and retainer from the fitting (Figure 7-11). Remove and discard the O-rings, spacer, and retainer from the fitting. Always use new O-rings, spacer, and retainer whenever a fuel line is separated.

Chrysler Models

The TBI systems used by Chrysler operate at two different fuel pressure levels:

* 1986 and later models, 14.7 psi
* 1985 and earlier models, 36 psi

To relieve the fuel pressure on these systems, first loosen the gas cap on the vehicle, then remove the

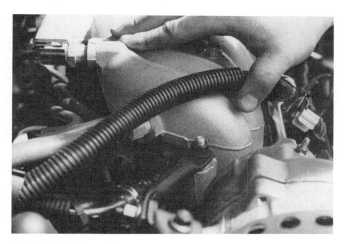

FIGURE 7-8   Alternate method of connecting a fuel gauge to check system pressure

FIGURE 7-9   Taking fuel pressure readings

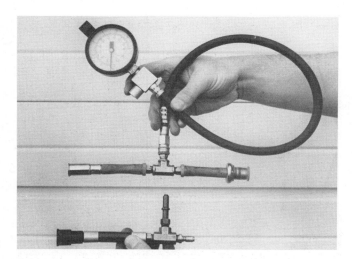

FIGURE 7-11 Quick-connect type fuel line fittings

FIGURE 7-13 Injector O-ring inspection

injector wiring harness. With the wiring harness free, ground one terminal on the connector and apply battery voltage (12 volts) to the other terminal. Apply the voltage for 10 seconds or less.

To check fuel pressure on a Chrysler TBI system, install a fuel pressure gauge between the fuel filter hose and throttle body (Figure 7-12). Release the system's fuel pressure before connecting the pressure gauge. Then start the engine and check the fuel pressure gauge. If the pressure is too low, move the gauge between the fuel filter hose and fuel line. Restart the car and check the gauge again. If the pressure is now correct, the fuel filter is at fault. If the same low pressure reading is obtained as before, try

gently squeezing the fuel return hose to see if the pressure increases. If it does, the pressure regulator is faulty. If the pressure reading remains unchanged, check the fuel pump. When servicing the injectors be sure to replace any O-rings or seals that are used with the injector (Figure 7-13).

Ford Models

Relieving the fuel pressure on Ford vehicles equipped with fuel injection is accomplished by a Schrader type valve. It is located on the fuel rail (MFI systems) or on the top of the fuel charging main

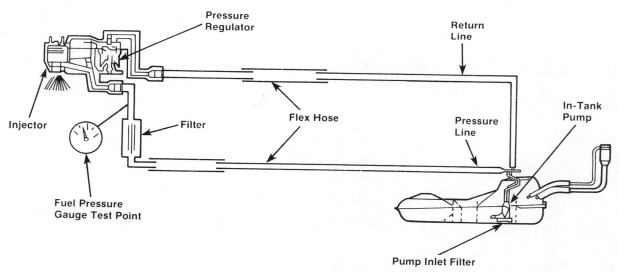

FIGURE 7-12 Checking fuel pressure on a Chrysler TBI system

body (on TBI systems) (Figure 7-14). The valve is also a convenient point to connect a fuel pressure gauge when testing the fuel system. A special test adapter is needed to go between the Schrader valve and the fuel pressure gauge. (An alternate method to relieve the pressure on a Ford is to disconnect the inertia switch [usually located in the trunk] and start the engine for about 15 seconds.)

If a fuel pressure gauge is not available, use a hand vacuum pump. Disconnect the vacuum hose from the pressure regulator and connect the pump to it (Figure 7-15). Apply a vacuum of about 25 inches of mercury to the regulator to vent the pressure into the fuel tank. This technique relieves the fuel pressure in the lines, but it does not clear the fuel from the lines. So have clean shop rags on hand when disconnecting any fuel line.

General Motors Models

On most General Motors models the fuel pressure can be relieved by removing the fuel pump fuse from the fuse block and starting the engine. The engine will stall when the fuel lines are empty. To be on the safe side, attempt to restart the engine for about 5 seconds. This will ensure that all pressure in the lines has been relieved. Some late-model GM cars have a pressure relief valve. It is usually located on or near the fuel rail. When testing the system's fuel pressure, connect a pressure gauge with a bleed valve to the pressure relief valve. Typical operating pressure for GM port injection systems is 30-psi nominal and 24 to 40 psi with the engine off (non-operating pressure). For GM throttle body injection systems, operating pressure is 11 psi nominal.

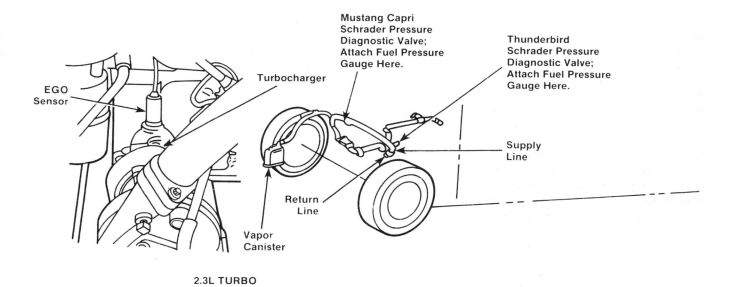

2.3L TURBO

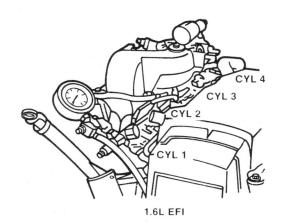

1.6L EFI

FIGURE 7-14   Fuel diagnostic valve location on Ford vehicles

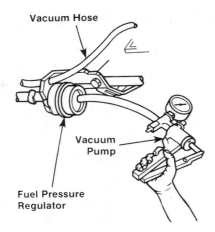

FIGURE 7-15   Relieving fuel system pressure using a hand vacuum pump

ADDITIONAL PRECAUTIONS

After the fuel injection system has been serviced, start the engine to allow the system's fuel pressure to build up and deliver fuel to the injectors.

WARNING: Never pour gasoline into the throat of the air intake in an attempt to start the car. This can cause a severe explosion and fire.

After servicing a fuel injector/fuel lines, it is sometimes necessary to bleed the lines of air. When this is done, fuel might leak or drip onto the shop floor in the vicinity of the engine. Be sure to wipe up all excess fuel before restarting the engine, and place the fuel-soaked shop rags in a safety can or otherwise designated container.

Some adjustments can be performed on a vehicle with an automatic transmission by running the engine and placing the shifting lever in DRIVE. If this is done, be sure the emergency brake is fully engaged and the driving wheels are properly blocked before attempting any fuel system adjustments with the engine running. Never stand in front of the vehicle when making the adjustments.

FUEL INJECTION PREVENTIVE MAINTENANCE

On the average, the fuel injection system is the least likely system to cause problems on a car. So the technician should not immediately assume that it is at fault when starting or running problems occur. Other vehicle systems, such as the battery, igni-

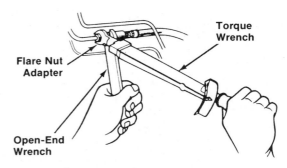

FIGURE 7-16   Tightening fuel line fittings

tion system, charging system, or starting system, can be at fault. These systems should be checked first before attempting to service the fuel injection system.

Fuel injection preventive maintenance involves periodic inspection of the system components and making the appropriate repairs if necessary. On a regular basis the technician or the owner/driver should inspect the system for the following:

- Fuel leaks
- Air leaks
- Dirty fuel filter
- Dirty air filter
- Good electrical connections
- General condition of all vacuum hoses
- Correct idle speed and fuel mixture
- Properly installed fuel lines and fuel injectors

LEAKS

A fuel leak or an air leak in a fuel injection system can be serviced in the same manner as those in a carbureted system. The only difference lies in the type and strength of the material that is used for the fuel lines. Fuel injection lines, being in a high-pressure situation, are usually made of a heavy steel tubing construction. If a fuel connection is loose, not only does it leak fuel but air will also be allowed to enter the system. This condition can cause erratic running headaches for the driver/owner. Always bleed the air from the fuel lines whenever they are serviced. When only fuel appears at the connection (no air bubbles accompanying the fuel) then the technician can begin to tighten the connection. Be sure to tighten the fittings according to the manufacturer's torque requirements (Figure 7-16).

The air induction system should also be checked for possible leaks. A small air leak can be a problem for the owner. The car will run poorly because of a lean air/fuel mixture. Air leaks are mainly determined by listening for a hissing sound. An alternate method is to use a soap solution or a carburetor cleaner spray on the suspected leaking area. Presence of air bubbles in the treated area indicate an air leak.

A vacuum leak can also cause erratic running problems and rough idling. Vacuum hoses are usually constructed of rubber or plastic. When subjected to engine heat, oil, and fuel vapors, the hoses deteriorate and crack. They can also become kinked or pulled loose during servicing. Make a thorough check of the condition of all vacuum hoses. Replace those hoses that are deteriorated, cracked, or otherwise suspected of being faulty. Be sure to follow the vacuum hose routing diagram (Figure 7-17) that is applicable to the car being serviced.

ELECTRICAL CONNECTIONS

All electrical connections to sensors and the fuel injectors should be checked to ensure that they are properly mated and tightly coupled (Figure 7-18). One way to test the connections is to gently wiggle or shake the connections to ensure they are tight. Also remove or break the electrical connections and check for corrosion. If corrosion is present, clean the male/female parts with an appropriate cleaner. Some connectors might require the use of a special conductive grease or fluid to prevent corrosion. Check the manufacturer's service manual to determine if the connector parts have to be coated

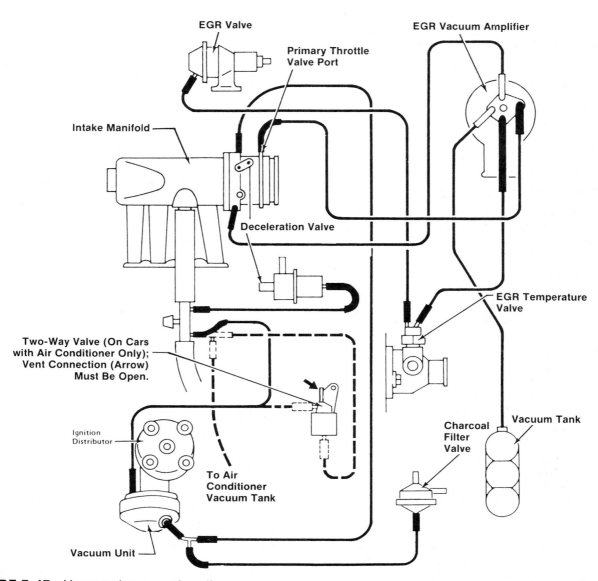

FIGURE 7-17 Vacuum hose routing diagram

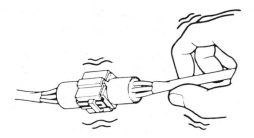

FIGURE 7-18 Checking electrical connectors for looseness

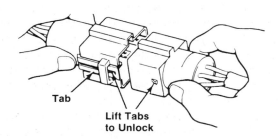

Tab

Lift Tabs to Unlock

FIGURE 7-19 Disconnecting electrical connectors

with a special anticorrosion application. Also, be sure to pull the connectors apart and check the condition of the metal pins and tabs (Figure 7-19).

Electrical connections to the fuel injectors should be periodically inspected. Injectors on an MFI system usually have a metal wire clip that securely holds the electrical connector to the injector. Use a screwdriver and gently pry the wire clip open while pulling on the connector to remove it (Figure 7-20). Never pull the wire harness—only the connector.

INSTALLATION/MOUNTING OF COMPONENTS

Fuel injectors can be installed by being pushed into a tight-fitting rubber grommet, or they can be secured to a mounting bracket by a bolt or nut (Figure 7-21). Occasionally check the mounting of the injectors by attempting to move it up, down, forward, and backward. Any excessive movement indicates the bolt or nut is loose or the rubber grommet is worn out or greatly expanded. Also check to ensure that the fuel rail and fuel lines are properly supported and secured. Engine vibration can loosen these parts very easily.

IDLE SPEED AND/OR MIXTURE ADJUSTMENTS

Periodic adjustment of idle speed and/or mixture is not necessary. If the adjustments are possi-

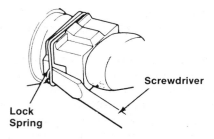

Lock Spring

Screwdriver

FIGURE 7-20 Removing the fuel injector connector

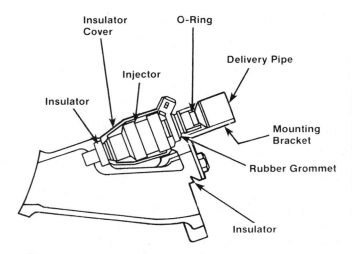

Insulator Cover

O-Ring

Injector

Delivery Pipe

Insulator

Mounting Bracket

Rubber Grommet

Insulator

FIGURE 7-21 Injector mounting

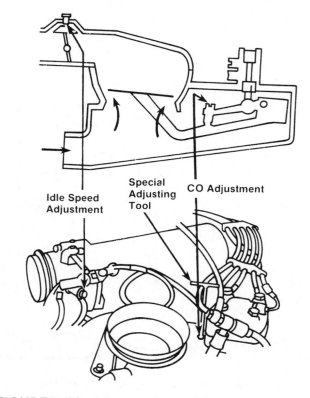

Idle Speed Adjustment

Special Adjusting Tool

CO Adjustment

FIGURE 7-22 Idle speed and CO adjustments

ble, they are usually performed only after or during major fuel injection servicing operations. Some vehicles have no such adjustments; the on-board computer electronically controls these parameters. Older vehicles have idle speed controls that are readily accessible. On new cars, however, these adjustment controls are usually hidden and might require special tools to prevent tampering with the fuel injection system (Figure 7-22). Mixture control adjustments on today's cars always require the use of an exhaust gas analyzer.

Troubleshooting

When driveability and performance problems occur in electronic fuel injection-equipped vehicles, it is often assumed that the injection system itself is at fault. Such an assumption at the outset can result in wasted time for the technician and the possibility of replacing good components. Therefore, when engine problems occur, be sure that all engine systems not associated with the fuel injection system are operating properly before attempting to troubleshoot the fuel injection system.

If all these systems are operating correctly, refer to the manufacturer's service manual for troubleshooting guides and/or charts. A troubleshooting chart is similar to a map with different directions at a crossroad. Start at the top and follow the directions downward. After following the beginning instructions, there will be two or more choices to make. Find the condition that matches the problem at hand. Then follow the chart downward again for further checks, explanations, or instructions for performing repairs (Figure 7-23). Manufacturer's service manuals for modern vehicles contain hundreds of such charts. A specific chart is needed for a certain vehicle, for varying engine and transmission combinations, and so on. Service manuals also list the meanings of abbreviations used by the manufacturer. Since no technician can memorize all of these charts, it is important to find and use the correct one in the appropriate service manual.

Troubleshooting guides, on the other hand, present remedial actions based on driveability and performance symptoms. The guides offer no testing procedures or recommendations. Their purpose is to assist the technician in locating a fault. Typical troubleshooting guides for a throttle body fuel injection system and a port-type system are given in Table 7-1 and Table 7-2.

ELECTRONIC DIAGNOSTICS

All of the major domestic manufacturers have an on-board diagnostic system for use on their spe-

cific types of electronic fuel injection (Figure 7-24). The diagnostic systems employed provide:

- Computer codes that can be read by the driver/owner or technician on the dashboard
- Computer codes that can be read on a tester by the technician to troubleshoot the fuel injection system
- Computer test commands that can exercise and test the fuel injection system

The on-board diagnostic system monitors at least twenty functions including the charging system, distributor, manifold absolute pressure, engine temperature, fuel system, and oxygen sensor. The testers used with the diagnostic system can troubleshoot the ECU, its sensors, and the fuel injection components that make up the system. The testers can also check many switches that send information to the computer or ECU (Figure 7-25).

The on-board diagnostic system is built into the ECU, which consists of a logic module and a power

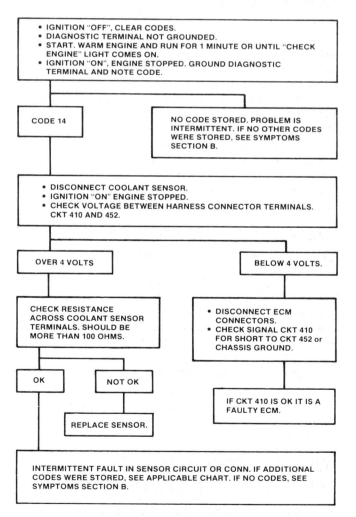

FIGURE 7-23 Typical troubleshooting charts

TABLE 7-1: TROUBLESHOOTING GUIDE FOR THROTTLE BODY FUEL INJECTION SYSTEMS

Condition	Possible Cause	Remedy
Preliminary checks		Check fuel system for fuel leaks.
		Check battery state of charge.
		Check all wiring and connections.
		Check cooling system level.
		Check ignition system.
		Check air cleaner and preheat system.
		Check fuel system pressure.
		Check fuel lines for restrictions.
		Check vacuum hoses for leaks and restrictions.
Hard start, cold or rough idle, cold	CTS	Check coolant level or replace sensor.
	Fuel pressure bleed down	Check for fuel leak or defective fuel pump.
	Leaking manifold gasket or base gasket	Replace defective gasket.
	MAT sensor	Replace defective MAT sensor.
	Wrong PCV valve	Replace PCV valve.
Stalling, hesitation, surging, hot or cold	CTS	Check coolant level or replace sensor.
	Lower fuel system pressure	Check fuel filter and fuel pump; service or replace as required.
	Restricted air intake system	Check air cleaner and preheat system; service or replace as required.
	TPS defective or not adjusted correctly	Check TPS; adjust or replace as required.
Hard start, hot	Bleeding injector	Inspect injector for dripping; service or replace as required.
	Leaking intake manifold gasket or base gasket	Replace defective gasket.
	MAP sensor	Check MAP sensor and vacuum hose; service or replace as required.
Rough idle, hot	MAP sensor	Check MAP sensor and vacuum hose; service or replace as required.
	CTS	Check coolant level or replace sensor.
	TPS	Check TPS; adjust or replace as required.
Stalling	ISC/IAC	Check idle speed control device; service or replace as required.
	TPS	Check TPS; adjust or replace as required.

TABLE 7-1: TROUBLESHOOTING GUIDE FOR THROTTLE BODY FUEL INJECTION SYSTEMS (CONTINUED)

Condition	Possible Cause	Remedy
Poor power	Dirty injector	Check injector spray pattern; clean or replace injector as required.
	Fuel pump	Check fuel pump pressure; replace fuel pump.
	Fuel pump pickup strainer	Check strainer; replace as required.
	Fuel filter	Check fuel filter; replace as required.

ABBREVIATIONS:

TBI:	Throttle Body Injector	ISC:	Idle Speed Control
CTS:	Coolant Temperature Sensor	IAC:	Idle Air Control
MAT:	Manifold Air Temperature Sensor	MAP:	Manifold Absolute Pressure Sensor
TPS:	Throttle Position Sensor	ACT:	Air Charge Temperature Sensor

TABLE 7-2: TROUBLESHOOTING GUIDE FOR PORT FUEL INJECTION SYSTEMS

Condition	Possible Cause	Remedy
Preliminary checks		Check fuel system for fuel leaks.
		Check battery state of charge.
		Check all wiring and connections.
		Check cooling system level.
		Check ignition system.
		Check air cleaner and preheat system.
		Check fuel system pressure.
		Check fuel lines for restrictions.
		Check vacuum hoses for leaks and restrictions.
Hard start, cold or rough idle, cold	CTS	Check coolant level or replace sensor.
	Fuel pressure bleed down	Check for fuel leak or defective fuel pump.
	Cold start injector	Check cold start injector; service or replace as required.
	Leaking manifold gasket or base gasket	Replace defective gasket.
	ACT/MAT Sensor	Replace defective ACT/MAT sensor.
	Wrong PCV valve	Replace PCV valve.
	Warm-up regulator	Replace warm-up regulator.
	Injector	Check injectors for variation in spray pattern; clean or replace injectors as required.
	Mass airflow sensor	Check airflow meter, fuel pump contacts.

TABLE 7-2: TROUBLESHOOTING GUIDE FOR PORT FUEL INJECTION SYSTEMS (CONTINUED)

Condition	Possible Cause	Remedy
	Pressure regulator	Check pressure regulator for setting and bleed down.
Hesitation or surging, hot or cold	CTS	Check coolant level or replace sensor.
	Low fuel system pressure	Check fuel filter and fuel pump; service or replace as required.
	Restricted air intake system	Check air cleaner and preheat system; service or replace as required.
	TPS defective or not adjusted correctly	Check TPS; adjust or replace as required.
	Mass airflow sensor	Check airflow meter, fuel pump contacts
	ACT/MAT sensor	Replace defective ACT/MAT sensor.
	Air leak in air intake system	Check gaskets, hoses and ducting; service or replace as required.
	Defective oxygen sensor	Replace oxygen sensor.
	Defective computer	Replace computer.
Hard start, hot	Bleeding injector	Inspect injector for dripping; service, or replace as required.
	Leaking intake manifold gasket or base gasket	Replace defective gasket.
	MAP sensor	Check MAP sensor and vacuum hose; service or replace as required.
	Pressure regulator	Check pressure regulator for setting and bleed down; service or replace as required.
Rough idle, hot	MAP sensor	Check MAP sensor and vacuum hose; service or replace as required.
	CTS	Check coolant level or replace sensor.
	TPS	Check TPS; adjust or replace as required.
	Injector	Check injector for variation in spray pattern; clean or replace injector as required.
	Oxygen sensor	Replace oxygen sensor.
	Defective computer	Replace computer.
	ISC/IAC	Check idle speed control device; service or replace as required.
Stalling	ISC/IAC	Check idle speed control device; service or replace as required.
	TPS	Check TPS; adjust or replace as required.

TABLE 7-2:   TROUBLESHOOTING GUIDE FOR PORT FUEL INJECTION SYSTEMS (CONTINUED)

Condition	Possible Cause	Remedy
	MAP Sensor	Check MAP sensor and vacuum hose; service or replace as required.
Poor power	Dirty injector	Check injector spray pattern; clean or replace injector as required.
	Fuel pump	Check fuel pump pressure; replace fuel pump.
	Fuel pump pickup strainer	Check strainer; replace as required.
	Fuel filter	Check fuel filter; replace as required.
	Pressure regulator	Check pressure regulator for setting and bleed down; service or replace as required.

ABBREVIATIONS:

CTS: Coolant Temperature Sensor
MAT: Manifold Air Temperature Sensor
TPS: Throttle Position Sensor
ISC: Idle Speed Control

IAC: Idle Air Control
MAP: Manifold Absolute Pressure Sensor
ACT: Air Charge Temperature Sensor

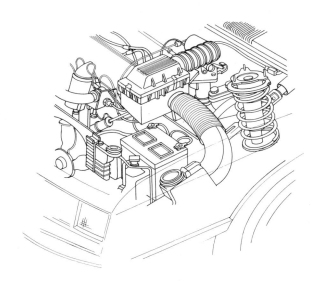

FIGURE 7-24   On-board diagnostic system

FIGURE 7-25   Typical EFI diagnostic tester

module (Figure 7-26). The logic module is really the brains of the operation. It evaluates input from sensors and switches, then sends some commands to solenoids and relays and others to the power module. The power module puts muscle behind the commands from the logic module. The power module regulates fuel flow at the injectors by controlling pulse width, and also regulates ignition timing.

Built into the logic module is a memory for on-board diagnostics (Figure 7-27). The memory records and stores any system trouble in the form of

fault or trouble codes. When a fault occurs, a trouble light illuminates on the tester. Turning the ignition switch on and off several times within a specified time interval will cause the tester light to begin flashing fault codes.

If the light flashes twice, pauses, then flashes twice again, it indicates code 22, a problem that could be interpreted to mean a defective temperature sensor circuit. Code 12 might be an indication that the battery feed to the logic module has been disconnected within the previous twenty to forty en-

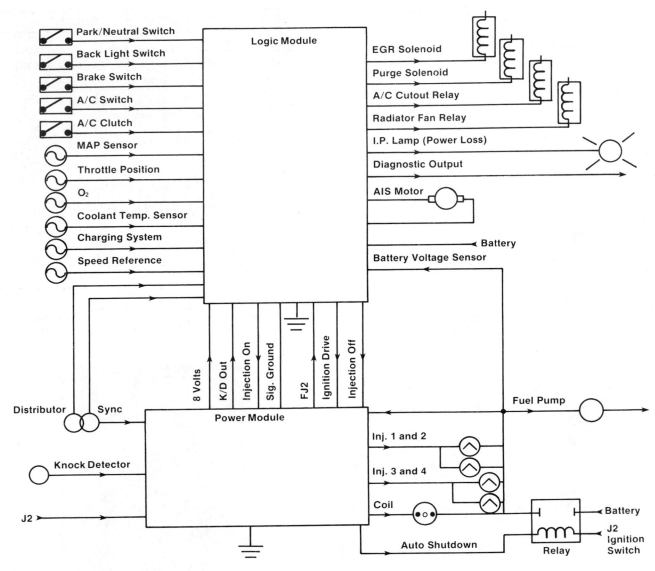

FIGURE 7-26 ECU and system diagnostics

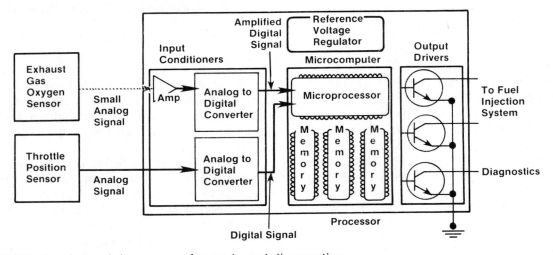

FIGURE 7-27 Logic module memory for on-board diagnostics

gine starts. And a code 55 could indicate the end of the message.

Variations in tester designs and the assignment of fault codes is different for each domestic manufacturer (Figure 7–28). The methodologies used to obtain the fault codes also vary from manufacturer to manufacturer.

Table 7–3 shows typical fault codes that are generated, indicating the area in which the problem exists. A code number 21, for example, could mean trouble in the oxygen sensor circuit. It does not always mean that the sensor is defective. It might be a broken wire, loose connector, or something else. The code number is localizing the possible area of the problem, and it is up to the technician to pinpoint the exact culprit. On some testers, if a fault code is

FIGURE 7–28   Tester connections and use with the on-board diagnostic system

Code	Circuit Affected or Possible Cause	Model 1988	Year 1989
11	Distributor reference circuit	X	X
12	Battery feed to logic module recently lost	X	X
13	MAP sensor circuit (vacuum)	X	X
14	MAP sensor circuit (electric)	X	X
15	Vehicle speed sensor	X	X
16	Loss of battery voltage		X
21	Oxygen sensor circuit	X	X
22	Coolant temperature circuit	X	X
23	Charge temperature circuit (turbo)	X	
24	Throttle position circuit	X	X
25	Automatic idle speed circuit	X	X
26	Fuel (peak injector; current not reached)		X
27	Fuel (no current in diagnostic transistor)		X
31	Purge solenoid	X	X
32	Power loss light circuit	X	X
33	A/C wide-open throttle circuit	X	X
34	EGR solenoid circuit	X	
34	Spare driver circuit		X
35	Fan relay circuit	X	X
36	Spare driver circuit		X
37	Shift indicator circuit (manual)		X
41	Charging system	X	X
42	Auto shutdown relay circuit	X	X
43	Ignition and fuel control interface	X	X
44	Logic module	X	
44	Battery temperature out of range		X
45	Overboost (turbo)	X	
46	Battery voltage high		X
47	Battery voltage low		X
51	Oxygen feedback system	X	
51	Closed loop latched lean		X
52	Closed loop latched rich		X
52 & 53	Logic module	X	
53	ROM bit sum fault		X
54	Distributor signal circuit	X	
55	End of test sequence	X	X
88	Start of test sequence	X	X

retained but the problem is only temporary, the logic module in the ECU will erase the code after a set number of ignition on/off cycles or employ some other time reference technique to reset the diagnostic system.

DIAGNOSING CHRYSLER'S FUEL INJECTION

In 1972, Chrysler became the first major auto manufacturer to offer electronic ignition as standard equipment. This seemed like a rather bold move, but once the advantages became evident, everyone began replacing mechanical breaker points with transistorized systems.

Four years later, Chrysler recorded another industry first with the introduction of a controversial electronic spark control system called Lean Burn. While the concept of burning a lean air/fuel mixture was not entirely revolutionary, the use of a microcomputer to control spark timing certainly was. Prior to Lean Burn, microcomputers were used in devices such as hand-held calculators, microwave ovens, and cash registers. After the debut of Lean Burn, however, the microcomputer was on its way to becoming a cornerstone in the design of automotive engine control systems.

Chrysler's engine control system is typical of most computer-controlled designs. However, there are enough differences, especially when it comes to system control, to warrant further review.

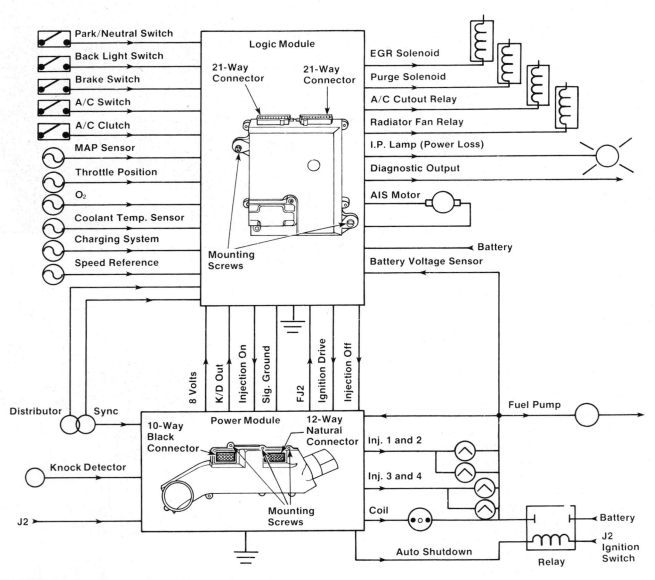

FIGURE 7-29 Chrysler's on-board diagnostic system

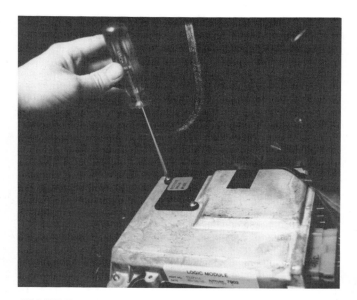

FIGURE 7-30   Logic module location

At the heart of Chrysler's computer control system is a digital preprogrammed microprocessor known as the logic module (LM) (Figure 7-29). From its location behind the right front kick pad (in the passenger compartment as shown in Figure 7-30), the LM issues commands affecting fuel delivery, ignition timing, idle speed, and operation of various emission control devices.

To accomplish all these tasks, the logic module operates in conjunction with a subordinate control unit called the power module (PM) (Figure 7-31), which controls the injector and ignition coil ground circuit (based on the LM's commands). It also supplies the ground to the automatic shutdown relay (ASD).

The ASD relay (Figure 7-32) controls the voltage supply to the fuel pump, logic module, injector, and coil drive circuits and is energized through the PM when the ignition switch is turned on. As a safety precaution, the PM will open the ASD relay circuit (by taking away the ground) after 1 second if it does not receive a signal from the distributor pickup indicating that the engine is being cranked or is running.

In addition to circuit control, the PM also contains a DC to DC voltage converter that reduces battery voltage used by the LM and Hall effect pickup to a regulated 8 volts.

The PM is located in the left front fender well behind the battery. Although this location makes access rather difficult, there are some logical reasons for its being there. First, the PM's switching transistor needs sufficient airflow to help keep it cool. Second, this remote location prevents "electri-

cal noise" from reaching and interfering with the logic module's operation.

OPERATION

Under normal operating conditions, the LM relies on information generated by several input devices or sensors to control vehicle operation. These data collectors include the MAP sensor, coolant temperature sensor, oxygen sensor, vehicle speed sensor, TPS, detonation sensor, charge temperature sensor, and a Hall effect pickup that provides data on engine rpm and crankshaft position. In addition to sensor input, the logic module also monitors the position of the brake switch, air-conditioner clutch

FIGURE 7-31   Power module location

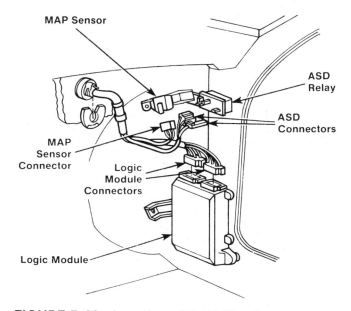

FIGURE 7-32   Location of the ASD relay

switch, neutral safety switch, heated back light switch, and wastegate control solenoid.

The logic module uses all this information to calculate the injector's pulse width, determine precise spark advance, maintain proper idle speed, govern turbo boost, control the EGR and purge solenoids, and direct the operation of the electric cooling fan (among other things). The LM even has a back-up plan, called *limp-in,* for when things go wrong.

The limp-in mode is nothing more than the computer's attempt to take control of vehicle operation when input from one of its critical circuits—the MAP, throttle position, coolant temperature, and/or charge temperature sensor—has been lost. If the computer sees a problem with the input from any of the previously mentioned devices, it will either work with fixed values in place of the failed sensor input or generate a modified value by combining two or more related sensor inputs.

To illustrate this last point, assume that the MAP sensor stops working (Figure 7-33). In place of the actual MAP measurement, the LM compensates by creating an "artificial" MAP signal from a combination of throttle position input and engine speed data. While this might not result in the most efficient operation, considering the alternatives, a modified MAP signal is better than no MAP signal.

In addition to the four sensors mentioned above, other problems that rate special consideration in Chrysler's computer control system include: incorrect voltage values at the waste gate control solenoid, excessive battery voltage (more than 1 volt above the desired control voltage), and a low battery voltage reading in the battery sensing circuit. By special consideration, the reference is made to the fact that any time a problem in one of these areas occurs, the LM will store a fault code identifying the problem circuit in its memory and turn on the power loss power limit light located in the instrument panel.

As for other computer-related problems (see the list of fault codes in Table 7-4), a fault code will set and store in memory, but the power loss/power limit light will not come on.

Once a fault code has been stored in memory, it will remain there until the battery is physically disconnected by the technician (at the quick-disconnect coupling) or the computer no longer sees a problem and erases the code. In the latter case, the computer is programmed to keep track of trouble conditions for twenty to forty engine starts. After that, if the problem is no longer detected, the computer will erase the code.

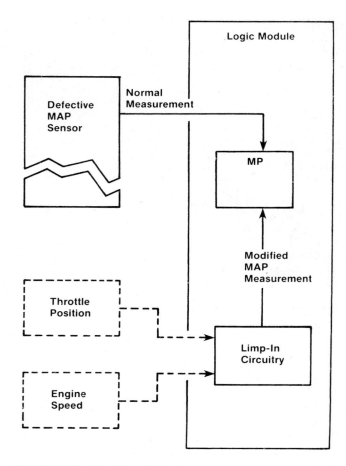

FIGURE 7-33 Use of the limp-in mode for a defective MAP sensor

TROUBLE CODES

The technician's preliminary diagnosis should include the following: a talk with the customer to find out what symptoms are present, a vehicle road test if possible, a thorough visual examination of all engine subsystems (including electrical connections, vacuum lines, and the like) and, of course, verification of the basics—ignition, fuel, and compression. After all that, if the problem persists, use the LM's on-board diagnostic capabilities.

Within the framework of Chrysler's driveability test procedure (which includes the LM's on-board diagnostic system) exists a variety of service codes and test modes, one of which—the fault code—has already been identified and discussed.

In addition to fault codes, indicator codes, actuation test mode (ATM) access codes, and diagnostic process codes have been developed to help guide the technician through Chrysler's diagnostic process. To understand what each of the code catego-

TABLE 7-4: CHRYSLER MPFI TROUBLE CODES

Code	Circuit Affected or Possible Cause	Power Loss Light (Run Mode Only)	Reset Required
11	Engine not cranked since battery disconnected.		
12	Memory standby power was recently lost.		
13	MAP sensor circuit (pneumatic)	On	Yes
14	MAP sensor circuit (electrical)	On	Yes
15	Vehicle speed sensor		
16	Loss of battery voltage sense	On	No
17	Engine running too cool		
21	Oxygen sensor circuit		
22	Coolant temperature sensor circuit	On	Yes
23	Charge temperature sensor circuit	On	Yes
24	Throttle position sensor circuit	On	Yes
25	AIS motor driver circuit		
26	Injector 1 circuit—these will open		Yes
27	Injector 2 circuit—shutdown relay		Yes
31	Purge solenoid driver circuit		
32	Not used		
33	A/C cutout relay driver circuit		
34	EGR solenoid driver circuit		
35	Fan control relay driver circuit		
36	Waste gate solenoid driver circuit	On	Yes
37	Baro read solenoid driver circuit		
41	Charging system—excess or no field current		
42	Auto shutdown relay driver		
43	Spark interface (logic to power module)		
44	Battery temperature sensor circuit		
45	Overboost shutoff		
46	Battery voltage too high	On	No
47	Battery voltage too low		
51	Closed loop fuel system latched lean		
52	Closed loop fuel system latched rich		
53	Rom bit sum fault		
54	Distributor sync pickup circuit		
55	End of message		

ries means, a closer look is needed to determine how and when each code is used.

1. *Indicator Codes.* These are two-digit numbers that tell the technician if certain conditions or sequences have occurred. For example, a code 88 at the beginning of the diagnostic test mode indicates the self-diagnostics have been properly activated and fault code transmission is about to begin. After all fault codes have been reported, a code 55 will appear indicating the end of the diagnostic test mode. As for the remaining indicator codes, a zero means the oxygen feedback system is running rich; a 01 indicates lean feedback system operation; 08 means the detonation sensor is detecting detonation; and 12 means that

the power feed to the battery has been disconnected some time within the last twenty to forty engine starts.

2. *ATM Test Codes.* These are two-digit numbers assigned to specific output circuits for the purpose of circuit identification. In other words, ATM test codes are the technician's way of telling the LM what actuators the technician wants to activate (such as canister purge, A/C cutout relay, EGR solenoid, and so on) during the circuit actuation test mode.

3. *Sensor Access Codes.* These allow the technician to selectively monitor the operation of various input devices. Like ATM test codes, sensor access codes are also two-digit numbers, but they are invalid and

of no use unless you are in the sensor access test mode.

TEST MODES

The five test modes are part of a total systematic approach that has been designed by Chrysler to help the technician determine the exact cause of a driveability problem. Taking shortcuts, failing to perform the test when requested, or ignoring these tests all are unwise.

The five test modes are:

1. Diagnostic
2. Circuit actuation
3. Sensor
4. Switch
5. Engine running

Diagnostic Test Mode

This mode is used to see if there are any fault codes stored in the on-board diagnostic system's memory. There are several ways to access fault codes. Either use the power loss/power limit light to flash out fault codes (for example, flash-flash-pause-flash equals code 21) or get a diagnostic readout box. If the former is chosen, all that is required to make the light start flashing is to turn the ignition key on-off-on-off-on within 5 seconds. However, be observant and prepared, because once the codes start transmitting, each will appear only once. Failure to observe the codes necessitates restarting this test mode.

The second choice is to use a diagnostic readout box (Figure 7-34). Chrysler's readout box, which is similar to Ford's STAR tester, is one option, but there are plenty of aftermarket equivalents available that do more and cost less.

The diagnostic readout box is needed to perform the circuit actuation test, sensor test, and engine running test. In addition, the following should also be on hand to properly diagnose the system: volt/ohmmeter, jumper wire, hand-held vacuum pump, timing light, fuel pressure gauge (high-pressure EFI type), and a service manual (Figure 7-35).

To obtain fault codes with the readout box, first locate the diagnostic connector located in the wiring harness next to the right front shock tower (see Figure 7-34).

After the readout device is plugged in, be sure that the read/hold switch on the readout box is in the READ position and turn the ignition switch on-off-on-off-on within 5 seconds to activate the test mode. If code 88 is indicated on start-up, it means the diagnostic mode is selected and fault codes will appear next. To stop the test for any reason once the fault codes start transmitting, place the read/hold switch in the HOLD position to freeze the display. When operation is to be resumed, simply move the switch back to the READ position. After all of the fault codes have been displayed, a code 55 will appear indicating the message is over.

From this point on, the remainder of the diagnostic tests are designed to identify the specific causes of the fault codes received during the diagnostic test mode.

FIGURE 7-34 Connection of the diagnostic readout box

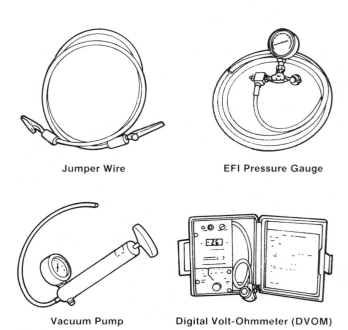

Jumper Wire

EFI Pressure Gauge

Vacuum Pump

Digital Volt-Ohmmeter (DVOM)

FIGURE 7-35 Additional equipment required to service on MPFI system

Circuit Actuation Test (ATM) Mode

To enter the ATM test, put the system into the diagnostic test mode (use the on/off ignition key sequence), make sure the READ/HOLD switch is in the READ position, and wait for code 55 to display. After that, press and hold the ATM switch until the desired ATM test code appears, then release the switch.

What happens next depends on the circuit selected. For example, if ATM code 04 was chosen, the radiator fan will turn on and off every 2 seconds for up to 5 minutes. Code 05, on the other hand, instructs the logic module to activate the air conditioner cutout relay. Like the radiator fan, if the cutout relay is working, it will cycle on and off at 2-second intervals for up to 5 minutes. To stop the test before the 5-minute cycle is up, press the TM switch to display the next code or turn off the ignition switch.

Some of the other ATM codes and the circuits they represent include (Figure 7-36):

01 Spark control (The ignition system will generate three sparks if the coil wire is held 1/4 inch from ground.)
02 Injector (Listen for a clicking sound.)
03 Automatic idle speed motor (cycles one step open and one step closed every 4 seconds)
06 Automatic shutdown (ASD) relay
07 Purge solenoid
08 EGR solenoid
09 Waste gate control solenoid
10 Baro read solenoid
11 Alternator field control

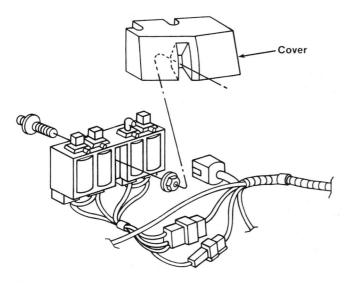

FIGURE 7-36   Some circuits that can be tested in the ATM test mode

Be aware that any time the ATM switch is pressed on the readout box, a fault code 42 will be generated. Do not worry about this code unless it turns up as a fault code and the ATM switch has not been pressed. Then treat it like any other fault code.

Sensor Test Mode

The sensor test is similar to the ATM test. But, instead of checking actuator output, the sensor test monitors select input circuits. To activate this test sequence, stay in the ATM test mode, hold the ATM switch down, and move the READ/HOLD switch to the HOLD position. The logic module will now use the readout box to display the output of the sensor selected. To move from sensor to sensor, press the ATM switch until the desired sensor access code appears.

Remember that in the sensor test mode all displayed readings, with the exception of the coolant temperature and battery voltage, must be divided by 10 to gain a true output reading. To obtain the correct temperature reading from the coolant temperature sensor, multiply the displayed reading by 10. Battery voltage will be displayed as an actual amount and needs no correction.

Switch Test Mode

The switch test is a quick and simple check designed to determine if specific switch inputs are being received by the logic module. Before entering the test, be sure all of the logic module input switches are turned off. To enter the switch test, put the system into the diagnostic test mode and wait for code 55 to be displayed and the power loss/power limit light to stop blinking. Once the test mode is selected and running, turn on each input switch, one at a time, and note the readout box display. If the display changes, it means the input is being received by the LM.

To use the power loss/power limit light for the switch test, enter the test mode following the same procedure as above (with the readout box), but instead of monitoring a display, simply watch the light. It should go on and off each time an input switch is turned on and off.

Engine Running Test Mode

This test is used to determine if the oxygen sensor feedback system is crossing over from rich to lean (and vice versa), and if the automatic idle speed motor and detonation systems are operational. With the diagnostic readout box connected, place the READ/HOLD switch in the READ position and start the engine. Once the oxygen sensor has reached operating temperature and the system is in closed loop, the display should start to alternate between 0 (lean) and 1 (rich) several times per second. If the reading does not change or response is slow, check the oxygen sensor circuit.

To check the automatic idle speed motor, idle the engine (in PARK or NEUTRAL), move the READ/HOLD switch to the HOLD position and note the engine rpm. It should increase to approximately 1500 rpm if the idle motor system is operating.

The last test in this sequence is designed to test the operation of the detonation sensor. With the engine speed above 1500 rpm, an 8 will appear in the display window next to the 0 or 1 (oxygen sensor indicator codes) when the detonation sensor is working. As an alternative to this last test, the technician can also tap on the intake manifold and watch the timing marks with a timing light. If the detonation sensor is working, the timing should retard with each knock.

After completing the driveability test modes to help locate and repair specific problems, clear the computer's memory. This task can be easily accomplished by disconnecting and reconnecting the quick disconnect coupling at the battery (the positive terminal). To verify that the memory has been cleared, recheck the fault codes. If you get a code 88-12-55 (if using the power loss/power limit light you will not get a code 88) with A/C or 88-12-33-55 without A/C, all is clear.

Once you have finished pulling codes, either start replacing parts piecemeal or follow the detailed, step-by-step logically developed troubleshooting sequences in the approproate manufacturer's service manual.

FUEL INJECTOR SERVICE

If any components must be removed or replaced in the fuel delivery system, always remember to depressurize the system first. This procedure is simple on Chrysler vehicles. After removing the fuel filler cap to relieve pressure in the tank, remove the electrical connection from an injector and, using two jumper wires, ground one injector terminal and connect the other to battery voltage. This will energize the injector, allowing pressure to bleed off. However, to avoid overheating the injector's sensitive solenoid coils, do not energize any one injector for more than 10 seconds.

If you are thinking about replacing an injector, the fuel rail must be removed first. Once the fuel rail assembly is unbolted, pull the rail so that the injectors come straight out of their ports. Next, remove

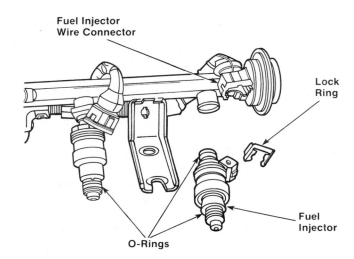

Fuel Injector Wire Connector

Lock Ring

Fuel Injector

O-Rings

FIGURE 7–37 Fuel injector service

the locking rings that secure the injectors to the rail and pull the injector straight out of the fuel rail receiver cup (Figure 7–37). If the injector is to be reused, always replace the injector O-rings, regardless of how they look.

To simplify injector installation, lubricate the new O-rings with a little ATF or engine oil and ease the injectors back into their fuel rail receiver cups. Replace the locking rings and reposition the fuel rail so that the injectors line up with their respective manifold slots. With a slight rocking motion, carefully seat each injector and retorque the fuel rail bolts to 250 inch-pounds.

Once the injectors are properly seated, check the installation with a flex head mirror. An improperly seated injector can result in a variety of problems ranging from a rough idle (small vacuum leak) to an overrevving condition (massive fuel and vacuum loss) that closely resembles a runaway diesel.

DIAGNOSING GM'S FUEL INJECTION

General Motor's computer command control (C-3) system offers the technician some definite diagnostic advantages over other domestic manufacturers, especially in the areas of data stream access and diagnostic testing capabilities.

The major components that make up the C-3 on-board diagnostic system (Figure 7–38) include a microcomputer called an electronic control module (ECM), input sensors, an assembly line communications link (ALCL), and a "check engine" or "service engine soon" warning light.

The primary job of the ECM is to control the subsystems (air/fuel ratio, ignition timing, EGR, canister purge, and the like) that make up the fuel control system. The ECM also contains built-in diagnostic circuits that enable it to detect and identify certain faults that can affect driveability.

The primary function of the sensors is to supply the computer with current information regarding the vehicle's operating condition. Refer to the accompanying component diagram shown in Figure 7-38 for information regarding sensor operation.

GM incorporates a warning light (Figure 7-39) into the C-3 system for two reasons. It is a signal to the vehicle operator that a detectable C-3 failure has occurred, and it can also be used by the service technician to aid in locating system malfunctions.

If a customer asks about the check engine light, assure him or her that it is not a panic light. On the other hand, do not downplay its significance either. Explain that it is simply a signal to remind the person to bring the vehicle in for a diagnostic inspection as soon as possible.

To make sure that the check engine or "CE" light is working, place the ignition switch in the RUN

FIGURE 7-39 GM's C-3 system warning light

position and observe the light. With the ignition switch on, the light should also be on. If the light does not come on, connect a test light between terminal A and D in the ALCL connector (Figure 7-40). If the test light glows with the key on, check for a

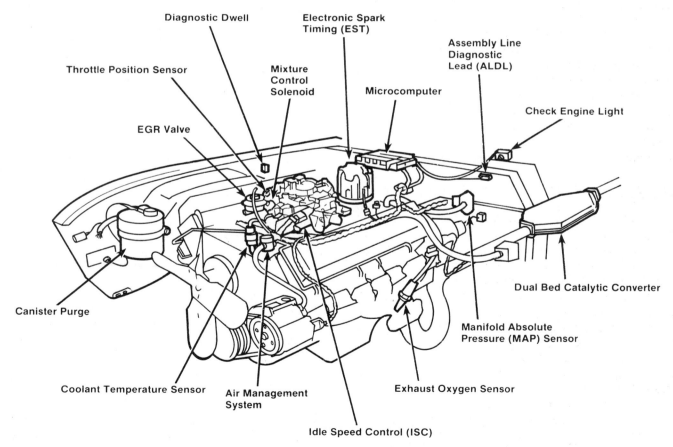

FIGURE 7-38 GM's computer command control system

FIGURE 7-40 GM's assembly line communication link (ALCL)

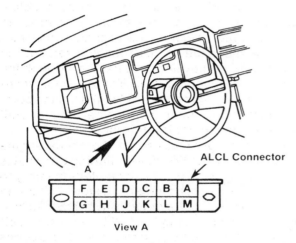

View A

FIGURE 7-41 ALCL connector pin assignments

burned out CE light bulb. If the bulb is all right, consult the service manual to find the cause of the inoperative light.

The ALCL is a connector that allows the technician to make certain diagnostic checks and read stored trouble codes (Figure 7-41). On most GM vehicles it is located under the dash. The ALCL was a five-pin connector (four pin on Chevette and T-1000) on 1981 models, but was changed to a twelve-pin design starting in 1982 (except Chevette and T-1000). Each pin or terminal is identified alphabetically starting with "A" at the upper right-hand position and ending with "M" at the lower left. Due to normal variations in each system's design, however, all pins are rarely used at once. When they are, here is what they mean:

- Terminal A is an ECM ground.
- Terminal B is the diagnostic "test" terminal.
- Terminal C is the air switching solenoid ground.
- Terminal D is the check engine light ground.
- Terminal E accesses serial data on fuel-injected vehicles.
- Terminal F is the TCC solenoid ground.
- Terminal G is a fuel pump test pin.
- Terminal K & L are BCM interfaces.
- Terminal M is for the level ride compressor.

By grounding or jumping different combinations of these test terminals, the technician can perform a variety of diagnostic checks from one central location.

SERVICE PRECAUTIONS

Before initiating a C-3 driveability diagnosis, some general service precautions should be observed. Keeping in mind that the computer and related components are very sensitive to voltage changes (spikes/surges), heat, and physical shock, here are a few precautions worth remembering.

To protect the system against damage from extreme voltage changes, do the following:

1. Never make or break electrical connections in any part of the C-3 wiring harness with the ignition switched on.
2. Do not charge the vehicle's battery with the battery cables still connected.
3. Avoid using a booster battery to jump start a vehicle with a C-3 system.

On a related note, extreme care should be taken to protect against damage resulting from static electrical charge. Recent studies have shown that voltage spikes caused by the discharge of "normal" static electricity can result in severe and permanent damage to a variety of electronic components, such as the digital dash elements and components of the microcomputer. Because tactic voltage is difficult to detect and virtually impossible to prevent, avoid handling or touching sensitive electronic parts unless both you and the part are properly grounded (allowing the static electricity to discharge).

To head off potential damage caused by overheating, always remove the computer prior to welding or exposure to the high temperatures generated by items such as paint drying lamps or baking ovens.

As a final precaution, if the computer must be removed for testing or service, protect it from sudden impact or physical shock while it is out of the vehicle. Also, to prolong the ECM's service life, make sure it is securely fastened into its special vibration-resistant mounts during reinstallation.

PRELIMINARY DIAGNOSTIC GUIDELINES

The number one mistake made by those who are inexperienced with computerized engine controls is to assume every driveability problem is related to the C-3 system. Jumping to conclusions can be very costly for both the technician (in terms of wasted time and comebacks) and the customer (saddled with expensive and unnecessary repairs).

Whenever a customer has a driveability complaint, the best way to avoid the temptation to immediately condemn the computer is to approach the vehicle as though the computer were not there. Take a few minutes to talk with the customer and find out what the symptoms are and under what conditions they occur. Start the diagnosis by looking for conventional problems first, such as vacuum leaks, fouled spark plugs, or arcing plug wires, and cover the basics—is there spark, fuel, and compression?

If these initial diagnostic analyses fail to locate the problem (assuming one exists), refer to a service manual that has complete diagnostic sequences and follow the diagnostic flow charts relating to the model line, engine design, and fuel system of the car being diagnosed. Skipping a step in the diagnostic sequence, using the wrong chart, or ignoring the manual are the best ways to misdiagnose a problem.

C-3 SYMPTOMS

If the engine is mechanically sound, all vacuum hoses and wires are properly routed, and the fuel delivery and ignition system check out, utilize the C-3 diagnostic capabilities when one of the following conditions exists:

- The check engine light is glowing steadily, flashing intermittently, or trouble codes are stored in the computer's memory.
- Incorrect ignition timing advance
- Idle problems such as stalling, wrong idle speed on ISC-equipped engines, or rough idle
- Rich or lean fuel condition, or symptoms that indicate carburetion problems such as surging, incorrect M/C dwell, or poor fuel economy
- Failure or malfunction of computer-actuated devices such as the canister purge, EGR valve, transmission torque converter clutch (TCC), or AIR system

Assuming the initial diagnosis has localized the C-3 system as being the possible suspect, continue the diagnosis with a careful visual inspection of the system affected (ignition, emission, fuel). Look for loose, dirty, or corroded connections and make sure there is a good ECM ground. If necessary, reroute all C-3 wires away from areas where electromagnetic induction emanating from plug wires and/or the coil could cause interference with or trigger false ECM signals. If everything appears to be in order, perform a diagnostic circuit check.

DIAGNOSTIC CIRCUIT CHECK

When certain types of faults occur in the C-3 system, a "check engine" warning light will come on and a trouble code will be stored in the computer's memory. If the light remains on while the vehicle is running, this indicates that a hard fault (problem present at the time of testing) exists. If the problem corrects itself or goes away (indicating an intermittent problem), a trouble code will still set, but the "check engine" light will go out 10 seconds after the problem is no longer detected. Trouble codes relating to intermittent problems will remain in the computer's memory for up to fifty engine starts or until power to the computer is interrupted.

Regardless of what causes the check engine light to come on, the first step in finding out what is wrong involves checking the computer's memory to see if any trouble codes have been stored. This can be accomplished in one of two ways. The quickest method is to simply plug the C-3 tester (scan tool) into the ALCL connector and read the trouble codes directly. If a C-3 tester is not available, rely on the "check engine" light to flash out codes.

To coax trouble codes out of any C-3 system using the "check engine" light method, simply jump terminal A and B together in the ALCL connector and turn on the ignition switch. The light should come on and immediately start flashing. These flashes are trouble codes. A single flash followed by a pause and then two more flashes (flash-pause-flash-flash) indicates code 12. Each code will be repeated three times before the next code is displayed. With the key on/engine off, code 12 is considered normal. If there are any other trouble codes stored in memory, they will be flashed out in a similar manner and in consecutive numerical order. See Table 7-5 for a list of typical GM trouble codes. When all stored codes have been displayed, code 12 will reappear and flash three times, signaling the end of code transmission.

TABLE 7–5: GM TROUBLE CODES

Code	Circuit Affected or Possible Cause	Code	Circuit Affected or Possible Cause
12	No distributor reference pulses to the ECM. This code is not stored in memory and will flash only while the fault is present. Normal code with ignition on, engine not running.	34	Vacuum sensor or Manifold Absolute Pressure (MAP) circuit. The engine must run up to 2 minutes at specified curb idle before this code will set.
13	Oxygen Sensor Circuit. The engine must run up to 4 minutes at part-throttle under road load before this code will set.	35	Idle speed control (ISC) switch circuit shorted. (Up to 70% TPS for over 5 seconds.)
14	Shorted coolant sensor circuit. The engine must run 2 minutes before this code will set.	41	No distributor reference pulses to the ECM at specified engine vacuum. This code will store in memory.
15	Open coolant sensor circuit. The engine must run 5 minutes before this code will set.	42	Electronic spark timing (EST) bypass circuit or EST circuit grounded or open.
21	Throttle Position Sensor (TPS) circuit voltage high (open circuit or misadjusted TPS). The engine must run 10 seconds at specified curb idle speed before this code will set.	43	Electronic Spark Control (ESC) retard signal for too long a time; causes retard in EST signal.
22	Throttle Position Sensor (TPS) circuit voltage low (grounded circuit or misadjsuted TPS). Engine must run 20 seconds at specified curb idle speed to set code.	44	Lean exhaust indication. The engine must run 2 minutes in closed loop and at part-throttle before this code will set.
23	M/C solenoid circuit open or grounded.	45	Rich exhaust indication. The engine must run 2 minutes in closed loop and at part throttle before this code will set.
24	Vehicle speed sensor (VSS) circuit. The vehicle must operate up to 2 minutes at road speed before this code will set.	51	Faulty or improperly installed calibration unit (PROM). It takes up to 30 seconds before this code will set.
32	Barometric pressure sensor (BARO) circuit low.	53	Exhaust Gas Recirculation (EGR) valve vacuum sensor has seen improper EGR vacuum.
		54	Shorted M/C solenoid circuit and/or faulty ECM.

Record any trouble codes received, but before proceeding, clear the computer's memory. This can easily be accomplished by pulling the ECM fuse or disconnecting the ECM feed wire at the battery for 10 seconds. Once cleared, start and run the vehicle (road test if necessary) for at least 5 minutes to see if the codes reset. It is not unusual to find a car with trouble codes that were left over from a previous repair. It is a good idea to start fresh and make sure any trouble codes indicated are valid.

After verifying the validity of all the trouble codes found, use the shop manual and refer to the troubleshooting procedures section. The tree charts have step-by-step instructions and the procedures needed to determine the exact cause of any hard fault condition.

When working with multiple codes, priority should be given to those in the 50s range since they indicate a problem in the ECM or PROM. Otherwise, check and repair codes in numerical sequence.

If no trouble codes are found, do not assume the C-3 system is all right until a system performance test is done and some of the essential sensor inputs checked. Many problems relating to C-3 operation will not set a trouble code. Problems of this type are often classified as "gray area problems" because they do not fit neatly into GM's diagnostic charts. For example, a misadjusted throttle position sensor (TPS) still supplies the computer with information, but it is not the right information. The circuit will meet all the self-diagnostic criteria (meaning no code will set), but as long as the ECM's decisions are

FIGURE 7-42   Use of the hand-held scan tool for diagnostic testing

based on incorrect data, the engine will not perform properly.

To diagnose situations such as the one mentioned above, a hand-held SCAN tool can be a valuable asset (Figure 7-42). With the SCAN tool, the actual voltage values generated by various C-3 sensors can be read and monitored. Once the actual voltage values are known, a comparison of the test results against specifications can be made.

TROUBLESHOOTING GM'S PFI SYSTEMS

When dealing with a fuel-related problem on a port injection system, the technician cannot watch the injector spray pattern. Fortunately, there are ways around this dilemma. An indirect indication of fuel to the cylinders can be made with the aid of an infrared analyzer (Figure 7-43). A high hydrocarbon (HC) reading during cranking means the fuel is getting to the cylinders. If an infrared unit is not handy or inclusive readings are obtained, further fuel system diagnosis is required.

Start the troubleshooting sequence on GM's PFI systems by checking the operation of the electric intank fuel pump. With the key on, listen for pump operation. The pump should run for 2 seconds and shut off. This 2-second energizing sequence is used to provide initial fuel pressure to the injectors for starting. If the ECM does not receive a distributor reference signal (indicating the engine is being cranked or has started), the pump is turned off. Once

FIGURE 7-43   Typical infrared analyzer

the car starts, the ECM will turn the relay on to power the pump.

If the pump does not run for 2 seconds with the key on, test for voltage with a test light or voltmeter. If an open circuit is indicated, check the fuel pump fuse, related wiring, and the pump relay mounted on the fire wall. A quick way to isolate a bad pump relay is to switch it with the air-conditioner relay right next to it (Figure 7-44).

As a backup system to the fuel pump relay, the pump can also be turned on by the oil pressure switch when lube pressure reaches about 4 psi. However, this will generally result in long cranking times. The technician, therefore, should immediately suspect a burnt relay whenever dealing with a hard start complaint.

PRESSURE CHECKS

Suppose the pump sounds as though it is working, but the car still does not start. At this point some fuel pressure checks are necessary. A fuel pressure gauge can be used to determine if fuel under suffi-

FIGURE 7-44 Confirming fuel pump relay operation

cient pressure (35 to 45 psi) is reaching the injectors. In the fuel rail of the multi-port system, a Schrader-type service fitting is provided for easy gauge hook-up (Figure 7-45).

Before hooking up a pressure gauge or performing any fuel system work, make sure the residual pressure is bled off first. The recommended procedure requires removing the fuel pump fuse and/or oil pressure switch connection (to kill the backup circuit) and running the car until it stalls. To be on the safe side, after the engine dies, try starting it one more time to confirm that it will not start.

In a no-start situation, the system might have pressure or it might not. To prevent the possibility of fuel spray, carefully connect the gauge while holding a rag around the service fitting. After connecting the gauge, bleed the air out of the gauge line. Since pressure is lost when the air is removed, the system

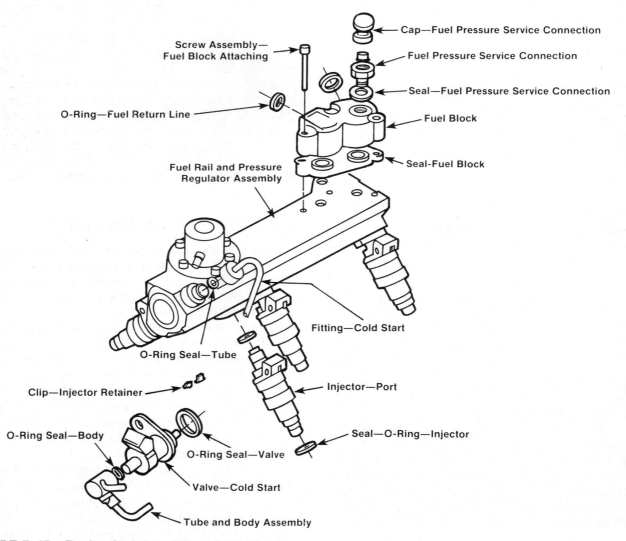

FIGURE 7-45 Engine fuel rial with cold start injector

will have to be repressurized. To build pressure, wait 10 seconds to allow the ECM to recycle, then turn the ignition key on to activate the fuel pump relay.

With the gauge hooked up, use it to visually verify pump operation. As soon as the ignition key is hit, pressure should build for 2 seconds and drop off slightly (3 to 4 psi) when the pump stops. Gauge pressure should read and hold at approximately 40 psi. A low reading can cause a no-start or poor running condition. Low readings can be caused by a blocked pickup tube sock, plugged fuel filter, faulty fuel pump, or kinked fuel line. If the pressure bleeds off too rapidly, suspect a faulty fuel pump check valve or open injector. Both of these will also cause a hard start condition.

In addition to checking pressure, a clogged filter, restricted line, or bad pump can be confirmed by performing a pump flow test. With system pressure relieved, disconnect the fuel inlet line and hold it in a suitable container (preferably a graduated one). Energize the pump by applying system voltage to the ALCL terminal "G." You should get a minimum of 1/2 pint in 15 seconds.

INJECTOR CHECKS

If the pump checks out, fuel flow and pressure are good, and the check valve is holding, the only things remaining on the fuel side to cause problems are the injectors. For a quick check, use a stethoscope to listen for the audible clicking or feel for the pulses that indicate injector operation. If you do not hear or feel anything, check for system voltage at each injector. If sufficient voltage is present, verify the supplied ground with a test light (Figure 7–46). Each time the injector is supposed to fire, the test

FIGURE 7–46   Troubleshooting a fuel injector

light should flash. Because the pulses are so close together, the flashes will appear as a rapidly flickering light. Perform this test on each injector individually.

If the injector's wiring circuit is good, check the injectors next. Injector windings can be tested for shorts, opens, or excessive resistance with an ohmmeter. Compare obtained resistance values to specs supplied in the shop manual. Do not assume that all injectors are the same. Any injector that has excessively high or infinite readings should be replaced.

After checking for specified resistance, perform an injector balance test. Figure 7–47 illustrates how the injector balance test is to be performed. Use of Kent-Moore tool J34730-3 (or equivalent) will simplify this procedure (Figure 7–48). Do not attempt to pulse the injectors with 12 volts. The injector coils will overheat and destroy the injector.

The tester is designed to pulse each injector the same amount for a controlled length of time while the technician monitors the pressure drop. Always reset the pressure after testing each injector. Look for inconsistencies in pressure readings from injector to injector.

Ideally, each injector should drop the same amount when opened. A variation of 1.5 to 2 psi or more is cause for concern. If there is no pressure drop or a low-pressure drop, suspect a plugged injector (assuming resistance is all right). A slight pressure drop indicates insufficient fuel delivery. On the other hand, a higher than average drop indicates a rich condition.

INJECTOR CLEANING

The relatively high cost of injectors suggests that cleaning should be tried first. Due to the inconsistency of today's gasoline blends, dirty fuel injectors are a common problem. To service a dirty injector, either clean or replace it.

Several varieties of injector cleaners are available (Figure 7–49). Some methods require the use of a pressure tank filled with chemical solvents and gasoline. These "on-the-car" systems generally connect directly into the fuel rail service fitting. Then start the car with the fuel pump disabled and the fuel lines blocked off. As the solvent/gasoline mixture runs through the injectors, the deposits should be dissolved without harming the oxygen sensor.

There are also ready-mixed alternatives. If prepackaged ready-mixed products are used, the pump still must be disabled and fuel lines blocked off. Because the cleaner comes in a ready-to-use single shot can, it only has to be hooked up to the service fitting. Then run the car until it dies; discard the can.

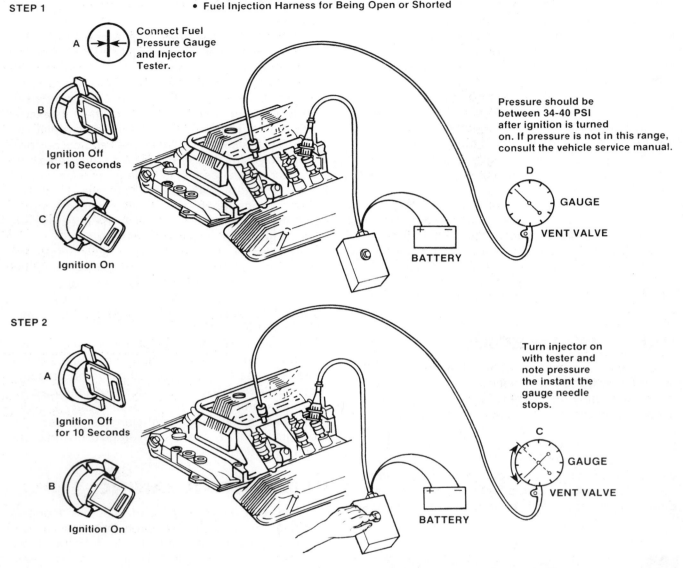

Before Performing This Test, Check the Following Items:
- Spark Plugs and Wires
- Compression
- Fuel Injection Harness for Being Open or Shorted

STEP 1

A — Connect Fuel Pressure Gauge and Injector Tester.

B — Ignition Off for 10 Seconds

C — Ignition On

Pressure should be between 34-40 PSI after ignition is turned on. If pressure is not in this range, consult the vehicle service manual.

D — GAUGE
VENT VALVE
BATTERY

STEP 2

A — Ignition Off for 10 Seconds

B — Ignition On

Turn injector on with tester and note pressure the instant the gauge needle stops.

C — GAUGE
VENT VALVE
BATTERY

STEP 3   Repeat test as in Step 2 on all injectors and record pressure drop on each.

Retest injectors that appear faulty. Replace any injectors that have a 1.5 to 2 PSI difference either (more or less) in pressure.

EXAMPLE

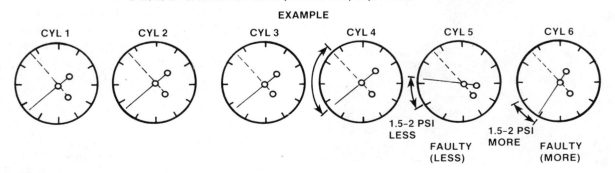

CYL 1   CYL 2   CYL 3   CYL 4   CYL 5   CYL 6

1.5-2 PSI LESS   FAULTY (LESS)   1.5-2 PSI MORE   FAULTY (MORE)

FIGURE 7-47   Port fuel injected system injector balance test

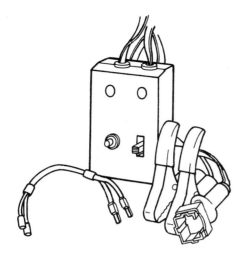

FIGURE 7-48 Typical fuel injector tester

After the injectors are cleaned, rerun the injector balance test and compare the results obtained before and after cleaning. If the problem was caused by dirty injectors and the cleaner worked, a pressure drop should be noted. If there is no pressure change, either try a stronger solvent or replace the injector.

Fuel pressure on PFI systems is automatically regulated in relation to manifold vacuum. A high vacuum signal (low manifold absolute pressure) will lower fuel pressure. As manifold vacuum drops off, absolute pressure increases, and fuel pressure should increase. With the engine running and the pressure gauge connected, a pressure change should be observed as the throttle is opened and closed.

As a final test to help isolate a troublesome injector, unplug the injectors one at a time while observing changes in engine speed. This test is similar to a cylinder balance test where plugs are individually shorted. Like a bad plug, a faulty or dirty injector will show up as a minimal rpm change. Compare the rpm drop across all injectors and look for consistency. To prevent the ECM from trying to compensate for any rpm drop, disconnect the idle air control (IAC) motor located in the throttle housing.

After completing any fuel system service, always make sure the portion of the ECM memory known as the integrator and block learn (I&BL) is cleared. This is accomplished by pulling the ECM fuse and waiting 10 seconds. This will neutralize the I&BL corrections and stop it from trying to compensate for problems already fixed.

DIAGNOSING FORD'S FUEL INJECTION

Some of the more common input and output devices found on a typical Ford Electronic Engine Control EEC-IV system are grouped according to

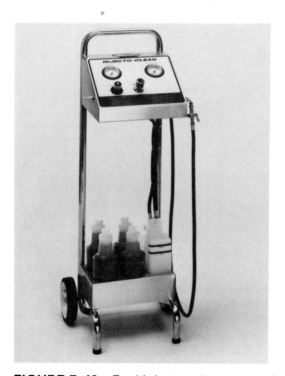

FIGURE 7-49 Fuel injector cleaners available on the market

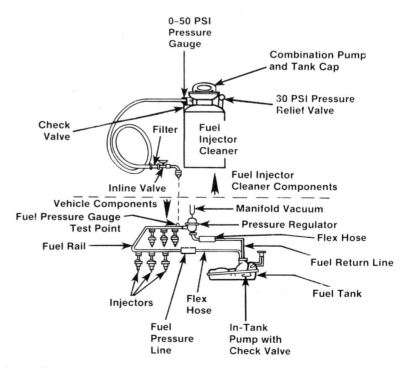

INPUTS

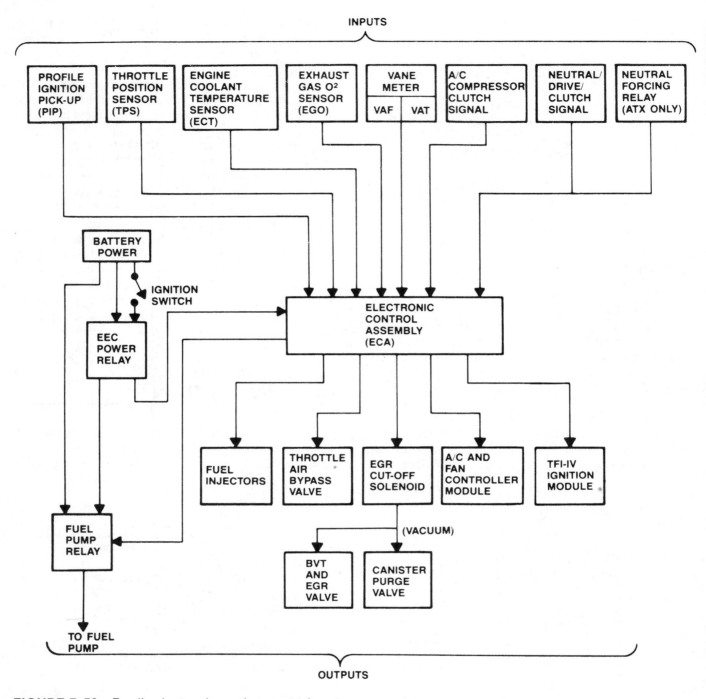

FIGURE 7-50 Ford's electronic engine control system

operating design and information type (input or output, Figure 7-50). There are three basic categories of input sensors: reference voltage, switches, and voltage-generating devices (Figure 7-51).

Reference voltage sensors include the throttle position sensor (TPS), manifold absolute pressure sensor (MAP), air charge temperature sensor (ACT), EGR valve position sensor (EVP), and the engine

coolant temperature sensor (ECT). On fuel-injected models, you can add a vane airflow (VAF) and vane air temperature (VAT) sensor.

Switches used in the EEC-IV system include the clutch engaged switch, idle tracking switch (ITS), power steering pressure switch (PSPS), neutral gear switch (NGS), and air conditioner clutch compressor signal switch (ACC).

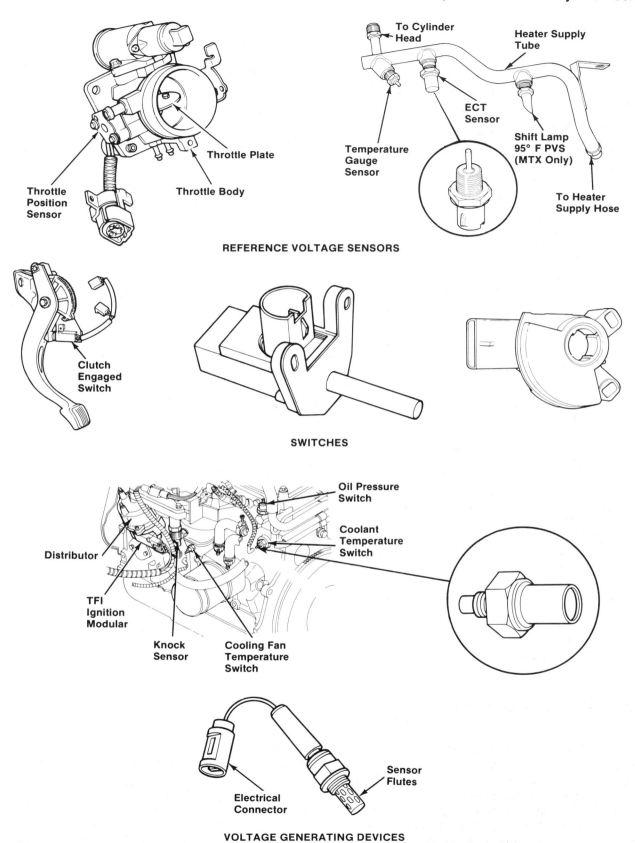

REFERENCE VOLTAGE SENSORS

SWITCHES

VOLTAGE GENERATING DEVICES

FIGURE 7-51 Types of input sensors

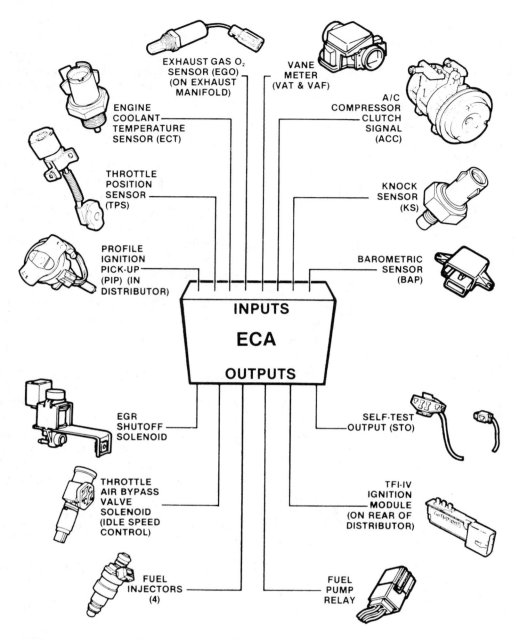

FIGURE 7-52 ECA—the heart of Ford's EEC-IV diagnostic system

Voltage-generating sensors capable of producing their own voltage signal include the exhaust gas oxygen sensor (EGO), the profile ignition pickup (PIP), and the knock sensor (KS).

Output devices can be divided into three categories: solenoids, electric motors, and controller modules.

Solenoids include the EGR shutoff solenoid, the canister purge solenoid (CANP), the carburetor feedback solenoid (FBC), the idle speed control solenoid (ISC), the throttle kicker solenoid (TKS), and the fuel injectors.

Electric motors used in the EEC system include the idle speed control motor (ISC), the fuel pump, and the cooling fan.

Controller modules are used to control more than one device. On the EEC-IV system, two controller modules are used. One is the air conditioner and cooling fan controller module and the other is the integrated controller module (ICM).

The electronic control assembly (ECA) is the heart of the EEC-IV system (Figure 7–52). The ECA is a two-chip microprocessor system in which all of the input/output and computational power resides on one chip, and the application's specific program memory resides on the other.

DIAGNOSTIC FEATURES

Like most computer-controlled systems, Ford's EEC-IV is endowed with self-diagnostic capabilities. By entering a mode known as self-test, the computer is able to evaluate the condition of the entire electronic system, including itself. If problems are found, they will show up as either hard faults (on-demand) or intermittent failures.

A hard fault means a problem has been found somewhere in the EEC system at the time of the self-test. An intermittent problem, on the other hand, indicates a malfunction occurred, such as a poor connection causing an open or short, but is not present at the time of the self-test. A feature known as the keep-alive memory (KAM) allows intermittent faults to be stored for up to twenty key on/off cycles. If the trouble does not reappear during that period, it is forgotten.

The ECA communicates its problems to the outside world by means of trouble codes that can be read only with the aid of special test equipment. If you are used to working on GM C-3 or Chrysler computer-controlled systems, do not bother looking for the equivalent of a flashing "service engine soon" light (or "check engine light") or "power loss" light. Ford does not use any type of flashing lights to signal failures.

To read trouble codes on the EEC-IV system, one of the following is needed: a STAR tester, SCAN tester, or analog (needle-type) voltmeter. The STAR (Self-Test Automatic Readout) tester is designed to read trouble codes on EEC-IV systems only (Figure 7-53). Once it is plugged into the diagnostic connector, it can do only one of two things: display trouble codes (digitally) or allow the technician to activate and deactivate the self-test mode.

A hand-held SCAN tester is definitely more versatile than the STAR. In addition to displaying trouble codes on Ford's systems, most scan tools are programmed to work on GM and Chrysler computer-controlled systems as well (Figure 7-54).

An ordinary analog voltmeter can also be used to perform the self-test (Figure 7-55). The voltmeter method is not nearly as convenient as the other testers, but once the self-test sequence is known, it makes little difference what is used. In this chapter, an analog voltmeter will be used to display the codes.

PRELIMINARY INSPECTIONS

When the complaint is presented, do not automatically assume the EEC-IV system is at fault. According to Ford, the key to making an accurate diagnosis of engine performance problems on a vehicle equipped with EEC-IV is to approach the vehicle as though it did not have the system. In other words, do

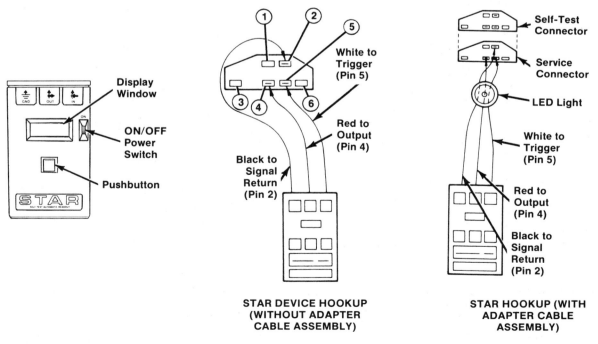

STAR DEVICE HOOKUP (WITHOUT ADAPTER CABLE ASSEMBLY)

STAR HOOKUP (WITH ADAPTER CABLE ASSEMBLY)

FIGURE 7-53 STAR tester for Ford diagnostics

FIGURE 7-54 SCAN tester

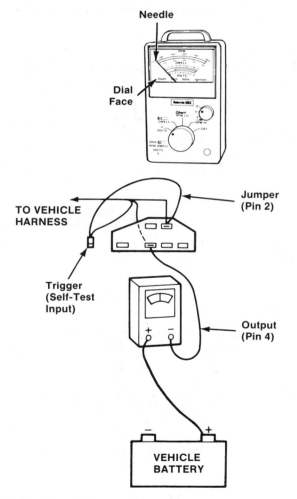

FIGURE 7-55 Typical voltmeter hookup for reading codes

not forget about the engine that is under all those sensors and actuators.

Start by looking for loose, corroded or damaged electrical connections. Next, check the condition and routing of all vacuum lines and fittings and take some vacuum readings with a vacuum gauge. Any type of leak in the air induction system, especially on fuel injected engines, will greatly affect engine performance. On turbo-equipped models, something as inconspicuous as a loose oil filler cap can cause the engine to run roughly at idle.

Finally, the technician should inspect the basics—fuel, ignition, and compression. No matter how sophisticated the electronic control system is, no computer has been developed that can compensate for an open plug wire, burned valve, or clogged fuel filter.

Then there is engine temperature to consider. Before entering the self-test mode, the engine has to be fully warmed up. If it is not, the computer will output a code 21 (ECT out of range) or code 41 (system always lean). When normal operating temperature is reached, the electric cooling fan will come on and the upper radiator hose will feel hot and pressurized. If the engine cools off too much between steps, a code 43 might be displayed, which means EGO sensor cool-down has occurred. To warm the sensor, start and run the engine for two minutes at 1500 rpm.

CHECKING THE BASICS

Base ignition timing should also be checked. To check base timing on EEC-IV systems, disconnect the single wire, in-line connector (spout connector) from the TFI module on the distributor, hook up a timing light in the normal fashion, and check for the required base spec (usually about 10 degrees BTDC) at the timing marks. Adjust if necessary and do not forget to reconnect the in-line connector. Leave the light hooked up because it will be needed during one of the self-test procedures.

To avoid sending a false reference signal to the ECA while it is in the self-test mode, turn off all electrical loads and keep the doors closed. Any changes in current draw during the test can confuse the ECA, causing it to set an erroneous code.

On Tempo/Topaz models equipped with the 2.3-liter HSC engine and automatic transmission, remove the rubber cap from the vacuum restrictor (VREST) that is located between the Thermactor air control valve (ACV) and the vacuum retard-delay valve (VRDV). If this is not done, a code 41 will be output to the tester.

FIGURE 7-56 Getting ready to run Ford's self-test diagnostic procedure

SELF-TESTS

Ford calls their ECC-IV's self-diagnostic procedure a self-test, which can be broken down into four parts:

1. *Key On/Engine Off (Figure 7-56).* This checks the system inputs for hard faults (malfunctions that occur during the self-test) and intermittent faults (malfunctions that occurred sometime prior to the self-test and were stored in memory). Sample output code formats are shown in Figure 7-57 and Figure 7-58.
2. *Computed Ignition Timing Check.* This checks the ECA's ability to advance or retard ignition timing. It is made while the self-test is activated and the engine is running (Figure 7-59).
3. *Engine Running Segment.* This checks the system's outputs for hard faults only.

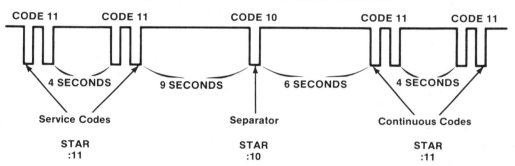

FIGURE 7-57 Key on/engine off self-test output code format

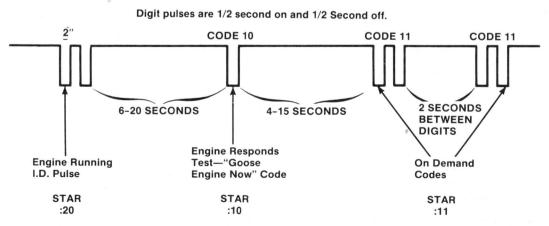

FIGURE 7-58 Key on/engine running self-test output code format

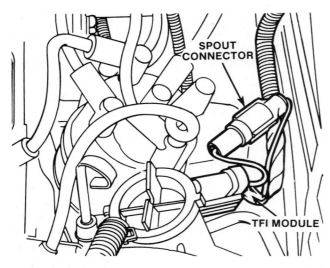

FIGURE 7-59 Ford's computed time check diagnostic procedure

4. *Continuous Monitoring Test (Wiggle Test).* This allows the technician to look for and set intermittent faults while the engine is running.

Within these four tests, there are six types of service codes.

1. On-demand
2. Memory
3. Separator
4. Dynamic response
5. Fast
6. Engine I.D.

On-demand codes are the same as hard faults. They mean something is wrong somewhere in the EEC-IV system at the time of the self-test. The term *on-demand* means the computer is being asked if a problem exists. Table 7-6 illustrates a typical listing

Code	Meaning	Code	Meaning
	TABLE 7-6: EEC-IV SYSTEM'S SELF-TEST CODES		
11	System "pass"	58	Idle tracking switch input too high (engine running test)
12	Rpm out of spec (extended idle)	61	ECT input too low
13	Rpm out of spec (normal idle)	63	TPS input too low
14	PIP was erratic (continuous test).	64	ACT (VAT) input too low
15	ROM test failed	65	Electrical charging over voltage
16	Rpm too low (fuel lean test)	66	MAF (VAF) input too low
17	Rpm too low (upstream/lean test)	67	Neutral drive switch—drive or accelerator on (engine off)
18	No tach	68	ITS open or AC on (engine-off test)
21	ECT out of range	72	No MPA change in "goose test"
22	MAP out of range	73	No TPS change in "goose test"
23	TPS out of range	76	No MAP (VAF) change in "goose test"
24	ACT out of range	77	Operator did not do "goose test"
25	Knock not sensed in test	81	Thermactor air bypass (TAB) circuit fault
26	MAF (VAF) out of range	82	Thermactor air diverter (TAD) circuit fault
31	EVP out of limits	83	EGR control circuit fault
32	EGR not controlling	84	EGR vent circuit fault
33	EVP not closing properly	85	Canister purge circuit fault
34	No EGR flow	86	WOT A/C cutoff circuit fault (all 3.8 L and 5.0 L Continentals)
35	Rpm too low (EGR test)	87	Fuel pump circuit fault
36	Fuel always lean (at idle)	88	Throttle kicker circuit fault (5.0 L)
37	Fuel always rich (at idle)	89	Exhaust heat control valve circuit fault
41	System always lean	91	Right EGO always lean
42	System always rich	92	Right EGO always rich
43	EGO cooldown occured.	93	Right EGO cooldown occurred
44	Air management system inoperative	94	Right secondary air inoperative
45	Air always upstream	95	Right air always upstream
46	Air not always bypassed	96	Right air always not bypassed
47	Up air/lean test always rich	97	Rpm drop (with fuel lean) but right EGO rich
48	Injectors imbalanced	98	Rpm drop (with fuel rich) but right EGO lean
51	ECT input too high		
53	TPS input too high		
54	ACT (VAT) input too high		
55	Electrical charging under voltage		
56	MAF (VAF) input too high		

of on-demand or self-test codes for Ford's EEC-IV system.

Memory codes mean a malfunction was noted sometime during the last twenty vehicle warm-ups, but is not present now. (If it were, it would be recorded as a hard fault.)

A separator code (10) indicates that the on-demand codes are over and the memory codes are about to begin. The separator code occurs as part of the key on/engine off segment of the self-test only.

When a code 10 appears during the engine running segment of the self-test, it is referred to as a dynamic response code. This is a signal to the technician to hit the throttle momentarily so that the ECA can verify the operation of its position and the manifold absolute pressure (MAP) sensor. Failure to respond to the dynamic code within 15 seconds after it appears will set a code 77.

Fast codes are of no value to the service technician. They are for factory use only and are transmitted about 100 times faster than even a STAR tester can read. On the voltmeter, fast codes will cause the needle to rapidly pulse between zero and 3 volts. On the STAR tester, the tester's LED light will flicker. Fast codes will appear twice during the entire self-test sequence, once at the very beginning of the key on/engine off test (right before the on-demand codes) and again after the dynamic response code (prior to hard fault transmission).

Engine identification codes are used to tell automated assembly line equipment how many cylinders the engine has. Two pulses indicate a four-cylinder, three pulses a six-cylinder, and four pulses identify the engine as an eight-cylinder model. Engine I.D. codes appear only at the beginning of the engine running segment.

READING THE TROUBLE CODES

Reading trouble codes with a voltmeter involves counting the sweeps and pauses of the analog voltmeter needle. All trouble codes are represented by two-digit numbers. Each two-digit number is represented by wide sweeps of the voltmeter needle (Figure 7–60). For example, three sweeps, a 2-second pause, then four sweeps is a code 34. If the ECA has recorded more than one problem, there will be a 4-second pause between one code and the next. Separator codes and dynamic response codes (both 10) are represented by one sweep (there is no sweep for zero) with a 6-second pause before and after.

All that is needed to activate the self-test mode is two jumper wires (each about 6 inches long) with male spade terminals and a voltmeter. Locate the large gray-colored self-test output connector under the hood, place one end of the first jumper lead in the #4 slot of the self-test connector, and clamp the

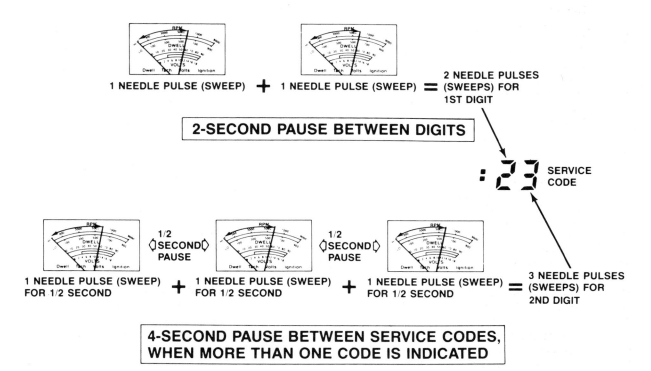

FIGURE 7–60 Analog voltmeter functional service code

other end to the negative voltmeter lead. Connect the positive voltmeter lead to the positive terminal of the battery. Bridge the #2 slot of the self-test connector and the self-test input connector with the other jumper (see Figure 7-55).

To begin the self-test, set the voltmeter on the 0 to 15 scale and place the ignition switch in the RUN position (do not start the vehicle). Immediately after the self-test is activated, the voltmeter needle will flutter rapidly between zero and 3 volts. These are the fast codes, which can be ignored.

After the fast codes are finished, the first usable code (on-demand) will be received. Once the codes start coming, it is easy to misinterpret or lose track of them. If the system is all right, the needle will sweep once, pause 2 seconds, then sweep again, indicating a code 11. This means the system has received a passing grade. In 4 seconds, the code will be repeated in case the technician missed it the first time.

If there is a problem that the computer has recognized, the code for that particular fault will be counted out on the meter. After each code is given once, the whole sequence is repeated. After a 6-second pause, the single sweep separator code (10) will appear, indicating it is time for memory codes. After another 6 seconds, the memory codes will start to count out and are read the same way as the on-demand codes. Once again, if the system passes, meaning no memory codes have been stored, a code 11 (repeated once) will be received.

After the key on/engine off test sequence is finished, start the engine and perform the computed ignition timing check. (This portion of the self-test must be performed within 2 minutes of self-test activation or the testing sequence will have to be started again.) The timing light should show base timing plus 20 degrees of advance. If it does not, refer to the shop service diagnosis manual.

If the computer proves it is capable of controlling spark advance, shut off the engine, remove the #2 jumper wire and timing light, then start it again and let it run at 1500 rpm for 2 minutes to warm the EGO sensor. Shut off the engine, reattach the jumper, wait 10 seconds, then start it again.

The engine running segment is now ready to begin. The first needle sweeps on the voltmeter are the engine I.D. codes pulsing out the cylinder codes. Within 6 to 50 seconds, the single-sweep dynamic response code will be received, requesting the technician to hit the throttle (within 15 seconds of code). After another 4 to 15 seconds, the fast codes will appear followed by the engine running on-demand codes. As in the key on/engine off segment, a code 11 means everything is fine, and it will be repeated in

4 seconds, signaling the end of the engine running sequence.

As each self-test segment is completed, it is a good idea to look up the codes received to find out to what specific problems they relate. Some codes require immediate attention, meaning a specific problem must be repaired before the technician can proceed to the next test segment. When tracing the cause of a specific code, follow the pinpoint test instructions as contained in the shop service diagnosis manual (or generic equivalent) before attempting to repair any faults.

The final self-test procedure is called the wiggle test, which derives its name from the action required to check for and set possible intermittent codes. In the continuous monitor test mode, the opportunity is given to recreate intermittent fault conditions by wiggling connectors, manipulating movable sensors or actuators, and heating and cooling thermistor-type sensors with a heat gun.

This procedure can be performed while the engine is off, running, or both. If you choose to look for intermittents while the engine is off, repair any codes as required (from the first test segments), then deactivate the self-test and place the ignition switch in the RUN position. With the voltmeter still connected, start "wiggling." If an intermittent problem is encountered, the voltmeter needle will deflect and a memory code will be set. After everything has been thoroughly shaken, tugged and pulled, reactivate the self-test (do not forget to turn off the key first) and perform the key on/engine off segment. The intermittent faults that were recorded during the wiggle test will read out at this time.

To perform the wiggle test while the engine is running, the self-test must remain activated. Approximately 2 minutes after the last code in the engine running segment has been displayed, the continuous monitor test mode will start. From this point on, the test is run the same as with the engine off.

DIAGNOSING BOSCH'S L-JETRONIC FUEL INJECTION SYSTEM

The original Bosch D-Jetronic system used a vacuum sensor to inform the electronics about the intake situation. However, soon a more accurate means of measuring the engine's air volume intake appeared: the airflow meter (also known as an air box or air vane). Using a spring-loaded rotating flap that turns a variable resistor, this device tells the

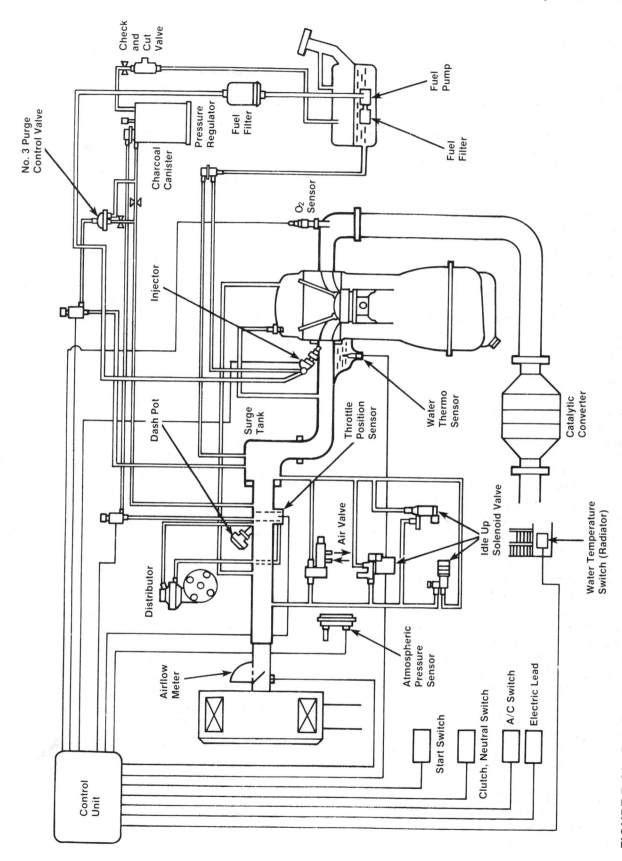

FIGURE 7-61   Bosch L-Jetronic system

computer how much of the atmosphere the power plant is ingesting at any particular moment.

It was named the L-Jetronic (the L stands for "Luftmengenmessung" aud Deutsch, meaning "airflow management"), although it has other names. Volkswagen, for instance, calls it AFC (Airflow Controlled) fuel injection. Regardless of what appellations the carmakers give for it, if it is port injection with an airflow meter, it belongs to the Bosch L-Jet design family. Later refinements include the addition of Lambda control (oxygen sensor feedback), a hot-wire mass air sensor (called LH-Jetronic, this system determines the actual mass of intake air corrected for density rather than just the simple cubic volume), and Motronic, which has integrated spark advance control.

The L-Jet is found on many late-model cars, both domestic and import. The Japanese are especially fond of it—the components might be labeled Nippondenso, but they are manufactured under a Bosch license. Figure 7-61 illustrates a typical Bosch L-Jetronic fuel injection system.

DIAGNOSTICS

In cases where the EFI is the problem, a good preliminary check is to listen to the injectors with a stethoscope. If an injector is quiet, remove its harness connector and check for voltage at one connector terminal and regularly occurring ground at the other. If this combination is found, the injector itself is bad (resistance across the terminals of a typical specimen should be 1.5 to 3 ohms). No power or ground means there is a broken wire in the harness branch to that cylinder. Remember, most L-Jets fire all injectors simultaneously, not one at a time as in some new sequential systems. If all the injectors are silent during cranking, that is the reason the engine will not start.

Fuel pressure testing is also one of the most basic procedures. Hook up the gauge, put a rag around the fitting to catch the spraying fuel, and run the pump, which will require the ignition to be switched on, then manually move the air vane; a common range is 35 to 40 psi. Check the specs for the particular vehicle. Start it up, let it idle, and the needle should drop to read between 28 to 31 psi.

If the vacuum hose is taken off the pressure regulator, the needle will read between 37 and 41 (Figure 7-62). Low pressure suggests a weak pump or clogged filter; too many psi might be due to a restricted fuel return line between the regulator and the tank. Fuel pressure regulators rarely go out of specs, but when they do, they can cause "ghosts,"

problems that throw off the diagnosis of other components. Also, they do not go bad as often as they are replaced. If a car is found that is hard to start when hot, suspect leakdown.

Obviously, zero pressure will result in a no-start. Check the pump fuse first, but if that is all right, there might be something else impeding electron flow. With the key on, reach inside the air box inlet and push the flap open. If the pumps starts running, the points in the box are all right. But maybe cranking vacuum is not enough to move the flap because of mechanical interference, a leak in the duct between the air box and the throttle body, or a backfire protection valve (if present) in the vane that is stuck open or blown out entirely. These should also be checked.

COLD IDLE PROBLEMS

Cold idle problems can usually be traced to the auxiliary air valve or its equivalent. The most straightforward check is to start the engine cold and pinch the hose that runs between the regulator and the manifold (Figure 7-63). Speed should drop. Let it run long enough to reach normal operating temperature, then squash the hose again. There should be little or no change in rpm. Also, the specified resistance across the terminals of the air valve's heating element is approximately 30 to 50 ohms.

Do a comprehensive exam of the system's electronics by probing the terminals of the ECU connector. Of course, the specific values and instructions for the vehicle in question will be needed, which can usually be found clearly stated in the service manual.

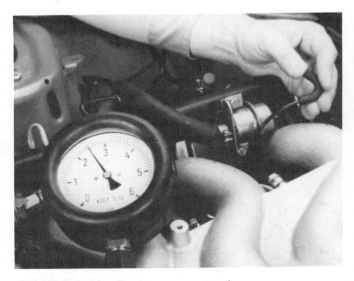

FIGURE 7-62 Fuel pressure testing

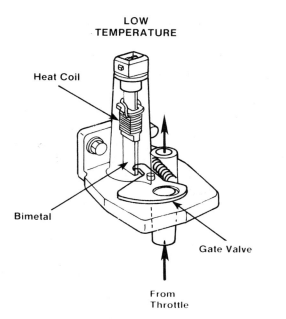
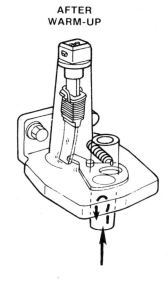

FIGURE 7-63   Cold idle diagnostics and auxiliary air valve

A lean mix that causes hesitation and surging is a relatively common problem, and clogged injectors or a bad temperature sensor that lets the ECU think the engine is warm when it is not is a likely cause. Some cars have come from the factory with a too-lean calibration, which the engineers often address with a service bulletin. For example, an electronic module is available for late 1970s Nissan Z-cars that is spliced into the coolant temperature sensor wire to alter the signal to the computer. For certain VW's that hesitate only during warm-up, there is a spacer that screws in between the sensor and the engine to slow the rate at which it heats up.

FLOW FAULTS

Clogged injectors can be found in several ways. One method requires the use of a device that triggers each solenoid for a precise amount of time. It is used in conjunction with a pressure gauge to compare the psi drop among injectors. If the fall-off is smaller with one than the others, the problem has been found.

Injector cleaning will get the flow rate where it should be in many cases, and it will also help insure a good spray pattern and sufficient atomization. The additives some oil companies put in gasoline might help avoid deposit buildup, but they are not really concentrated enough to blow away what is already there. You can get cleaning agents in bulk for use in various cleaning setups or in one-shot aerosols.

Most systems work pretty well. If aerosol is chosen, get a quality brand that has enough pressure in the can to keep the engine operating properly.

It often makes sense to clean the EFI as a preliminary to troubleshooting. It is inexpensive and effective. Also, the problem might disappear, eliminating the need for engaging in the mental discipline of diagnosis. Actually, this procedure can be considered part of a modern tune-up.

DIAGNOSTIC SUMMARY

Some miscellaneous tips that apply to servicing all airflow meter L-Jets include

- Whenever a broken air flap mechanism is found, make sure all the shrapnel is cleaned out.
- If binding or roughness is felt as the air vane is slowly moved through its range of travel, trying to fix it is a waste of time.
- In cases where the airflow meter's variable resistor appears to be worn through or has burned spots, a new box is the only solution.
- Resistance between terminals E2 and VS of the airflow meter should be 20 ohms with the vane in the CLOSED position and 1,000 ohms in the OPEN position.
- Never try to adjust the air vane return spring tension. It can only be adjusted with special equipment.

REVIEW QUESTIONS

1. Fuel exits from the injector nozzle as
   _____ .
   a. gas
   b. liquid drops
   c. fine mist
   d. liquid stream

2. Fuel injection pressures are _____ .
   a. higher than pressure in a carburetor
      system
   b. lower than cylinder pressure
   c. about atmospheric pressure
   d. lower than atomspheric pressure

3. Fuel injectors are inserted into the manifold
   head by _____ .
   a. pressing into place, sealing with an O-ring
   b. screwing into the cylinder like a spark plug
   c. torquing the injector against a gasket
   d. giving 1/2 turn to lock the injector

4. On a GMC vehicle equipped with a TBI system, idle speed can be adjusted
   _____ .
   a. by the technician, by altering the throttle
      opening
   b. by the technician, by altering the amount
      of air bypassing the throttle
   c. only by the ECM (electronic control
      module)
   d. by the technician, by adjusting the stepper
      motor and pintle valve

5. What should a technician do to clear a
   flooded engine on a car equipped with a
   GMC TBI system?
   a. Stop cranking immediately.
   b. Floor the accelerator pedal and continue
      to crank.
   c. Relax throttle pressure to about 50
      percent.
   d. Continue to crank without using throttle
      pressure.

6. Technician A says that fuel injected systems
   can operate with fuel pressures of 10 psi.
   Technician B says that fuel injected systems
   operate with fuel pressure as high as 85 psi.
   Who is correct?
   a. Technician A

   b. Technician B
   c. Both A and B
   d. Neither A nor B

7. Before attempting work on any fuel injection
   system, what should the technician do?
   a. Talk to the owner about running
      problems.
   b. Test-drive the vehicle.
   c. Check for fuel delivery.
   d. Refer to the manufacturer's service manual
      for troubleshooting procedures and
      charts.

8. Technician A says the ignition switch can be
   on when connecting or disconnecting any
   fuel injector system electrical connects.
   Technician B says that electronic parts and/
   or modules are immune to grounding or
   shorting. Who is correct?
   a. Technician A
   b. Technician B
   c. Both A and B
   d. Neither A nor B

9. Technician A says that fuel pressure can be
   relieved by energizing the cold start valve for
   about 10 seconds. Technician B says that the
   pressure can be relieved by carefully cracking the fuel line loose on the control pressure
   regulator with a clean shop rag wrapped
   around the connection. Who has the safest
   method?
   a. Technician A
   b. Technician B
   c. Both A and B
   d. Neither A nor B

10. The sequence for testing fuel pressure is to
    check _____ .
    a. cold control pressure first, hot control
       pressure next, and rest pressure last
    b. rest pressure first, cold control pressure
       next, then hot control pressure
    c. only hot control pressure
    d. check cold control pressure first, then hot
       control pressure

11. Running the engine with the fuel pump fuse
    or relay removed until the engine stalls is one
    way of relieving fuel pressure.
    a. True
    b. False

12. On Chrysler TBI systems, a pressure gauge is installed between the fuel filter hose and the throttle body after releasing system pressure. When the engine is started, the gauge indicates very low pressure. The gauge is then installed between the fuel filter hose and the fuel line, and the engine is started again. The same low-pressure reading is observed again. Technician A says the fuel filter is defective. Technician B says the pressure regulator is defective. Who is correct?
    a. Technician A
    b. Technician B
    c. Both A and B
    d. Neither A nor B

13. On-board diagnostic systems provide _____ .
    a. computer codes that can be read on the dashboard
    b. computer codes that can be read on a tester
    c. computer commands to exercise the fuel injection system
    d. all of the above

14. On Chrysler's MPFI Diagnostic System, the diagnostic test mode is used to see if there are any fault codes stored in the on-board system's memory. Technician A says only a diagnostic readout box can detect the codes. Technician B says the power loss/power limit light on the instrument panel can be used. Who is correct?
    a. Technician A
    b. Technician B
    c. Both A and B
    d. Neither A nor B

15. An electric fuel pump on a GMC port injection system is not working. A way to isolate the bad pump is to _____ .
    a. check the fuel pump fuse
    b. check the fuel pump wiring
    c. switch the fuel pump relay with the air conditioner relay and check pump operation
    d. all of the above

16. Fuel pressure on a port injection system is automatically regulated in relation to manifold vacuum. A high vacuum signal (low manifold absolute pressure) will _____ .

a. increase fuel pressure
b. lower fuel pressure
c. equalize fuel pressure
d. not affect fuel pressure

17. When using Ford's EEC-IV diagnostic system, what tester must be used to read the trouble codes?
    a. STAR tester
    b. SCAN tester
    c. analog (needle-type) voltmeter
    d. all of the above
    e. none of the above

18. A car equipped with AFC (airflow control) L-Jetronic fuel injection will not start. Technician A says this is caused by a blown fuel pump fuse. Technician B says this is caused by an air box leak. Who is correct?
    a. Technician A
    b. Technician B
    c. Both A and B
    d. Neither A nor B

19. Which of these could be the cause of high fuel consumption on a fuel injected car?
    a. fuel pressure too high
    b. cracked sensor vacuum hose
    c. restricted muffler
    d. all of the above

20. A car with a TBI system is repeatedly stalling. Technician A says the idle speed control device is defective. Technician B says the throttle position sensor needs adjustment. Who is correct?
    a. Technician A
    b. Technician B
    c. Both A and B
    d. Neither A nor B

21. Technician A says that a fuel fire can be extinguished using a water hose. Technician B says that a Class A fire extinguisher should be used on fuel fires. Who is correct?
    a. Technician A
    b. Technician B
    c. Both A and B
    d. Neither A nor B

22. Which of the following manufacturers uses a Schrader type valve for relieving fuel pressure?

a. AMC
b. Ford
c. Chrysler
d. All of the above

23. The fuel injection system is the most likely system to cause problems on a car. Inspect the fuel system before suspecting other systems.
    a. True
    b. False

24. A trouble code from the onboard diagnostics indicates that the oxygen sensor is faulty. Technician A says the oxygen sensor should be immediately replaced. Technician B says that the sensor should be tested and the wiring inspected first. Who is correct?
    a. Technician A
    b. Technician B
    c. Both A and B
    d. Neither A nor B

25. If the power module of Chrysler's computer control system does not receive a synchronization signal from the distributor within 1 second after the ignition switch is turned on,

the _____ will turn off the fuel and ignition systems.
    a. MAP sensor
    b. fuel pressure regulator
    c. ASD relay
    d. logic module

26. Which of the following is not an output of Chrysler's logic module?
    a. ignition coil
    b. EGR solenoid
    c. AIS motor
    d. purge solenoid

27. To check for a plugged injector you should _____ .
    a. check the injector ground with a test light
    b. listen for injector operation with a stethoscope
    c. check for available voltage at the wiring harness connector
    d. perform an injector balance test.

28. Which company developed the L-Jetronic fuel injection system?
    a. General Motors
    b. Nippondenso
    c. Robert Bosch
    d. Motorcraft

CHAPTER EIGHT

DIESEL FUEL INJECTION

Objectives

After reading this chapter, you should be able to:
- Describe the purpose, construction, and operation of a diesel fuel injection system and its components.
- Diagnose basic diesel fuel injection problems.
- Follow accepted general precautions prior to and during diesel fuel injection servicing.
- Describe the operation and construction of an inline injection pump.
- Describe the operation and construction of a distribution injection pump.
- Follow general corrective procedures to repair a diesel fuel injection system.

This chapter will discuss the basics of diesel fuel injection and provide servicing procedures for the system. Servicing will cover preventive, diagnostic, and corrective measures.

GENERAL PRINCIPLES OF OPERATION

The mechanical systems of a diesel engine and a gasoline engine are very similar, although diesel engine components are of heavier construction than those of a gasoline engine. The heavier construction in the diesel engine accommodates the highly compressed air pressures and the higher fuel injection pressures utilized by this type of engine.

Systems similar in both types of engines include:

- Starting
- Charging
- Cooling
- Lubricating
- Fuel delivery

The two types of engines differ in their ignition systems and their fuel metering systems. In contrast to the gasoline engine, ignition in a diesel engine takes place as the fuel meets the superheated air near the end of the compression stroke. Fuel is forced into this highly compressed air with a high-pressure fuel pump through an injector nozzle. The

timing of fuel injection in a diesel engine has the same effect on fuel economy and power as ignition timing has on a gasoline engine. Figure 8-1 illustrates a typical diesel fuel injection system.

Gasoline engines are self-governing because the air entering the engine is throttled by the use of a carburetor butterfly valve or by the throttle body in a fuel injected system. The driver can change engine speed by manipulating the throttle pedal.

Because of this method, in most cases, the gasoline engine has no need for a governor. If such an engine is equipped with a governor mechanism, it is for the sole purpose of limiting the maximum road speed of the vehicle and the engine rpm to prevent engine abuse or avoid poor fuel economy.

The diesel engine, on the other hand, operates with an excessive amount of unthrottled air throughout its operating speed range. The fuel injection system is separately controlled from the airflow system. When a diesel engine is started, the air is unthrottled and remains so as the driver opens the fuel control mechanism or throttle linkage to allow the injection system to mix or inject more fuel into the cylinders. Without some form of fuel control, the diesel engine can accelerate very rapidly and self-destruct, especially if the throttle were to be placed in the full-fuel position and left there with no mechanism to regulate the fuel input.

Therefore, a governor is required on diesel engines to regulate the fuel input and prevent engine stalling at the low-speed end and from overspeeding at the high-speed end. The governor controls the

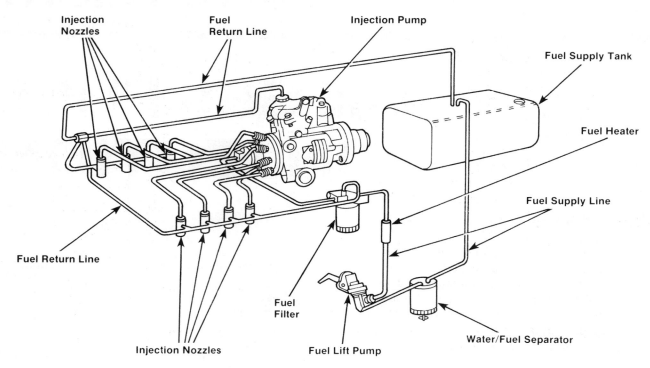

FIGURE 8-1   Diesel fuel-injection system

speed of the engine. In most diesel engines, airflow is not throttled so engine speed is determined by the amount of fuel injected. In passenger car diesel engines, fuel flow is directly controlled by the driver except for governor control at the two extremes of the speed range. It is called a min-max governor.

At minimum speed or idle, the governor controls pump injections so the engine gets just enough fuel to keep it running. At maximum rpm, the governor controls pump injection to limit fuel flow to keep the engine from overspeeding. At intermediate speeds, the pump responds to movements of the accelerator pedal so the driver can control car speed directly. During coasting, the governor can cut off fuel delivery for economy. See Figure 8-2 for the operation of a typical governor.

In comparing gasoline and diesel engine operation, the following characteristics about diesel engines should be noted:

1. Power is developed by expanding gases.
2. Fuel is efficiently converted into heat and mechanical energy due to a higher rate of expansion of gases and slower vaporization of fuel.
3. The air intake system has no throttle valves.
4. Air is compressed only during the compression stroke and yields a high compression ratio.

5. Fuel is injected into the combustion chamber at a precise time in relation to piston position and stroke.
6. Fuel is injected at extremely high pressures—up to 15,000 psi or higher.
7. There is no ignition system—fuel is ignited by the heat of the compressed air in the cylinder.
8. The amount of fuel injected and the duration of injection are precisely controlled.
9. They have higher air to fuel ratio than gasoline engines—up to 60 to 1 as compared to a maximum of 18 to 1 for gasoline engines.
10. They make more noise than gasoline engines.

The diesel engine does not use a throttle plate mechanism between the air cleaner and the air intake valve. A charge of intake air is supplied to the combustion chamber on each intake stroke of the combustion cycle. This arrangement allows the diesel engine to operate in an efficient, fuel economy mode. Typically, the intake air on a diesel engine has a compression ratio of about 22 to 1, a pressure approximating 500 psi, and an operating temperature approaching 1000 degrees Fahrenheit. A fuel charge injected into this volatile air environment will automatically ignite; therefore, there is no need for a

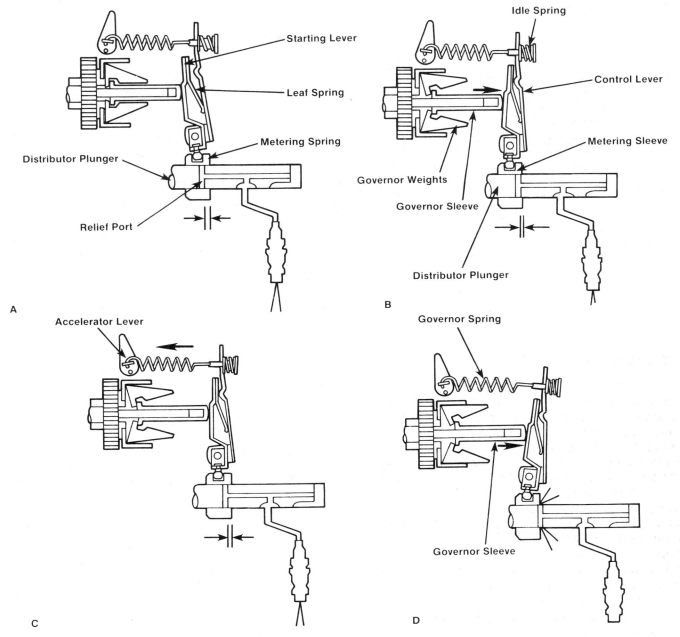

FIGURE 8-2 Govenor operation in a typical diesel fuel injection system: (A) During starting, metering sleeve is pushed farthest from BDC, which results in a longer injection period; (B) as engine speed increases, the governor flyweights push the governor sleeve to the right, which moves the metering sleeve to the left and reduces injection duration. (C) Depressing the accelerator moves the metering sleeve to the right, which causes more fuel to be injected. (D) At maximum speed, the engine's speed is governed by limiting fuel delivery as during idle.

spark plug/coil/distributor system typical of a gasoline engine. The fuel charge is injected into the combustion chamber at high pressure. This allows for a metered quantity of fuel to be delivered as a fine mist or spray into the chamber. The design and adjustment of the fuel injector used will determine the system's injection pressure. Figure 8-3 illustrates a typical schematic diagram of a diesel fuel injection system.

The output power of a diesel engine is directly proportional to the fuel charge injected into the combustion chamber. At an idle speed, a small fuel

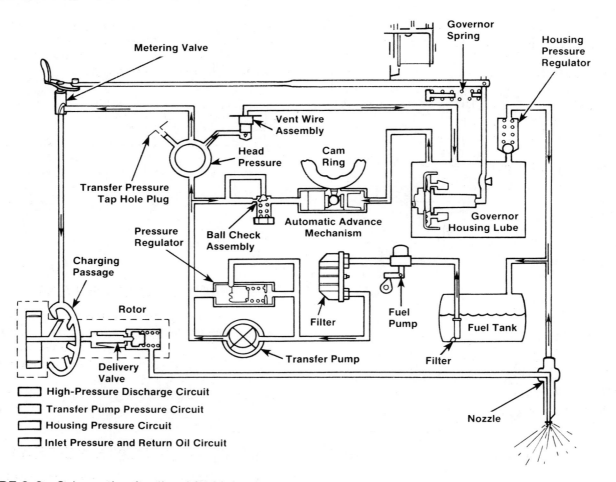

High-Pressure Discharge Circuit
Transfer Pump Pressure Circuit
Housing Pressure Circuit
Inlet Pressure and Return Oil Circuit

FIGURE 8-3 Schematic of a diesel fuel injection system

charge is used that has a very lean mixture (a higher air ratio than fuel). As more fuel is injected, the mixture becomes rich, engine speed increases, and output power increases. The amount of fuel that can be injected is determined by the amount of air drawn into the engine on the combustion intake stroke. In all types of diesel engines, however, the amount of fuel to be injected is limited by a governor (see Figure 8-2). This device limits the air/fuel ratio for a diesel engine up to a maximum of 10 to 1 by weight. The air/fuel ratio of a typical gasoline engine is approximately 14.7 to 1 by weight.

Combustion in a diesel engine occurs in four sequential stages or periods: the delay period, the uncontrolled burning period, the controlled burning period, and the after burning period (Figure 8-4). When the fuel is first injected into the combustion chamber, there is an initial delay as the fuel changes from a liquid state to a vapor or gas state. This liquid-to-vapor conversion is necessary so that the fuel will burn. The delay period is followed by a period of uncontrolled burning of the fuel already injected

into the chamber. This period is followed by a controlled burning period as the injector continues to feed fuel into the combustion chamber. If and when the fuel injection stops, all the remaining fuel in the chamber will continue to burn until it is consumed. This stage is known as the *after burning period*. In the diesel engine these periods of combustion simultaneously occur in different parts of the combustion chamber during injection.

The combustion of diesel fuel is not identical under all conditions. Combustion is affected by:

- Injector timing
- Rate of fuel injection
- Length of time (or pulse width) of injection
- Position of injector nozzle
- Injection pressure
- Vaporization of fuel
- Distribution of fuel in the combustion chamber

It must be remembered that the diesel engine has no ignition system that is typically found in gasoline engines. The moment fuel is injected into the cylinder it is ignited. Therefore, the fuel injection pump utilized by the diesel engine system must inject the fuel at the precise instant when the piston is at the correct position on the compression stroke. When this occurs is determined by such factors as ignition lag time of the fuel, engine temperature, engine speed and load, and exhaust emission considerations.

Diesel injection pumps are driven from the engine crankshaft by means of gears or by a positive, toothed drive-type timing belt. If a belt is used, it is a positive, toothed drive type. The injection pump in conjunction with the accelerator system utilizes a governor to control engine speed. The accelerator linkage interacts with the governor spring to determine engine speed. Without the governor and an air throttling system, the diesel engine would not idle at proper speeds. The engine could also increase in speed to the point of self-destruction.

TYPES OF DIESEL FUEL SYSTEMS

Four types of diesel fuel systems are used:

- Pump controlled
- Unit injection
- Common rail
- Distributor pump

PUMP CONTROLLED SYSTEM

Each cylinder is equipped with a high-pressure plunger and metering mechanism. The plungers are designed to provide the correct metering of fuel under high pressure to each cylinder injector at the correct time in the combustion cycle. The high-pressure plungers are cam operated. The effective stroke of the fuel pump can be changed by rotating the plunger in its barrel.

UNIT INJECTION SYSTEM

This system is similar to the pump controlled system except that the high-pressure pumping and metering mechanisms are an integral part of the fuel injector. Each cylinder is equipped with a cam-operated injector that can be adjusted to control the metering of fuel for each pump stroke.

COMMON RAIL SYSTEM

This type of system uses a high-pressure fuel pump that is connected to a common fuel rail. Each cylinder fuel injector is connected to the common fuel rail. A cam, pushrod, and rocker arm device are used to actuate the fuel injectors. Adjusting the length of the pushrod mechanism determines the injector's open time and therefore the amount of fuel to be delivered to the combustion chamber.

DISTRIBUTOR PUMP SYSTEM

This is the most commonly used system with several types of distributor pumps. The more common types include the Bosch distributor pump and the Stanadyne pump. These types of distributor pumps are covered in more detail in this chapter.

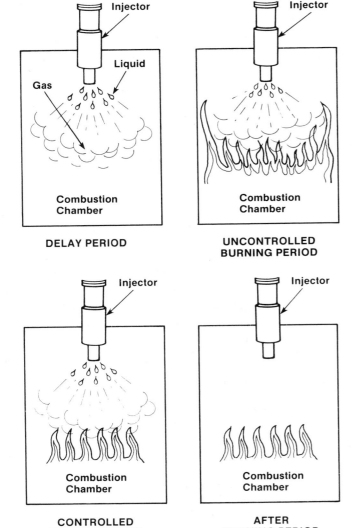

FIGURE 8-4   Four stages of diesel combustion

DIESEL INJECTION PUMPS

Diesel fuel injection pumps are designed to deliver a precise fuel charge to the combustion chamber through an injector at a specific crankshaft angle usually at the end of a compression stroke. This unique operation controls ignition timing and the metering of the diesel fuel. The fuel injection pump used in the system is synchronized or timed to the engine's crankshaft by drive gears. The pump is designed to create or build up high fuel pressure (15,000 psi to as high as 30,000 psi) so that diesel fuel is injected into the combustion chamber as a fine mist or spray. The fuel spray will rapidly evaporate when it encounters the high-pressure, high-temperature air charge in the combustion chamber and allow for automatic ignition of the fuel. In order for the engine to operate smoothly, the following fuel injection conditions must exist:

- Pressure must be available at the precise instant of injection into the chamber.
- Injection rate must be equal for all cylinders so there will be equal power pulses from each cylinder.
- Pressure must shut off at the precise instant to control the total amount of fuel to be injected.

The fuel lines that carry the highly pressurized diesel fuel between the injection pump and fuel injector are manufactured from special, thick-walled steel tubing. The steel fuel lines are always of the same length on a diesel engine so that each cylinder receives an identical or equal fuel charge. During operation the steel fuel lines are completely full of fuel. If the fuel line contains 0.001 cc of fuel, then 0.001 cc of fuel will be delivered to the fuel injector because liquid diesel fuel will not compress under the high pump pressure.

A typical diesel fuel injection pump (Figure 8-5) uses a plunger that moves in a barrel that is similar to a steel bushing. The clearance between the plunger and barrel is very small. Each plunger/barrel assembly is lapped to allow the plunger to freely slide in the barrel. This lapped fit prevents internal pump parts from being interchanged between fuel injectors. The plunger-to-barrel clearance is small enough to prevent fuel from leaking between the two parts during the injection cycle, thereby ensuring that the correct fuel charge is injected into the chamber. At the same time, the plunger-to-barrel clearance is adequate to allow the fuel to lubricate the parts during operation. The plunger is moved in the barrel by a cam and in turn forces the fuel through the injector. The mechanism is recharged with fuel under pressure by a

feed pump via the fuel charging ports. The feed pump in some systems is sometimes referred to as a transfer pump. After the fuel charge has been delivered, the plunger is returned to its original position, sometimes with the aid of a return spring.

Three types of feed or transfer pumps are used today: vane, gear, and piston (Figure 8-6). The feed pump can be mounted externally or internally to the fuel injection pump housing. In some systems, a diaphragm fuel pump, similar to those used on gasoline engines, is used to deliver the diesel fuel from the fuel tank to the feed pump.

Today's diesel fuel injection pumps may also be equipped with the following:

- Automatic pump start-stop devices
- Automatic fuel delivery correction to compensate for engine power, altitude, and so on
- Pump speed regulation devices
- Temperature-controlled fuel starting devices

The automotive diesel engine uses two types of diesel fuel ignition pumps: a multiple plunger pump often referred to as an in-line injection pump and a distributor or rotary injection pump.

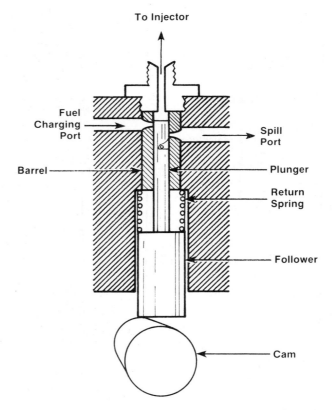

FIGURE 8-5 Typical diesel fuel injection pump

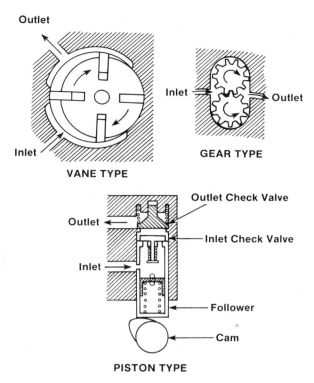

FIGURE 8-6   Types of diesel fuel feed pumps

In-Line Injection Pump

The multicylinder in-line injection pump (Figure 8-7) has a plunger and barrel assembly for each engine cylinder. The assemblies are grouped together in one housing that resembles cylinders in the block of an in-line engine. A high-pressure fuel line connects each pump assembly to one injector (Figure 8-8).

In operation a cam shaft moves the pump plunger toward the incoming fuel charge. The plunger in turn pushes or forces the fuel charge into the injector. Each plunger in the pump has its own or separate cam lobe on the cam shaft. A spring force keeps the roller tappet riding on the cam lobe and also serves to move the plunger downward on the fuel charging stroke of the injection pump. Figure 8-9 illustrates a sectional view of an in-line injection pump.

Low pressure fuel from the feed pump encompasses the barrel of the injection pump. The barrel is designed to have one or more holes through its structure. These holes are opened and closed as the plunger slides up and down the barrel. This operation is similar to the opening and closing of the

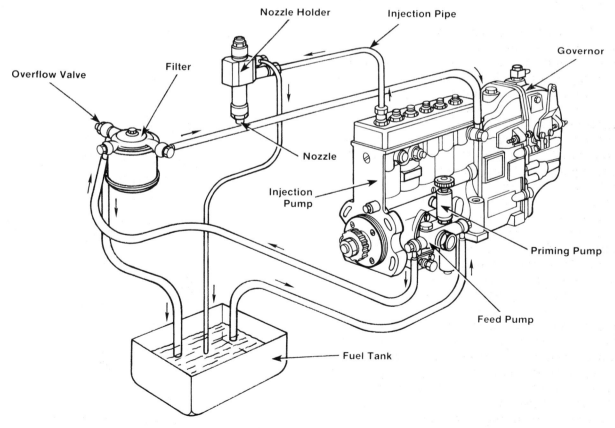

FIGURE 8-7   Typical multicylinder in-line injection pump

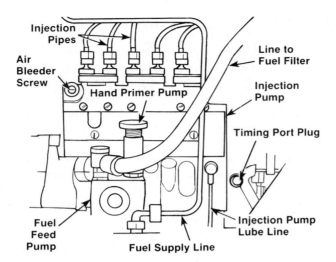

FIGURE 8-8 High-pressure lines connect each pump assembly to one injector.

intake fuel ports on a two-stroke cycle engine. The in-line pump delivers or pumps an equal amount of fuel on each stroke and is only useful for constant-speed, constant-load diesel engines.

For all engine speeds or loads, each pump plunger has a constant mechanical stroke. The shape of the plunger near or at its delivery end also determines the effective stroke of the plunger. Cut into the exterior surface of the plunger is a curved groove called a helix (Figure 8-10). Some designs might also use a groove or hole to connect this helix to the tip of the plunger. A flat surface on one side of the plunger (near the cam end) fits or mates with a flat surface in the center hole of a gear. The plunger

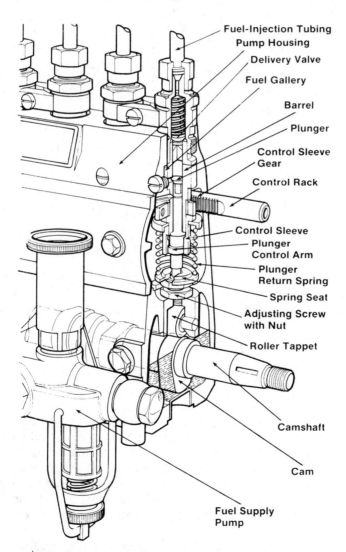

FIGURE 8-9 Sectional view of an in-line injection pump

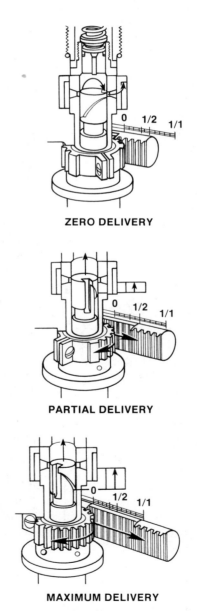

FIGURE 8-10 Typical diesel pump plunger

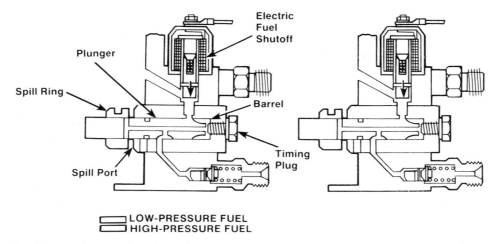

LOW-PRESSURE FUEL
HIGH-PRESSURE FUEL

FIGURE 8-11   Pump plunger in a barrel

moves up and down freely through the gear and rotates when the gear rotates. The teeth on the gear mesh with a control rack. The control rack movement is what causes the plunger to rotate and change the alignment of the plunger helix with the holes in the barrel (Figure 8-11).

Fuel enters the pump and fills it when there is an opening through the barrel into the pump chamber. This opening is controlled by the position of the plunger. With fuel in the pump chamber, the plunger will move through its stroke and block or close the holes in the barrel. The fuel that is trapped in the chamber is then pressurized and forced out through the injector nozzle. The fuel is injected until another part of the helix groove in the plunger again reaches the holes in the barrel. Any trapped fuel in the pump chamber can then be passed back to the fuel supply side through these holes in the barrel. As this happens, fuel pressurization immediately decreases and stops fuel injection. When the barrel holes are used in this manner to drop fuel pressure, they are referred to as *spill ports*. The operation of this type of pump is illustrated in Figure 8-12. The helix groove can be aligned at different positions in relation to the spill ports by rotating the plunger. The technician therefore can adjust the amount of fuel injected into the combustion chamber by rotating the plunger.

The single rack concept is used to control the position of all the injectors simultaneously. Injector plungers are shown in two different rotated positions in Figure 8-13. The stop screws on the rack are adjusted to control the engine idle speed. Depressing the driver's accelerator pedal moves the rack through a governor to rotate the plungers and increase engine speed beyond idle speed. Figure 8-14 illustrates a typical governor mechanism that is used to limit engine speed on diesels. When the diesel

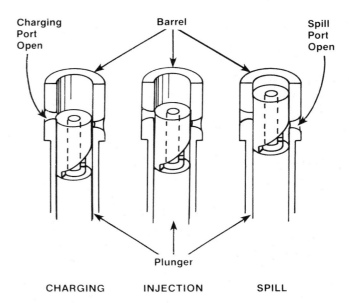

CHARGING   INJECTION   SPILL

FIGURE 8-12   Operation of a typical injection pump

engine attains an overspeed condition, the governor will automatically move the control rack. This in turn rotates the plungers, which reduces the amount of fuel injected into the combustion chamber. An overspeed condition can be encountered when a vehicle is going down a steep grade. In this case the control rack can be up against the idle stop. The accelerator is used to determine the speed of the engine between engine idle and maximum speed, and the governor is used when the maximum speed limit is exceeded.

A diesel engine, like the gasoline engine, produces the most power and best fuel economy when ignition is properly timed with respect to the com-

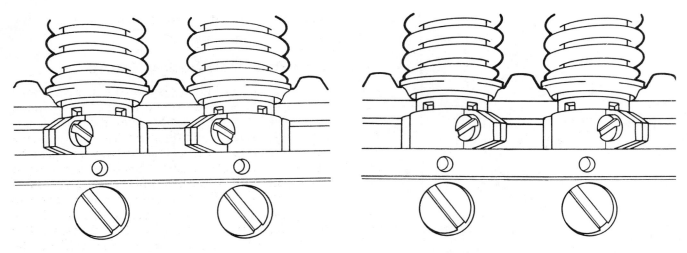

FIGURE 8–13 Rotating the injectors to control the amount of fuel injection

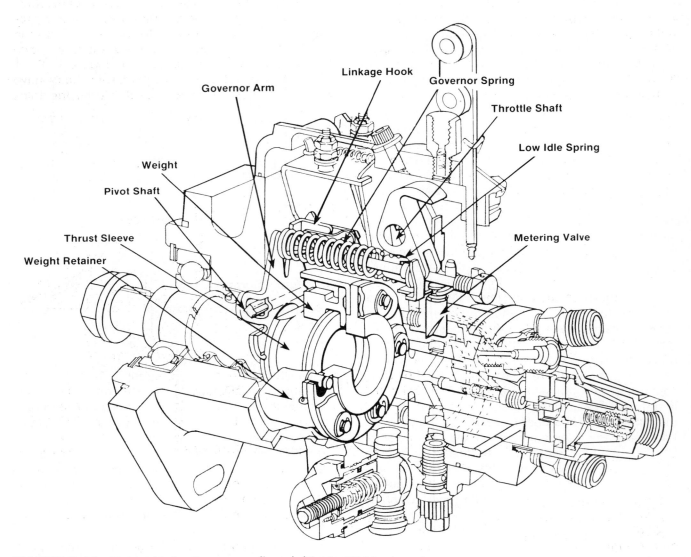

FIGURE 8–14 Typical injection pump flyweight governor

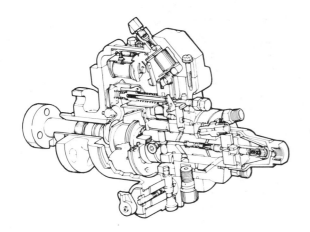

FIGURE 8–15 Typical distributor injection pump

bustion cycle. In diesel engines, combustion takes place immediately after fuel injection starts. On automotive diesel engines, fuel injection pumps have a timing-advance mechanism to advance the start of fuel injection as engine speed increases.

The in-line injection pump used on automotive diesel engines is lubricated with engine oil. The pump does not have a seal around the cam drive end

bearing so the oil returns to the engine through the bearing.

Distributor Injection Pump

A typical distributor injection pump is illustrated in Figure 8–15. Like the in-line injection pump, it too is driven by the engine's crankshaft through the use of timing gears. The timing gears drive a rotor on the pump that is fitted into a close-tolerance bore in the hydraulic head of the pump. Drilled passageways to carry fuel are incorporated in both the pump rotor and hydraulic head. One type of pump uses two plungers that operate in a cross-drilled bore in the distributor rotor; another type of pump uses only a single plunger. The two-plunger concept allows the plungers to move outward and opposite to each other when the pump chamber is filled with fuel. The plungers are forced together by internal cam lobes. When the plungers are together they effectively reduce the chamber size and force the fuel into the injectors. Typical cam plunger movement in a distributor injection pump is shown in Figure 8–16.

Both the rotor and the hydraulic head have holes in their structure. As the rotor turns, the holes

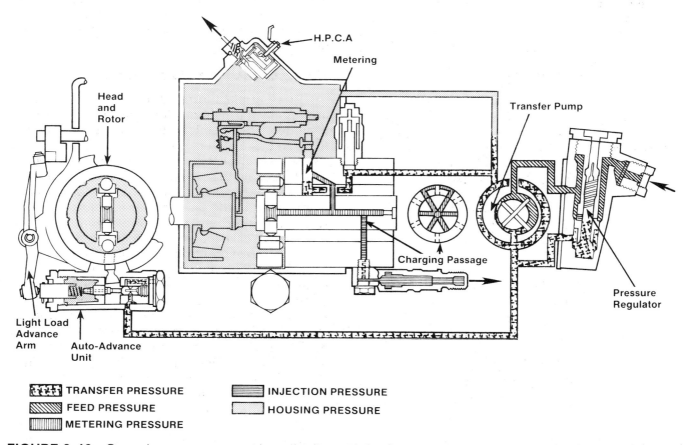

▨ TRANSFER PRESSURE	▤ INJECTION PRESSURE
▧ FEED PRESSURE	☐ HOUSING PRESSURE
▥ METERING PRESSURE	

FIGURE 8–16 Cam plunger movement in a distributor injection pump

align with each other and allow fuel to flow between the rotor and the head. When the holes are misaligned, the fuel flow stops. The holes in both components are referred to as *ports.* Two types of ports are utilized: charging ports and discharging ports. When the charging ports are aligned, the pump chamber fills with fuel. When the discharging ports are aligned, fuel injection takes place.

A vane-type transfer pump (Figure 8-16) draws and pushes fuel through passages in the hydraulic head to a fuel metering valve. The size of the opening in the metering valve is controlled by the accelerator pedal acting through a governor. A small opening in the valve is maintained at idle speed. Depressing the accelerator causes the valve opening to increase. When maximum engine speed is attained or an overspeed condition exists, the governor takes over and automatically begins to close the valve.

Fuel from the metering valve is routed to the charging ports on the distributor injection pump as shown in Figure 8-17. As the fuel enters the pump, it forces the plungers outward or away from each other. At idle speed the metering valve is almost closed so very little fuel is routed to the pump chamber. This forces the plungers slightly outward, partially charging the chamber. Depressing the accelerator pump causes more fuel to enter the chamber and forces the plungers further apart. As the rotor turns, the charging ports on the rotor and the hydraulic head misalign, effectively closing the ports. Continued

rotation of the rotor will then align the discharge ports of both components. At this time, the plungers will be forced or pushed inward by the rollers contacting the cam lobes. As the plungers move together, the fuel is forced out of the pump chamber into the fuel lines to the injectors. Fuel is prevented from dripping into the combustion chamber after the injection cycle by a delivery valve. The delivery valve, located in the rotor, ensures a sharp fuel cutoff after the injection cycle.

The quantity of fuel delivered by the distributor injection pump is directly proportional to the amount of fuel entering the pump chamber through the metering valve. Fuel injection timing is controlled by an internal time-advance mechanism. Timing is altered by rotating an internal cam ring in the rotor and head assembly. A pin in the auto-advance unit (see Figure 8-16) is connected to the internal cam and is located between a spring and piston. Fuel pressure from the transfer pump pushes the piston toward the spring and moves the pin. Movement of the pin in turn moves the cam in an advance direction. As the engine speed increases, transfer pump fuel pressure increases, and the timing advances accordingly.

Unused injection fuel from the transfer pump is vented back to the fuel tank through the governor housing. It is also used to lubricate and cool the internal components of the distributor injection pump.

INJECTION NOZZLES

The fuel injection nozzle is designed to vaporize and direct the metered fuel into the combustion chamber. The injection pump forces the required fuel into the injection nozzle at the precise time it is needed. The design of the combustion chamber will usually dictate the type of nozzle used, the droplet size, and the spray pattern required for optimum combustion in the given time frame and space. A typical injection nozzle is shown in Figure 8-18.

The typical injection nozzle has small openings so that the pressure can build up under some operating conditions. A spring-loaded needle valve in the injector nozzle keeps the opening of the nozzle closed until the pressure reaches the operating level. The parts of a typical fuel injection nozzle are shown in Figure 8-19. The fuel pressure opens the nozzle valve and the spring closes it. Fuel from the injection pump enters and pressurizes the fuel lines and the pressure chamber. When the force on the lift area is greater than the set spring force on the spindle, the needle valve will lift off its seat and rest with its upper shoulder against the face of the holder body. Fuel in a mist or spray-type pattern is then forced out into the combustion chamber. The pat-

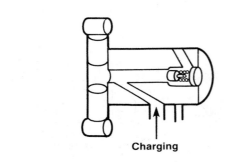

Charging

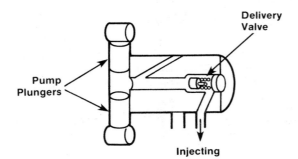

Delivery Valve

Pump Plungers

Injecting

FIGURE 8-17 Charging and injecting cycles for a distributor injection pump

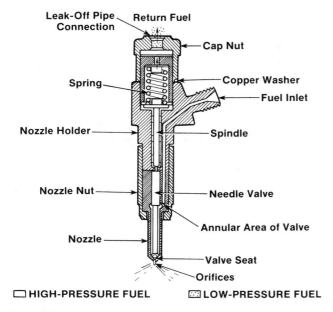

FIGURE 8-18   Typical injector nozzle

tern is determined by the type of tip used on the injector nozzle.

Only the tip of the injection nozzle protrudes into the combustion chamber. Two types of injector tips are used on diesel engines: the hole type and the pintle type, as shown in Figure 8-20. Open combustion chambers use the hole type, but it is also used on a few engines with divided combustion chambers. The pintle type tip is used only on engines with a divided combustion chamber.

Hole Type

The holes in this type of tip are drilled so that the fuel will spray into an open portion of the engine's combustion chamber. The holes are equally spaced around the tip and are approximately 0.007 inches in diameter. In some design applications, the tip will have holes only on one side. Because the holes in the tip are so small, fuel injection pressure can be as high as 30,000 psi during injection.

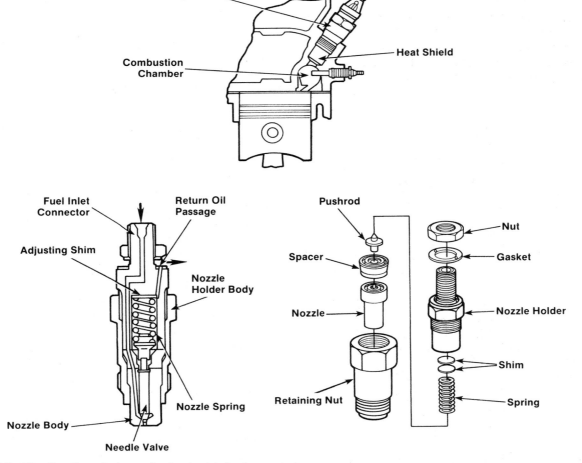

FIGURE 8-19   Sectional view of a typical injector nozzle

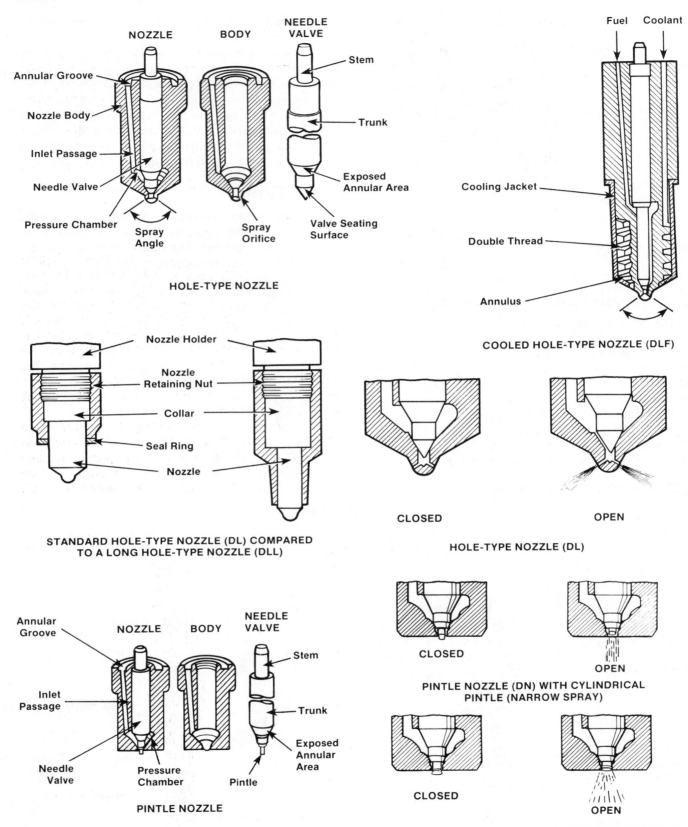

FIGURE 8–20 Sectional views of hole-type and pintle-type injection nozzles

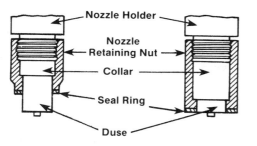

STANDARD PINTLE NOZZLE   ROLLER-TYPE NOZZLE

STANDARD PINTLE NOZZLE

FIGURE 8–20   (continued)

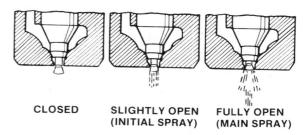

CLOSED   SLIGHTLY OPEN   FULLY OPEN
             (INITIAL SPRAY)   (MAIN SPRAY)

THROTTLING PINTLE NOZZLE (PILOT INJECTION)

Pintle Type

This type of injector tip has a single hole with a pin in it, often referred to as a pintle. Fuel sprays from the space between the pin and the hole and produces a narrow cone-shaped spray pattern. The shape of the fuel injection pattern is solely determined by the shape of the pintle. The pintle-type tip has a larger opening than the hole-type tip. Therefore, the pintle-type tip can operate only at pressures between 1,100 and 1,800 psi. Some design applications use a throttling pintle where a large amount of fuel is sprayed at the start of injection. Then, as the pintle moves up, it gradually reduces the fuel flow until it is completely stopped.

STARTING DEVICES

Since diesel fuel does not vaporize or ignite as readily as gasoline, starting devices are required on diesel engines. These include:

* Glow plugs
* Fuel heaters
* Engine block heaters

Glow plugs (Figure 8–21), located in the combustion chamber, preheat the air and fuel during cranking (or prior to cranking depending on the system design). Figure 8–22 shows a cross-sectional view of a typical glow plug arrangement. Engine block heaters are used in colder climates to heat the engine's coolant, which in turn keeps the cylinder block and head at a temperature suitable for starting purposes. A fuel heater preheats the fuel electrically before it reaches the filter. The heater, usually thermostatically controlled, is a resistance type designed to heat the fuel before it enters the filter. This reduces the possibility of wax plugging the filter, which usually occurs when the fuel temperature is 20 degrees Fahrenheit or lower.

A glow plug is a low-voltage heating element that is inserted into the combustion chamber on the intake manifold (Figure 8–23). The plug is usually controlled by the ignition switch. However, other design applications might have a separate on-off

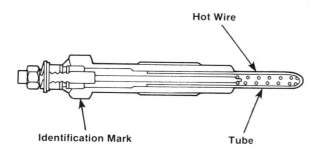

FIGURE 8–21   Typical glow plug

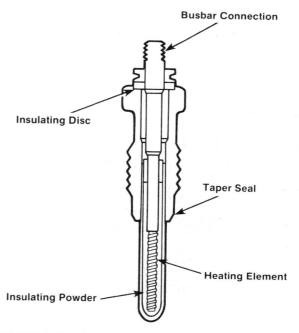

FIGURE 8–22   Cross-sectional view of a glow plug

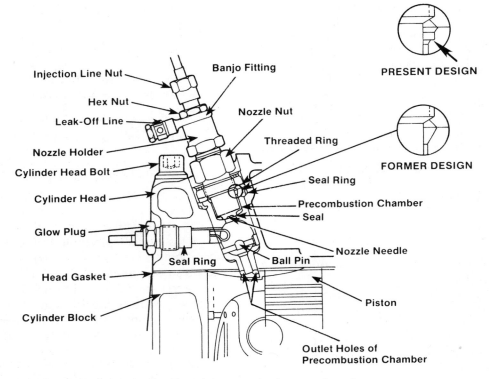

FIGURE 8-23 Injector tip and glow plug locations within the prechamber

switch within the passenger compartment. The glow plug is energized only until the air in the combustion chamber is adequately heated and can support ignition at start-up. The engerized or heating period for a glow plug is directly dependent upon how fast the plug can heat up and what the ambient temperature of the combustion chamber is. Most glow plugs work off of the car's 12-volt system.

A control module is the heart of the glow plug system (Figure 8-24). The module senses the engine's coolant temperature and de-energizes the glow plugs when the coolant is sufficiently heated after start-up. Other design variations might use a cycling type of relay to pulse the glow plugs on and off. This type of system usually has a manual on-off switch to control the cycling relay.

GENERAL SERVICE PRECAUTIONS

Diesel fuel system service is somewhat more specialized than gasoline system service. Before attempting to service a diesel fuel injection system, the technician should follow or perform certain servicing precautions. The most important guideline is to

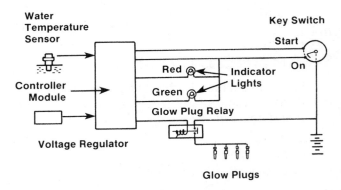

FIGURE 8-24 Electrical schematic of glow plug system

follow all normal personal and shop safety practices. Preventing accidents or hazardous conditions is just as important, if not more so, than repairing the faulty fuel system.

In addition, the following special precautions should be observed:

- Never disconnect any fuel line(s) while the engine is still running or hot.
- Ensure all disconnected fuel lines and fittings are properly capped.

- Whenever disconnecting a fuel line fitting from a double nut connection always use two wrenches to prevent the lines from twisting and cracking open.
- Never bend, twist, or damage the fuel injection lines.
- Maintain a clean work area and ensure all parts are clean. This is especially important when the system is assembled. Dirt or foreign matter can cause complete failure of the injection system. Abrasives damage friction surfaces severely, and extremely close tolerance can be destroyed by particles of dirt.
- Use only clean, filtered, compressed air to clean parts after they are washed.
- Use only filtered, temperature-controlled testing fluids for injector and injector pump service.
- The diesel fuel injection system utilizes very high pressures, on the order of 15,000 psi. Such pressure levels when released will penetrate the skin very easily and cause serious personal damage and possibly infection. Avoid contact with fuel that is under injection pressure.
- Use a face mask and filter to avoid prolonged inhalation of fuel or test fluid vapors.
- Ensure that the correct specifications are being used for the system being serviced. Refer to the appropriate manufacturer's service manual for:
  —Specifications
  —Installation procedures
  —Alignment/adjustment procedures
  —Timing procedures

DIESEL FUEL INJECTION SERVICING

Troubleshooting a diesel fuel injection system is similar to troubleshooting a gasoline fuel injection system. First, the diesel engine must be in good mechanical condition. The engine must also have good compression or the compressed air will not become hot enough to ignite the fuel. Second, the electrical system must be in good condition to enable the engine to crank rapidly and allow the glow plugs to get hot. Third, the fuel injection system must be operating correctly.

The first person to notice an engine problem is the driver/owner. The technician should ask the driver/owner specific questions until the exact nature of the problem is clear to both of them.

Typical diesel engine problems include:

- Overheating
- Abnormal engine knock
- Loss of power
- Smokey exhaust
- High fuel consumption

It is possible for these problems to be caused by the fuel injection system, but they can also be caused by the other engine systems.

Overheating problems are caused by the same conditions that cause a gasoline engine to overheat—low coolant level, slipping fan belt, defective head gasket, or a thermostat stuck closed.

When too much fuel is injected into the cylinder too early in the combustion cycle, engine knock will result. If engine knock is noted on all cylinders, it could indicate incorrect injection pump timing. Normally, engine knock will occur only on one or two cylinders and signals that the applicable injector is at fault. Incorrect operation of an injector will result in a much larger uncontrolled burning period and produce noticeable engine knock.

A restricted air inlet system or fuel system can be the cause of a loss of power. A check should be made to ensure that injector nozzle tips are clean and opened, that the fuel lines are not plugged, that the car filter is clean, and that the engine compression is correct.

A common problem with diesel engines is a smokey exhaust. This is caused by incomplete combustion of the fuel. At idle speeds this is more pronounced because the combustion chamber might still be cool and will tend to quench combustion before all the fuel is consumed. At higher engine speeds, if the smokey exhaust condition is still noted, the air/fuel mixture is too rich. Possible causes could be a faulty EGR valve, faulty injection pump, or an incorrect injection pump timing.

The injection of too much fuel results in poor fuel economy. A fuel leak in the system also produces the same symptom. High fuel consumption is always caused by low compression and incorrect injection pump timing.

Table 8-1 is a typical troubleshooting or diagnostic chart for a diesel fuel injection system.

FUEL FILTER REPLACEMENT

To protect the fuel injection system of a diesel engine, a good filtration is essential. The clearances between injector parts require close control of solids in the fuel. To avoid erosion and wear, solids must not pass between closely mated injector parts. In

TABLE 8-1: DIESEL FUEL INJECTION DIAGNOSIS

Condition	Possible Cause	Remedy
Engine will not crank.	Loose or corroded battery cables	Check all cable connections and tighten or clean as needed.
	Discharged battery	Recharge the battery and check the charging system output.
	Inoperative starter	Check voltage and amperage to starter.
Engine cranks slowly and will not start.	Loose or corroded battery cables	Check all cable connections and tighten or clean as needed.
	Undercharged battery	Recharge the battery and check the charging system output.
	Wrong engine oil	Drain oil and refill with recommended type.
Engine cranks normally, but will not start.	Incorrect starting procedure	Use recommended starting procedure.
	Inoperative glow plug system	Check glow plugs and electrical feed to glow plugs. Replace as needed.
	No fuel to nozzle	Loosen the nozzle line at the nozzle. Wipe the connection dry, and crank for 5 seconds. If fuel does not flow from the line, correct the restriction. Retighten the connection.
	Inoperative fuel shut-off solenoid	Repeatedly turn the ignition switch on and off. There should be a clicking noise, indicating that the solenoid is working. If not, connect a voltmeter and make a reading. If the voltage is less than 9 volts, check the circuit. If it is more than 9 volts, replace the solenoid.
	Restricted fuel filter	Loosen the fuel line going to the filter. If fuel sprays from the fitting while cranking the engine, replace the filter.
	Incorrect pump timing	Check and adjust the timing.
	No fuel to injection pump (if equipped with external supply pump)	Loosen the fuel line coming out of the filter. If fuel sprays from the filter while cranking the engine, check for a restricted fuel line or fuel tank strainer. If no fuel sprays from the filter, replace the injection pump.
	Inoperative fuel supply pump	Remove the inlet hose to the fuel pump. Connect a hose from the fuel pump to a container, then loosen the line going to the filter. If fuel does not spray from the fitting, replace the fuel supply pump.
	Restricted fuel line or fuel tank strainer	Check the fuel line for dents and kinks. If no restriction is discovered, remove the tank and check the strainer.
	Plugged fuel return line	Disconnect the line at the injection pump and connect a hose from the pump to a container. If the engine starts, find and correct the restriction in the fuel return circuit.

TABLE 8-1:   DIESEL FUEL INJECTION DIAGNOSIS (CONTINUED)

Condition	Possible Cause	Remedy
	Wrong or contaminated fuel	Flush the fuel system and add the correct fuel.
	Faulty nozzle(s)	Remove the nozzle(s) and test with a nozzle tester. Replace as needed.
	Low compression	Check the cylinder compression.
Engine starts, but will not keep running at idle.	Incorrect slow-speed idle	Adjust idle to specifications.
	Inoperative fast-idle solenoid	With the engine cold, check that the solenoid moves to hold the injection pump lever in the fast-idle position. If not, check the fast-idle circuit.
	Glow plugs turn off too soon.	Check glow plugs and electrical feed to glow plugs. Replace as needed.
	Restricted fuel return system	Disconnect the fuel return line at the injection pump and connect a hose from the pump to a container. If the engine idles normally, correct the restriction in the fuel return line circuit. If the idle is not normal, remove the return line check valve and make sure it is not blocked.
	Incorrect pump timing	Adjust the timing to specifications.
	Limited fuel to injection pump	Check the fuel filter, fuel lines, and fuel supply pump. Repair or replace as needed.
	Wrong or contaminated fuel	Flush the fuel system and add the correct fuel.
	Low compression	Check the cylinder compression.
	Injection pump malfunction	Remove and repair injeciton pump.
	Fuel shut-off solenoid closes	Adjust the ignition switch.
Excessive surge at light throttle, under load	Injection pump timing incorrect	Adjust timing to specifications
	Injection pump housing pressure too high	Use the recommended service procedure, as specified by the manufacturer.
	Restricted fuel filter.	Check fuel pump pressure at inlet and outlet sides of the filter. Replace the filter if necessary.
	Torque converter clutch is engaging too soon (if applicable)	Refer to automatic transmission service manual.
Engine starts and idles roughly, with excess noise and/or smoke, but clears up after warm-up.	Incorrect starting procedure Injection pump timing	Use the recommended starting procedure. Adjust timing to specifications.
	Air in system	Attach a section of clear plastic tubing on the fuel return line. If bubbles appear when the engine is cranking or running, a leak is present on the suction side of the fuel system.
	Inoperative glow plug(s)	Locate and replace defective glow plug(s).

TABLE 8-1: DIESEL FUEL INJECTION DIAGNOSIS (CONTINUED)

Condition	Possible Cause	Remedy
	Malfunctiong nozzle(s)	Clean or replace malfunctioning nozzle(s).
	Inoperative HPCA	Check the HPCA circuit.
Rough engine idle	Incorrect idle speed	Adjust idle to specifications.
	Air in system	Attach a section of clear plastic tubing on the fuel return line. Repair the leak and purge all air from the system.
	Incorrect injection pump timing	Adjust timing to specifications.
	Leaking nozzle lines	Inspect the lines at both the pump and nozzles. Tighten or replace the lines.
	Low fuel pressure between the fuel filter and injection pump	Check the fuel pressure and repair any restrictions that are found.
	Malfunctioning nozzle(s)	Clean or replace malfunctioning nozzle(s).
	Low cylinder compression	Perform a compression test to determine cylinder pressure.
Engine misfires above idle, but idles fine.	Restricted fuel filter	Check filter and replace if necessary.
	Incorrect injection pump timing	Adjust the timing.
	Wrong or contaminated fuel	Flush the fuel system and add the correct fuel.
Engine will not return to idle.	External linkage binding or misadjusted	Free the linkage and adjust or replace as necessary.
	Fast-idle system malfunction.	Check fast-idle adjustment and circuit.
	Internal injection pump malfunction	Remove the pump and repair or replace as necessary.
Fuel leaks onto the ground, but the engine is not malfunctioning.	Loose or broken fuel line or connection, or both	Inspect the fuel system and make any necessary repairs.
	Defective injection pump seal(s)	Replace the seal(s).
Noticeable loss of engine power	Restricted air intake	Check the air cleaner element and passages.
	Injection pump timing not set to specifications.	Adjust pump timing.
	EGR malfunction (if applicable)	Refer to EGR/EPR service manual.
	Restricted or damaged exhaust system	Check the exhaust system and make any necessary repairs.
	Restricted fuel filter	Check the filter and replace if necessary.

TABLE 8-1: DIESEL FUEL INJECTION DIAGNOSIS (CONTINUED)

Condition	Possible Cause	Remedy
	Restricted fuel supply from the fuel tank to the injection pump	Examine the fuel supply system. Locate the restriction and make the necessary repairs.
	Restricted fuel tank strainer	Remove the fuel tank and check the strainer.
	Restricted fuel return line circuit	Locate the restriction and make the necessary repairs.
	Plugged fuel tank vacuum vent in the fuel cap	Remove the cap. If loud hissing is heard, the vent is plugged and the cap must be replaced. (A slight hissing is normal.)
	Wrong or contaminated fuel	Flush the fuel system and add the correct fuel.
	External compression leaks	Check for compression leaks at all nozzles and glow plugs. If a leak is found, tighten the nozzle(s) or glow plugs.
	Faulty nozzle(s)	Remove the nozzle(s). Clean or replace as needed.
	Low cylinder compression	Check the compression to determine the cause.
Rapping noise from one or more cylinders; sounds like rod bearing knock.	Nozzle(s) sticking open or very low nozzle opening pressure	Remove the nozzle from the noisy cylinder(s). Test the nozzle and clean or replace as needed.
	Piston hitting cylinder head	Replace the malfunctioning parts.
	Mechanical problem	Refer to appropriate section in service manual.
Overall combustion noise (above normal noise level) and excessive black smoke	Timing not set to specifications.	Adjust timing to specifications.
	EGR malfunction (if applicable)	Refer to EGR/EPR service manual.
	Injection pump housing pressure out of specifications.	Adjust to specifications.
	Injection pump internal malfunction	Remove pump for repair.
Engine noise—internal or external	Faulty fuel pump, water pump, generator, valve train, bearings, etc.	Repair or replace as needed.
Engine overheats.	Coolant system leak, oil cooler system leak, or coolant recovery system malfunction	Locate and repair any leaks. Check the coolant recovery jar, hose, and radiator cap.
	Slipping or damaged belt	Replace or adjust as needed.
	Thermostat stuck closed	Replace the thermostat.
	Leaking head gasket	Replace the head gasket.
Instrument panel oil warning light comes on during idle.	Restricted oil cooler or cooler line	Repair the restriction.
	Low oil pump pressure	Refer to appropriate section in service manual.

NOTE: Whenever fuel is being sprayed during test procedures, be careful to always direct the fuel away from ignition sources.

addition to controlling solid contaminants, fuel filters also present a barrier to water that might be present in the fuel tank.

Diesel fuel, as it comes from the refinery, is usually free of contaminants. As it is transferred and handled, however, the chances of contamination increase. The main contaminants are water, dirt, bacteria, and wax crystals.

- *Water Contamination.* When the fuel tank is not kept filled, warm, moisture-laden air condenses on the cooler inside the metal wall of the fuel tank. The result is water contamination of the fuel. This type of contamination can also occur with careless filling of a fuel tank during a heavy rain or snowfall. Water in the diesel fuel can cause a tip to blow off a fuel injector or the seizing of a plunger and bushing, seriously damaging the fuel injectors and pump.
- *Dirt Contamination.* Contamination from dirt and foreign matter can be the result of careless filling of the fuel tank. It can happen when a hose or filler cap is dropped and not cleaned off before being used. As a result, dirt, mud, or grass clinging to the hose nozzle or cap can be washed into the tank by the flowing fuel.
- *Bacteria Contamination.* This form of fuel contamination occurs when microorganisms, which are always present in the air, begin to grow in the interface, the area between the fuel and any water in the tank. The dark, nonturbulent nature of the fuel tank, the viscous fuel, and a small amount of water at the tank bottom can provide ideal bacteria growth conditions. As bacteria, yeasts, molds, and fungi multiply, they form a slime that can easily clog fuel filters, retard fuel flow, and reduce engine efficiency. Bacterial contamination can best be controlled through the use of a fuel-soluble biocide. The biocide blends with the fuel, destroys bacteria, and prevents regrowth.
- *Wax Crystals or Fuel Cloud Point.* Fuel cloud point is the temperature at which wax crystals will form in diesel fuel, making it appear cloudy. As temperatures drop, these wax-like crystals will gradually increase in size and number. Eventually the crystals can clog the fuel filter and lead to engine shutdown. During cold weather engine operation, the cloud point should be 10 degrees Fahrenheit below the lowest expected outside temperature. This will prevent wax crystals from

clogging the fuel filters. Chemical flow improvers, to minimize clouding, are normally added to diesel fuels for use in cold climates. Without the use of chemical flow improvers, frequent replacement of fuel filters is necessary. Locating the fuel filters in the engine compartment for heating purposes will help keep crystals from clogging the filters.

Diesel Fuel Filtering System

The filters used on diesel fuel systems, in general, fall into two categories; single and two-filter systems. Some engine manufacturers use only one filter, putting it between the fuel tank and the fuel transfer pump (Figure 8-25). Since the filter is on the suction side of the pump, the pressure drop across the filter is limited. The filter for single-filter systems has sufficient dirt-holding capacity (with a low pressure drop) to operate for the recommended service interval.

Diesel fuel systems using two fuel filters are common (Figure 8-26). This arrangement locates

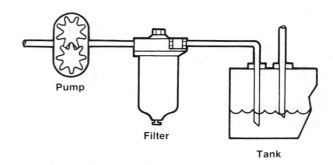

FIGURE 8-25 Single-filter system

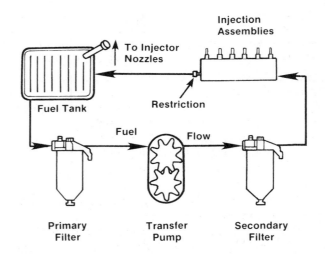

FIGURE 8-26 Two-filter system

the *primary* filter on the suction side of the transfer pump, protecting it from large solid contaminants. Most of the water in the fuel is also removed by this filter. The *secondary* filter is located between the transfer pump and the fuel injectors. Since the filter is located on the discharge side of the pump, the pressure drop across the secondary filter can be much higher than on a filter located on the suction side of the pump. This allows the secondary filter to be made in a compact size using a fine filtering media. This filter controls the size of particles allowed to pass into the fuel injectors. It also stops any water that might have passed through the primary filter.

Replacing Diesel Filters

The diesel fuel filter is usually one of two types:

- Spin-on separate fuel filter units
- Self-contained units with filter elements

Spin-on Fuel Filters. To replace a spin-on fuel filter unit, proceed as follows:

1. To service a spin-on fuel filter, simply remove the used filter. If the filter has a flat-bottom type housing (Figure 8–27), it can be removed using an oil filter wrench.
2. Hand tighten the new spin-on filter until the gasket touches the filter assembly. Then use an oil filter wrench and tighten the filter the number of turns called out on the filter.

Self-Contained Fuel Filters. To replace a fuel filter element in a self-contained unit, proceed as follows:

1. Drain fuel, dirt, and water from the filter housing sump by opening the petcock or removing the drain plug (Figure 8–28).
2. Loosen the center stud bolt located at the top of the filter assembly; discard gaskets; remove filter housing and/or cover and element together.
3. Remove the used element from the filter housing; discard the element.
4. Wipe the inside of the filter housing clean of fuel, residue, or other deposits.
5. Place a new gasket on the center stud bolt (Figure 8–29).
6. Place a new gasket in the filter cover assembly (Figure 8–29).
7. Place a new fuel filter element in the filter housing; assemble the housing and cover and tighten the center stud bolt to the

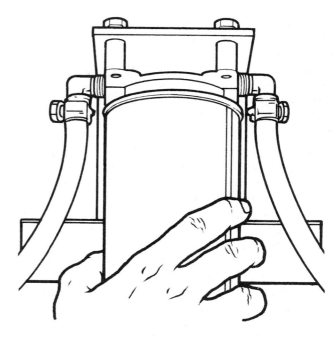

FIGURE 8-27   AC spin-on filters, extruded housing, and flat-bottom type housing

torque called out on the housing or in the instructions (Figure 8-30).
8. Close petcock or replace drain plug in filter housing.

Most service manuals usually provide detailed procedures on bleeding the air from the fuel system. In some cases, a hand priming pump is required for this procedure (Figure 8-31).

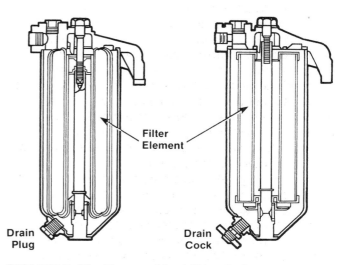

FIGURE 8-28   Types of fuel filter drains

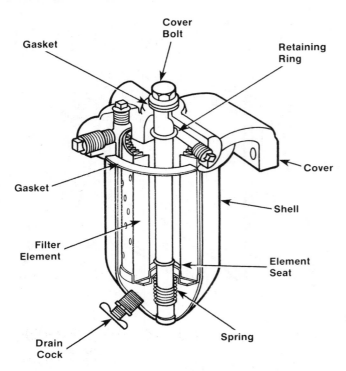

FIGURE 8-29 Typical fuel filter assembly

REMOVING WATER FROM THE SYSTEM

As just mentioned a common diesel fuel problem is water in the fuel. Some vehicles are equipped with a water separator in the fuel system. A fuel water separator works much like a filter, except that the element is replaced by a baffle. When contaminated fuel enters the separator, the heavier water settles at the bottom, while the lighter fuel rises to the top. The separator requires no servicing other than draining of the water (Figure 8-32).

A water separator can also be installed as aftermarket equipment. Some diesel systems are equipped with a combination fuel pickup—fuel level indicator—water sensor as illustrated in Figure 8-33. This device will detect water in the system when it reaches a level of 1 to 1-1/2 gallons in the tank. The device will illuminate a dashboard light to signal the driver. The vehicle can be driven some distance under this condition before water is drawn into the fuel system. However, it should be drained as soon as possible. To drain water from the system, remove the fuel return hose, usually located at the front of the engine (Figure 8-34). Remove the fuel tank filler cap, and pump or siphon the water from the previously disconnected line. Use clear plastic line to determine when uncontaminated fuel begins to flow. When the water is drained, replace the filter cap and the return hose.

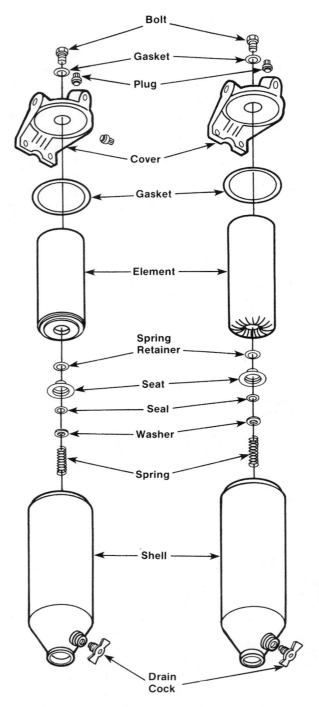

FIGURE 8-30 Typical fuel filter detail

CAUTION: Be extremely careful when working with open fuel lines. Hot exhaust system parts, sparks, or open flames can cause fires. Keep readily available a fully charged fire extinguisher capable of extinguishing class B fires.

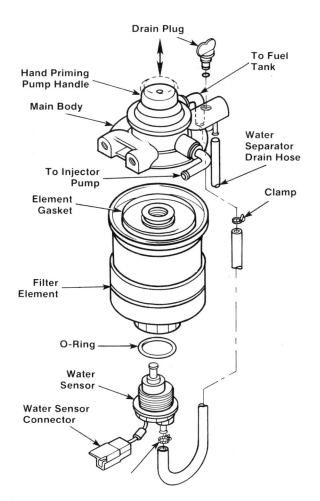

FIGURE 8-31   Using a hand priming pump on a diesel fuel filter

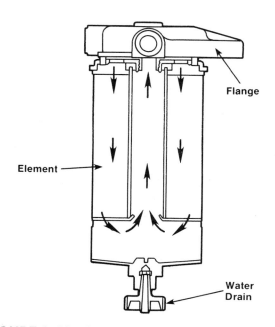

FIGURE 8-32   Diesel fuel filter/water separator

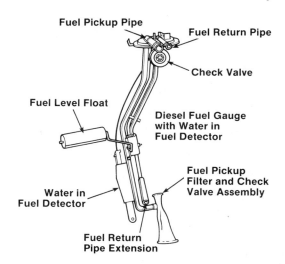

FIGURE 8-33   A combination fuel pickup/fuel level indicator/water sensor

INJECTION NOZZLE SERVICE

The diesel injection nozzle is an important part of the diesel fuel injection system. The injection pump develops very high pressures (800 to 3,000 psi). These high pressures are routed through thick steel fuel lines to the injection nozzles (Figure 8–35).

Rough idle, knock on one or more cylinders, smokey exhaust, and loss of power can be caused by a faulty injector. The injector valve can stick, leak, or the tip can be plugged or corroded.

To identify a faulty injector, first loosen the high-pressure fitting on the injector fuel line (Figure

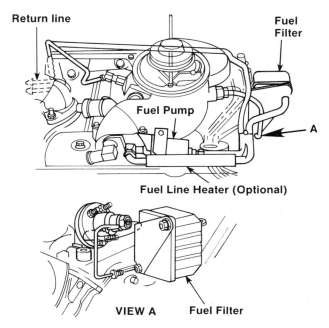

FIGURE 8-34   Location of diesel fuel lines

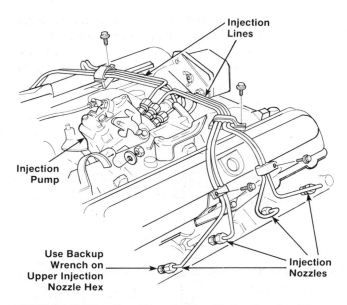

FIGURE 8-35 Routing of diesel injection fuel lines

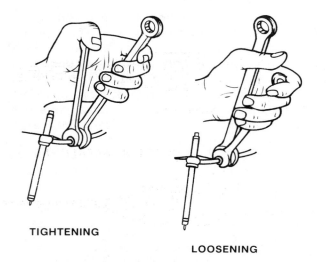

TIGHTENING

LOOSENING

FIGURE 8-36 Preferred method of loosening fuel line fittings

8-36). This will prevent the injector from operating and spraying fuel into the combustion chamber. Then set the engine to fast idle. Loosen each fitting in turn, one at a time. An injector is faulty if the engine speed does not change when the appropriate fitting is loosened. This action is the same as grounding one spark plug at a time to locate a weak cylinder on a gasoline engine. A general procedure is presented here to remove, replace, and test an injection nozzle. Be sure to check the manufacturer's service manual for more detailed and specific servicing procedures.

CAUTION: Be sure the nozzles and lines are clean before proceeding with nozzle removal. Any contaminants or foreign material in the fuel system can cause problems.

1. Clean the area in the vicinity of the injector(s) that are to be removed. This will prevent dirt and foreign materials from contaminating the fuel system.
2. Disconnect the fuel line(s) from the injectors to be serviced. Cap all fuel lines and fittings to prevent contamination.
3. Follow the procedures in the service manual and remove the injector.
4. Disassemble and clean the injector (Figure 8-37). Do not interchange parts from one injector to another.
5. Carefully inspect all parts for wear and damage. Refer to the specifications given

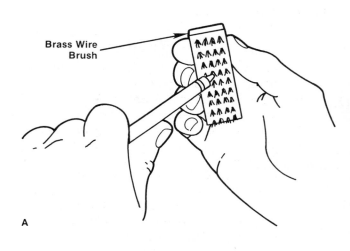

Brass Wire Brush

A

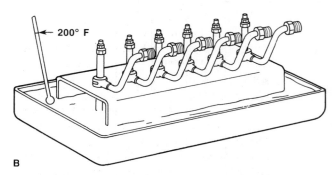

← 200° F

B

FIGURE 8-37 Cleaning the injector nozzle: (A) using a soft brass bristle brush to clean injector nozzle tip; (B) injectors in rack with nozzles immersed in special cleaning fluid at a controlled temperature to help loosen deposits.

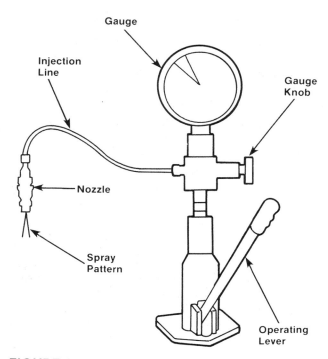

FIGURE 8-38   Nozzle tester

in the manufacturer's service manual to determine wear tolerances.

6. Reassemble the injector as per the procedures given in the service manual.

7. To test the injector, install it in a nozzle stand similar to the one shown in Figure 8-38. Be sure the correct test third is used and is at the proper temperature.

8. Testing and adjusting procedures (depending on make and model of injector) must be performed to achieve the following:
   • specified nozzle-opening pressure
   • correct spray pattern
   • proper chatter
   • no nozzle drip at specified fuel pressure

9. Carefully follow the equipment manufacturer's test procedures to check for proper spray pattern (Figure 8-39).

10. While rapidly moving the operating lever on the nozzle tester, listen for a creaking sound. This indicates the nozzle is clean and in good condition.

11. Open the gauge knob on the nozzle tester slightly and observe the gauge while slowly operating the lever. The gauge should indicate the opening pressure of the nozzle. This gauge valve should be compared to the specifications given in the manufacturer's service manual. On some types of

injectors the nozzle opening pressure can be adjusted as shown in Figure 8-40.

12. Again, slowly operate the lever on the nozzle tester and bring the pressure to 200 psi and observe the nozzle tip.

ACCEPTABLE NOZZLE SPRAY PATTERNS

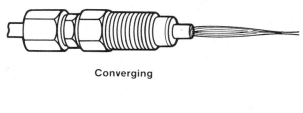

Converging

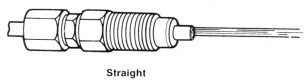

Straight

Narrow Cone

UNACCEPTABLE NOZZLE SPRAY PATTERNS

Hosing

Partial Cone

Wide Cone

FIGURE 8-39   Spray patterns for a diesel fuel injection nozzle

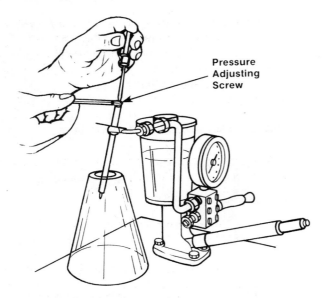

FIGURE 8-40 Adjustment of the nozzle opening pressure

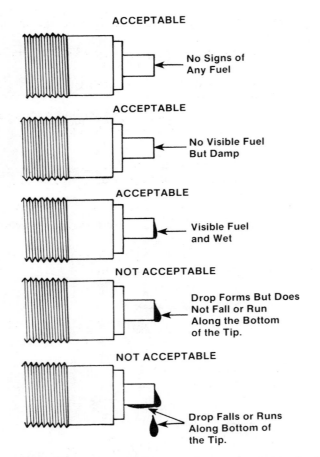

ACCEPTABLE

No Signs of Any Fuel

ACCEPTABLE

No Visible Fuel But Damp

ACCEPTABLE

Visible Fuel and Wet

NOT ACCEPTABLE

Drop Forms But Does Not Fall or Run Along the Bottom of the Tip.

NOT ACCEPTABLE

Drop Falls or Runs Along Bottom of the Tip.

FIGURE 8-41 Acceptable and unacceptable fuel leakage from an injection nozzle

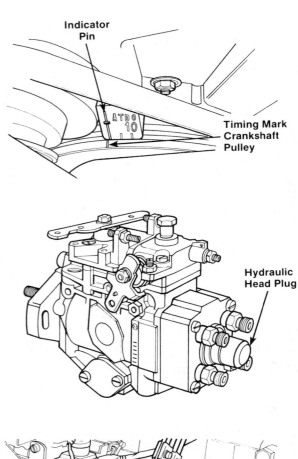

Indicator Pin

Timing Mark Crankshaft Pulley

Hydraulic Head Plug

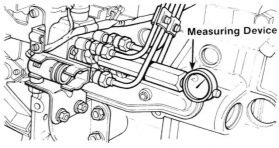

Measuring Device

FIGURE 8-42 Adjusting the injection pump timing: (A) position crankshaft as specified; (B) remove hydraulic head plug; (C) insert special dial-type measuring tool. Then position and tighten the pump mounting to achieve the specified pump plunger timing position.

CAUTION: Do not touch the tip or place your hand near it. The pressure from the spray can cause the fluid to penetrate the skin and cause blood poisoning.

Figure 8-41 shows acceptable and unacceptable leakage amounts for one type of nozzle. Refer to the manufacturer's service manual for proper specifications and recommendations.

13. Prior to reinstalling the injector(s), be sure that there is no gasket material remaining in the injector mounting hole. Clean all carbon deposits from the injector opening and the precombustion chamber. Remove all old O-rings and use only new O-rings and gaskets where required.

14. Install the injector(s) into the cylinder head. Tighten the injector to the manufacturer's recommended torque. Be sure that there is a positive seal around the injector.

15. Connect all fuel lines. Tighten the fuel lines to the injectors according to the manufacturer's recommended torque values.

16. If necessary, bleed all air from the fuel system.

17. Start the engine and inspect for fuel leaks. Tighten all loose connections.

INJECTION PUMP TIMING ADJUSTMENT (FIGURE 8-42)

The procedure for adjusting the timing of the injection pumps is as follows:

1. Put the crankshaft in the specified position.
2. Remove the hydraulic head plug from the pump.
3. Connect a special dirt-type measuring tool to the pump.
4. Position and tighten the pump mounting to achieve the specified pump plunger timing position as noted on the measuring device.

REVIEW QUESTIONS

1. A water fuel separator works on which of the following principles?
   a. relative weight difference and water filtering
   b. water filtering and trapping water
   c. water is heavier than fuel and fuel penetrates filters more easily
   d. all of the above

2. The fuel charge in a diesel engine is ignited by the _____ .
   a. injection nozzle
   b. glow plug
   c. compressed air
   d. advanced timing mechanism

3. After removing and replacing a diesel fuel line, the technician must do which of the following procedures before starting the engine?
   a. change the fuel filter
   b. bleed the air from the fuel lines
   c. adjust injection timing
   d. drain water from the water separator

4. The governor on a diesel fuel injector system is used to control the _____ .
   a. injection timing
   b. injection timing advance
   c. fuel metering mechanism
   d. all of the above

5. The primary function of a diesel injection pump is to _____ .
   a. force a precise metered amount of fuel through the injector
   b. build up a low fuel pressure source
   c. control fuel atomization
   d. all of the above

6. A pintle type injection nozzle is used only on _____ .
   a. engines with open or divided combustion chambers
   b. engines with divided combustion chambers
   c. engines with open combustion chambers
   d. on diesel as well as gasoline engines

7. A hole type injection nozzle _____ .
   a. operates at twice the pressure level of a pintle type nozzle
   b. can be used on engines with open combustion chambers as well as divided combustion chambers
   c. has small holes that are equally spaced around the diameter of the tip
   d. all of the above

8. In cold climates, the diesel engine's coolant system is heated by a _____ .
   a. glow plug
   b. fuel heater
   c. engine block heater
   d. none of the above

9. Glow plugs _____ .
   a. preheat the air and fuel on start-up
   b. are used on a continuous basis to heat the air and fuel in colder climates

c. use very high voltage heating elements
d. preheat the engine's coolant system

10. An automobile equipped with a diesel fuel injection system is experiencing excessive engine knocking. Technician A says this is caused by a restricted air inlet system. Technician B says that too much fuel is being injected too early in the combustion cycle. Who is correct?
   a. Technician A
   b. Technician B
   c. Both Technician A and B
   d. Neither Technician A nor B

11. A diesel fuel injected engine is experiencing loss of power at highway speeds as well as during city driving. Technician A says that the air inlet system is restricted either by a duty air filter or a plugged air passage. Tech-

nician B says that the culprit is a restricted or plugged injection nozzle. Who is right?
   a. Technician A
   b. Technician B
   c. Both Technician A and B
   d. Neither Technician A nor B

12. To drain water from a diesel fuel injection system Technician A removed the fuel return hose and fuel tank filter cap and siphoned the water from the disconnected fuel return hose. Technician B removed the fuel return hose and fuel tank filter cap and siphoned the water from the auxiliary output on the fuel filter. Who performed the correct procedure?
   a. Technician A
   b. Technician B
   c. Both Technician A and B
   d. Neither Technician A nor B

CHAPTER NINE

EXHAUST SYSTEMS

Objectives

After reading this chapter, you should be able to:

- Explain the three basic functions of a vehicle's exhaust system.
- Explain the function of exhaust system components, including exhaust manifold, gaskets, exhaust pipe and seal, catalytic converter, muffler, resonator, and clamps, brackets, and hangers.
- Explain the importance of proper exhaust system operations to vehicle performance and personal safety.
- Properly perform an exhaust system inspection.
- Properly service and/or replace exhaust system components.
- Explain how a turbocharger operates.
- Properly service a turbocharger.

A vehicle's exhaust system has three basic functions:

1. Carry away the poisonous, lethal gases from the passenger compartment
2. Clean the exhaust emissions
3. Muffle the sound of the engine

The system starts at the exhaust manifold and includes the exhaust pipe, a catalytic converter, muffler, tail pipe, and sometimes a resonator (Figure 9–1). All the parts of the system are designed to conform to the available space of the car's undercarriage and yet be a safe distance above the road.

A vehicle can be equipped with either a single or a dual exhaust system. Most four- and six-cylinder engines use a single exhaust system (Figure 9–2). Many V-type engines use a dual exhaust system for increased power, performance, and volumetric effi-

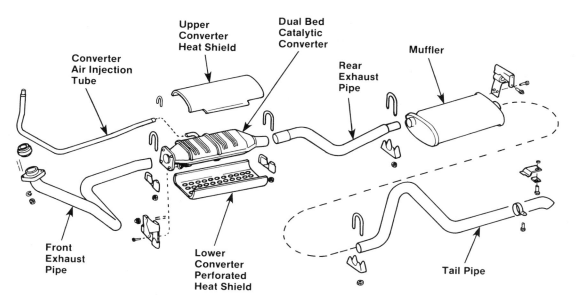

FIGURE 9–1 Typical exhaust system components

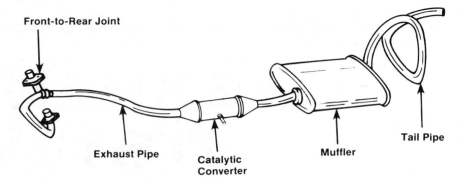

FIGURE 9-2   Single exhaust system

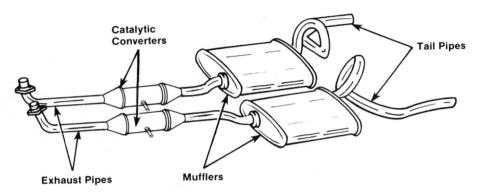

FIGURE 9-3   Dual exhaust system

ciency. A dual exhaust system has a separate system of pipes, mufflers, and converters for each bank of cylinders (Figure 9-3).

WARNING: When inspecting or working on the exhaust system, remember that its components get very hot when the engine is running and contact with them could cause a severe burn. Also, always wear safety glasses or goggles when working on the system.

EXHAUST SYSTEM COMPONENTS

An overview of the components of the exhaust system was given in Chapter 1. The following is a more detailed look at the function of various components in the system.

EXHAUST MANIFOLD

The exhaust manifold is designed to collect high-temperature burned gases and guide them from the cylinder head of the engine to the exhaust manifold. Exhaust gas temperature depends on the power produced by the engine. Under full-power conditions, the exhaust manifold will become red hot, causing a great deal of expansion. At idle, the exhaust manifold is just warm, causing little expansion. In passenger car operation, the engine will normally be run under light-load, part-throttle conditions, where the manifold temperatures will not make the manifold red hot. Most modern exhaust manifolds are bolted to the engine and made of:

- *Cast Iron.* The majority of exhaust manifolds are still made of cast iron. This material is inexpensive and easy to machine, but is heavy.
- *Cast Nodular Iron.* Nodular steel is stronger and more durable than conventional cast iron. Manifolds can be made lighter with nodular iron because its strength allows less material to be used. The disadvantage of nodular iron is that it is difficult to machine.
- *Fabricated Steel.* Fabricated steel is lighter than iron and easier to assemble into tubes for tubular headers. Unfortunately, steel is also more costly than either conventional cast iron or cast nodular iron.

In-line or straight-line engines, those with all the cylinders in a row, have one exhaust manifold (Figure 9–4). Engines of the V-type have an exhaust manifold on each side of the engine (Figure 9–5). An exhaust manifold will have either three, four, or six passages (depending on the type of engine) at one end. These passages blend into a single passage at the other end, which connects to an exhaust pipe. From here, the flow of exhaust gases continues to the back of the car.

Exhaust systems are designed for the engine-chassis combination. The exhaust system length, pipe size, and silencer are designed to make use of the tuning effect of the gas column resonating within the exhaust system. Tuning occurs when the ex-

haust pulses are emptied into the manifold between the pulses of other cylinders. Proper tuning of the exhaust manifold tubes can actually create a vacuum that helps to draw exhaust gases out of the exhaust system improving volumetric efficiency. Separate, tuned exhaust headers (Figure 9–6) can also improve efficiency by preventing exhaust flow from one cylinder from interfering with the exhaust flow from another cylinder. Cylinders next to one another may release exhaust gas at about the same time. When this happens, the pressure of the exhaust gas from one cylinder can interfere with the flow from the other cylinder (Figure 9–7). With separate headers, the cylinders are isolated from one another, interference is eliminated, and the engine breathes

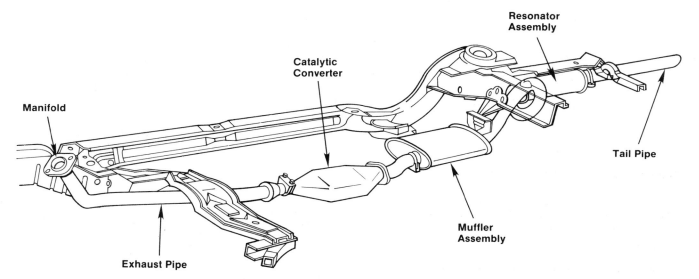

FIGURE 9–4   In-line engines have one exhaust manifold.

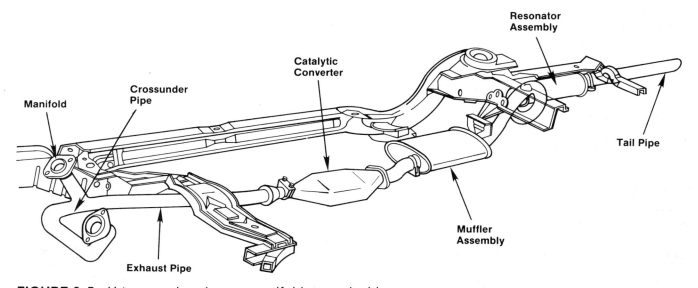

FIGURE 9–5   V-type engines have a manifold on each side.

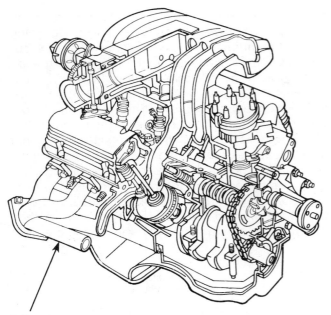

Tuned Exhaust Headers

FIGURE 9-6 Efficiency can be improved with tuned exhaust headers.

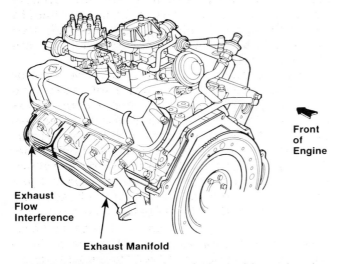

Exhaust
Flow
Interference

Exhaust Manifold

Front
of
Engine

FIGURE 9-7 Cylinders interfering with each other's exhaust flow

better. The problem of interference is especially common with V-8 engines, because they have a lot of cylinder overlap. V-6 engines have less cylinder overlap.

Four-cylinder engines often have an exhaust overlap problem with the first and second or third and fourth cylinders (Figure 9-8). While the overlap is usually not as severe as with a V-8 engine, the problems caused can be bad, because half of the engine's cylinders are affected.

Some exhaust manifolds are designed to go above the spark plug, whereas others are designed to go below. The spark plug and carefully routed ignition wires are usually shielded from the exhaust heat with sheet metal heat deflector shields. Often, a stove for the automatic choke is placed in the exhaust manifold.

To make exhaust gases less harmful, fresh air can be injected into the exhaust manifold. This helps to oxidize the unburned fuel still remaining in the exhaust gases. Metal connector pipes and fittings to carry and distribute fresh air are screwed into the exhaust manifold (Figure 9-9). Air injection systems are covered fully in Chapter 10; their service is described in Chapter 11.

Another control in the exhaust manifold setup is the exhaust gas recirculation (EGR). This device is

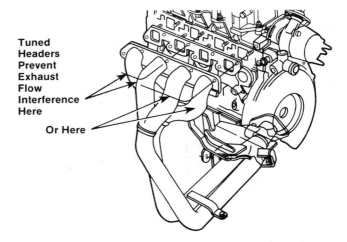

Tuned
Headers
Prevent
Exhaust
Flow
Interference
Here

Or Here

FIGURE 9-8 Tuned exhaust headers in a four-cylinder engine.

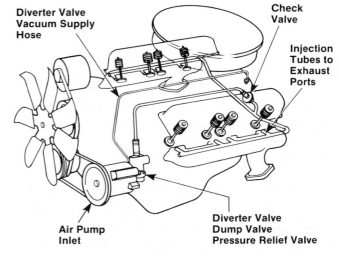

Diverter Valve
Vacuum Supply
Hose

Check
Valve

Injection
Tubes to
Exhaust
Ports

Air Pump
Inlet

Diverter Valve
Dump Valve
Pressure Relief Valve

FIGURE 9-9 Air injection system

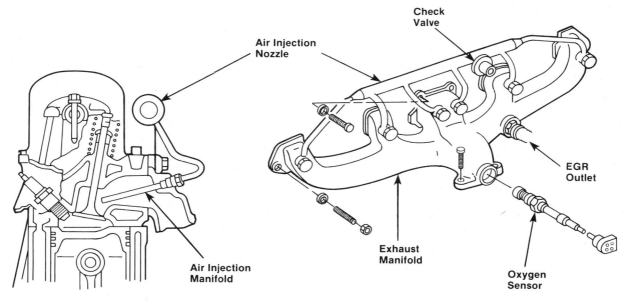

FIGURE 9-10 Emission controls on an exhaust manifold

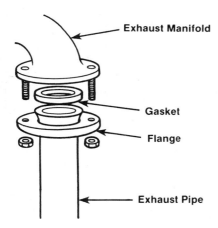

FIGURE 9-11 Manifold connection to exhaust pipe

used to lower combustion temperatures for control of oxides of nitrogen. The EGR process routes small amounts of exhaust gas into the intake air/fuel charge. EGR is covered in Chapters 10 and 11. The location of the emission controls in a typical exhaust manifold is illustrated in Figure 9-10.

Exhaust Manifold Gaskets

The exhaust manifold gasket seals the joint between the head and exhaust manifold (Figure 9-11). Many new engines are assembled without exhaust manifold gaskets. This is possible because new manifolds are flat and fit tightly against the head without leaks. During use, the exhaust manifolds go through many heating/cooling cycles. This causes stress and some corrosion in the exhaust manifold. Removing the manifold will usually distort the manifold slightly so it is no longer flat enough to seal without a gasket. Exhaust manifold gaskets are normally used to eliminate leaks when exhaust manifolds are reinstalled.

Some exhaust manifold gaskets have a perforated steel face on one side and a soft face on the other. The steel face is positioned toward the exhaust manifold on the engine. This is done because the exhaust manifold expands when it is heated during operation. As it expands, it moves on the head. This design seals the exhaust manifold while allowing it to move on the metal surface of the gasket, just as it does on the metal surface of the head in the absence of a gasket. A new gasket uses a ceramic-based facing on perforated steel. The ceramic-based facing does an excellent job of sealing high-temperature engines found on late-model engines. Most heavy-duty engines use a separate embossed shim gasket at each port, usually with a single bead embossment. A double bead embossment is used in more critical sealing applications.

In-line engines that have the intake and exhaust manifold on the same side of the head often use a combination gasket. This gasket has a perforated steel core with the facing on one side at the exhaust ports and facings on both sides at the intake ports. Special steps must be taken when installing these gaskets to prevent intake or exhaust leaks. Instructions packaged with the gaskets describe the required installation steps to produce a good seal.

EXHAUST PIPE AND SEAL

The exhaust pipe is metal pipe—either aluminized steel, stainless steel, or zinc-plated heavy-gauge steel—that runs under the vehicle between the exhaust manifold. On cars built before 1975, the exhaust pipe was connected directly to the muffler.

On most V-type engines, an exhaust pipe resembling the shape of a Y is used. The two ends at the top of the Y are connected to the two exhaust manifolds used on V-type engines. The bottom of the Y connects to the catalytic converter (Figure 9–12). A crossover pipe (Figure 9–13), which can be routed behind, above, or below the engine, is frequently used to connect the two sides of the V-type engine to the exhaust pipe.

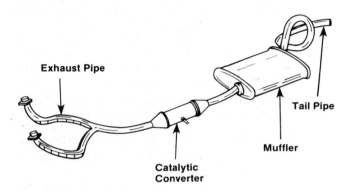

FIGURE 9–12 V-type engines use an exhaust pipe in the shape of a Y.

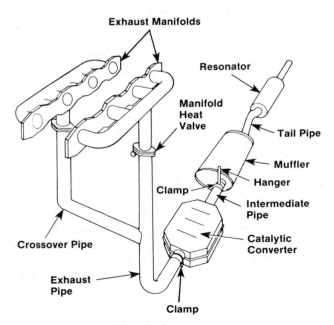

FIGURE 9–13 Crossover pipe connects the V-type engine to the exhaust pipe.

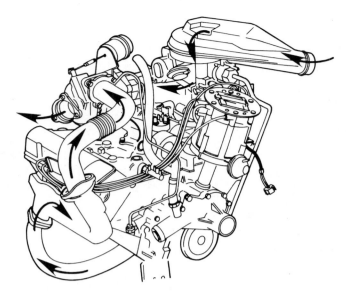

FIGURE 9–14 Route of intake and exhaust flow through a turbocharged engine

On some high-performance V-type engines, a dual exhaust system uses separate components on each side. An exhaust pipe extends from both side exhaust manifolds and runs along each side of the vehicle to its own converter/muffler system.

With most turbocharged engine exhaust connections, the exhaust gases that drive the turbocharger must be routed from the exhaust pipe into and out of the exhaust turbine housing (Figure 9–14). Various other turbocharger/exhaust pipe arrangements are shown in Figure 1–30 (see Chapter 1).

Some high-performance or supercharged engines have a combination of the exhaust manifold and the exhaust pipe custom-formed from thin chrome-plated steel tubing. Called a *header,* this design permits an exhaust manifold design that will handle the large volumes of exhaust gas produced when the engine is operating at a high speed, such as during a race. But, for normal, legal driving speeds, headers do not do anything to improve engine performance. In addition, they have a relatively short life and are costly. Because of headers' appearance of power and speed, some vehicle owners have them installed inspite of their cost.

Some type of emission controls such as the exhaust gas recirculation (EGR) valve, the early fuel evaporation (EFE), or a heat-riser is usually mounted between the exhaust manifold and the exhaust pipe. Full details on the operation and servicing of these emission controls are given in Chapters 10 and 11. In an electronic emission-control system, an oxygen sensor is installed in the exhaust pipe. This

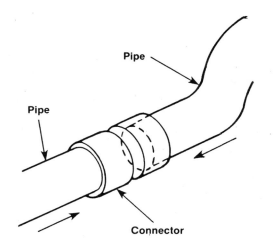

FIGURE 9-15 Exhaust pipe coupling

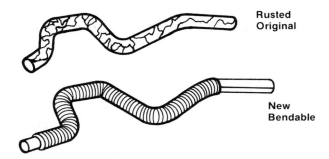

FIGURE 9-16 Bendable pipe

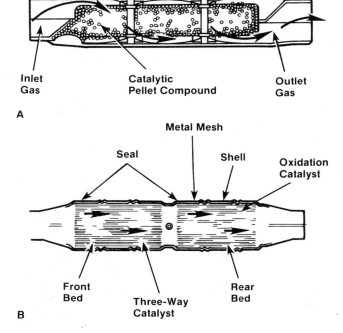

FIGURE 9-17 (A) Pellet catalytic converter; (B) monolithic catalytic converter

device is used to sense the amount of oxygen in the exhaust gas and to send signals to an electronic control unit to change the mixture of air and fuel being delivered to the engine. The operation and service of an oxygen sensor is given in Chapter 12.

Intermediate, or connecting zinc-plated steel or stainless steel pipes, are used at any point past the exhaust pipe to join the exhaust system components (Figure 9-15). As shown in Figure 9-16, bendable pipe is available to replace rusted originals.

CATALYTIC CONVERTERS

Catalytic converters have been the biggest change in exhaust systems over the years. They were added in the mid-1970s to reduce emissions, but carmakers quickly discovered that they also did part of the job of silencing the exhaust. As a result, mufflers have become lighter and simpler.

The catalytic converter is located ahead of the muffler in the exhaust system. The extreme heat in the converter oxidizes the exhaust emissions that flow out of the engine. As described in Chapters 10 and 11, there are several types of catalytic converters. The most common are the following:

- *Pellet catalytic converter.* Uses hundreds of small beads that act as the catalyst agent (Figure 9-17A).
- *Monolithic catalytic converter.* Uses a ceramic block shaped like a honeycomb or beehive that is coated with a special chemical that helps the converter act on the exhaust gases (Figure 9-17B).
- *Minicatalytic converter.* Provides a close coupled converter that is either built in the engine exhaust manifold or located next to it (Figure 9-18). It is primarily used on four-

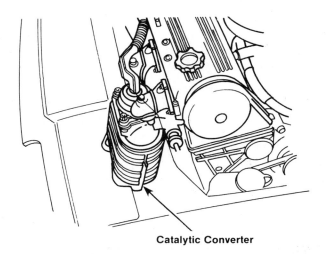

FIGURE 9-18 Minicatalytic converter

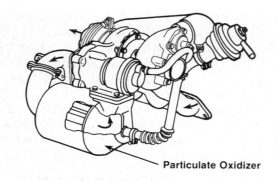

Particulate Oxidizer

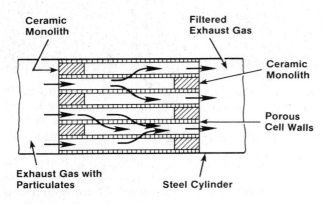

Ceramic Monolith

Filtered Exhaust Gas

Ceramic Monolith

Porous Cell Walls

Exhaust Gas with Particulates

Steel Cylinder

FIGURE 9–19 Particulate catalytic converter

cylinder vehicles, or on larger vehicles as a second converter.

- *Particulate oxidizer catalytic converter.* Uses a monolithic element, located between the exhaust manifold and turbocharger, to react to the particulates of a diesel engine (Figure 9–19). Particulates are solid particles of carbon-like soot that are emitted from the diesel engine as black smoke.

MUFFLERS

The muffler (Figure 9–20) is a cylindrical or oval-shaped component, generally about 2 feet long, mounted in the exhaust system about midway or toward the rear of the car. Consisting of a series of baffles, chambers, tubes, and holes to break up, cancel out, or silence the pressure pulsations that occur each time an exhaust valve opens. Two types of mufflers are commonly used on passenger vehicles (Figure 9–21):

1. *Reverse-flow mufflers.* These mufflers change the direction of the exhaust gas flow through the inside of the unit. This is the most common type of muffler found on passenger cars.

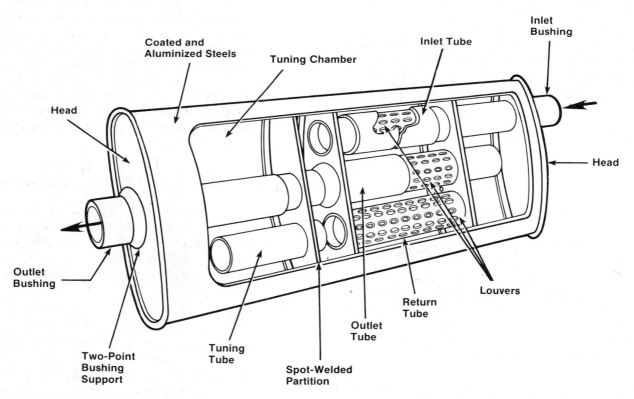

Coated and Aluminized Steels

Tuning Chamber

Inlet Tube

Inlet Bushing

Head

Head

Outlet Bushing

Two-Point Bushing Support

Tuning Tube

Spot-Welded Partition

Outlet Tube

Return Tube

Louvers

FIGURE 9–20 Typical muffler

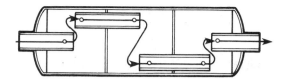

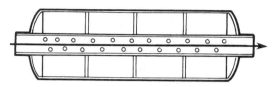

A Reverse-Flow Muffler

B Straight-Through Muffler

FIGURE 9–21 (A) Reverse-flow muffler; (B) straight-through muffler

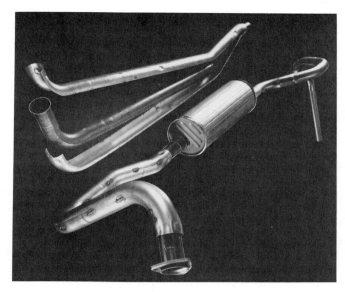

FIGURE 9–22 The number of exhaust parts made of stainless steel is increasing.

2. *Straight-through mufflers.* These mufflers permit exhaust gases to pass through a single tube. The tube has perforations that tend to break up pressure pulsations, but they are not as quiet as the reverse-flow type.

Though mufflers still reduce noise as the exhaust gases pulse through them, there have been several important changes in recent years in their design. Most of these changes have been centered at reducing weight and emissions, improving fuel economy, and simplifying assembly. Many also affect the technician who repairs or replaces the system. These changes include:

- *New Materials.* More and more mufflers are being made of aluminized and stainless steel (Figure 9–22). Car designers, trying to make exhaust systems lighter, sometimes call for thinner metal for some components. Be sure the muffler is of high-quality steel.

 SHOP TALK ——————

Remember that stainless steel pipe tends to break rather than rust out, and it might take some tact to explain to the customers why a cracked pipe that does not look rusted is actually worn out. Also, if low-quality components are used where the car had stainless or aluminized steel, you are likely to end up with a premature rustout and an unhappy customer.

- *Double-Wall Design.* Retarded engine ignition timing that is used on many small cars

tends to make the exhaust pulses sharper, so many small cars now use a double-wall exhaust pipe to better contain the sound and reduce pipe ring.

- *Rear-Mounted Mufflers.* More and more often, the only space left under the car for the muffler is at the very rear. This means that the muffler runs cooler than before and is much more easily damaged by condensation in the exhaust system. This moisture, combined with nitrogen and sulfur oxides in the exhaust gas, forms acids that rot the muffler from the inside out. Many more mufflers are being produced with drain holes drilled in one of the heads.

 SHOP TALK ——————

Many customers do not always like the drain holes because they allow a small amount of gas and noise to escape, but they are one of the cheapest and most effective ways to fight corrosion. Do not plug the drain holes with a sheet metal screw.

- *Back Pressure.* Even a well-designed muffler will produce some back pressure in the system. Back pressure reduces an engine's volumetric efficiency, or ability to "breathe." Excessive back pressure caused by defects in a muffler or other exhaust system part can

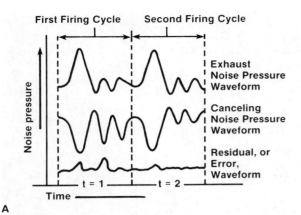

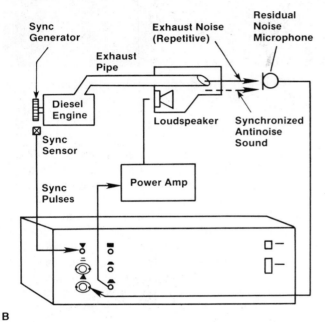

FIGURE 9-23 (A) An electronic muffler might seem far-fetched, but the production of a mirror image, out-of-phase waveform can cancel out sound. (B) Schematic shows how sensors, microphones, and speakers are teamed up to alternate noise.

slow or stop the engine. However, a small amount of back pressure can be used intentionally to allow a slower passage of exhaust gases through the catalytic converter. This slower passage results in more complete conversion to less harmful gases.

- *Electronic Mufflers.* Basically, sensors and microphones pick up the pattern of the pressure waves an engine emits from its exhaust pipe (Figure 9-23A). This data is analyzed by a computer, which instantly produces a mirror-image pattern of pulses and sends it to

speakers mounted near the exhaust outlet, creating contrawaves that cancel out the noise (Figure 9-23B). While electronic mufflers are not produced in models at present, several manufacturers suggest that they be used by the mid-1990s.

RESONATOR

On some vehicles, there is an additional muffler, known as a *resonator* or silencer, to further reduce the sound level of the exhaust. This is located toward the back end of the system and generally looks like a smaller, rounder version of a muffler. The resonator is constructed like a straight-through muffler and is connected to the muffler by an intermediate pipe.

Because of vehicle manufacturers' concern with weight and cost, resonators have been eliminated on most cars except for the larger models.

TAIL PIPE

The tail pipe is the end of the pipeline carrying exhaust fumes to the atmosphere beyond the back end of the car. On cars equipped with a resonator, the resonator can be located in the middle of the tail pipe. In fact, the resonator on some cars is an integral part of the tail pipe, forming a one-piece unit. If one or the other is damaged, the entire unit must be replaced.

In most cases, the tail pipe opens at the rear of the vehicle below the rear bumper. In some cases, it opens at the side of the vehicle just ahead of or just behind the rear wheel. Decorative tail pipe exhaust extensions in either chrome or black are available (Figure 9-24). Most extensions are fastened to the tail pipe with double-lock screws. The downward exhaust spout deflects gases down and away from the vehicle.

 SHOP TALK _____

Do not use mufflers on the street without proper tail pipes.

CLAMPS, BRACKETS, AND HANGERS

Clamps, brackets, and hangers are used to properly join and support various exhaust system components as well as to help isolate noise by preventing its transfer through the frame or body to the passenger compartment. Clamps help to secure exhaust system parts to one another. The pipes are

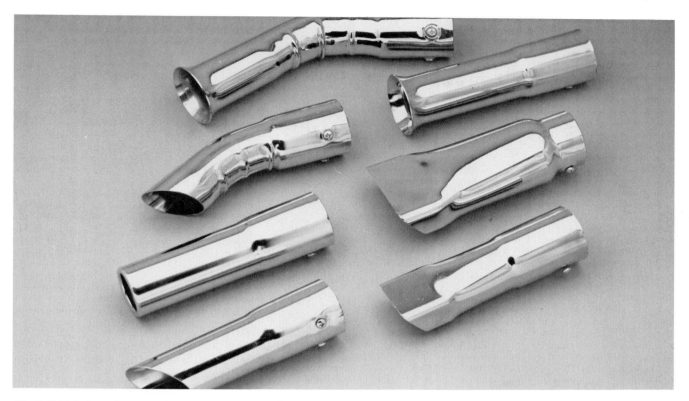

FIGURE 9-24 Decorative exhaust extensions

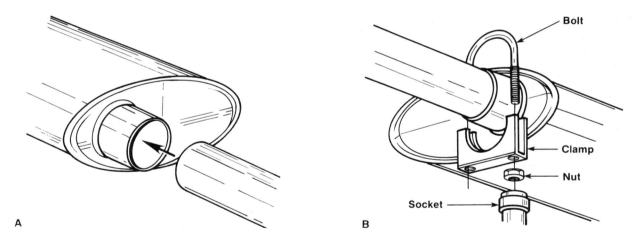

A

B

FIGURE 9-25 (A) One pipe slips inside the other. (B) Clamps secure exhaust system parts to each other.

formed in such a way that one slips inside the other (Figure 9-25). This design makes a close fit. A U-type clamp usually holds this connection tight. Another important job of clamps and brackets (Figure 9-26) is to hold pipes to the bottom of the vehicle.

New hanger arrangements have been necessary due to automobile design. For instance, transverse mounted engines rock back and forth on their mounts, instead of rocking from side to side. With-

out some sort of hinge in the exhaust system, this rocking motion would cause the tail pipe to move up and down—as much as a foot or two on some cars. There are several solutions: Honda uses flex pipe, VW relies on convoluted pipe, and GM has a slick spring-loaded spherical bushing (Figure 9-27).

Another type of pipe connection is a ball joint (Figure 9-28), a connection with a rounded, ball-shaped part that fits into a matching rounded, hol-

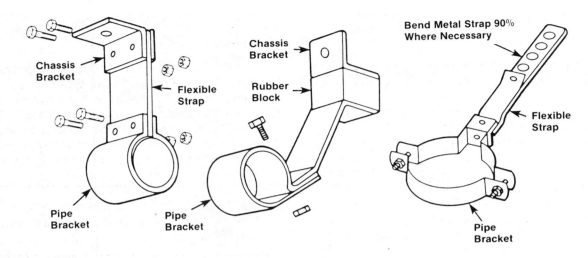

FIGURE 9-26 Clamp assemblies

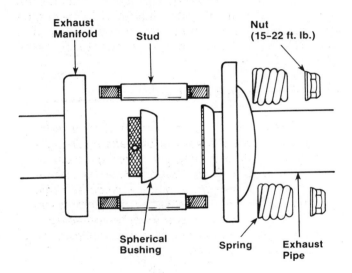

FIGURE 9-27 GM spherical bushing

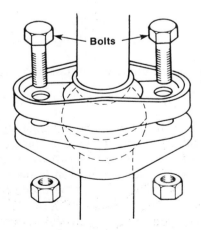

FIGURE 9-28 Ball joint exhaust pipe manifold connection

low socket. Ball joints are used where connections cannot be made in a straight line. Nuts and bolts are used to hold the ball joint together. Ball joints are used extensively in front-wheel drive vehicles to join the exhaust pipe to the manifold.

Various types of hangers are available. Some exhaust systems are supported by doughnut-shaped rubber rings between hooks on the exhaust component and on the frame or car body (Figure 9-29). Others are supported at the exhaust pipe and tail pipe connections by a combination of metal and reinforced fabric hanger. Both the doughnuts and the reinforced fabric allow the exhaust system to vibrate without breakage that could be caused by a hard physical connection to the vehicle's frame.

Exhaust system pipes can be welded together. Welding, of course, is a process in which heat is used to melt metal and allow the molten portions to flow together. When cooled, a welded joint is as strong as, or sometimes stronger than, a single piece of metal. The welding torch can also be used to cut exhaust system pipes.

WARNING: Do not attempt to use welding equipment without proper training and safety equipment. The light caused by welding can cause permanent eye damage or even blindness. Further, the heat from welding can ignite fuel vapors or damage parts.

Some OE is a fully welded exhaust system. By welding instead of clamping, carmakers save the weight of overlapping joints as well as that of clamps. It does not sound like much, but it adds up.

Figure 9-30 shows a fully welded system that is stainless steel from top to bottom.

EXHAUST SYSTEM SERVICE

Exhaust system components are subject to both physical and chemical damage. Any physical damage to an exhaust system part that causes a partially restricted or blocked exhaust system usually results in loss of power or backfire up through the carburetor. In addition to improper engine operation, a blocked or restricted exhaust system causes increased vehicle noise and air pollution. Leaks in the exhaust system caused by either physical or chemical (rust) damage could result in illness, asphyxiation, or even death. Remember that vehicle exhaust fumes can be *very* dangerous to your health.

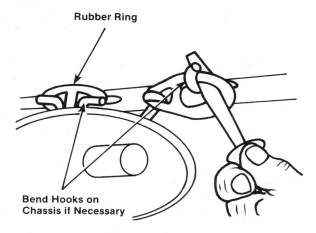

FIGURE 9-29 Exhaust pipe hanger

EXHAUST SYSTEM INSPECTION

Most parts of the exhaust system, especially the exhaust pipe, muffler, and tail pipe, are subject to rust and cracks. But these parts are prone to other malfunctions besides corrosion and heat. Clamps break or work loose, permitting parts to separate. They also fail because of impact with ruts and potholes in the road and rocks thrown up by the wheels.

WARNING: On all inspection and repair work, be sure to wear safety glasses or equivalent eye protection.

Before making a visual inspection, listen closely for the hissing or rumbling sound that indicates the beginning of exhaust system failure. With the engine idling, slowly move along the entire system and listen for leaks. It is generally quite easy to locate the source of the leak.

 SHOP TALK

Some experienced exhaust system technicians locate the source of leaks by using a length of rubber tubing and holding it close to the car to magnify any sound. Move the rubber tubing near and around the suspected leak. When found, the sound will be loud and clearly defined. In some cases, simply replacing a missing bolt or nut or tightening a loose clamp can stop the leakage.

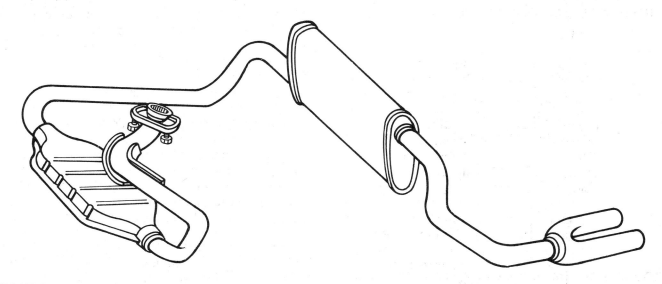

FIGURE 9-30 Fully-welded system made entirely of stainless steel

FIGURE 9-31 Checking a muffler with a hammer

FIGURE 9-32 Checking pipes for kinks and dents

WARNING: When using either leak detection method, be very careful. Remember that the exhaust system gets very hot. Do not get your face too close when listening for leaks.

The vehicle should be safely raised and supported (see Chapter 2). After the exhaust system has cooled thoroughly, the system can be inspected as follows:

 1. With a flashlight or trouble light, check for the following:
   - Holes and road damage
   - Discoloration and rust
   - Carbon smudges
   - Bulging muffler seams
   - Interfering rattle points
   - Torn or broken hangers and clamps

2. Sound out the system by gently tapping the pipes and muffler with a hammer or mallet (Figure 9-31). A good part will have a solid metallic sound. A weak or worn out part will have a dull sound. Listen for falling rust particles on the inside of the muffler. Mufflers usually corrode from the inside out, so it might be worn out on the inside but cannot be seen from the outside. Remember that some rust spots might be only surface rust. Suspected deep rust should be jabbed with an old screwdriver. If the screwdriver penetrates the rusty spot, the part is defective and will fail soon.

3. Grab the tail pipe (when it is cool) and try to move it up and down and from side to side. There should be only slight movement in any direction. If the system feels wobbly or loose, check the clamps and hangers that fasten the exhaust to the vehicle.

4. Check all pipes for kinks and dents that might restrict the flow of exhaust gases (Figure 9-32). A bent, kinked, or dented pipe should be replaced immediately because this restriction causes excessive back pressure and poor engine performance.

5. Take a close look at each connection. Check the exhaust manifold to exhaust pipe connection. If any white powdery deposits are found, the bolts could be loose or the gasket might be leaking. Check for loose connections at the muffler by pushing up on the muffler slightly.

6. If a visual inspection does not identify a partially restricted or blocked exhaust system, perform the following test:
   - Attach a vacuum gauge to the intake manifold.
   - Connect the tachometer.
   - Start the engine and observe the vacuum gauge. It should indicate a vacuum of 16 to 21 inches of mercury.
   - Increase the engine speed to 2000 rpm and observe the vacuum gauge. Vacuum will decrease when the speed is increased rapidly, but it should stabilize at 16 to 21 inches of mercury and remain constant. If the vacuum remains below 16 inches of mercury, the exhaust system is restricted or blocked.

7. Catalytic converters can overheat. This can be seen in a bluish or brownish discoloration of the outer stainless steel shell. In

addition, undercoating or paint above and near the converter might appear blistered or burned. Details on inspecting catalytic converters can be found in Chapter 11.

A troubleshooting chart for exhaust systems is given in Table 9-1. A diagnosis of catalytic converter and emission control device problems is given in Chapter 11.

Before replacing any exhaust system components, the following steps *must* be taken to prepare the job:

 1. Make sure the system has cooled before working on it. Working on a cooled engine avoids burned arms. It is a good idea to wear long-sleeved shirts for protection.
2. If replacing the exhaust pipe, remove the battery ground to avoid short-circuiting the electrical system.
3. Depending on the vehicle's make and model, removing the rear wheel might make the job easier. Consult the vehicle's service manual for any special precautions.

TABLE 9-1: EXHAUST SYSTEM PROBLEM DIAGNOSIS

Condition	Possible Cause	Remedy
Heat control valve noisy	Loose, weak, or broken antirattle spring or shaft at heat control valve	Replace antirattle spring or heat control valve.
	Thermostat broken	Replace thermostat.
Excessive exhaust system noise	System components striking body or chassis	Reposition components.
	Broken or loose clamps or brackets	Repair or replace.
	Leaks at manifold or pipe connections	Torque clamps or leaking connections to specifications.
	Burned-out or blown-out muffler	Replace muffler.
	Burned-out or rusted-out pipe	Replace pipe.
	Exhaust pipe leaking at manifold flange	Torque nuts to specifications.
	Exhaust manifold cracked or broken	Replace manifold.
	Leak between manifold and cylinder head	Torque studs to specifications or replace gaskets.
	Restriction in muffler or tail pipe	Remove restriction if possible or replace as necessary.
Leaking exhaust	Leaks at pipe joints	Reseal joints with exhaust system sealer. Tighten clamp bolts securely. If leaks persist, replace pipes.
	Rusted out pipes	Replace.
	Damaged or improperly placed gasket at exhaust pipe/exhaust manifold joint	Replace.
	Rusted out muffler	Replace.
Restricted exhaust	Kinked or plugged pipe	Repair or replace pipe.
Engine hard to warm up or will not go to normal idle	Heat control valve frozen in OPEN position	Free up manifold heat control valve using a suitable solvent or penetrating oil.
	Blocked crossover passage in intake manifold	Remove restriction or replace intake manifold.

4. Before attempting to remove the old parts, soak all connections and threads on all the nuts, bolts, clamps, hangers, brackets, and connections with a good quality penetrating oil.
5. Check the old system's routing for critical clearance points so that the new system or component is properly installed. Inspect the clearance at the following:
   - Steering assembly and stabilizer bars
   - Shocks and springs
   - Transmission shift and clutch linkage
   - Frame and crossmembers
   - Brake lines
   - Gas tank and fuel lines

REPLACING EXHAUST SYSTEM COMPONENTS

When replacing exhaust system components, it is important that original equipment parts (or their equivalent) are used to ensure proper alignment with other parts in the system and provide acceptable exhaust noise levels. When replacing only one component in an exhaust system, it is not always necessary to take off the parts behind it.

Exhaust system component replacement might require the use of special tools (Figure 9–33) and/or welding equipment.

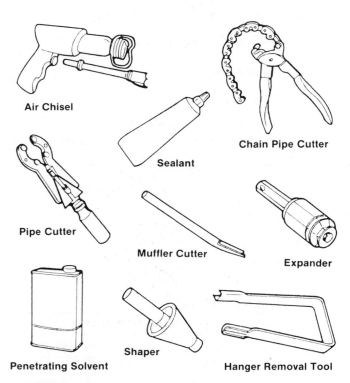

Air Chisel

Sealant

Chain Pipe Cutter

Pipe Cutter

Muffler Cutter

Expander

Penetrating Solvent

Shaper

Hanger Removal Tool

FIGURE 9–33 Exhaust system tools

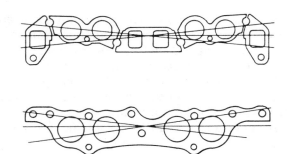

FIGURE 9–34 Exhaust manifold surfaces are checked for warpage.

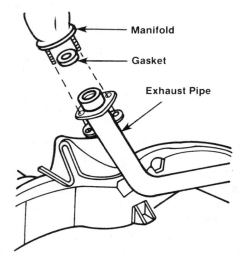

Manifold

Gasket

Exhaust Pipe

FIGURE 9–35 Leaking gaskets and seals are found most often between the manifold and cylinder head.

Exhaust Manifold and Exhaust Pipe Servicing

As mentioned, the manifold itself rarely causes any problems. On occasion, an exhaust manifold will warp because of excess heat. A straightedge and feeler gauge can be used to check the machined surface of the manifold across the area indicated in Figure 9–34.

Another problem—also the result of high temperatures generated by the engine—is a cracked manifold. This usually occurs after the car passes through a large puddle and cold water splashes on the manifold's hot surface. If the manifold is warped beyond manufacturer's specifications or is cracked, it must be replaced. Also, check the exhaust pipe for collapse. If there is damage, repair it. These repairs should be done as directed in the vehicle's service manual.

Replacing Leaking Gaskets and Seals. The most likely spots for leaking gaskets and seals are between the exhaust manifold and the cylinder head and between the exhaust pipe and the exhaust manifold (Figure 9–35).

When installing exhaust gaskets, carefully follow the chemical recommendations on the gasket package label and instruction forms. Read through all installation steps before beginning. Take note of any of the original equipment manufacturer's recommendations in service manuals that could affect engine sealing. This is especially important when working with aluminum components. Manifolds warp more easily if an attempt is made to remove them while still hot. Remember heat expands metal, making assembly bolts more difficult to remove and easier to break.

Follow the torque sequence in reverse to loosen each bolt. Then repeat the process again to remove the bolts. This minimizes the chance of components warping.

Any debris left on the sealing surfaces increases the chance of leaks. A good gasket remover will quickly soften the old gasket debris and dry adhesive for quick removal. Carefully remove the softened pieces with a scraper and a wire brush. Be sure to use a nonmetallic scraper when attempting to remove gasket material from aluminum surfaces.

Inspect the manifold for irregularities that might cause leaks, such as gouges, scratches, or cracks. Check for surface flatness with a straightedge and feeler gauge, and machine only enough to make the surface flat. Replace any parts that are cracked or badly warped. This will insure proper sealing of the manifold.

Due to high heat conditions, it is important to retap and redie all threaded bolt holes, studs, and mounting bolts. This procedure insures tight, balanced clamping forces on the gasket. Lubricate the thread with a good high-temperature antiseize lubricant. This will give a good thread lubrication for the proper clamping force. It will also keep the threads from seizing under the high temperatures. Use a small amount of contact adhesive to hold the gasket in place. Align the gasket properly before the adhesive dries. Allow the adhesive to dry completely before proceeding with manifold installation.

Install the bolts finger tight. Tighten the bolts in three steps—one-half, three-quarters, and full torque—following the torque tables in the OEM installation manual or gasket manufacturers' instructions. Torquing is usually begun in the center of the manifold, working outward in an X pattern.

To replace a damaged exhaust pipe, disconnect it at the exhaust manifold and catalytic converter. Support the converter to keep it from falling. Carefully remove the oxygen sensor if there is one. Re-

move any hangers or clamps holding the exhaust pipe to the frame. Unbolt the flange holding the exhaust pipe to the exhaust manifold. When removing the exhaust pipe, check to see if there is a gasket. If so, discard it and replace it with a new one. Once the joint has been taken apart, the gasket loses its effectiveness. Pull the front exhaust pipe loose and remove it.

Although most exhaust systems use a slip joint and clamps to fasten the pipe to the muffler, a few use a welded connection. If the vehicle's system is welded, cut the pipe at the joint with a hacksaw or pipe cutter (Figure 9–36). The new pipe need not be welded to the muffler; an adapter, available with the pipe, can be used instead. But when measuring the length for the new pipe, allow at least 2 inches for the adapter to enter the muffler. If the pipe has to be expanded, it could be done as shown in Figure 9–37.

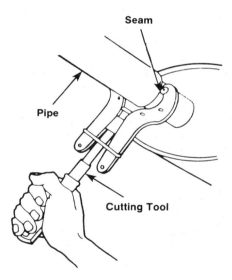

FIGURE 9–36 Cutting an exhaust pipe

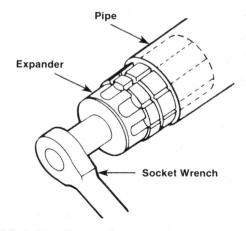

FIGURE 9–37 Expanding an exhaust pipe

The old exhaust pipe might be rusted into the converter opening. Attempt to collapse the old pipe by using a cold chisel or slitting tool and a hammer (Figure 9–38). While freeing the pipe, try not to damage the muffler inlet. It must be perfectly round to accept the new pipe.

Slide the new pipe into the muffler (some lubricant might be helpful), and attach the front end to the manifold. The pipe must fit at least 1-1/2 inches into the converter and muffler. A new gasket must be used at the manifold, as previously described. Before tightening the connectors, check the system for alignment. When it is properly aligned, tighten the clamps.

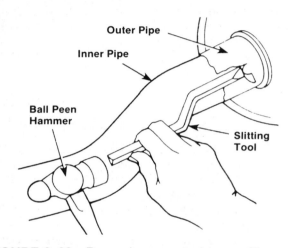

FIGURE 9–38   Removing a rusted-on muffler

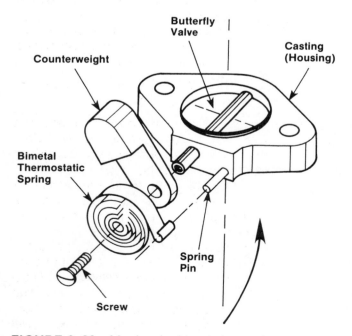

FIGURE 9–39   Mechanical heat riser valve

When removing the exhaust pipe, be careful with the emission controls. If the vehicle is equipped with a mechanical heat riser valve, it should be sprayed with graphite oil. Also be sure that the counterweight is in balance (Figure 9–39). To do this, use pliers to move the counterweight and valve shaft slowly and carefully. Graphite oil is applied liberally during this process until the shaft moves freely and easily.

 SHOP TALK _____

Graphite oil is a thin lubricant containing graphite particles. After the engine starts, the oil will burn off, but the graphite, an excellent dry lubricant, will remain.

Muffler Removal and Installation. When replacing a typical muffler, proceed as follows:

 1. Loosen the clamps and hangers from the muffler (Figure 9–40).
2. Pull the assembly back and let it hang. Tap the old muffler free (Figure 9–41) to remove it from the system.
3. Before installing the new muffler, coat the inside of both the inlet and outlet with muffler sealant. Also, coat the outside of the pipe coming into the muffler and the tail pipe. The muffler sealant forms a permanent seal that prevents sound and gas leakage.
4. Make sure the muffler outlet arrow (if there is one) is pointed in the right direction (Figure 9–42).

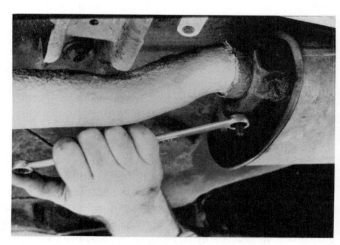

FIGURE 9–40   Removing the old muffler

FIGURE 9-41 Tapping the old muffler free from the exhaust pipe

FIGURE 9-43 Installing the rear muffler hanger

FIGURE 9-42 Installing a new muffler

5. Connect the muffler inlet to the exhaust or intermediate pipe first. The pipe should go 1-1/2 to 2 inches into the muffler inlet.

6. Rotate the muffler to its proper alignment. Tighten the front clamp snugly. Install the rear muffler hanger and place the muffler outlet into the hanger to support it (Figure 9-43). Insert the tail pipe into the muffler and clamp snugly.

7. Snug tighten the other clamps, hangers, and brackets on the remaining parts that have been loosened. The assembly is now loosely installed.

8. Check that all connections are inserted fully and that the components are properly aligned.

9. Make sure all exhaust components have adequate clearance. Heat expansion will cause the system to grow as much as 1 inch. Also, engine torque might decrease clearance along the left side of the system when the engine is under load.

10. If the system is properly installed, tighten all connections securely.

11. With the transmission in NEUTRAL, start the engine.

12. Accelerate and decelerate. Listen for noise or rattles.

13. If the shop has adequate ventilation, stuff a shop rag into the tail pipe and visually check for leaks. Remove the shop rag after the inspection.

14. After the exhaust system is thoroughly warmed up, retighten all connections and recheck critical clearances.

When replacing major sections of the entire exhaust system, proceed as follows:

1. The exhaust system is primarily secured at the manifold. So, dismantle the old system from the rear forward. Be careful not to damage any parts to be reused.

2. Raise the body of the vehicle to gain clearance at the rear axle. Safety stands can be used to support components during removal and installation.

3. When removing the exhaust pipe at the manifold, use caution not to break the manifold studs, if applicable. Use plenty of nut-loosening lubrication and, if needed, nut cutters to avoid damaging the studs. Also, do not drop the heat riser valve—this will damage it.

SHOP TALK _____

An easy way to break off rusted nuts is to tighten them instead of loosening them. Sometimes a badly rusted clamp or hanger strap will snap off with ease. Sometimes the old exhaust system will not drop free of the body because a large part is in the way, such as the rear end or the transmission support. Use a large cold chisel, pipe cutter, hacksaw, muffler cutter, or chain cutter to cut the old system at convenient points to make the exhaust assembly smaller.

CAUTION: Be sure to wear work gloves to prevent cutting your hands on rusted metal parts.

---

4. Make sure that any components that are to be reused are in good condition. Remove any burrs, corrosion, and scale at connections. Remove clamp grooves and restore roundness, as needed. Clean all gaskets and ball and socket surfaces to ensure a gas tight seal.
5. Install the new system from front to rear.
6. Check that the heat riser valve and its anti-rattle spring are working properly. A malfunctioning heat riser must cause noise. If in doubt, replace the heat riser.
7. Change the manifold studs, if applicable.
8. Install a new, clean gasket. Slipping a rubber band over the manifold studs to hold the gasket is a good way to free the hands to connect the exhaust pipe. (The rubber band will disintegrate.)
9. Connect the exhaust pipe to the manifold. Use new brass nuts to connect it to the studs. Or, use new bolts. Tighten these connections snugly (Figure 9-44).
10. Install the remaining components. Because correct alignment of the entire system is very important, snug tighten connections so that the system will be loosely installed (Figure 9-45). Use muffler sealant to make sure all connections are sealed properly.

To complete the job follow steps 9 to 15 for the installation of the muffler.

When the components are reassembled, there will be a number of common joints, such as the one between the tail pipe and the muffler. Before clamping these parts together, coat them with exhaust system joint sealer. The sealer, rather than the clamp, prevents leaks. Tighten all nuts or bolts to the specifications given by the carmaker.

Go over the entire system to be sure each hanger and clamp bolt or nut is securely fastened. The exhaust system must be aligned so there is no stress on parts and no part can bang against any area of the car.

CAUTION: Be sure no exhaust part comes into direct contact with any section of the body, fuel lines, fuel tank, or brake lines.

TURBOCHARGERS

As mentioned earlier in this chapter, on some vehicles, the turbocharger can be considered part of the exhaust system since it is driven by the high velocity of the engine exhaust gases. These exhaust gases cause the turbocharger shaft assembly to ro-

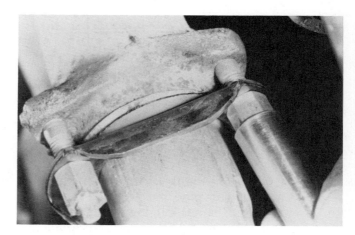

FIGURE 9-44 Fastening the manifold studs

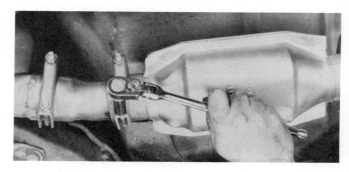

FIGURE 9-45 Fastening together the remaining exhaust system parts

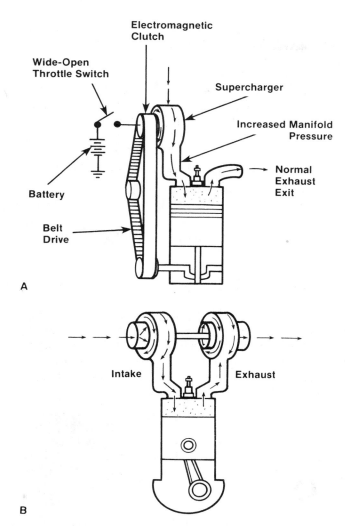

FIGURE 9–46 Two devices for increasing an engine's ability to take in air: (A) belt-driven supercharger and (B) exhaust-driven turbocharger.

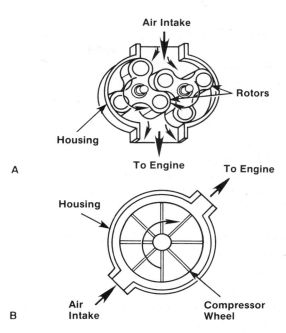

FIGURE 9–47 Two types of superchargers: (A) positive displacement and (B) centrifugal

tate at speeds of 1,000 to 30,000 revolutions per minute.

By definition, a turbocharger is a type of supercharging device that utilizes an exhaust-driven turbine to generate intake air pressure. To comprehend what this means, think back to the fundamentals of engine operation given in Chapter 1. Remember that the power generated by the internal combustion engine is directly related to the compression force exerted on the air/fuel charge. In other words, the greater the compression (within reason), the greater the output of the engine.

Two approaches can be used to increase engine compression. One is to modify the internal configuration of the engine to increase the basic compression ratio. This has been accomplished in many ways including by the use of such things as domed

or high top pistons, altered crankshaft strokes, or changes in the shape and structure of the combustion chamber design.

Another, less expensive way to increase mixture compression (and engine power) without physically changing the shape of the combustion chamber is to simply increase the quantity of the intake charge. By precompressing the intake mixture before it enters the cylinder, more air and fuel molecules can be packed into the combustion chamber. Keep in mind that any time the amount of the air/fuel mix that enters the cylinder (within reason) is increased, there is a substantial increase in power.

The process of artificially increasing the amount of airflow into the engine is known as *supercharging*. The ways to accomplish this are by a mechanical or belt-driven supercharger or an exhaust-driven turbocharger (Figure 9–46).

A supercharger, or blower, as it is commonly called, is an air or air/fuel mixture pressurizing pump that is mechanically driven from the engine crankshaft through belts, gears, or chains. There are two primary categories of superchargers: positive displacement and centrifugal. In automotive applications, the positive displacement type is the more commonly used design (Figure 9–47A). With a positive displacement pump, the amount of charge per pump revolution is essentially the same for each revolution of the engine regardless of speed. Conversely, the output of a centrifugal pump (Figure 9–47B) is speed related. Because of the inherent

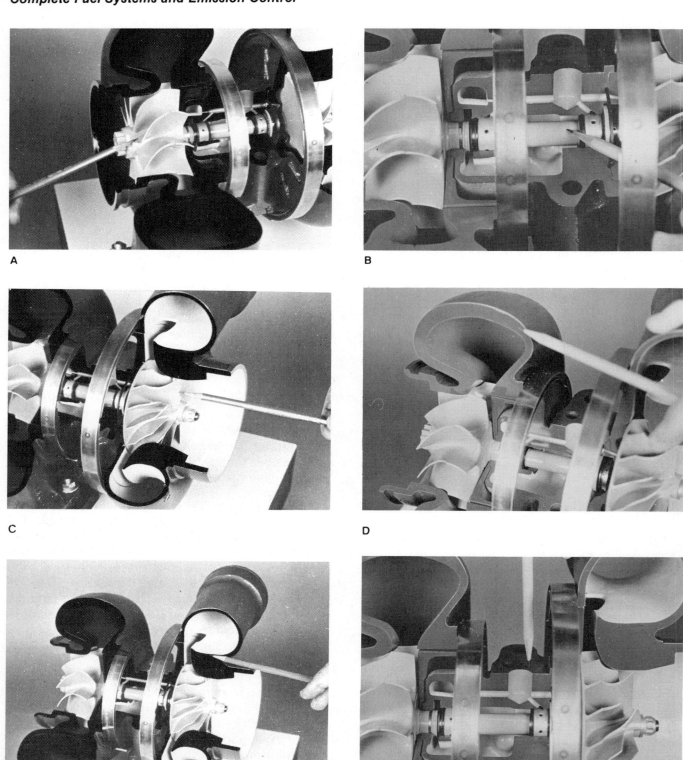

FIGURE 9-48 Components of a turbocharger: (A) turbine or hot wheel; (B) shaft; (C) compressor or cold wheel; (D) turbine housing; (E) compressor housing; (F) center housing; (G) turbine seal assembly; (H) compressor seal assembly

G

H

FIGURE 9-48   continued

design constraints and the basic inefficiencies involved in using the engine to drive the supercharging unit (operating the blower can cost up to 8 to 10 percent of the engine's power output), mechanically driven superchargers are rarely used on production passenger car applications. They are being experimented with by some auto manufacturers and show promise for the future.

A turbocharger, on the other hand, is a type of air-compressing device that is similar in design and appearance to the centrifugal supercharger pump. However, there is one very obvious difference: The turbocharger does not require a mechanical connection between the engine and the pressurizing pump to compress the intake gases. Instead, it relies on the rapid expansion of hot exhaust gases exiting the cylinder to spin turbine blades (hence the name turbocharger). Because exhaust gas is a waste product, the energy developed by the turbine is said to be free since it theoretically does not rob the engine of any of the power it helps to produce.

TURBOCHARGER OPERATION

A typical turbocharger, usually called a *turbo,* consists of the following components (Figure 9-48):

- Turbine or hot wheel
- Shaft
- Compressor or cold wheel
- Turbine housing
- Compressor housing
- Center housing (This component contains the bearings, the turbine seal assembly, and the compressor seal assembly.)

The turbocharger is located to one side of the engine, usually close to the exhaust manifold (Figure 9-49). An exhaust pipe runs between the engine exhaust manifold and the turbine housing to carry the exhaust flow to the turbine wheel. Another pipe connects the compressor housing intake to an injector throttle body or a carburetor (Figure 9-50), and another conducts compressed air to the intake manifold.

Inside the turbocharger, an exhaust-driven turbine wheel (hot wheel) is attached via a shaft to an intake compressor wheel (cold wheel). Each wheel is encased in its own spiral-shaped housing that serves to control and direct the flow of exhaust and intake gases. The shaft that joins the two wheels rides on bearings (generally the free-floating type), which are part of a bearing lubrication and rotational housing cartridge.

FIGURE 9-49   The turbocharger is usually located close to the exhaust manifold

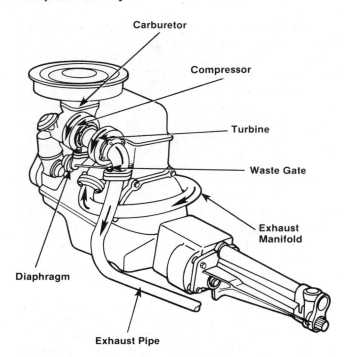

FIGURE 9-50 Turbocharger installation

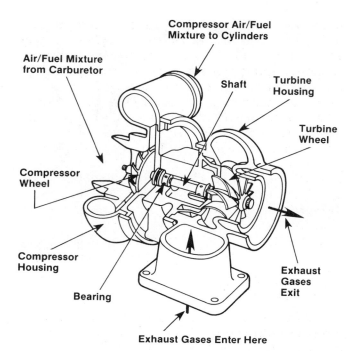

FIGURE 9-51 Cutaway of a turbocharger shows the turbine wheel, the compressor wheel, and their connecting shaft.

The air compressing process starts when exhaust gas enters the turbine housing (Figure 9-51). When the engine's speed and load are high enough (generally above 2000 rpm), the force of the exhaust flow is directed through a nozzle against the side of the turbine wheel. As the hot gases hit the turbine wheel causing it to spin, the specially curved turbine fins direct the air toward the center of the housing where it exits. This action creates a flow called a *vortex,* which can be likened to the action of water going down a drain. Once the turbine starts to spin, the compressor wheel (shaped like a turbine wheel in reverse) also starts to spin, causing air to be drawn into the center where it is caught by the whirling blades of the compressor and thrown outward by centrifugal force. From there the air exits under pressure through the remainder of the induction system on its way to the cylinder.

Under normal atmospheric conditions, air is drawn into the engine at a maximum of 14.7 psi at sea level. A turbocharged engine, however, is capable of pressurizing the intake charge above and beyond nature's limits. *Turbo boost* is the term used to describe the positive pressure increase created by a turbocharger. For example, 10 psi of boost means the air is being fed into the engine at 24.7 psi (14.7 psi atmospheric plus 10 pounds of boost).

Even though most turbocharged engines are fortified to withstand the added pressures generated by the turbo, there is a limit to the beating they can take. So to help minimize the risk of internal engine

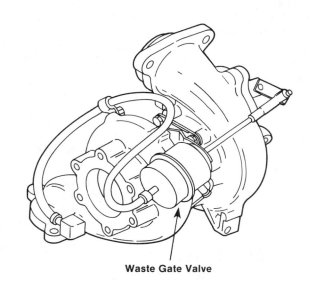

FIGURE 9-52 Typical waste gate assembly

damage resulting from too much boost, all turbochargers use some means of boost control to limit their output—the most effective and widely used method is the waste gate. The waste gate manages turbo output by controlling the amount of exhaust gas that is allowed to enter the turbine housing. A typical waste gate assembly consists of a poppet valve (located on the exhaust side), control diaphragm, and signal hose (Figure 9-52). The control diaphragm and poppet valve (waste gate) are physically connected by a mechanical link. During normal vehicle operation, the poppet valve is held closed by calibrated spring pressure acting on the control diaphragm. The other side of the diaphragm is exposed to intake manifold pressure via the signal hose. When boost pressure starts to exceed a preset level, the spring's opposing pressure is overcome, causing the diaphragm to move. This movement in turn forces the poppet valve open and allows some of the exhaust gases (depending on the amount of valve opening) to bypass the turbine housing. As long as the valve is open, the turbocharger's speed cannot increase, keeping boost pressure constant.

Other types of control devices that are commonly used with a turbocharged engine include:

- Intercooling devices
- Spark retard systems

Intercooling, or aftercooling as some experts argue, is the process of removing heat from the air or air/fuel charge after it has been compressed. There are two reasons intercoolers have gained considerable attention from car makers in recent years. First, intercooling is an effective means of controlling the charge temperature and thus detonation (in some cases, the final air charge temperature can be reduced by as much as 200 degrees Fahrenheit). Second, the cooler, denser air that results increases the amount of air and fuel that can be packed in per charge.

Some pressure will be lost as the compressed air is cooled, but the charge will continue to increase in density. This increased density allows a greater mass of air to flow through the engine at a given intake pressure. With a higher density charge (more air molecules), a proportionately greater amount of fuel can be burned, producing an overall increase in power. As an added benefit, the lowered charge temperature will significantly reduce the heat rejection requirements of the engine (cooling system load) and, as mentioned, help prevent destructive detonation. The intercooler itself is actually nothing more than a heat exchanger that uses air or liquid as a cooling medium (Figure 9-53). On automotive applications, the air-to-air type is by far the most common design because it is easier to locate and requires less plumbing than a typical air-to-liquid style.

Retarding spark timing is another often used method of controlling detonation on turbocharged engines. Unfortunately, any time the ignition is permanently retarded to prevent detonation, power is lost, fuel economy suffers, and the engine tends to run hotter. Because of these tradeoffs, most systems use knock-sensing devices to retard timing only when detonation is detected.

The design capacity of the turbocharger is dependent on engine size and type, which is determined by the vehicle manufacturer. Control devices limit the degree of turbocharging to prevent detonation and engine damage. In fact, some turbocharg-

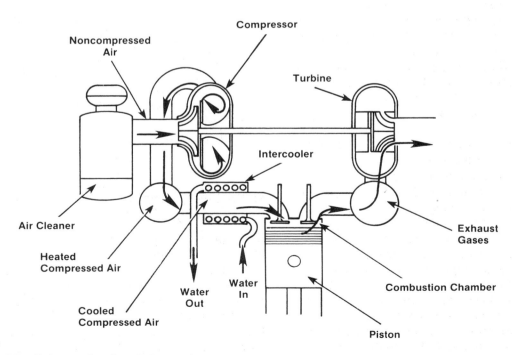

FIGURE 9-53 Schematic of an intercooler

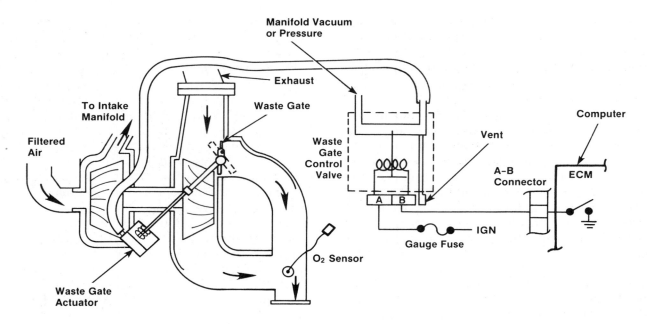

FIGURE 9-54 Computer-controlled turbocharging system

ing systems on computer-controlled vehicles use an electronic control unit to operate the waste gate control valve through sensor signals (Figure 9-54). The computer output is conducted (sends current) to a vacuum control solenoid. The vacuum solenoid controls the engine vacuum sent to the waste gate diaphragm. Several sensors, including the anti-knock sensor (which detects preignition and knock), speed sensor, and oxygen sensor, provide data to the computer. Information on the operation of these control devices can be found in Chapter 12.

TURBOCHARGER INSPECTION

The purpose of system inspection is to identify the reason for failure so repair can be made before installing a new unit. Common symptoms that might indicate possible turbocharger trouble are:

1. Engine lacks power
2. Black smoke
3. Blue smoke
4. Excessive engine oil consumption
5. Noisy operation of the turbocharger

But before making any inspection of the turbocharger, remember that there is a very good chance that the problem lies outside the turbocharger. Here is what to look for:

- *Intake System.* Any restriction or leak in the system limits the volume of air the turbo can

deliver. Check for air filter blockage or dirt buildup, then check all intake plumbing from air filter to turbo, including the clamps. Next, check the plumbing from turbo to intake manifold and be sure the manifold is properly torqued to the cylinder head.
- *Exhaust System.* Again, leaks and restrictions can have a major impact. Be sure the exhaust manifolds are properly torqued to the heads and that the turbo is properly torqued to the manifold. A leaking or improperly installed turbo/manifold gasket could be the culprit. Finally, check the entire exhaust system downstream from the turbo, including muffler and catalytic converter, for restrictions. Exhaust blockage causes sluggish and reduced engine performance.
- *Carburetion/Fuel Injection.* All the air a turbo can deliver will not help if it is not mixed with the right amount of fuel. Check the function of carburetion, fuel injectors, and fuel pump. Make certain that all fuel lines and distribution blocks are unrestricted and that the entire fuel system is free of air-entry leaks (Figure 9-55).
- *Waste Gate.* This device, also called a *turbo actuator,* must operate properly to prevent loss of performance/economy and possible engine damage. And remember, any attempt to increase boost pressure by modifying or removing the waste gate can easily allow an engine and turbocharger to rev to destruction.

To inspect a turbocharger, start the engine and listen to the sound the turbo system makes. As a technician becomes more familiar with this characteristic sound, he or she will be able to identify an air leak between the compressor outlet and engine or an exhaust leak between engine and turbo by a higher pitched sound. If the turbo sound cycles or changes in intensity, the likely causes are a plugged air cleaner or loose material in the compressor inlet ducts or dirt buildup on the compressor wheel and housing.

After listening, check the air cleaner for a dirty element. If in doubt, measure for restrictions per engine manufacturer's shop manual. Next, with the engine stopped, remove the ducting from the air cleaner to turbo and look for dirt buildup or damage from foreign objects. Check for loose clamps on compressor outlet connections and check the engine intake system for loose bolts, leaking gaskets, and so forth. Then, disconnect the exhaust pipe and look for restrictions or loose material. Examine the engine exhaust system for cracks, loose nuts, or blown gaskets. Rotate the turbo shaft assembly. Does it rotate freely? Are there signs of rubbing or wheel impact damage? Axial shaft play is end-to-

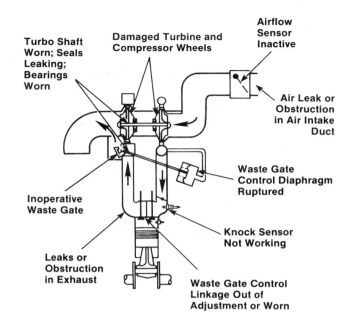

FIGURE 9-56   Any of these problems with a turbocharger system will affect the operation of a turbocharged engine.

end movement and radial shaft play is side-to-side movement. There is normally side-to-side play; however, if this play is sufficient to permit either of the wheels to touch the housing when the shaft is rotated by hand, then there is excessive wear. Any of the problems shown in Figure 9-56 could affect the operation of the turbo.

If all these checks fail to isolate the problem, it is time to inspect the turbocharger. Before taking a turbo apart, it is a good idea to mark the locations of all the housings and components that make up the whole turbo assembly (Figure 9-57). This will ensure that everything can be reassembled in its proper place.

Once the turbine housing (at both ends) is open to view, the technician is ready to make a diagnosis. Inspect the compressor and turbine wheels very closely (Figure 9-58). Typical damage and wear are shown in Figure 9-59. Look for bent, damaged, cracked, and/or excessively worn blades. Check for impact damage due to excessive shaft movement, contaminant injestion, or improper lubrication. Spin the turbine by hand, as mentioned earlier, paying specific attention to the amount of resistance that can be felt as the blades move (assuming they do move). The resistance should be even and not extreme, and there should be no binding or sticking. In short, there should be no indication of rubbing or contact at all.

Bearing clearance is critical and should be checked according to the manufacturer's specified

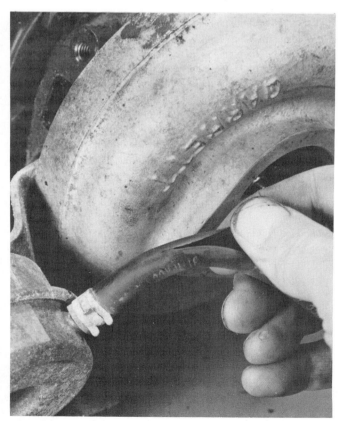

FIGURE 9-55   Checking the turbo for leaks

FIGURE 9-57 Before taking a turbo apart, mark the major housing and component relationships.

FIGURE 9-58 Look for bent, cracked, and/or excessively worn blades.

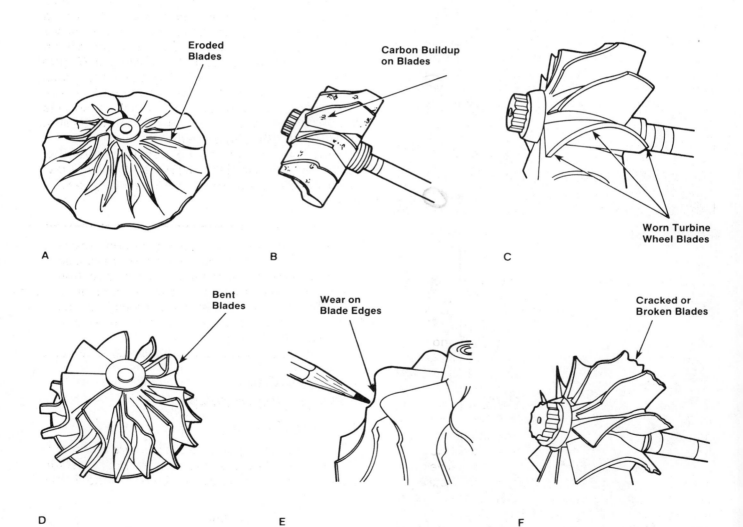

Eroded Blades

Carbon Buildup on Blades

Worn Turbine Wheel Blades

A

B

C

Bent Blades

Wear on Blade Edges

Cracked or Broken Blades

D

E

F

FIGURE 9-59 Typical turbine and compressor wheel damage and wear: (A) Extreme erosion from sand striking the compressor wheel; (B) turbocharger seal leak allowed oil into the turbine where it burned; (C) wear caused by heavy blade contact after bearing failure; (D) damaged compressor wheel due to ingestion of soft material; (E) worn compressor wheel caused by dirt or other abrasive; (F) blades damaged by foreign objects.

FIGURE 9-60 Checking full floating shaft bearings

FIGURE 9-61 Checking shaft movement

procedures for both radial and axial clearance. Figure 9-60 shows the full floating shaft bearings being checked for total axial movement. As a quick test before hooking up the measuring gauge, grasp the center shaft at both ends (never rock it up and down from one end only) and try moving it up and down. Shaft movement should not exceed 0.003 to 0.006 inch unless otherwise stated (Figure 9-61).

Visually inspect all hoses, gaskets, and tubing for proper fit, damage, and wear. Do not underestimate the importance of a broken or deteriorated hose in a system that depends on vacuum and pressure for proper operation. Check the negative pressure, or air cleaner side, of the intake system for vacuum leaks. This is one area where propane works well. As described in Chapter 4, propane is good for checking potential leak areas. Infrared analyzers (see Chapter 11) are an excellent means of checking for this problem.

On the pressure side of the system you can check for leaks by using soapy water. After applying the soap mixture, look for bubbles to pinpoint the source of the leak.

Leakage in the exhaust system upstream from the turbine housing will also affect turbo operation. If exhaust gases are allowed to escape prior to entering the turbine housing, the reduced temperature and pressure will cause a proportionate reduction in boost and an accompanying loss of power. If the waste gate does not appear to be operating properly (for example, too much or too little boost), check to make sure the connecting linkage is operating smoothly and not binding. Also, check to make sure the pressure sensing hose is clear and properly connected.

Waste Gate Service

If the waste gate is not operating properly, boost pressure can be either too little or too much. Waste gate malfunction can usually be traced to carbon buildup, which keeps the unit from closing or causes it to bind. A defective diaphragm or leaking vacuum hose can result in an inoperative waste gate (Figure 9-62). But, before condemning the waste gate, check the ignition timing, the spark retard system, vacuum hoses, and (if used) the knock sensor, the oxygen sensor, and the computer itself to be sure each one is operating properly.

When testing or checking the waste gate and its control, always carefully follow the procedures given in the vehicle or turbo manufacturer's service manual.

CAUTION: When removing carbon deposits from turbine and waste gate parts, never use a hard metal tool or sandpaper. Remember that any gouges or scratches on these metal parts can cause severe vibration or damage to the turbocharger. To clean these parts, use a soft brush and a solvent.

COMMON TURBOCHARGER PROBLEMS

The turbocharger, with proper care and servicing, will provide years of reliable service. Most turbocharger failures are caused by one of the following reasons:

- Lack of lubricant
- Ingestion of foreign objects
- Contamination of lubricant

As can be noted proper lubrication—clean oil and a fresh oil fiter—is the most important factor

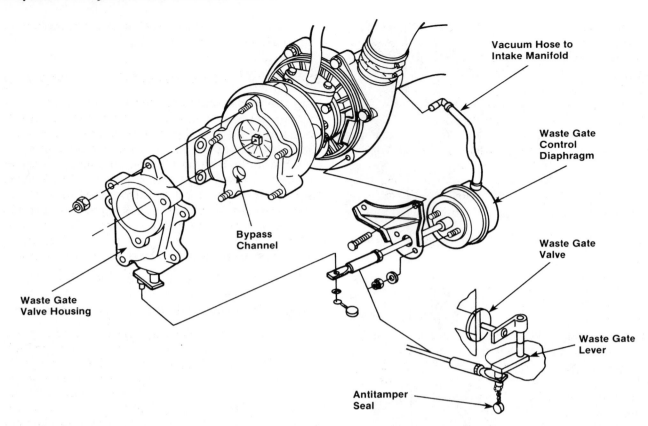

FIGURE 9-62 Diaphragm defects or vacuum hose leakage can affect waste gate operation.

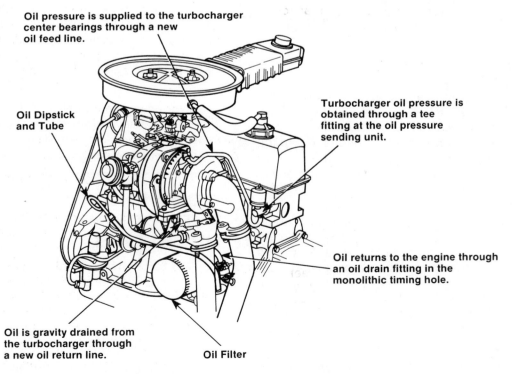

FIGURE 9-63 External components of a turbocharger lubrication system

determining the service life of a turbocharger. The main shaft of a turbo rotates within its twin bearings at up to sixty times the crankshaft speed under maximum load, greater than 2,000 revolutions per second. At those speeds, contaminated lubricant or insufficient lubrication can destroy a turbocharger in less time than it takes to describe it.

To ensure maximum life for a turbocharger the oil and oil filter must be changed at least as often as recommended by the vehicle or engine manufacturer (Figure 9-63); more often when the vehicle is operated off-road or in similar, dirty environments. And, as with a nonturbocharged engine, the oil filter should be changed at each oil change.

Proper oil level is equally important. Low level increases oil temperature and can lead to momentary interruption of oil flow to the turbocharger's bearings. At speeds up to 150,000 rpm, a moment is all it takes. The bearing surface area in any turbocharger is relatively small, especially compared to the large bearings on the slower-spinning crankshaft. The heat turbo bearings must dissipate is much greater. Oil lubricating the turbocharger is contaminated by high temperatures and must be changed more often than the oil in a conventional engine.

While preventive maintenance is not required for the turbocharger itself, oil lines leading to and from the turbo should be checked periodically for leaks or seepage and tightened as required.

Some common turbocharger problems are covered in the following section.

Black or Blue Smoke

If the engine is giving low power and emitting black smoke, check for the following:

- Dirty air cleaner
- Loose ducting connections
- Engine over-fueled

If the engine is emitting blue smoke and oil consumption is high, check the air cleaner for restrictions per the engine manufacturer's shop manual. Higher than normal air cleaner restriction can cause compressor oil seal leakage. With the engine stopped, remove the turbo ducts and check the shaft assembly for free rotation, damage to wheels, or rubbing against housing walls. Next, check the oil drain line for restriction or damage, which can cause seal flooding and leakage. Also check for high crankcase pressure. If in doubt, measure the crankcase pressure, which must be within the engine manufacturer's specifications. Finally, loosen the exhaust manifold duct and check for oil in the engine exhaust. If oil is present, see the engine manual for appropriate repairs.

Excessive Engine Oil Consumption

If there is no black or blue smoke, check the air cleaner for restrictions per engine manufacturer's shop manual. Check the compressor discharge duct for loose connections, and check crankcase pressure, which must be within the manufacturer's specifications. Check the turbo shaft assembly for free rotation. Also check for evidence of wheel rubbing on housing walls. If there is rubbing, it is possible to feel it when rotating the shaft while pulling or pushing on it. Check the oil supply line for damage or restriction. If no fault is found, consult the engine manual for further troubleshooting procedures.

Noisy Turbocharger

Every engine has its own unique sound or noise level. Any change in the sound/noise level can indicate a turbocharger or engine problem and should be investigated immediately. If the noise changes to a higher pitch, look for an air leak between the air cleaner and engine or a gas leak in the exhaust system between turbo and engine. Noise-level cycling can indicate a plugged air cleaner, restricted air inlet before the turbo, or dirt buildup in the compressor housing or on the compressor wheel.

With the engine running, uneven noise and vibration can indicate a malfunction in the shaft wheel assembly. If a problem is indicated, shut the engine down immediately and repair it. If the turbocharger is functional, check the air system (see Chapter 3) for the following:

- Air cleaner restriction
- Hose clamps tight
- Intake manifold gaskets
- Cracked or deteriorated hoses

With the engine running at idle the following can be performed:

- Air tube and air cleaner/turbo connections can be checked by spraying exterior surfaces lightly with starting fluid; an engine speed increase indicates a leak.
- Air leaks between turbo and engine can be checked by feel and by application of lightweight oil or soap suds on the crossover tube, connections, and hoses. Look for bubbles.

Exhaust gas leaks between the engine block and inlet to the turbo will create a noise-level change

TABLE 9-2: TURBOCHARGER TROUBLESHOOTING GUIDE

Condition	Possible Causes Code Numbers	Remedy Description by Code Numbers
Engine lacks power	1, 4, 5, 6, 7, 8, 9, 10, 11, 18, 20, 21, 22, 25, 26, 27, 28, 29, 30, 37, 38, 39, 40, 41, 42, 43	1. Dirty air cleaner element 2. Plugged crankcase breathers 3. Air cleaner element missing, leaking, not sealing correctly; loose connections to turbocharger 4. Collapsed or restricted air tube before turbocharger 5. Restricted-damaged crossover pipe, turbocharger to inlet manifold
Black smoke	1, 4, 5, 6, 7, 8, 9, 10, 11, 18, 20, 21, 22, 25, 26, 27, 28, 29, 30, 37, 38, 39, 40, 41, 43	6. Foreign object between air cleaner and turbocharger 7. Foreign object in exhaust system (from engine, check engine) 8. Turbocharger flanges, clamps, or bolts loose. 9. Inlet manifold cracked; gaskets loose or missing; connections loose
Blue smoke	1, 2, 4, 6, 8, 9, 17, 19, 20, 21, 22, 32, 33, 34, 37, 45	10. Exhaust manifold cracked, burned; gaskets loose, blown, or missing 11. Restricted exhaust system
Excessive oil consumption	2, 8, 15, 17, 19, 20, 29, 30, 31, 33, 34, 37, 45	12. Oil lag (oil delay to turbocharger at start-up) 13. Insufficient lubrication 14. Lubricating oil contaminated with dirt or other material
Excessive oil turbine end	2, 7, 8, 17, 19, 20, 22, 29, 30, 32, 33, 34, 45	15. Improper type lubricating oil used 16. Restricted oil feed line 17. Restricted oil drain line
Excessive oil compressor end	1, 2, 4, 5, 6, 8, 19, 20, 21, 29, 30, 33, 34, 45	18. Turbine housing damaged or restricted 19. Turbocharger seal leakage 20. Worn journal bearings
Insufficient lubrication	8, 12, 14, 15, 16, 23, 24, 31, 34, 35, 36, 44, 46	21. Excessive dirt buildup in compressor housing 22. Excessive carbon buildup behind turbine wheel 23. Too fast acceleration at initial start (oil lag)
Oil in exhaust manifold	2, 7, 17, 18, 19, 20, 22, 29, 30, 33, 34, 45	24. Too little warm-up time 25. Fuel pump malfunction 26. Worn or damaged injectors 27. Valve timing
Damaged compressor wheel	3, 4, 6, 8, 12, 15, 16, 20, 21, 23, 24, 31, 34, 35, 36, 44, 46	28. Burned valves 29. Worn piston rings 30. Burned pistons 31. Leaking oil feed line
Damaged turbine wheel	7, 8, 12, 13, 14, 15, 16 18, 20, 22, 23, 24, 25, 28, 30, 31, 34, 35, 36, 44, 46	32. Excessive engine pre-oil 33. Excessive engine idle 34. Coked or sludged center housing 35. Oil pump malfunction 36. Oil filter plugged

Drag or bind in rotating assembly	3, 6, 7, 8, 12, 13, 14, 15, 16, 18, 20, 21, 22, 23, 24, 31, 34, 35, 36, 44, 46	37. Oil-bath-type air cleaner:
		  • Air inlet screen restricted
		  • Oil pullover
		  • Dirty air cleaner
		  • Oil viscosity low
		  • Oil viscosity high

Worn bearings, journals, bearing bores	6, 7, 8, 12, 13, 14, 15, 16, 23, 24, 31, 35, 36, 44, 46	38. Actuator damaged or defective 39. Waste gate binding 40. Electronic control module or connector(s) defective 41. Waste gate actuator solenoid or connector defective
Noisy	1, 3, 4, 5, 6, 7, 8, 9, 10, 11, 12, 13, 14, 15, 16, 18, 20, 21, 22, 23, 24, 31, 34, 35, 36, 37, 44, 46	42. EGR valve defective 43. Alternator voltage incorrect 44. Engine shut off without adequate cooldown time
Sludged or coked center housing	2, 11, 13, 14, 15, 17, 18, 24, 31, 35, 36, 44, 46	45. Leaking valve guide seals 46. Low oil level

and reduced performance. Check the exhaust system for:

- Manifold gasket leakage
- Manifold retaining bolt tightness
- Cracked or porous manifold
- Compressor discharge ducting
- Turbo inlet gasket leakage
- Turbo inlet flange bolt tightness

As mentioned earlier in this chapter, exhaust gas leakage can be detected by carbon deposits or heat discoloration in the area of the leak.

It is also important to check the turbo shaft for looseness and look for wheel rubbing or impact damage to blades from foreign material. If rubbing or impact damage is found, remove and replace the turbocharger.

Table 9–2 gives a summary of the most common turbocharger problems and recommended remedies.

REPLACING A TURBOCHARGER

If the turbocharger is faulty it can be replaced with a new or rebuilt unit. Always follow the exact replacement procedure given in the service manual, but the following steps are a good general guide:

1. Remove the old turbocharger by loosening the bolts, connections, and gaskets.
2. Make sure that the unit is the correct one for the engine. The part can be checked against the service manual.
3. Install the new gaskets and seals as needed, and be sure that they are properly positioned.
4. Install the new or rebuilt unit and torque the fasteners to the service manual specifications and recommended sequence.
5. Spin the impeller-turbine to check for binding before installing the unit on the engine.
6. Change the engine oil and flush the oil lines. If oil problems caused the failure of the original unit, check the oil supply pressure in the feed line to the turbo.

Once the new or rebuilt unit is installed, the turbo should be started up as described in the following section.

Turbo Start-Up and Shutdown

Oil supply to a turbocharger's shaft bearings is, as mentioned previously, critical to long life for this high-speed component. After replacement of a turbocharger, or after an engine has been unused or stored, there can be a considerable lag after engine start-up before the oil pressure is sufficient to deliver oil to the turbo's bearings. To prevent the problem, which can lead to premature turbo failure, follow these simple steps:

1. When installing a new or remanufactured turbocharger, make certain that the oil inlet and drain lines are clean before connecting.
2. Be sure the engine oil is clean and at the proper level.
3. Fill the oil filter with clean oil to minimize cranking time.
4. Leave the oil drain line disconnected at the turbo and crank the engine without starting until oil flows out of the turbo drain port.
5. Connect the drain line, start the engine, and operate at low idle for a few minutes before operating at higher speeds.

After an oil and filter change, crank the engine without starting until the oil pressure gauge indicates normal oil pressure. Alternate: run the engine at low idle until a steady oil pressure is obtained. This same procedure should be followed if the engine has not been operated for some time.

Instruct the driver as to the proper shutdown procedure. If he or she shuts down from high speed, the turbo will continue to rotate after the engine oil pressure has dropped to zero, which can cause bearing damage. Also, advise the driver about oil lag. Allow 30 seconds for the oil flow to become established before running up a high rpm. Ask the driver if engine oil and oil filter are changed at recommended intervals; review the proper lube oil and filter change interval with the operator. Contaminated oil can cause sludge buildups within the turbo. Check the oil drain outlet for sludge buildup with the oil drain line removed. Failure to follow these steps can result in bearing failure on the turbo's main shaft. Remember, shaft rotation speed on modern turbochargers can easily exceed 2,000 revolutions per second, so momentary oil flow interruption is likely to lead to turbocharger failure.

REVIEW QUESTIONS

1. Which of the following statements is not a basic function of a vehicle's exhaust system?
   a. carry away poisonous gases from the passenger compartment
   b. muffle the sound of the engine
   c. enhance carburetor performance
   d. clean exhaust emissions

2. The exhaust manifold gasket seals the joint between the exhaust manifold and the
   _____ .
   a. cylinder head
   b. engine block
   c. muffler
   d. resonator

3. Some high-performance engines have a combination of the exhaust manifold and the exhaust pipe called a(n) _____ .
   a. dual-exhaust system
   b. header
   c. exhaust gas recirculation
   d. catalytic converter

4. Other than reducing engine noise, what other functions does a muffler serve?
   a. reduces weight
   b. improves fuel economy
   c. reduces emissions
   d. all of the above

5. What exhaust system component, mainly used on large model cars, further reduces the sound level of the exhaust?
   a. manifold
   b. exhaust pipe
   c. resonator
   d. header

6. What exhaust system component carries exhaust fumes to the atmosphere?
   a. muffler
   b. tail pipe
   c. resonator
   d. manifold

7. A blocked or restricted exhaust system can cause _____ .
   a. increased vehicle noise
   b. increased air pollution
   c. improper engine operation
   d. all of the above

8. When inspecting an exhaust system, Technician A first makes a visual inspection. Technician B first listens for sounds that might indicate the beginning of exhaust system failure. Who is right?
   a. Technician A
   b. Technician B
   c. Both A and B
   d. Neither A nor B

9. Before replacing any exhaust system component, Technician A soaks all old connections with a penetrating oil. Technician B checks the old system's routing for critical clearance points. Who is right?
   a. Technician A
   b. Technician B
   c. Both A and B
   d. Neither A nor B

10. A vehicle's manifold is warped beyond the manufacturer's specifications. Technician A replaces it. Technician B rebuilds it. Who is right?
    a. Technician A
    b. Technician B
    c. Both A and B
    d. Neither A nor B

11. A customer complains of an engine that is hard to warm up or will not return to normal idle. Technician A checks the heat control valve. Technician B checks for a blocked crossover passage in the intake manifold. Who is right?
    a. Technician A
    b. Technician B
    c. Both A and B
    d. Neither A nor B

12. Which of the following is not a possible cause of excessive exhaust system noise?
    a. broken or loose clamps
    b. exhaust pipe leak
    c. restrictions in muffler or tail pipe
    d. blocked crossover passage in the intake manifold

13. When replacing major sections of the entire exhaust system, Technician A dismantles the old system from the rear forward. Technician B dismantles the old system from the front to back. Who is right?
    a. Technician A
    b. Technician B
    c. Both A and B
    d. Neither A nor B

14. What problem in the exhaust manifold could cause a white powdery deposit at the exhaust pipe connection?
    a. loose gasket
    b. missing gasket
    c. leaking gasket
    d. all of the above

15. Which of the following can cause exhaust system component failure?
    a. rust
    b. impact with ruts
    c. rocks thrown up by the tires
    d. all of the above

CHAPTER TEN

EMISSION CONTROL SYSTEMS

Objectives

After reading this chapter, you should be able to:
- Explain the two types of emission control systems that are used in modern vehicles.
- Describe a PCV system, naming its components and explaining its function in emission control.
- Describe an EGR system and its role in emission control.
- Explain the operation of the fuel evaporative emission control system.
- Differentiate between carburetor fuel bowl emission control and air temperature emission control.
- Explain the operation of a basic air injection system used in most vehicles.
- Identify the importance of the catalytic converter.
- Describe the various methods of modifying engines to control vehicle emissions.
- Explain the operation of a diesel emission system.

Automobile manufacturers have been working toward reduction of automotive air pollutants since the early 1950s, when emissions first were related to Los Angeles pollution. Governmental interest developed around the same time.

In late 1959, California established the first standards for automotive emissions. In 1967, the Federal Clean Air Act was amended to provide for the establishment of federal standards to apply to motor vehicles.

The first source of emissions to be brought under control was the crankcase. Positive crankcase ventilation systems (PCV) to route these vapors back to the engine intake manifold were developed and incorporated into 1961 cars and light trucks sold in California (Figure 10-1). These systems were installed on all cars nationwide beginning with the 1963 models.

Control of unburned hydrocarbons and carbon monoxide in the engine exhaust was the next major development. An Air Injection Reactor (AIR) system was built into cars and light trucks sold in California in 1966. Other systems, including the Controlled

Combustion System (CCS), were developed and used nationwide in 1968. Further improvements in years following improved combustion to reduce hydrocarbon and carbon monoxide emissions.

Fuel vapors from the gasoline tank and the carburetor float bowl were brought under control with the introduction of evaporation control systems. These systems were first installed in 1970 model cars sold in California and in most domestic-made cars nationwide beginning with 1971 models.

Most vehicle manufacturers started to provide emission control systems that reduced oxides of nitrogen as early as 1970. The Exhaust Gas Recirculation system used on some 1972 models was used extensively for 1973 models when federal standards for oxides of nitrogen took effect.

Present ultimate government goals call for a 98 percent reduction of unburned hydrocarbons, a 97 percent reduction of carbon monoxide, and a 90 percent reduction of oxides of nitrogen compared to precontrolled cars.

One of the most important developments for getting to lower emission levels has been the avail-

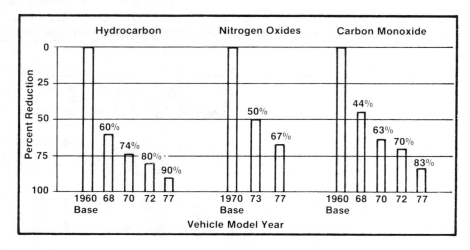

FIGURE 10-1 Average yearly reduction of automotive exhaust emission in the U.S. since federal standards were established.

ability and use of unleaded gasolines. Beginning with 1971, cars have been designed to operate on unleaded fuels.

Removing lead from gasoline brings some immediate benefits. It eliminates the emission of lead particles to the atmosphere from automobile exhausts. It increases spark plug life—important from an emission standpoint. It avoids formation of lead deposits in the combustion chambers which tend to increase hydrocarbon emissions.

The catalytic converter, a later development, provided a means for oxidizing the carbon monoxide and hydrocarbon emissions in the engine exhaust, a process that lowers the amount of these pollutants. Beginning with the 1975 model year, passenger cars and light trucks have been equipped with converters.

The three kinds of dangerous emissions that are being controlled in gasoline engines today are:

1. *Unburned Hydrocarbons (HC).* These are particles, usually vapors, of gasoline that have not been fully burned. They are present in the exhaust and in crankcase vapors. Of course, any raw gas that evaporates out of the tank or carburetor is classed as HC.
2. *Carbon Monoxide (CO)* is a poisonous chemical compound of carbon (part of gasoline and oxygen from the air). It forms in the engine when the fuel burning (combustion) is less than complete. CO is found in the exhaust principally, but can also be in the crankcase.
3. *Oxides of Nitrogen (NOx; pronounced "nox").* Various compounds of nitrogen and oxygen, both present in the air used

for combustion, are formed in the cylinders during combustion and are part of exhaust gas. They become part of the tail pipe emissions if not reduced in the exhaust system.

Federal standards for these pollutants are given in Table 10-1. The exceptions to these standards are a few high-altitude western states and California. Because there is less oxygen at high altitudes to promote combustion, emission standards at high-altitudes are slightly less strict. California's standards allow less pollution than federal standards.

Two basic types of emission control systems are used in modern vehicles:

1. *Precombustion Control Systems.* Precombustion systems are by far the more efficient of the two types of control systems. It is better, easier, and cheaper not to produce emissions in the first place than it is to control or destroy them after they have been formed. Most of the pollution control systems used today prevent emissions from being created in the engine, either during or before the combustion cycle. The following are common precombustion control systems:

   • *Positive Crankcase Ventilation (PCV).* First used in the early 1960s, the PCV system removes gases that blow by the pistons into the crankcase. Originally, these gases (HC, CO, and NOx) were vented to the air by a road draft tube. Now, they are recirculated to the induction system.

TABLE 10-1:   Federal Standards*—New Vehicle Summary

Year	Test Procedure[2]	Hydrocarbons	Carbon Monoxide	Oxides Of Nitrogen	Particulates[3]	Evaporative[4] Hydrocarbon
Prior	7-mode	850 ppm	3.4%	1000 ppm	—	—
to	7-mode	11 gpm	80 gpm	4 gpm	—	—
controls	CVS-75	8.8 gpm	87.0 gpm	3.6 gpm	—	—
1968–69	7-mode					
	50-100 CID	410 ppm	2.3%	—	—	—
	101-140 CID	350 ppm	2.0%	—	—	—
	over 140 CID	275 ppm	1.5%	—	—	—
1970	7-mode	2.2 gpm	23 gpm	—	—	—
1971	7-mode	2.2 gpm	23 hpm	—	—	6 g/test[5]
1972	CVS-72	3.4 gpm	39 gpm	—	—	2 g/test
1973–74	CVS-72	3.4 gpm	39 gpm	3.0 gpm	—	2 g/test
1975–76	CVS-75	1.5 gpm	15 gpm[6]	3.1 gpm	—	2 g/test
1977[7]	CVS-75	1.5 gpm	15 gpm	2.0 gpm	—	2.0 g/test
1978–79	CVS-75	1.5 gpm	15 gpm	2.0 gpm	—	6.0 g/test
1980	CVS-75	0.41 gpm	7.0 gpm	2.0 gpm	—	6.0 g/test
1981	CVS-75	0.41 gpm	3.4 gpm[8]	1.0 gpm[9] [10]	—	2.0 g/test
1982[11]	CVS-75	0.41 gpm	3.4 gpm[8]	1.0 gpm[9] [10]	0.6 gpm	2.0 g/test
		(0.57)	(7.8)	(1.0)[9]	—	(2.6)
1983[11]	CVS-75	0.41 gpm	3.4 gpm	1.0 gpm[9]	0.6 gpm	2.0 g/test
		(0.57)	(7.8)	(1.0)[9]	—	(2.6)
1984	CVS-75	0.41 gpm	3.4 gpm	1.0 gpm[9]	0.6 gpm	2.0 g/test
1985 and later	CVS-75	0.41 gpm	3.4 gpm	1.0 gpm	0.2 gpm	2.0 g/test

*The following standards, up to 1975, apply only to gasoline-fueled light-duty vehicles. Standards for 1975 and later apply to both gasoline-fueled and diesel light-duty vehicles.

[1]Standards do not apply to vehicles with engines less than 60 CID from 1968 through 1974.

[2]Different test procedures, which vary in stringency, have been used since the early years of emission control. The appearance that the standards were relaxed from 1971 to 1972 is incorrect. The 1972 standards are actually more stringent because of the greater strictness of the 1972 test procedure.

[3]Applies only to diesels.

[4]Evaporative emissions determined by carbon trap method through 1977, SHED procedure beginning in 1978. Applies only to gasoline-fueled vehicles.

[5]Evaporative standard does not apply to off-road utility vehicles for 1971.

[6]Carbon monoxide standard for vehicles sold in the State of California is 9.0 gpm.

[7]Cars sold in specified high altitude countries required to meet standards at high altitude.

[8]Carbon monoxide standard can be waived to 7.0 gpm for 1981-82 by the EPA Administrator.

[9]Oxides of nitrogen standard can be waived to 1.5 gpm for innovative technology or diesel.

[10]Oxides of nitrogen standard can be waived to 2.0 gpm for American Motors Corporation.

[11]Standards in parentheses apply to vehicles sold in specified high altitude counties. Vehicles eligible for a carbon monoxide waiver for 7.0 gpm at low altitude are eligible for a waiver to 11 gpm at high altitude.

gpm: grams per mile
CID: cubic inch displacement
CVS-72: constant volume sample cold start test
CVS-75: constant volume sample test, includes cold and hot starts
7-mode: 127-second driving cycle test
ppm: parts per million.

- *Engine Modification System.* This system appeared in the 1960s as a group of engine modifications designed to improve combustion and reduce HC-CO in the exhaust. It included a heated primary air system, carburetor design changes, engine "breathing" refinements, and some spark timing controls.
- *Evaporative Control Systems.* In the 1960s, evaporative control systems were used to trap raw gas vapors in the fuel tank (and later carburetor bowl) and route them to the air cleaner when the engine ran. In the 1970s the system was refined to a sealed housing system to better control emissions and purge them to the intake manifold in specific engine models.
- *Exhaust Gas Recirculating (EGR) Systems.* EGR is strictly a control for NOx

in the exhaust gases. It reduces NOx by diluting the air/fuel mixture with some exhaust gas, which does not burn. This reduces peak combustion chamber temperature, so less NOx is formed.

2. *Post-Combustion Control Systems.* Post-combustion control systems clean up the exhaust gases after the fuel has been burned. Secondary air or air injector systems appeared in the mid-1960s. Their function is to pull fresh air into the exhaust to reduce HC and CO to harmless water vapor and carbon dioxide by chemical (thermal) reaction with oxygen in the air. In the 1970s, catalytic converters were added to help this process. Some catalysts now reduce NOx as well as HC-CO.

Most emission control devices have some kind of engine vacuum control. These controls are used to block out the operation in certain modes:

- Systems that put air or fuel vapors into the air/fuel induction system do not operate at idle, when the engine is cold, and sometimes when the intake air is cold. The exception is PCV, which operates whenever the engine is running.
- EGR systems are usually blocked out at idle, since mixture dilution then causes rough idle. They have various controls to vary the gas flow in other models.
- Thermactor (secondary air) systems are usually blocked out from the exhaust when it is rich (idle, deceleration, cold engine, WOT, and the like) to prevent backfire and damage to the catalysts.

The control devices are principally vacuum diaphragms (Figure 10–2). Engine vacuum from various sources is switched on or off to the diaphragms. The switching devices can be in the air cleaner (controlled by inlet air temperature), in the water jacket (controlled by coolant temperature), or incorporated into solenoids that are controlled by the electronic control assembly (ECA). The principal sources for the vacuum muscle are manifold vacuum, spark port or "S" vacuum, and "E" port (EGR) vacuum. "S" and "E" vacuum sources are used to lock out certain systems at curb idle when they are essentially zero. Venturi vacuum is used only occasionally.

Chapter 1 contained an overview of the components that compose the emission control system. In various other chapters of this book, specific fuel emission controls have been fully discussed. This chapter covers components required by federal and state emission agencies. The following chapter in the book details servicing and maintenance of various parts of typical emission systems. Chapter 12 covers electronic engine controlled emission systems.

PCV SYSTEMS

The crankcase of an engine must "breathe" to purge itself of fuel and water vapors. If these vapors are not purged, they can thin the oil or form sludge. Water forms in the crankcase when a cold engine condenses it from the air. In fact, it has been estimated that every gallon of gasoline burned forms more than a gallon of water. During the last part of the engine's combustion stroke, some unburned fuel and products of combustion—water vapor, for instance—leak past the engine's piston rings into the crankcase. This leakage results in the following:

- *High Pressures in the Engine Combustion Chamber.* This condition is created by the normal compression stroke in the engine under operation.
- *Necessary Working Clearance of Piston Rings in Their Grooves.* Without normal ring clearance, the engine's piston rings do not have room to expand from heat, which is created by normal engine operation, and seal properly against the cylinder walls.
- *Normal Shifting of Piston Rings in Their Grooves.* Sometimes this shift lines up the clearance gaps of two or more rings, which is a normal condition. As the piston rings continue to turn in their grooves, the situation corrects itself.
- *Reduction in Piston Ring Sealing Contact Area.* This occurs as the piston moves up and down in the cylinder.

Leakage into the engine crankcase is called *blowby*. Blowby must be removed from the engine before it condenses in the crankcase and reacts with the oil, which forms sludge. Sludge, if allowed to circulate with engine oil, will corrode and accelerate wear of pistons, piston rings, valves, bearings, and other internal working parts of the engine.

Because the air/fuel mixture in an engine never completely burns, blowby carries some unburned fuel into the crankcase. If unremoved, the unburned fuel dilutes the crankcase oil. When oil is diluted with gaoline, it does not lubricate the engine properly, which causes excessive wear.

Combustion gases that enter the crankcase are removed by a positive crankcase ventilation system,

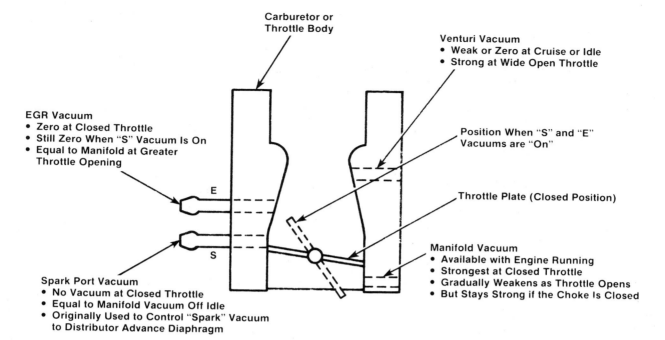

FIGURE 10-2   Emission control devices that are used principally are vacuum diaphragms.

which uses engine vacuum to draw fresh air through the crankcase.

This fresh air, which dissipates the harmful gases, enters through the air filter on top of the carburetor or through a separate PCV breather filter located on the inside of the air filter housing (Figure 10-3).

Because the vacuum supply for the PCV system is from the engine's intake manifold, the airflow through this system must be controlled in such a way that it varies in proportion to the regular air/fuel ratio being drawn into the intake manifold through the carburetor. Otherwise, the additional air that is drawn into the system would cause the air/fuel mixture to become too lean for efficient engine operation.

To summarize, the positive crankcase ventilation system (Figure 10-4) has two major functions:

1. It prevents the emission of blowby gases from the engine crankcase to the atmosphere. These gases were once vented through a road draft tube. Now they are recirculated to the engine intake and burned during combustion.
2. It scavenges the crankcase of vapors that could dilute the oil and cause it to deteriorate or that could build undesirable pressure in the crankcase. Fresh air from the air cleaner mixes with these vapors and makes them flow to the air/fuel intake.

Thus, the PCV system benefits the vehicle's driveability by

*   Eliminating harmful crankcase gases
*   Reducing air pollution
*   Promoting fuel economy

Recirculated gases in the system are a combustible mixture that becomes fuel for the engine when added to the air/fuel mixture entering the intake manifold from the carburetor or fuel injectors. Consequently, an inoperative PCV system could shorten the life of the engine by allowing harmful blowby gases to remain in the engine, causing corrosion and accelerating wear.

TYPES OF PCV SYSTEMS

Since 1961, there have been two types of positive or forced crankcase ventilation systems (Figure 10-5):

1. Open PCV system
2. Closed PCV system

When first introduced, all PCV systems were open. That means the air is drawn through a conventional open breather cap and released through the PCV valve into the intake manifold, where it goes into the combustion chambers and is burned.

The closed PCV system, on the other hand, draws from the clean side of the air cleaner through a connecting hose to the breather cap. All 1968 and

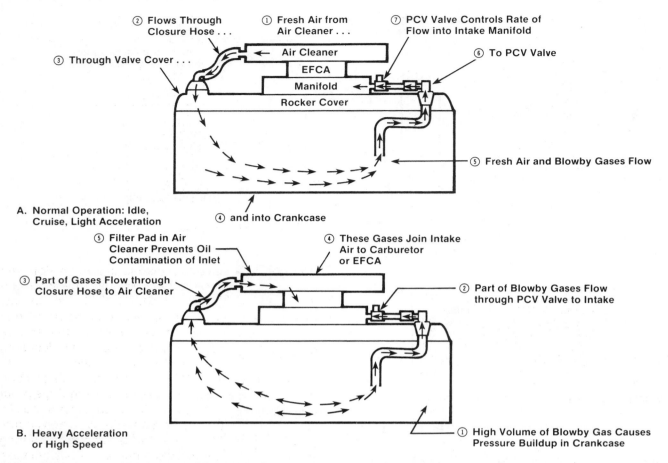

② Flows Through Closure Hose . . .

① Fresh Air from Air Cleaner . . .

⑦ PCV Valve Controls Rate of Flow into Intake Manifold

③ Through Valve Cover . . .

⑥ To PCV Valve

Air Cleaner

EFCA

Manifold

Rocker Cover

⑤ Fresh Air and Blowby Gases Flow

A. Normal Operation: Idle, Cruise, Light Acceleration

④ and into Crankcase

⑤ Filter Pad in Air Cleaner Prevents Oil Contamination of Inlet

④ These Gases Join Intake Air to Carburetor or EFCA

③ Part of Gases Flow through Closure Hose to Air Cleaner

② Part of Blowby Gases Flow through PCV Valve to Intake

B. Heavy Acceleration or High Speed

① High Volume of Blowby Gas Causes Pressure Buildup in Crankcase

FIGURE 10-3   (A) Fresh air flow during normal operation: idle, cruise, and light acceleration; (B) during heavy acceleration or high speed.

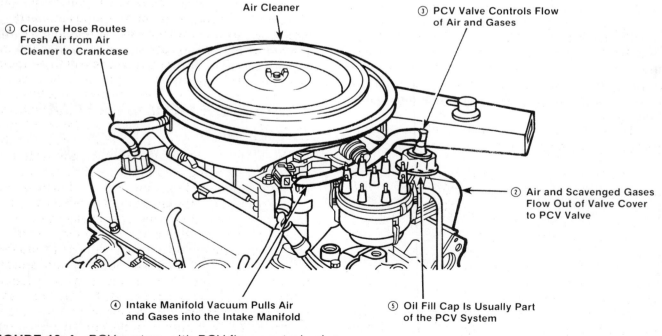

① Closure Hose Routes Fresh Air from Air Cleaner to Crankcase

Air Cleaner

③ PCV Valve Controls Flow of Air and Gases

② Air and Scavenged Gases Flow Out of Valve Cover to PCV Valve

④ Intake Manifold Vacuum Pulls Air and Gases into the Intake Manifold

⑤ Oil Fill Cap Is Usually Part of the PCV System

FIGURE 10-4   PCV system with PCV flow control valve

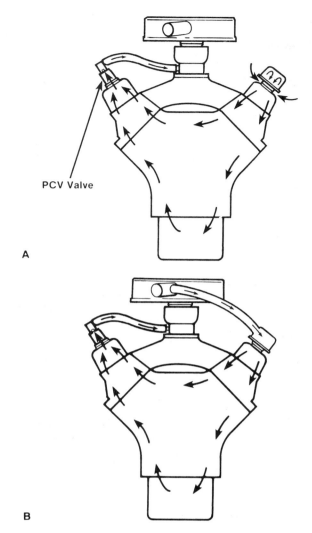

A

B

FIGURE 10-5 (A) Open PCV system; (B) closed PCV system

newer engines are designed with closed PCV systems; the crankcase is not vented in any way to the outside atmosphere.

PCV VALVE

The PCV valve is usually located in the hose or tube that is connected between the engine valve cover and a vacuum source fitting at the intake manifold below the carburetor or injectors. Depending on the particular engine, the actual location of the PCV valve will vary. It might be inserted with a rubber grommet in the engine valve cover, or it might be closer to the intake manifold with rubber hoses connected to each end of the valve.

The PCV valve itself is a variable-orifice device containing a contoured plunger, spring, orifice washer, and washer housed in a one-piece body

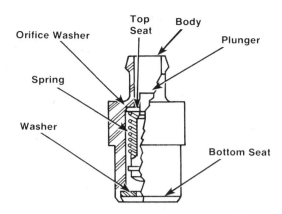

FIGURE 10-6 Typical PCV valve

(Figure 10-6). Valve dimensions, spring tension, and internal dimensions vary according to the engine on which they are used to produce the desired airflow requirements. Because of this, it is important that the repair technician checks the vehicle's service manual to get the PCV valve specifically designed for the particular engine being serviced.

In operation, the PCV valve reacts to intake manifold vacuum, thereby regulating the amounts of air and blowby gases allowed to combine with the air/fuel mixture in the intake manifold. With no vacuum applied to the PCV valve, the tension of the spring forces the plunger to close against the bottom seat. Excessive crankcase pressure can push the plunger off the seat and force the blowby gases through the valve.

Vacuum applied to the valve will pull the plunger off the bottom seat, overcoming spring tension (Figure 10-7A). This allows the valve to regulate blowby gases from the crankcase. As applied vacuum increases, the PCV flow also increases. With no vacuum, the valve is against the bottom seat. As vacuum increases to a specified level, the plunger is lifted off the bottom seat, allowing a maximum flow of blowby gases at approximately 3.0 inches mercury (Figure 10-7B). After this point, the vacuum will pull the plunger into the top seat. However, the design is such that the plunger cannot be pulled far enough into the top seat to block the flow of blowby gases. This allows the contoured plunger to reduce the flow of gases in proportion to the opening allowed by the size of the top seat to the contour of the plunger.

An intake manifold backfire will seat the tapered valve against the PCV valve housing and prevent backfiring into the crankcase as shown in Figure 10-7C. This is an important safety feature because a backfire could cause an explosion in the intake manifold. If the flame should enter the crankcase, serious damage would occur. To prevent this, the piston

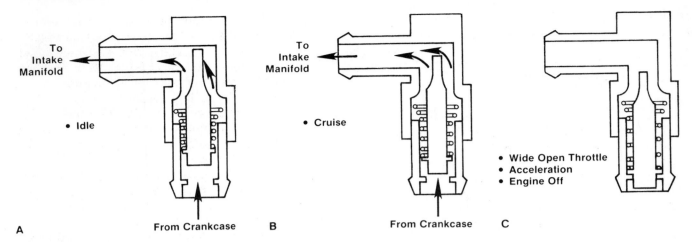

A • Idle

B • Cruise

From Crankcase

• Wide Open Throttle
• Acceleration
• Engine Off

From Crankcase **C**

FIGURE 10-7 PCV operation

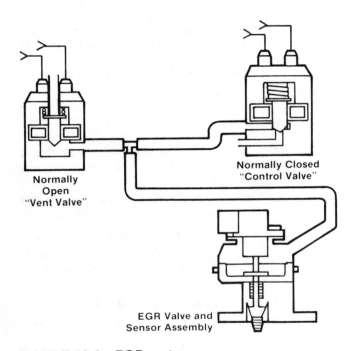

Normally Open "Vent Valve"

Normally Closed "Control Valve"

EGR Valve and Sensor Assembly

FIGURE 10-8 EGR system

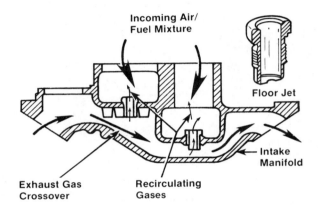

Incoming Air/ Fuel Mixture

Floor Jet

Intake Manifold

Exhaust Gas Crossover

Recirculating Gases

FIGURE 10-9 Jets allow a calibrated amount of exhaust gas to be drawn into the intake system.

of nitrogen (NO$_x$) emitted by the exhaust system. Oxides of nitrogen are compounds of two elements or substances that are the main components of air: oxygen (about 20 percent) and nitrogen (about 80 percent). The oxygen is used in burning the fuel, and ideally, the nitrogen just passes through into the exhaust. However, if the combustion temperatures in the engine are very high, a chemical reaction takes place between nitrogen and oxygen in the engine's combustion chamber.

This chemistry produces compounds of the two, called oxides of nitrogen (NO$_x$). The best way to lower combustion temperature and reduce NO$_x$ emissions, without compromising the engine's fuel economy and performance, is to recirculate exhaust gases back into the combustion chamber.

TYPES OF EGR SYSTEMS

When first introduced to emission control, there were two types of EGR systems that regulated the

closes the valve completely by seating on the end of the valve chamber away from the manifold. Most PCV systems have a screen on the end of the clear air hose in the air cleaner as a backfire safety device.

EXHAUST GAS RECIRCULATION SYSTEMS

The exhaust gas recirculation (EGR) system (Figure 10-8) is used to reduce the amount of oxides

amount of exhaust gases entering the intake system. One was the floor jet system, which utilizes two orificed jets built into the intake manifold. The other was the valve-controlled system, which utilizes a vacuum-operated valve.

Floor Jet System

As shown in Figure 10-9, there is an opening from the exhaust passage to the intake manifold, which is provided by the jets to allow a calibrated amount of exhaust gas to be drawn into the intake system. The amount of exhaust gas drawn into the intake system is regulated by the size of the orifices in the jets.

Valve Controlled System

This system uses a vacuum operated EGR valve to regulate the exhaust gas flow into the intake manifold. The exhaust gas is channelled to the valve through exhaust crossover passages under the intake manifold. (In some in-line engines, the exhaust gas is routed to the valve through an external tube.) Typical mounting of the EGR valve is either on a plate under the carburetor or directly on the manifold. Figure 10-10 illustrates how the basic valve system operates:

1. The EGR valve is a vacuum operated, flow control valve. On most systems, it is attached to a carburetor spacer.

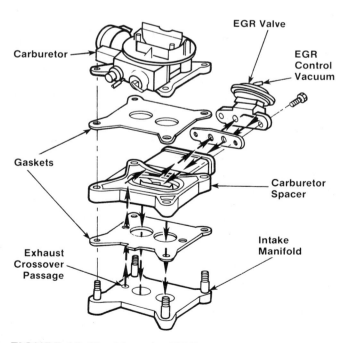

FIGURE 10-10   How the EGR system works

2. The carburetor spacer is sandwiched between the carburetor and intake manifold.
3. Gaskets are used above and below the spacer to seal the EGR system and the carburetor-to-manifold air/fuel flow.
4. Exhaust gases are admitted to the spacer through a small exhaust crossover passage in the intake manifold. These gases flow through the spacer to the inlet port of the EGR valve.
5. If the EGR valve is opened by control vacuum at the diaphragm, exhaust gases flow through the valve and back to another port of the spacer.
6. Here, the exhaust gas mixes with the air/fuel mixture leaving the carburetor and then into the intake manifold. The effect is to dilute or lean out the mixture so that it still burns completely but with a reduction in combustion chamber temperatures. In an ignition-type system, the EGR system operates in basically the same manner as a carburetor.

In summary, the effects of exhaust in gas recirculation system are:

- Nitrogen in the air does not form as much NO_x, but simply goes out the exhaust.
- Improved combustion
- Combustion pressure is maintained at a level the prevents spark knock or detonation.

In the mid-1970s, the floor jet system was abandoned by the auto industry and the valve controlled system became the universally accepted system.

TYPES OF EGR VALVES

Several types of control valves are used in EGR systems. The more common types are:

- *Ported EGR Valve.* The ported or diaphragm-operated valve has a stem that pulls the valve off its seat as the vacuum applied to the diaphragm port increases (Figure 10-11). There are several styles of ported EGR valves (Figure 10-12). Some typical designs are:
  — Poppet Type. The valve stem pulls a valve poppet off its seat when vacuum is applied to the diaphragm. The poppets are various shapes. Also, there might be a flow restrictor in the valve body inlet port to regulate flow.
  — Tapered Stem. The valve end of the stem is tapered. Flow rate depends on how far the stem end is pulled out of the seat.

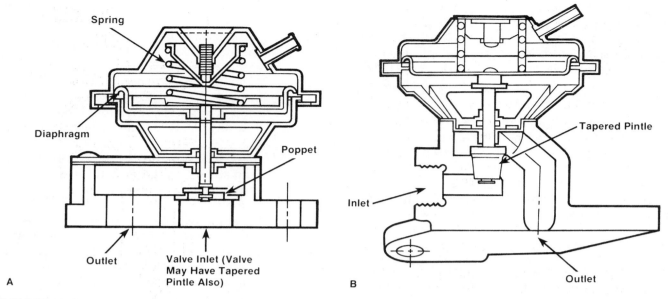

FIGURE 10-11   Ported EGR valves: (A) base entry; (B) side entry

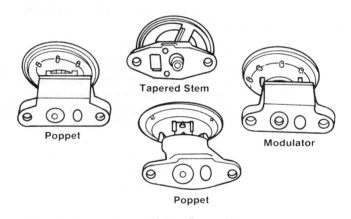

FIGURE 10-12   Types of EGR valves

— Modulator. A modulator valve has an extra disk valve on the stem below the main valve. This disk will restrict flow when a high vacuum is applied to the diaphragm. In all ported design EGR valves, the inlet port is located at the valve and the outlet is opposite to it. The valve opens when vacuum is applied to the spring end of the diaphragm. The spring calibrates the amount the valve opens by balancing against the force of vacuum. When vacuum is released, the spring pulls the valve closed and there is no EGR flow. The valve units can have either base entry or side entry installation. Both entry system operates in the same way.

Remote Back Pressure Transducer EGR Valve. This system utilizes an externally

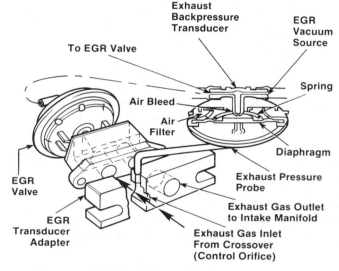

FIGURE 10-13   Remote back pressure transducer EGR valve

mounted, remote back pressure transducer, which uses the exhaust back pressure as a signal to modulate or control vacuum to the EGR valve diaphragm. The transducer receives an exhaust back pressure signal through the exhaust inlet (Figure 10-13). The exhaust back pressure is high, forcing the transducer diaphragm up and directing EGR vacuum to the EGR valve. At this time, the EGR vacuum is high enough to open the EGR valve allowing the exhaust gas to flow to the intake manifold for recirculation.

• *Positive Integral Back Pressure Transducer EGR Valve.* The positive back pressure EGR valve uses an exhaust back pressure signal to assist in the opening and closing of the EGR valve. The integral transducer receives an exhaust back pressure signal through the hollow shaft (Figure 10-14A), which exerts a force on the bottom of the transducer diaphragm opposed by the light spring. As exhaust back pressure increases, the control valve closes, shutting off airflow through the valve (Figure 10-14B). Vacuum then builds up in the vacuum chamber until the spring force holding the EGR valve closed is overcome. This allows the exhaust gas to enter the intake manifold for recirculation.

• *Negative Integral Back Pressure Transducer EGR Valve.* The negative back pressure EGR valve shown in Figure 10-15 uses an exhaust back pressure signal to assist in the opening and closing of the EGR valve. During light engine loads, the ported vacuum signal is applied to the main vacuum chamber, partially opening the EGR valve. This allows manifold vacuum to bleed into the seat area and travel up the hollow valve stem. The manifold vacuum then acts on the transducer diaphragm to open the air bleed. With the ported vacuum signal bled off, the diaphragm be-

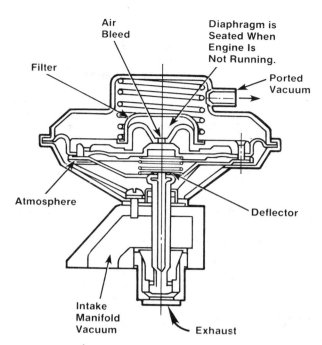

FIGURE 10-15 Negative integral back pressure transducer EGR valve

gins to close. At heavier engine loads, the manifold vacuum signal is very weak. The exhaust back pressure is strong enough to open the valve and enter the intake manifold.

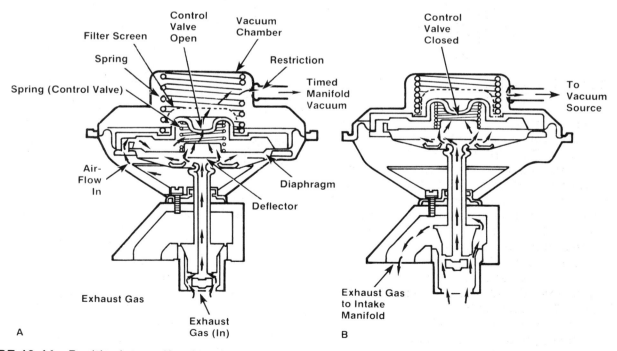

FIGURE 10-14 Positive integral back pressure transducer EGR valve (A) control valve open; (B) control valve closed

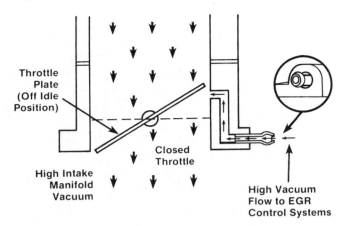

FIGURE 10-16   EGR vacuum port

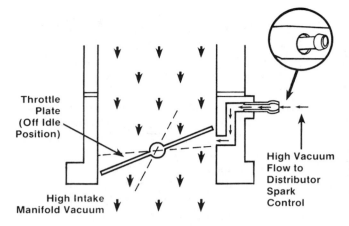

FIGURE 10-17   Spark vacuum port

Since the manifold vacuum signal is weak, the air bleed closes and allows ported vacuum to control EGR valve operations.

EGR VACUUM SOURCES

Vacuum used to operate and control the EGR valve is supplied by one of three possible ports in the carburetor body: the EGR vacuum port, spark vacuum port, or venturi vacuum port.

EGR Vacuum Port

The EGR vacuum port is used to supply vacuum to the EGR valve, which is located on the carburetor above the throttle plate. The EGR port is open to manifold vacuum when the throttle plate is opened (Figure 10-16). A high vacuum is available to the EGR valve, permitting exhaust gas recirculation. At closed throttle, the EGR port is cut off from manifold vacuum, closing the EGR valve and restricting the flow of exhaust gases. At full throttle, the vacuum signal at the EGR port is usually too weak to maintain EGR flow, closing the EGR valve and restricting exhaust gas recirculation.

Spark Vacuum Port

The spark vacuum port is used to supply a vacuum signal to the distributor spark control systems. In some applications the spark vacuum port is also used to supply EGR vacuum. The throttle plate is open and a high vacuum signal is available to the EGR valve and the distributor spark control systems (Figure 10-17). At closed throttle, the spark port is cut off from manifold vacuum, closing the EGR valve and restricting the flow of exhaust gases. At full throttle, the vacuum signal at the spark port is usually too weak to maintain EGR flow, closing the EGR valve and restricting exhaust gas recirculation.

Venturi Vacuum Port

The venturi vacuum port uses the venturi in the carburetor as a source to supply EGR vacuum. The throttle plate is open (Figure 10-18). This creates an airflow through the venturi, pulling fuel out of the main discharge nozzle and producing a weak vacuum signal. The weak vacuum signal is sensed at the venturi vacuum port but is too weak to open the EGR valve. To produce a strong enough vacuum signal, the vacuum amplifier is used. This increases the vacuum signal and opens the EGR valve permitting exhaust gas to flow. At closed throttle, the airflow is greatly reduced in the venturi, closing the EGR valve and restricting the flow of exhaust gases. At full throttle, venturi vacuum is high. However, manifold vacuum is low. This supplies the vacuum amplifier with a weak vacuum signal, closing the EGR valve and restricting the flow of exhaust gases.

EGR with Electronic Control. Some engines, beginning in 1978, are equipped with electronic engine control. On these engines, a special

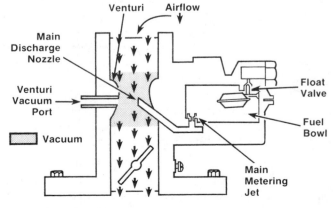

FIGURE 10-18   Venturi vacuum port

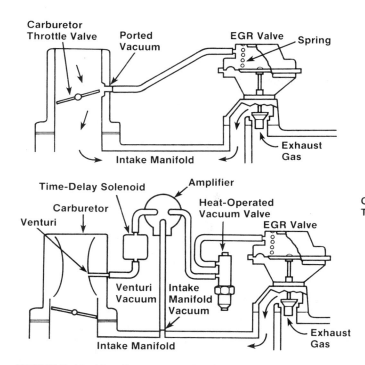

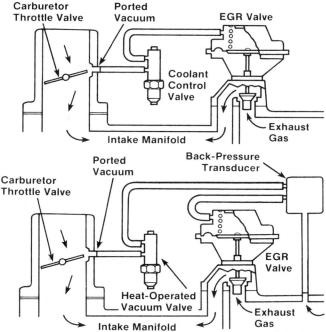

FIGURE 10-19 Summary of EGR variations

design EGR valve is adapted to the electronic control. It has a position sensor to tell the control how far the valve is open. Vacuum (or air) can be ported to the diaphragm to change the opening. This is done through two electronically controlled solenoid valves. A detailed explanation of the operation of the electronic system can be found in Chapter 12. A summary of EGR variations is illustrated in Figure 10-19.

EGR CONTROLS

EGR systems are usually designed to operate when the engine reaches operating temperature and when the engine is operated under conditions other than idle or wide-open throttle. Various controls have been incorporated into EGR systems to control the operation of the EGR valve. In some applications, cold engine EGR lockout and wide-open throttle EGR lockout are used. Basically, cold EGR lockout is required to keep the EGR valve closed during cold engine operation. Wide-open throttle EGR lockout might be required to keep the EGR valve closed when the engine is under maximum load. The following are various controls that are directly related to the EGR system:

- *Thermal Vacuum Switch (TVS)*. Senses the air temperature in the carburetor air cleaner to control vacuum to the EGR valve (Figure 10-20). When the engine reaches operating

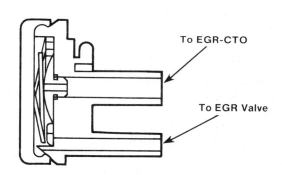

FIGURE 10-20 Thermal vacuum switch

temperature, the TVS opens to supply vacuum to the EGR valve. This opens the EGR valve for exhaust gas recirculation.

- *Ported Vacuum Switch (PVS)*. Senses the coolant temperature to control vacuum to the EGR valve (Figure 10-21). The PVS operates in the same manner as the TVS except it senses the coolant temperature instead of the air temperature. That is, the PVS function is to cut off vacuum to the EGR valve when the engine is cold and connects the vacuum to the EGR valve when the engine is warm.

- *Coolant Controlled Exhaust Gas Recirculation (CCEGR)*. Operation of the CCEGR is the same as the PVS, but is used by a different manufacturer and labeled differently.

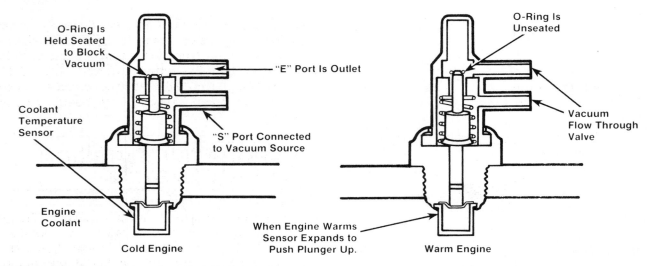

FIGURE 10-21 How the two-port PVS switch works

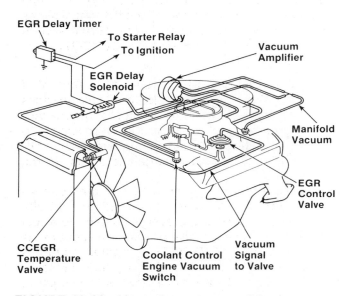

FIGURE 10-22 Venturi vacuum amplifier

- *Venturi Vacuum Amplifier (VVA).* This control is used with some EGR systems so that the carburetor venturi vacuum can control the EGR valve operation (Figure 10-22). Venturi vacuum is more desirable because it is proportioned to the airflow through the carburetor. Since it is a relatively weak vacuum signal, the VVA is used to convert it to a strong enough signal to operate the EGR valve. Manifold vacuum is used in the VVA system for the strength and venturi vacuum for the control signal (Figure 10-23).
- *EGR Delay Timer Control.* Some vehicles are equipped with an EGR delay system, which consists of an electrical timer that is con-

nected to an engine-mounted solenoid. Together, the purpose of the delay timer and solenoid is to prevent EGR operation for a predetermined amount of time after warm engine start-up. On cold engine start-ups, the delay timer is overriden by the TVS and/or PVS valve.
- *Early Fuel Evaporation/Thermal Vacuum Switch (EGR—EFE/TVS).* In most common applications the EFE uses a valve that increases the exhaust gas flow under the intake manifold during cold engine operation through a crossover passage to heat up the incoming air/fuel charge (Figure 10-24). The

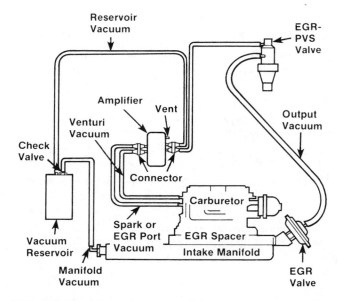

FIGURE 10-23 A typical EGR system with a vacuum amplifier

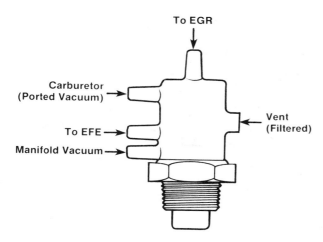

FIGURE 10-24 Early fuel evaporation/thermal vacuum switch

EFE is vacuum operated and controlled by a TVS that applies vacuum to the EFE valve when the coolant temperature is low. Once the engine reaches operating temperature, the TVS will block off vacuum to the EFE and direct it to the EGR valve for EGR operation.

- *Wide-Open Throttle Valve (WOT).* The WOT may be used in some applications where it is desirable to cut off EGR flow at wide-open throttle (Figure 10-25).

The various EGR system controls described represent some of the common controls currently

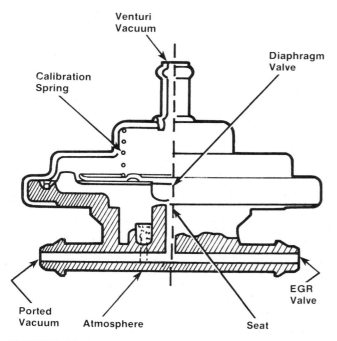

FIGURE 10-25 Wide-open throttle valve

used by automobile manufacturers. Control devices used in the various systems might be labeled differently but actually complete the same function within the EGR system. For further information on EGR system controls it is advisable to consult the service manual that pertains to each vehicle being tested or serviced.

EVAPORATIVE EMISSION CONTROL SYSTEM

The fuel evaporative emission system (Figure 10-26) reduces the amount of raw fuel vapors (HC) that are emitted into the air from the fuel tank and carburetor (if so equipped). While different manufacturers have slight variations and names for their evaporative emission control (EEC) systems they are very similar in operation. Since the first systems were used nationwide in the early 1970s, several refinements have been added (Figure 10-27). Current systems include:

- A special filler design to limit the amount of fuel that can be put in the tank. This provides an air space of 10 to 12 percent of tank volume for heat expansion and for vapors to collect until they can be removed. When filling a gas tank, do not attempt to "top off" the fuel.
- A pressure/vacuum relief fuel tank cap instead of a plain vented cap (Figure 10-28). This type of cap prevents damage to a fuel tank from excessive pressure or vacuum. As fuel cools, it contracts. A partial vacuum is formed in the tank. Atmospheric pressure acting on the walls of a tank could collapse it. A vacuum relief valve opens to allow outside air into the tank. This action balances pressure against the outside surfaces of a tank. When a fuel tank is heated by outside temperatures, fuel and vapors expand, producing pressure that could rupture fuel tank seams. Normally, this pressure buildup is slight and pressure is dissipated throughout a vapor recovery system. However, if hoses or tubes become clogged or accidentally pinched, pressure could damage a tank. A pressure-relief valve opens to allow excessive pressure to escape.
- A vapor separator in the top of the fuel tank (Figure 10-29). This device permits fuel vapors to pass into connecting hoses. The vapor/fuel separator collects droplets of liquid fuel and directs them back into the tank.

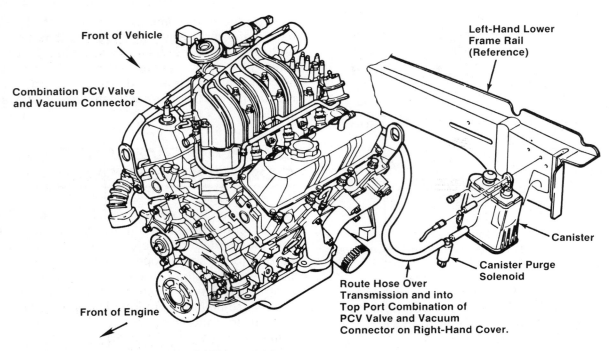

Front of Vehicle

Left-Hand Lower Frame Rail (Reference)

Combination PCV Valve and Vacuum Connector

Canister

Canister Purge Solenoid

Route Hose Over Transmission and into Top Port Combination of PCV Valve and Vacuum Connector on Right-Hand Cover.

Front of Engine

FIGURE 10-26 Typical fuel evaporative emissions system

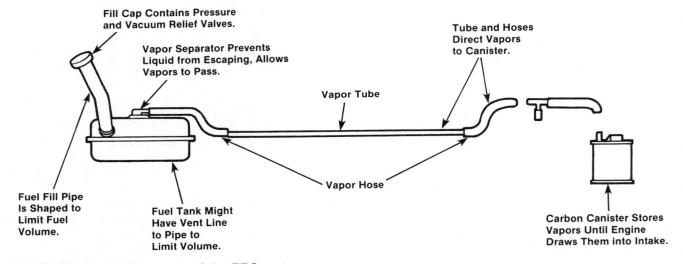

Fill Cap Contains Pressure and Vacuum Relief Valves.

Tube and Hoses Direct Vapors to Canister.

Vapor Separator Prevents Liquid from Escaping, Allows Vapors to Pass.

Vapor Tube

Vapor Hose

Fuel Fill Pipe Is Shaped to Limit Fuel Volume.

Fuel Tank Might Have Vent Line to Pipe to Limit Volume.

Carbon Canister Stores Vapors Until Engine Draws Them into Intake.

FIGURE 10-27 Refinements of the EEC system

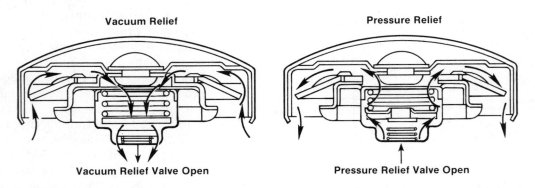

Vacuum Relief

Pressure Relief

Vacuum Relief Valve Open

Pressure Relief Valve Open

FIGURE 10-28 Sealed fuel tank cap

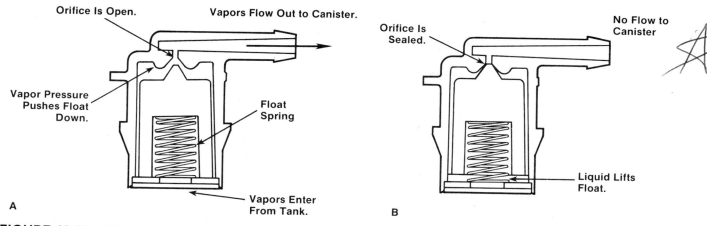

FIGURE 10-29 (A) Normal operation of vapor separator; (B) with liquid in separator

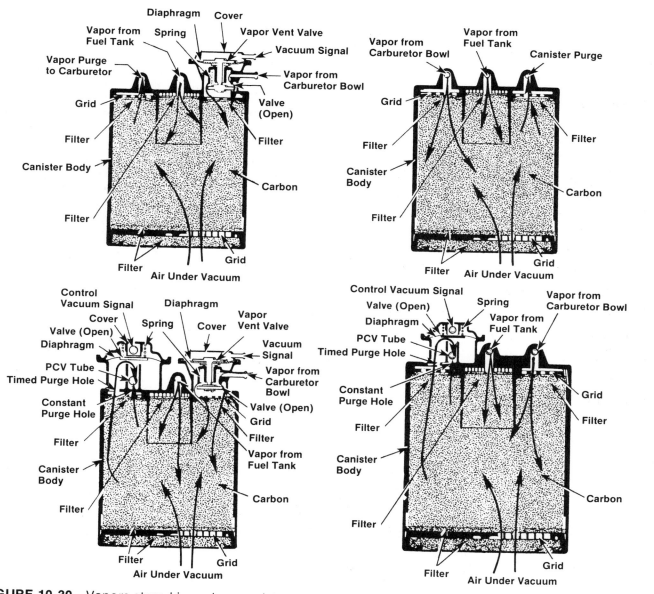

FIGURE 10-30 Vapors stored in carbon canister

- A domed fuel tank in which the upper portion is raised. Fuel vapors rise to this upper portion and collect.
- Check, or one-way, valves keep vapors confined. When an engine runs, vacuum and/or electrically operated valves open. Fresh air and vapors are drawn into the intake manifold.
- Hoses and tubes connect parts of a vapor-recovery system. Special fuel/vapor rubber tubing must be used. Ordinary rubber hoses would rot and crack quickly from the effects of fuel. Metal tubing is used under a vehicle.

The most obvious part of the evaporative emission control system is the charcoal (carbon) canister in the engine compartment. This canister (Figure 10-30) that stores the vapors is located in or near the engine compartment. It is connected to the air cleaner, carburetor, or throttle body so as to purge the vapors when the engine is running.

- Two canisters sometimes used with dual tanks, dual fuel bowls or large capacity tanks.
- Storage medium is activated carbon (charcoal).
- Dust cap and vapor purge connections are interchangeable.
- Canister purge control is usually accomplished with a vacuum-operated or electrically operated valve. The valve opens during specific operating conditions: when the engine can accept the extra fuel vapor.
- Fuel bowl vent control is part of the system on carbureted engines (Figure 10-31). The carburetor bowl is vented to the bowl vent line. Bowl vapors thus flow into the canister when the purge is not operating. Several types of values are used on various engines to control the bowl venting.

In 1978, new government emission standards led to a group of changes called Sealed Housing Evaporative Determination (SHED) or Evaporative Emission SHED System. The principal features of SHED, which have carried through to current models, are:

- Vapors from the gas tank and carburetor fuel bowl are routed to and stored in the carbon canister.
- The carbon canister is purged while the engine is running.

Fuel vapors are absorbed onto the surfaces of the charcoal granules. When the car is restarted, vapors are sucked into the carburetor or intake manifold to be burned in the engine. Canister purging varies widely with make and model. In some instances a fixed restriction allows constant purging whenever there is manifold vacuum. In others, a

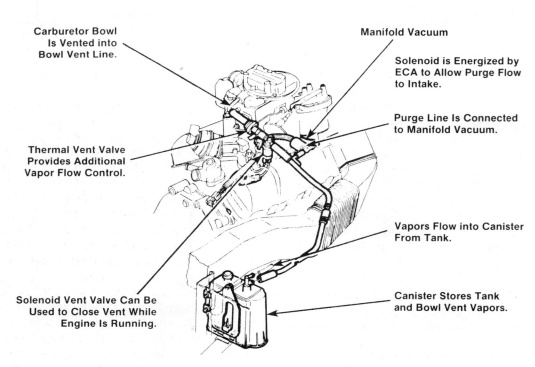

Carburetor Bowl Is Vented into Bowl Vent Line.

Manifold Vacuum

Solenoid is Energized by ECA to Allow Purge Flow to Intake.

Purge Line Is Connected to Manifold Vacuum.

Thermal Vent Valve Provides Additional Vapor Flow Control.

Vapors Flow into Canister From Tank.

Solenoid Vent Valve Can Be Used to Close Vent While Engine Is Running.

Canister Stores Tank and Bowl Vent Vapors.

FIGURE 10-31 Typical carbureted engine vapor control

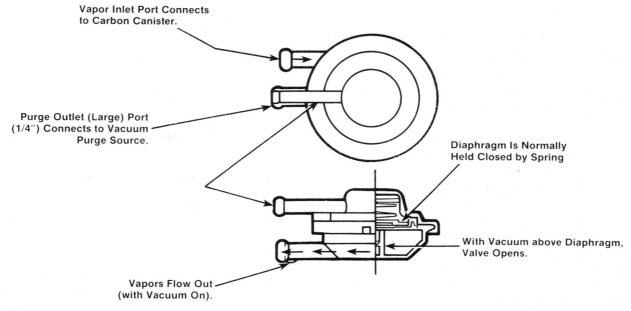

FIGURE 10-32 Canister purge valve operation

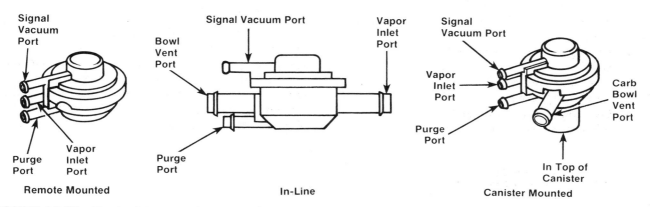

FIGURE 10-33 Typical purge valve mountings

staged valve provides purging only at speeds above idle. Generally, the canister purge valve is normally closed (Figure 10-32). It opens the inlet to the purge outlet when vacuum is applied. Some units incorporate a thermal-delay valve so the canister is not purged until the engine reaches operating temperature. Purging at idle or with a cold engine creates other problems, such as rough running and increased emissions because of the additional vapor added to the intake manifold. Typical purge valve mountings are shown in Figure 10-33.

CARBURETOR FUEL BOWL EMISSION CONTROL

Most carburetor fuel bowls are vented to the air horn. This is necessary with the engine running to maintain the same pressure in the fuel bowl as downstream of the air cleaner. Otherwise fuel metering would vary with pressure changes due to fuel level in the bowl or airflow through the filter.

When the engine is shut down, vapors in the bowl can be emitted through the air filter and air cleaner to the atmosphere. To minimize these emissions, the SHED evaporative control incorporated a large diameter fuel bowl vent line to the carbon canister. It is large enough to provide an easier flow path for vapors to the canister than through the air horn vent (or other external vent).

The evaporative fuel bowl vent is tied into the canister purge line. It directs fuel bowl vapors into the canister with the engine off when the canister purge is not operating. If the system has an in-line or canister-mounted purge valve, the valve has a port for the bowl vent connection (Figure 10-34).

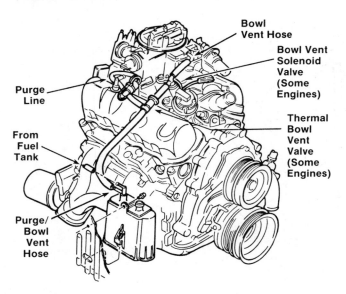

Bowl Vent Hose

Bowl Vent Solenoid Valve (Some Engines)

Thermal Bowl Vent Valve (Some Engines)

Purge Line

From Fuel Tank

Purge/ Bowl Vent Hose

FIGURE 10-34 In an evaporative fuel bowl vent with an inline or canister-mounted purge valve, the valve has a port for the bowl vent connection.

There are two general types of fuel bowl vent valves:

1. *Vacuum Bowl Vent Valve.* The vacuum-operated bowl vent valve is a normally open valve, which closes when the engine is running. Manifold vacuum closes the valve to prevent back suctioning of the carburetor bowl, as illustrated in Figure 10-35.

2. *Vacuum/Thermal Bowl Vent Valve.* The vacuum/thermal-operated bowl vent valve controls the flow of fuel vapors between the carburetor and the carbon canister. The valve responds to temperature and vacuum to provide purge control, as shown in Figure 10-36.

Fuel bowl vent vapor line solenoids or valves are employed to prevent vapor back flow from the canister when the engine is off. The canister purge shutoff valve (CSOV) functions identically to the vacuum

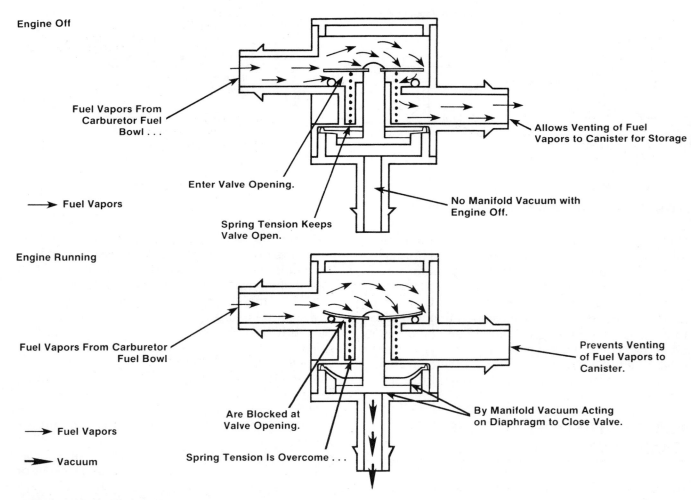

Engine Off

Fuel Vapors From Carburetor Fuel Bowl . . .

⟶ Fuel Vapors

Enter Valve Opening.

Spring Tension Keeps Valve Open.

Allows Venting of Fuel Vapors to Canister for Storage

No Manifold Vacuum with Engine Off.

Engine Running

Fuel Vapors From Carburetor Fuel Bowl

⟶ Fuel Vapors

⟶ Vacuum

Are Blocked at Valve Opening.

Spring Tension Is Overcome . . .

Prevents Venting of Fuel Vapors to Canister.

By Manifold Vacuum Acting on Diaphragm to Close Valve.

FIGURE 10-35 Valve is closed by manifold vacuum.

350

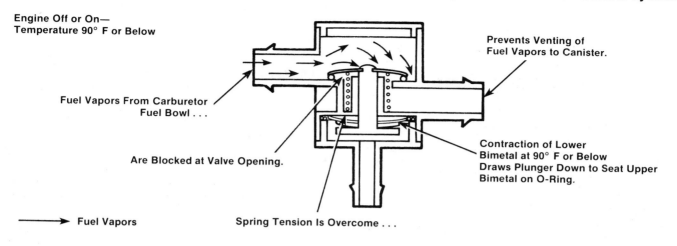

Engine Off or On—
Temperature 90° F or Below

Fuel Vapors From Carburetor
Fuel Bowl . . .

Are Blocked at Valve Opening.

Prevents Venting of
Fuel Vapors to Canister.

Contraction of Lower
Bimetal at 90° F or Below
Draws Plunger Down to Seat Upper
Bimetal on O-Ring.

Spring Tension Is Overcome . . .

⟶ Fuel Vapors

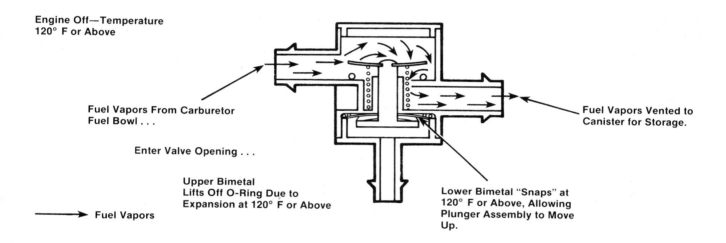

Engine Off—Temperature
120° F or Above

Fuel Vapors From Carburetor
Fuel Bowl . . .

Enter Valve Opening . . .

Upper Bimetal
Lifts Off O-Ring Due to
Expansion at 120° F or Above

Fuel Vapors Vented to
Canister for Storage.

Lower Bimetal "Snaps" at
120° F or Above, Allowing
Plunger Assembly to Move
Up.

⟶ Fuel Vapors

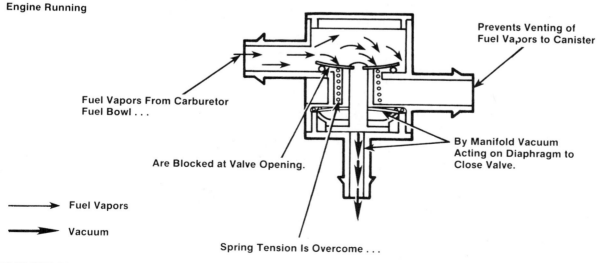

Engine Running

Fuel Vapors From Carburetor
Fuel Bowl . . .

Are Blocked at Valve Opening.

Prevents Venting of
Fuel Vapors to Canister

By Manifold Vacuum
Acting on Diaphragm to
Close Valve.

Spring Tension Is Overcome . . .

⟶ Fuel Vapors

⟶ Vacuum

FIGURE 10-36 Vacuum/thermal bowl vent valve responds to temperature and vacuum, providing purge control.

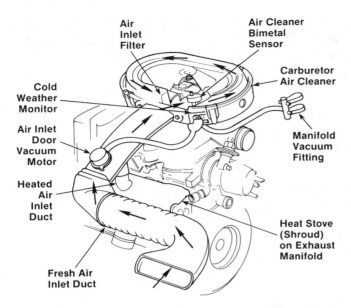

FIGURE 10-37 Typical system with conventional air cleaner

bowl vent valve. It opens and closes the purge line to the intake manifold. The CSOV operates in reverse order of the canister purge valve.

AIR TEMPERATURE EMISSION CONTROL

Hydrocarbon and carbon monoxide exhaust emissions are highest when the engine is cold. However, warm combustion air improves the vaporization of the fuel in the carburetor, fuel injector body, or intake manifold. This makes the fuel burn, therefore reducing HC and CO emissions in the exhaust. Three systems are used on various gasoline engines to heat the inlet air and/or the air/fuel mixture.

1. *Heated Air Inlet.* This is used on all carbureted and central fuel injected engines. The idea originated with MC emission control.
2. *Manifold Heat Control Valves.* These valves are used on engines with a manifold heat riser (see Chapter 3).
3. *Early Fuel Evaporation (EFE) Heater.* Its function is basically the same as the manifold heat control valve, except that a heater warms the air.

HEATED AIR INLET

A heated air inlet control, as previously stated, is used on gasoline engines with carburetion or central

fuel injection. It is not used with turbocharging or ported fuel injection. This system controls the inlet air temperature on the way to the carburetor or fuel injection body. By warming the air as necessary, it reduces HC and CO emissions by improved fuel vaporization and faster warm-up.

The principal components (Figure 10-37) and functions of a conventional air cleaner system are:

- *Heated Air Inlet Duct.* Air that has been warmed by the heat stove (shroud) on the intake manifold to the snorkel of the air cleaner is directed by this duct.
- *Air Inlet Door Vacuum Motor.* A flapper door inside the snorkel to admit manifold heated air or fresh air (or a mixture of both) into the air cleaner is controlled by this motor.
- *Air Cleaner Bimetal Sensor.* This regulates vacuum to the vacuum motor to determine the position of the air door. It is sensitive to air cleaner temperature.
- *Cold Weather Modulator (CWM).* Vacuum to the motor is trapped by the CWM if manifold vacuum drops off due to the throttle opening while the air cleaner is cold.

The air cleaner bimetal sensor is installed in the air cleaner body (or air horn) so that it senses the air cleaner temperature. Depending on the calibration, the sensor can be set to operate at 75 degrees Fahrenheit, 90 degrees Fahrenheit, or 105 degrees Fahrenheit. It is what controls the operating modes shown in Figure 10-38.

Some vacuum control systems use a retard delay valve (Figure 10-39) instead of a cold weather modulator. The difference is that the retard dely valve traps the vacuum for a few seconds when the throttle opens. Its function is the same: to prevent a change in the air door position if vacuum drops off because the throttle opens.

With a remote air cleaner, the functions are the same. As illustrated in Figure 10-40, however, the components are located differently.

- The bimetal sensor is in the air horn assembly.
- The air inlet door vacuum motor and door are in the air cleaner assembly instead of the inlet snorkel.

The bimetal sensor (Figure 10-41) is installed in the air cleaner body. It senses air cleaner temperature and starts to bleed off vacuum near its setting. The sensing element is a bimetal spring, which is linked to a sensing valve. Several temperature settings are available. The sensor is calibrated to provide a specific output vacuum to the air door motor as it warms to its temperature setting. The calibration is based on 16 inches of source vacuum. When

vacuum starts to bleed off, the reduced vacuum allows the air door motor to modulate; that is, to mix fresh air with heated air to maintain the calibrated air temperature.

COLD START-UP
• Strong Manifold Vacuum
• Engine at Fast Idle

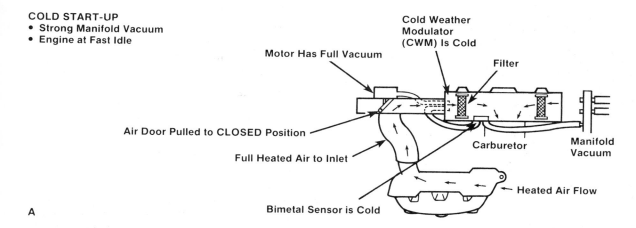

MODULATING PARTIAL WARM-UP
• Strong to Moderate Vacuum
• Air Cleaner Temperature Near Setting

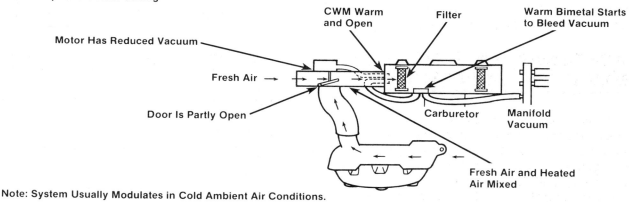

Note: System Usually Modulates in Cold Ambient Air Conditions.

HOT ENGINE/HOT AMBIENT AIR
• Strong to Moderate Vacuum
• Air Cleaner Above Temperature Setting

FIGURE 10-38   Air cleaner bimetal sensor at (A) cold start-up; (B) modulating partial warm-up; (C) hot engine/hot ambient air

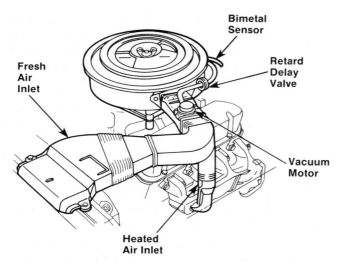

FIGURE 10-39 Conventional air cleaner with retard-delay valve

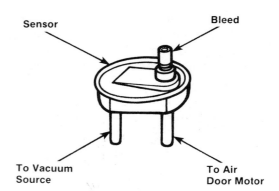

FIGURE 10-41 Air cleaner bimetal sensor operation

MANIFOLD HEAT CONTROL VALVES

The exhaust manifold heat control valve is used to route exhaust gases to warm the intake manifold heat riser when the engine is cold. This heats the air/fuel mixture in the intake manifold and improves ventilation. The result is reduced HC and CO emissions in the exhaust. The two general types of valves are:

1. *Vacuum-Operated* (Figure 10-42). A vacuum-operated valve is used on some V-8 engines. It is bolted between the left-side exhaust manifold and exhaust pipe. The vacuum diaphragm is connected to manifold vacuum through a ported vacuum switch (PVS). On EEC-controlled engines, the vacuum system may also include an electric solenoid-operated vacuum valve.

2. *Thermostat-Operated* (Figure 10-43). A thermostat-operated valve is used mostly on 6-cylinder engines. It has the same function as the vacuum-operated valve. It operates as follows:

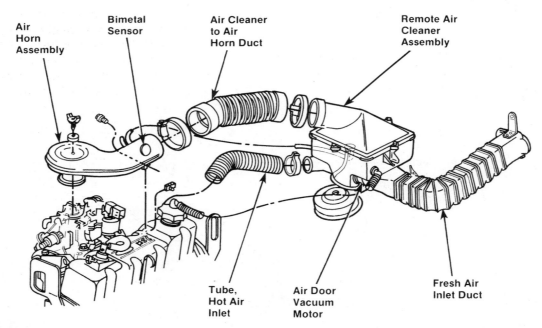

FIGURE 10-40 Remote air cleaner

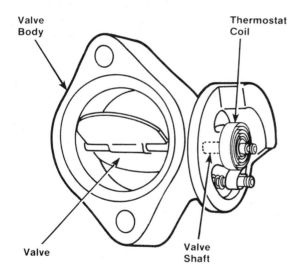

FIGURE 10-42   Vacuum-operated valve

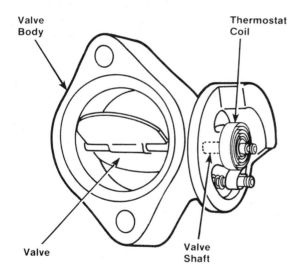

FIGURE 10-43   Thermostat-operated valve

- Cold Engine. The thermostat closes the valve to block exhaust gas flow from the manifold. The gas is forced to flow up through the heat riser and then to the exhaust pipe.
- Warm Engine. The thermostat opens the valve to a position that seals off the heat riser passage. Exhaust gases flow directly to the exhaust pipe.

On some V-8 engines, a more complicated manifold heat control valve, called a power heat control valve (Figure 10-44), is used. It works similarly to the vacuum-controlled EFE, but is designed specifically to work with a minicatalyst as well as to preheat the air/fuel mixture for improved cold engine driveability. A vacuum actuator keeps the power heat control valve closed during warm-up. All right-side exhaust gas is forced up through the intake manifold crossover to the left side of the engine. Then, all exhaust gas from the engine passes through a miniconverter just down from the left manifold. This converter warms up rapidly because it is small and close to the engine. Its rapid warm-up reduces exhaust emissions. As the engine and main converter warm up, a coolant-controlled engine vacuum switch (CCEVS) closes, which cuts vacuum to the actuator and allows the valve to open. Exhaust gas flows through both manifolds into the exhaust system and main converter.

EARLY FUEL EVAPORATION (EFE) CONTROL

The early fuel evaporation (EFE) heater contains a resistance grid that heats the mixture from the primary venturi of the carburetor (Figure 10-45). Its purpose is the same as a manifold heat control valve: to improve vaporization in a cold engine. The heater operates for about the first two minutes, per-

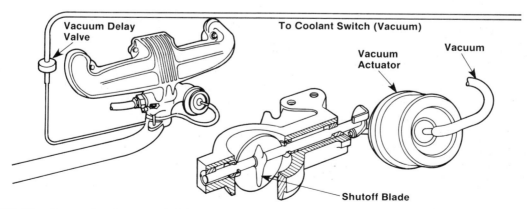

FIGURE 10-44   Power heat control valve

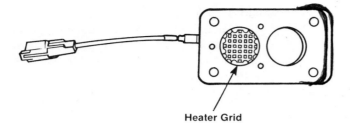

Heater Grid

FIGURE 10-45 EFE heater resistance grid

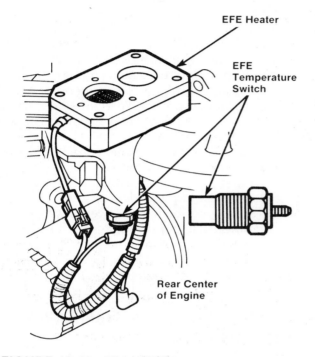

EFE Heater

EFE Temperature Switch

Rear Center of Engine

FIGURE 10-46 The EFE temperature switch is mounted on the engine.

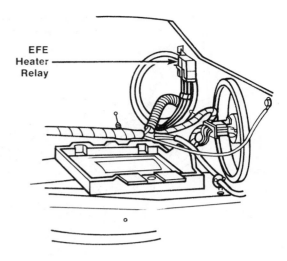

EFE Heater Relay

FIGURE 10-47 The EFE heater really is usually mounted on the body of the vehicle.

powers the EFE heater when the temperature switch is cold and closed. After the engine has warmed up and EFE is no longer needed, the relay is de-energized and the grid heater is turned off. The EFE heater relay is usually mounted on the body of the vehicle (Figure 10-47).

AIR INJECTION SYSTEMS

One of the earliest components used to control HC and CO was the air pump. Formal names of these systems include air injection reaction (AIR) by GM, Thermactor Emission (TE) by Ford, the air guard by American Motors, and air injection system by Chrysler. The name used by auto mechanics in the field was smog pump.

The basic air injection system, regardless of name, forces air from an air pump, through air manifolds, into the cylinder head exhaust ports or exhaust manifold. The fresh air or oxygen promotes after burning of any combustibles remaining in the exhaust gases flowing from the combustion chamber. Carburetor calibrations are generally slightly richer, which also permits improved driveability.

Earlier in this chapter it is stated that the air injection was a post-combustion emission system. Actually the typical air injector has elements of both pre- and postcombustion systems. That is, the air injection system introduces extra air (oxygen) into the exhaust stream of the engine. In most cases, combustion in the cylinders is limited by the amount of oxygen available to sustain burning; so when extra air is introduced into the hot exhaust system, all

mitting leaner choke calibrations for improved emissions without cold driveaway problems.

The basic EFE system is similar from one engine to the next. In addition to the grid heater, EFE has two other important components:

1. *Coolant Temperature Switch.* The EFE temperature switch or solenoid is mounted on the engine, usually on the bottom of the intake manifold (Figure 10-46). The switch is closed when its temperature is below a specified temperature (generally between 130 to 150 degrees Fahrenheit). The switch opens as the engine coolant temperature goes above the specified temperature (see the service manual for the exact temperature).

2. *EFE Heater Relay.* The temperature switch controls the EFE relay or valve. It

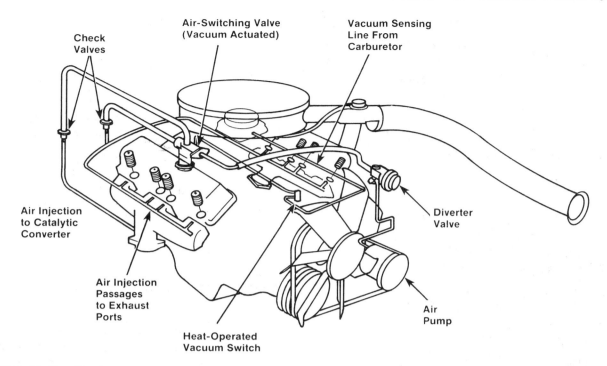

Check
Valves

Air-Switching Valve
(Vacuum Actuated)

Vacuum Sensing
Line From
Carburetor

Diverter
Valve

Air Injection
to Catalytic
Converter

Air Injection
Passages
to Exhaust
Ports

Heat-Operated
Vacuum Switch

Air
Pump

FIGURE 10-48   Engine-driven air pump

the remaining fuel is oxidized, or burned. Since air injection helps the process of combustion, it qualifies as a precombustion device. But the fact that the combustion process is continued in the exhaust system (and requires additional components) qualifies air injection as a postcombustion system. However it is classified, air injection is an effective system that economically reduces HC and CO emissions.

The typical air injector system used in most vehicles consists of a belt-driven vane pump (Figure 10-48); a vacuum-operated diverter valve to vent pump air to the atmosphere during deceleration so the combustion of a rich fuel mixture and oxygen does not cause backfiring; a pressure relief valve that allows excess pump output to escape; a one-way check valve that allows air into the exhaust but prevents exhaust from entering the pump in the event the belt breaks; and the hoses and nozzles necessary to distribute and inject the air (Figure 10-49).

On most air pump-equipped models, the system has the ability to switch airflow from the exhaust manifold to the catalytic converter. During engine warm-up, air is injected directly into the exhaust manifold. But once the engine is warm, the extra air in the manifold would affect EGR operation, so air injection is switched downstream to the converter, where it aids the converter in oxidizing emissions.

A vacuum-operated valve does the actual switching. Vacuum to the valve is controlled by a thermal

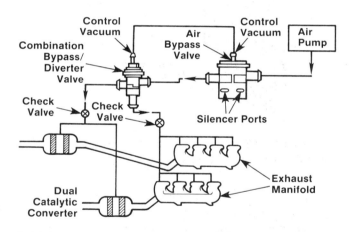

Control
Vacuum

Combination
Bypass/
Diverter
Valve

Air
Bypass
Valve

Control
Vacuum

Air
Pump

Check
Valve

Check
Valve

Silencer Ports

Dual
Catalytic
Converter

Exhaust
Manifold

FIGURE 10-49   Typical managed air system

vacuum switch. When the coolant is cold, it signals the switching valve to direct air to the exhaust manifold. Then when the engine warms to normal operating temperature, the thermal vacuum switch signals the switching valve to reroute the air to the converter. A relief valve in the switching valve vents extra air to the atmosphere. When air pump output rises with engine speed, the air pressure overcomes the relief valve spring pressure.

An antibackfire valve prevents backfire in the exhaust. Two types of antibackfire valves are used, both of which are controlled by intake manifold vacuum. All late-model engines use an air bypass system

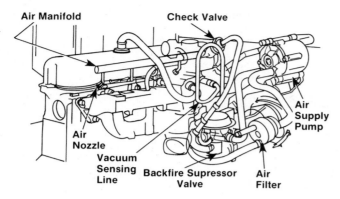

FIGURE 10-50 The deceleration valve should prevent exhaust backfire during deceleration.

FIGURE 10-50 Air injection system with a bypass type of backfire-suppressor valve

(Figure 10-50), which momentarily bypasses or dumps the air pump output at the start of deceleration when manifold (engine) is highest. The diverter or deceleration bypass valve (Figure 10-51) directs airflow through a muffling device, which is usually mounted on the air pump. After one or two seconds, the valve closes and again supplies air to the exhaust manifold.

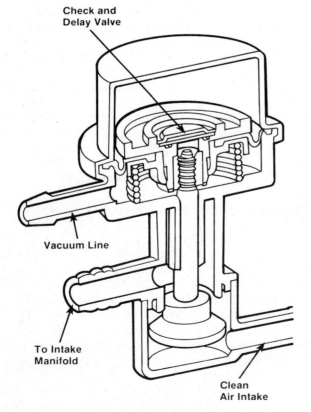

FIGURE 10-51 The deceleration valve should prevent exhaust backfire during deceleration.

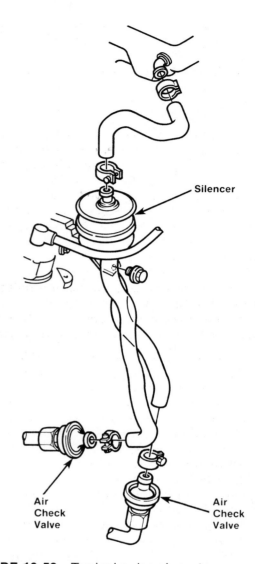

FIGURE 10-52 Typical pulse air system

In some newer systems, there is a vacuum differential valve (VDV), which is used in conjunction with an air bypass valve. The VDV shuts off the signal to the bypass valve momentarily whenever intake manifold vacuum rises or drops sharply. When the vacuum signal is gone, a spring inside the bypass opens a vent port and dumps the air pressure. Under normal vacuum, the vent spring is overriden and air flows into the exhaust ports.

Older model engines and many imports use an air gulp type antibackfire valve, which accomplishes the same task in a slightly different manner. The gulp valve also monitors engine vacuum and, when it gets a strong signal, instead of dumping the air into the atmosphere, it routes pump air into the intake manifold, leaning the mixture enough to prevent unwanted explosions in the exhaust system. Some gulpers do not use pump air. They simply open a passage to the air cleaner on deceleration and let the engine, with its high vacuum, take as deep a breath as it can to lean the mixture.

In some engines in the late 1970s the aspirator air system replaced the mechanical air pump (Figure 10-52). This system draws air into the system from the air cleaner. The aspirator valve, which contains a spring-loaded one-way valve, is located in a hose that runs between the air cleaner and the exhaust manifold. Negative pressure pulses in the exhaust system cause the aspirator valve to open and allow fresh air to mix with the exhaust gases. When positive pulses occur in the exhaust, the one-way aspirator valve closes to prevent the hot exhaust gases from backing up in the hoses.

Several other pulse air systems are in use. In one operation, the system uses the natural pulses already mentioned in this chapter to "pull" air from the air cleaner into the catalytic converters. The typical pulsed air system operates in the following manner:

- An air check valve and silencer (Figure 10-52) are connected in-line between the air cleaner and the catalytic converter.
- When pressure in the exhaust system is greater than the pressure in the air cleaner, the reed valve (Figure 10-53) inside the air check valve closes.
- When the pressure in the exhaust system is less than the pressure in the air cleaner, the reed valve opens and air is drawn into the catalytic converter.
- The incoming oxygen reduces the hydrocarbons and carbon monoxide content of the exhaust gases by continuing the combustion of unburned gases in the same manner as the conventional system (with an air pump).

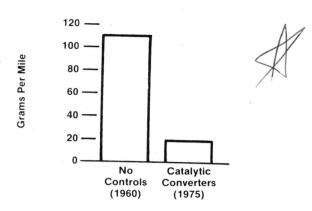

FIGURE 10-54   Reduction in pollutants

CATALYTIC CONVERTERS

Catalytic converters are the most effective devices for controlling exhaust emissions. Their placement in the exhaust system began on new cars in 1975. In that year, a reduction in the allowable emission levels set by the EPA came into effect (Figure 10-54). Until 1975, carmakers had done an effective job of controlling emissions by the use of other systems—auxiliary air injection systems, exhaust gas recirculation systems, and positive crankcase ventilation. But controlling emissions with these systems meant lean mixtures and exotic ignition timing, which often severely penalized power and fuel economy.

When catalytic converters (CC) were introduced (Figure 10-55), much of the emission control job could be taken out of the engine and moved into the exhaust system. This change allowed manufacturers to retune the engine for better performance and improved fuel economy.

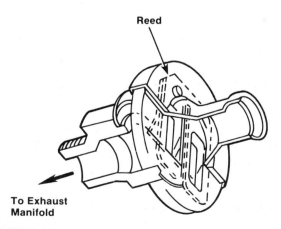

FIGURE 10-53   Typical air check valve

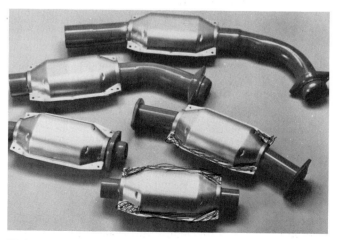

FIGURE 10-55   Catalytic converters

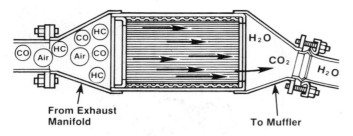

FIGURE 10-56   Conventional catalytic converter

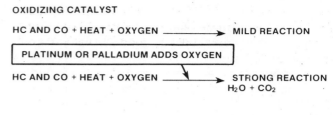

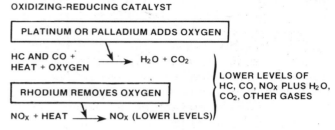

FIGURE 10-57   Effects of (A) two-way oxidizing catalyst and (B) three-way oxidizing catalyst on exhaust gases

A catalyst is a substance that causes a chemical reaction in other elements without actually becoming a part of the chemical change itself and without being used up or consumed in the process. An automatic CC is a device that uses catalysts to cause a change in the elements of the waste exhaust gases as they pass through the exhaust system.

The catalyst elements used in CC are platinum, palladium, and rhodium. These elements are used alone or in combination to change the undesirable CO, HC, and NO_x into harmless water vapor, carbon dioxide, nitrogen, and oxygen (Figure 10-56).

These rare metal elements thinly coat a substrate supported inside the converter shell. The neutral substrate is a ceramic material designed to withstand the high temperature of the exhaust gases and the additional heat that is caused by the chemical changes of the catalytic reaction. It is also designed so that a very large surface—several thousand square yards—is exposed to the exhaust gases.

The NO_x, CO, and HC molecules enter the converter and contact and attach themselves to the rare metals that cover the substrate. When heat is applied, the catalytic process causes the molecules to rearrange themselves into less harmful elements. Following are descriptions of the converter chemical processes:

- *Oxidation Process.* The process through which HC and CO are reduced to relatively harmless byproducts is called *oxidation* (Figure 10-57A). Two-way converters use platinum and palladium as the chemical catalysts. For the converter to produce a chemical reaction, oxygen must be added to the exhaust gases. The extra oxygen is provided by an air pump or a pulse-air system. When the exhaust gases pass through the converter with the oxygen, oxidation takes place, reducing the harmful gases to water and carbon dioxide, which are harmless parts of the natural environment. Platinum and palladium enter into the chemical reaction process, but are not in themselves used up, hence the

reason the converter should not wear out and indeed last the lifetime of the car. (But the converter can be destroyed as described in Chapter 11).

- *Reduction Process.* The process through which the NO_x is reduced is called *reduction* (Figure 10-57B). Reduction is the removal of oxygen from the exhaust gases (actually, from any compound). This action takes place in the converter when the gases pass through the catalysts of platinum and rhodium (some use all three catalysts, i.e., platinum, palladium, and rhodium) reducing the NO_x to harmless nitrogen and oxygen.

The converter itself is located either under the floor of the automobile or in the exhaust manifold (Figure 10-58).

TYPES OF CATALYTIC CONVERTERS

There are two basic designs of catalytic converters:

1. *Monolithic Design.*   This design, used on Chrysler and Ford (Figure 10-59), contains a ceramic honeycomb-shaped substrate, or carrier, on which the precious metal catalyst is deposited. The catalyst is usally platinum and/or palladium. This substrate is housed in a stainless steel casing that is insulated to produce a skin temperature of between 250 and 300 degrees Fahrenheit.

2. *Pelletized Design.*   This design (Figure 10-60), used in GM and AM vehicles, is

wide and flat (looks like a water bottle) and contains a substrate made up of thousands of tin porous beads or pellets that are coated with the same precious metals as those

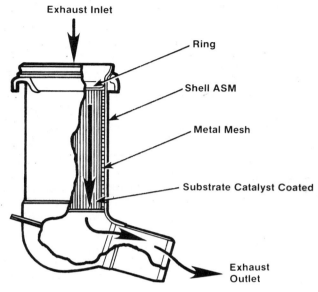

FIGURE 10-58   Manifold catalytic converter

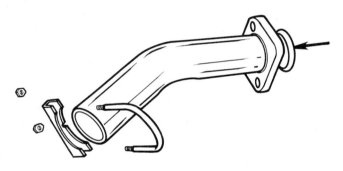

FIGURE 10-59   Monolithic design

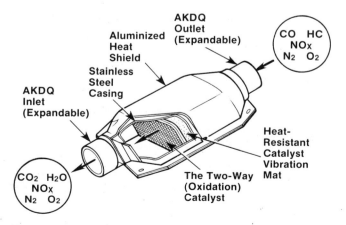

FIGURE 10-60   Pelletized design

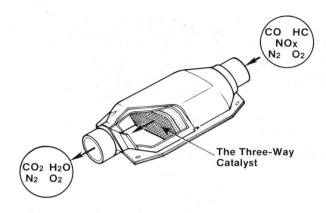

FIGURE 10-61   Dual-bed converter

found on the honeycomb or the monolithic design. This pellet design CC is housed in a stainless steel casing, which is also insulated.

There are three types of catalytic converters used on vehicles:

1. Single-bed, two-way, introduced on vehicles in the mid-1970s
2. Three-way
3. Dual-bed, three-way. The three-way converters were introduced on cars in the late 1970s.

A single-bed converter has all of the catalytic materials in one chamber (Figure 10-60). A dual-bed converter has a reducing chamber for breaking down NO_x and an oxidizing chamber for oxidizing HC and CO. The exhaust passes through the reducing chamber first (Figure 10-61). Additional oxygen is introduced into the oxidizing chamber to make it more efficient. A two-way converter oxidizes HC and CO; a three-way converter oxidizes HC and CO and reduces NO_x (Figure 10-62).

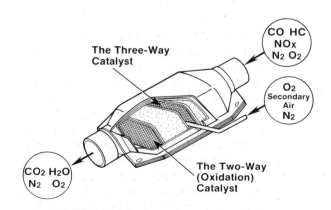

FIGURE 10-62   Three-way converter

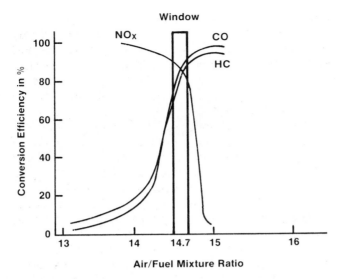

FIGURE 10-63   Air/fuel mixture ratio

In the three-way converter, the air/fuel ratio must be maintained by the computer controls in a closed-loop, feedback system for both carbureted and fuel-injected systems. It is important to understand that prior to computer controls, there was no way to reach and maintain 14:7:1 fuel ratios (Figure 10-63). The best air/fuel ratio for controlling HC, CO, and NOx emissions and, at the same time, provide good engine performance, driveability, and fuel economy is called a *stoichiometric mixture*. For this to happen, the air/fuel ratio must be maintained (see Chapter 11).

 SHOP TALK

A stoichiometric air/fuel ratio provides just enough air and fuel for the most complete combustion, resulting in the least possible total amount of leftover combustible materials: oxygen, carbon, and hydrogen. This maximizes the production of CO_2 and H_2O and minimizes the potential for producing CO, HC, and NOx. It also provides good driveability and economy. The most power is obtained from an air/fuel ratio of about 12.5 to 1, although the best economy is obtained at a ratio of about 16 to 1. Air/fuel ratios this far from stoichiometric, however, are not compatible with the three-way catalytic converter. The leaner mixtures also increase engine temperature as a result of a slower burn rate.

The closed-loop fuel control system with its computer, sensors, and actuators controls the air/

fuel ratio under all driving conditions (see Chapter 12). The system is continually checking and rechecking very rapidly for the entire time the vehicle is operated. This action is correcting the fuel mixture about 10 times per second. Keep in mind that there are times during vehicle operation when the system is in open loop. You have to know whether or not the system is in closed loop or open loop when diagnosing or making tests. The shop service manual, as usual, is of utmost importance for the modern repair technician.

It is important to keep in mind that CC cannot regulate oxidation or reduction. The process depends on how much air and fuel are available. The more efficiently the other systems control emissions coming out of the engine, the better and longer the converter operation will be.

Some original equipment manufacturers find that the catalytic converter is more efficient if it is located close to the exhaust manifold where the exhaust temperatures are higher. These higher temperatures age the catalyst faster, reducing its service life. By installing a secondary catalytic converter (Figure 10-64) farther from the exhaust manifold, the original equipment manufacturers combine the fast light-off time with longer durability. In addition, some manufacturers find that dividing the catalyst volume into two containers (converters) divides the concentration of the high temperatures emitted from the converters and dissipates this heat faster. When the undercarriage space confinement does not permit the installation of one large converter, two smaller converters can be installed. Often the front converters will be three-way units, which reduce the NOx and inject air into the exhaust pipe. The rear converter is used to improve the oxidation, acting as a two-way unit. Figure 10-65 illustrates typical methods connecting a catalytic converter in an in-line and V-8 engine arrangement.

As stated in Chapter 2, one major element of the converter system is the fuel inlet restrictor (Figure 10-66). This restrictor is used to prevent the use of leaded gasoline. Remember that the use of only unleaded fuel is necessary for the proper operation of the converter.

ENGINE MODIFICATION PROGRAMS

One of the automotive industry's best programs for the control of vehicle emissions is engine modifications. These include engine and intake manifold modifications that have been made by the manufacturers to improve fuel vaporization, intake gas flow

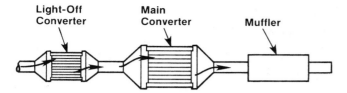

FIGURE 10-64 Separate catalytic converters

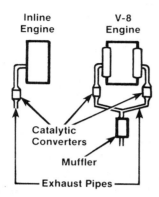

FIGURE 10-65 Typical methods of connecting inline and V-8 engines

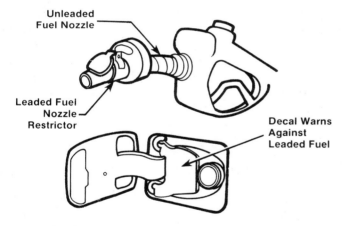

FIGURE 10-66 Fuel inlet restrictor

and turbulence, and exhaust gas flow. In addition, exhaust gases can be mixed with an intake charge to decrease burning temperatures. Also, revised timing patterns and control modifications have greatly reduced CO, NO_x, and HC emissions.

ENGINE AND INTAKE MANIFOLD MODIFICATIONS

Modern engines are being made, including reducing compression ratio, to reduce combustion pressure and temperatures, valve-timing variations, and combustion chamber design. Combustion

chamber alterations include eliminating flat quench surfaces to show burning and keep NO_x down. By eliminating the cooler surfaces of quench areas, HC emissions are dropped. Piston design is important. If the rings are higher or the piston top is tapered toward the top ring, there is less volume in which unburned HC can hide, and this further reduces HC.

Keeping the chamber hot reduces deposits, which provide still more places for unburned HC to hide. All these are minimal changes, but the game of reducing emissions is made up of these. Keeping compression ratio down reduces peak pressures, which reduces NO_x. Actually, the reducing of emissions by engine modification is a delicate balancing act. The major complexity facing engine designers is the engine's tendency to produce more NO_x whenever HC and CO are being reduced, and vice versa.

The service technicians can keep up-to-date on engine and manifold modifications by reading automotive trade publications and studying manufacturers' service manuals.

IGNITION TIMING MODIFICATIONS

Carburetors are set up with leaner mixtures in the part-throttle range. Mechanical limiters are placed on idle-mixture adjustment screws to prevent excessively rich idle mixtures. Dashpots and throttle retarders, plus special idle setting solenoids, are used. And the choke is designed to come off very quickly after the engine starts. Inlet air is heated to ensure good fuel vaporization and distribution.

Decelerations create very high manifold vacuum unless special controls are used. With a closed throttle, so much exhaust is sucked back into the intake manifold that the air/fuel mixture is diluted (leaned). Borderline firing occurs, causing missing and, consequently, high emission concentrations of unburned hydrocarbons.

Several controls can be used singly or in combinations:

- Shut off the fuel flow so there will be no unburned hydrocarbons because no fuel will be entering the manifold. This creates a bump when the fuel is turned back on near normal idling manifold vacuum.
- The fuel cutoff solenoid helps prevent dieseling after the engine is shut off. A spring-loaded needle valve at the end of the solenoid extends to block the idle discharge passage when the ignition is turned off and the solenoid deenergizes. When the ignition is turned on, the solenoid energizes and the

363

needle valve retracts to uncover the idle discharge passage. If the solenoid or wiring fails, the idle discharge passage will remain covered after starting and the engine will not idle.

- Supply a richer mixture to ensure burning. This was done by a deceleration valve and special deceleration air/fuel feed circuit. In fact, as many as three systems are used on some engines to control emission during deceleration. The coasting air valve (CAV) and the air switching valve (ASV) work together to inject additional air into the carburetor during deceleration. Also, a deceleration spark advance system (DSA) advances ignition timing by switching full manifold vacuum to the vacuum advance diaphragm during deceleration.
- Retard throttle closing to avoid high vacuum buildup. Although commonly used, it reduces the braking effect that would have been obtained from the engine during deceleration with a closed throttle.
- Due to the extremely lean idle mixtures it has been necessary to increase idle speeds from the 500 rpm range to a 700 to 750 rpm range. The main function of the throttle opener (idle-up) is to increase idle speed when an electrical or A/C load is placed on the engine. When idle speed is too low, the engine

can vibrate and shudder. Emissions also increase.

Distributors are set up with more retard at idle and part-throttle. Various switches, valves, and other controls are used to provide advance or retard as required to meet emission. Spark advance systems, for example, have been on distributors since the earliest gasoline engines. Until electronic engine controls, most distributors have had both a vacuum-operated and a mechanical (centrifugal) spark advance mechanism. About the time of improved combustion refinements to the engine to reduce exhaust emissions, various controls began to be used on the vacuum advance for better emission controls (Figure 10-67):

- In general, retarding (or delaying) the spark timing improves combustion by increasing combustion temperature and allowing leaner fuel mixtures.
- Some controls have bettered the overall emissions level by causing a faster warm-up of a cold engine. This has allowed reducing the choke come-off time so that the engine runs lean faster.
- Retarding the spark during deceleration or curb idle also reduces richness and improves emission levels.

Spark retard is an effective means of reducing HC. This is because exhaust temperatures are in-

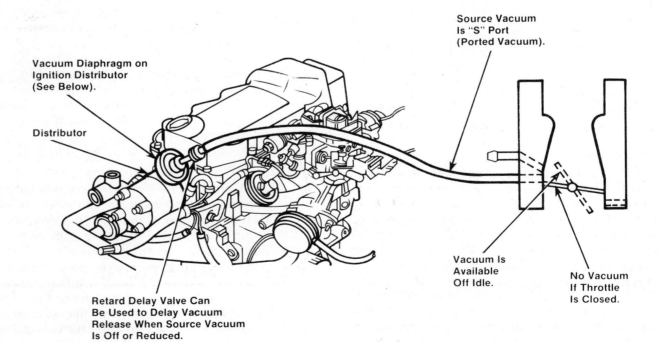

Vacuum Diaphragm on Ignition Distributor (See Below).

Distributor

Source Vacuum Is "S" Port (Ported Vacuum).

Vacuum Is Available Off Idle.

No Vacuum If Throttle Is Closed.

Retard Delay Valve Can Be Used to Delay Vacuum Release When Source Vacuum Is Off or Reduced.

FIGURE 10-67 Ported vacuum spark advance control

creased so the hydrocarbons are burned completely. One of the most effective spark retard delay systems employs a thermal vacuum valve (TVV). When the engine coolant is cold, the TVV remains closed. The only vacuum path to the distributor is through the spark retard delay valve. On hard acceleration, the manifold vacuum decreases instantly, and the spark retard delay valve traps vacuum in the distributor advance for 4 seconds. Cold driveability is improved by gradually retarding the vacuum advance. Warm engine coolant opens the TVV and supplies manifold vacuum directly to the vacuum advance without going through the spark retard delay valve (Figure 10-68).

New engine modification controls to check emission problems are being tested by automobile engineers. Of course, no engine can use them all. Check the vehicle's service manual to see what ones are used on the engine being serviced. See early chapters and Chapter 12 for further information on ignition timing controls.

All major auto manufacturers have some form of engine modification:

- American Motors—Engine Mod System
- Chrysler—Cleaner Air System (CAS)
- Ford—Improved Combustion System (IMCO)
- General Motors—Controlled Combustion System (CCS)

DIESEL ENGINE EMISSION SYSTEM

As mentioned in Chapter 2, one of the advantages of diesel fuels is their low emission levels. Many diesel engine systems include a PCV control and catalytic converter, but the most important and, in some cases, the only emission control, is EGR. Figure 10-69 illustrates the GM V-6 diesel engine, which operates, according to the service manual, in the following manner.

The vacuum regular valve (VRV), which is mounted on the injection pump, modulates vacuum from the vacuum pump. At idle, vacuum is highest; at wide-open throttle, vacuum decreases to zero. At idle, the EGR valve is open to its maximum; at wide-open throttle, the EGR valve is closed. A vacuum modulator valve (VMV) further modulates the amount of the EGR valve opening: the VMM allows for a vacuum increase in the EGR valve as the throttle is closed (up to the VMV switching point). Between the VRV and torque converter clutch-operated (TCC) solenoid, a response vacuum reducer (RVR) is used to allow the EGR valve to change

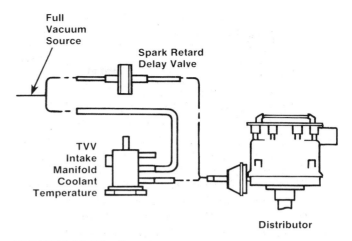

FIGURE 10-68 Spark retard delay valve

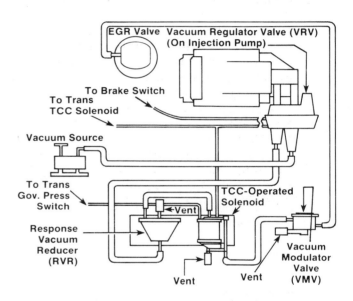

FIGURE 10-69 Diesel engine EGR system as used on a GM V-6 diesel engine

position quickly as the position of the throttle is changed. To block vacuum to the EGR valve whenever the TCC is applied, a solenoid is placed between the RVR and VMV. The solenoid is fed 12 volts from the TCC switch portion of the VRV and is grounded through the transmission's governor pressure switch.

REVIEW QUESTIONS

1. Which of the following systems help to control harmful emissions from vehicle exhaust?
   a. positive crankcase ventilation system
   b. air injection reactor system

c. controlled combustion system
d. all of the above

2. What component, available on vehicles since 1975, provides a means for oxidizing the carbon monoxide and hydrocarbon emissions in the engine exhaust?
   a. EGR
   b. catalytic converter
   c. PCV
   d. NOx

3. Which of the following is generally not classified as a type of precombustion control system?
   a. air injector system
   b. PCV
   c. EGR
   d. evaporative control system

4. Leakage into the engine crankcase is called _____.
   a. exhaust
   b. water vapor
   c. sludge
   d. blowby

5. Blowby that condenses in the crankcase and reacts with the oil forms _____.
   a. hydrocarbons
   b. NOx.
   c. sludge
   d. carbon monoxide

6. Combustion gases that enter the crankcase are removed by a(n) _____.
   a. PCV system
   b. evaporative control system
   c. EGR
   d. catalytic converter

7. Exhaust gas recirculation systems reduce what harmful exhaust emission?
   a. carbon monoxide
   b. hydrocarbons
   c. NOx
   d. all of the above

8. Vacuum used to operate and control the EGR valve is supplied by _____.
   a. the spark vacuum port
   b. a port in the carburetor body
   c. the venturi vacuum port
   d. all of the above

9. What component of an evaporative emission control system, located in or near the engine compartment, stores raw fuel vapors until it is purged?
   a. catalytic converter
   b. charcoal canister
   c. blowby
   d. fuel bowl

10. Which of the following systems does not heat the inlet air and/or air/fuel mixture in order to reduce HC and CO emissions in the exhaust?
    a. manifold heat control valves
    b. heated air inlet
    c. early fuel evaporation heater
    d. evaporative emission control

11. A smog pump is used mainly on older model vehicles with what type of emission control system?
    a. air injection system
    b. evaporative emission control
    c. PCV system
    d. EGR system

12. What is a substance that causes a chemical reaction in other elements without being affected in any way itself?
    a. blowby
    b. converter
    c. catalyst
    d. oxidizer

13. What type of fuel is necessary for proper operation of the catalytic converter?
    a. diesel fuel
    b. regular gasoline
    c. kerosene
    d. unleaded fuel

14. When inspecting an emission control system, Technician A looks for the catalytic converter under the floor of the vehicle. Technician B looks in the exhaust manifold. Who is right?
    a. Technician A
    b. Technician B
    c. Both A and B
    d. Neither A nor B

15. The most important and sometimes the only emission control on diesel engine systems is _____.
    a. an air injector system
    b. exhaust gas recirculation
    c. carburetor fuel bowl emission control
    d. evaporative emission control

CHAPTER ELEVEN

EMISSION CONTROL SYSTEMS SERVICE

Objectives

After reading this chapter, you should be able to:
- Explain the Federal Clean Air Act and how it affects the technician.
- List the information included on a Vehicle Emission Control Information (VECI) Label.
- Explain the information coded into the Vehicle Identification Number (VIN).
- Explain the function of and the diagnosis and service procedures for the positive crankcase ventilation system.
- Explain the function of and the diagnosis and service procedures for the exhaust gas recirculating system.
- Explain the function of and the diagnosis and service procedures for the evaporative emission control system.
- Explain the function of and the diagnosis and service procedures for air temperature emission controls.
- Explain the function of and the diagnosis and service procedures for the air inspection system.
- Inspect and replace a catalytic converter.
- Explain the operation and uses of the exhaust emission analyzer.
- Explain the operation of the spark advance system.

The Federal Clean Air Act makes the repair technician responsible for the emission control systems. The law requires service technicians to restore emission control systems to their original design. The law prescribes penalties for repair shops and technicians who alter emission control systems or fail to restore them to their original design.

- Damaged or faulty parts must be replaced with good parts. Eliminating damaged parts to avoid replacement parts is against the law.

- Using damaged or faulty parts that prevent proper operation of the emission control system is also against the law.
- Proper repairs of the emission control system must be made to manufacturers' specifications.
- All replacement parts for emission control systems must satisfy the original design requirements of the manufacturer.

VEHICLE EMISSION CONTROL INFORMATION LABEL

To help the repair technician in servicing the emission system, vehicle manufacturers are required to install Vehicle Emission Control Information (VECI) labels (Figure 11-1). The labels, located in the engine compartment, contain important emission specifications and setting procedures, as well as a vacuum hose schematic with emission components identified. The VECI label should be checked for up-to-date information whenever the engine or emission systems are serviced.

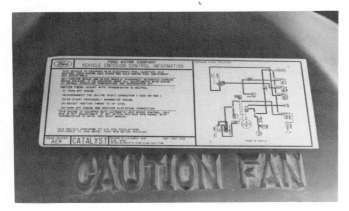

FIGURE 11-1 Typical VECI decals

Information on a VEC label includes

- Engine size
- Label code
- Adjustment procedure
- Engine adjustment specifications
- Area of certification
- Label part number
- Emission component and vacuum hose schematic
- Evaporative emission system
- Exhaust emission system

Information about the exhaust emission system is further broken down into a letter/number code that identifies the following:

- Certification year
- Division
- Displacement
- Vehicle class and standards
- Fuel metering
- Catalyst description
- Engine family suffix code (describes emission system)
- Check sum digit

Another important bit of emission information can be found on the Engine Code Information label (Figure 11–2). This label displays the engine calibration number which is frequently used to determine proper engine application for Technical Service Bulletins. Also included are Emission Control Application Charts for all the vehicles. These charts describe the types of fuel, ignition, EGR, Thermactor, idle speed control, and electronic engine controls used on all vehicles.

Vacuum hoses on modern emissions systems are usually color coded (Figure 11–3). Decals showing the proper hose routing and connection are required by law on modern cars and can be very helpful in reinstallation.

VEHICLE IDENTIFICATION NUMBER

Another important source of information for the service technician to know, especially when ordering parts, is the Vehicle Identification Number (VIN), which is the vehicle's serial number. By law, the VIN must appear inside the windshield on the driver's side.

The car service manual has a diagram showing where the engine production code is found. Usually this code shows as a letter or group of letters in the VIN. The adoption of the standard 17-digit code of

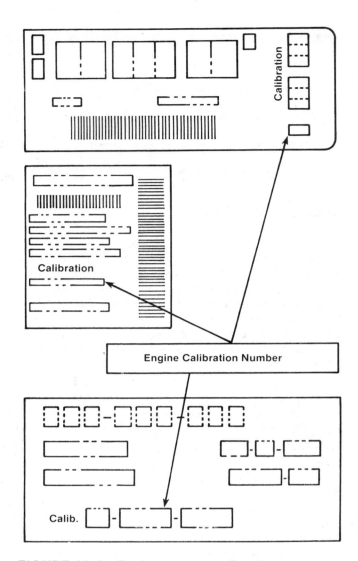

FIGURE 11–2   Typical engine calibration tabs

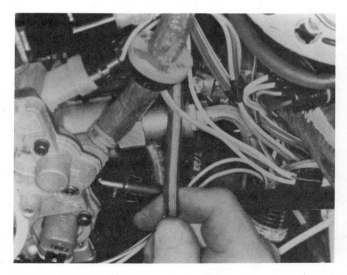

FIGURE 11–3   Color-coded vacuum hoses

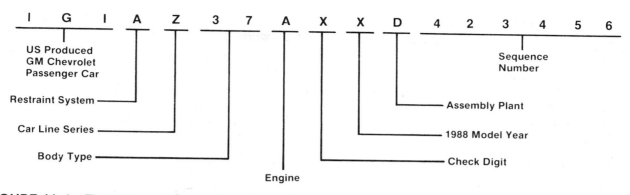

FIGURE 11-4 The VIN contains useful information for the technician.

letters and numbers is mandatory beginning with 1981 vehicles. The standard VIN (Figure 11-4) of the United States National Highway Transportation and Safety Administration, Department of Transportation is being used by all manufacturers of vehicles both domestic and foreign.

The VIN number code includes the following information:

- Country of origin
- Manufacturer
- Make and type
- Series code
- Body types
- Engine codes
- Check digit
- Model year

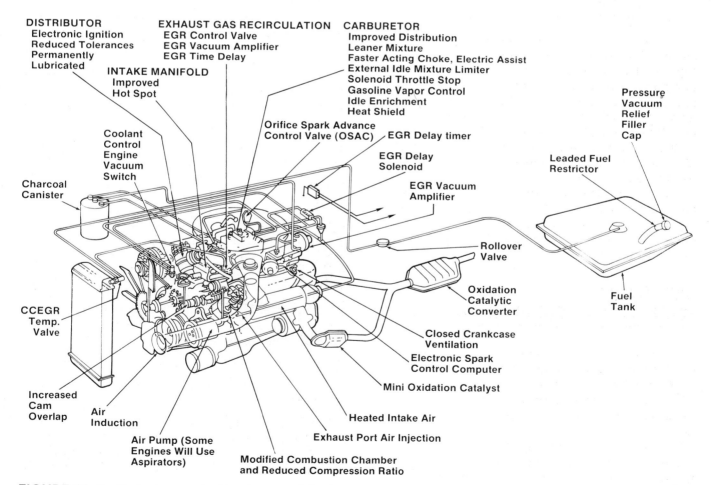

FIGURE 11-5 Emission control systems and devices

TABLE 11-1: EMISSION SYSTEM MAINTENANCE INTERVALS

Maintenance Services	Maintenance Intervals Time or Mileage, Whichever Occurs First*					
	4/6	12/12	24/24	36/36	48/48	60 Mo.
Air Cleaner Element			R		R	
Thermostatically Controlled Air Cleaner	1	1	1	1	1	1
Choke—For Free Operation	1	1	1	1	1	1
Bolts, Carburetor to Manifold—Torque	A					
Filter—Carburetor or Fuel Pump		R	R	R	R	R
Spark Plugs Unleaded Fuels		R	R	R	R	R
Spark Plugs Leaded Fuels	R 6,000-Mile Intervals					
Engine Compression			1		1	
Distributor Points, Cam Lubricator		1	R	1	R	1
Distributor Cap—Clean			1		1	
Spark Plug Wires			1	1	1	1
Timing, Points (Dwell), Idle Speed	A	A	A	A	A	A
CEC Valve or Vacuum Advance Solenoid & Hoses		1	1	1	1	1
Idle Stop Solenoid		1	1	1	1	1
ECS Fuel & Vapor Lines—Leaks			1		1	
ECS Canister—For Cracks or Damage			1		1	
Filter—ECS Canister			R		R	
PCV Valve & Filter			R		R	
PCV System—Hose, Connections		1		1		1
AIR Pump & Engine Belts	A		1		1	1
AIR System Hoses & Connections			1		1	
Manifold Heat Valve	1	1	1	1	1	1
Coolant—Level & Protection	1 at 6-Month Intervals		R		R	
Radiator Core Exterior—Clean	1	1	1	1	1	1
Fuel Cap, Tank, & Lines for Leaks			1		1	
Engine Oil	R	4-Month or 6,000-Mile Intervals				
Engine Oil Filter	R	R Every Other Oil Change				
Automatic Transmission Fluid	Check Level at Eng. Oil Changes		R		R	

*Months, Miles in thousands

EMISSION SYSTEM ABBREVIATIONS
ECS: Evaporation Control System
AIR: Air Injection Reactor
PCV: Positive Crankcase Ventilation

MAINTENANCE OPERATIONS
A: Adjust
R: Replace
I: Inspect, Correct—Replace If Necessary

- Plant code
- Plant sequence number

As with any other vehicle system, maintenance and service are required on emission controls. The technician must be able to carefully inspect, test, and use proper diagnostic procedures to find the problem. Although the general techniques in this chapter will help to make these tasks easier, remember to always follow the manufacturer's recommendations and step-by-step procedures given in the appropriate service manual. Never intentionally disconnect, bypass, or disable any emission control devices. Such practices are prohibited by law and can result in fines and/or loss of licenses or certifications. In other words, tampering with emission control systems can cause a licensed or certified mechanic to lose his or her license or certification and be fined.

In many states, vehicles must pass a strictly monitored emission inspection. Failure to pass means that the vehicle cannot be registered. In California, for example, tests and inspections are performed regularly to detect high emission levels and/or tampering with controls. Emission control equipment that has been tampered with must be returned to OE condition.

Specific emission control devices (Figure 11-5) that should be inspected and checked include:

- PCV systems
- EGR systems
- EMC systems
- Thermostatically controlled air cleaner
- Air injector system
- Spark timing control system
- Catalytic converter

Emission control systems and devices are often part of other systems and therefore are not serviced in isolation. All systems involved with emission control and control devices should be considered when diagnosing and servicing these items. For information on emission control system diagnosis and service, refer to chapters 3, 4, 5, and 6.

MAINTENANCE SCHEDULE

A service technician can help customers keep their cars efficient in both performance and emission control by recommending scheduled maintenance or other needed work shown in Table 11-1.

POSITIVE CRANKCASE VENTILATION

As described in Chapter 10, a positive crankcase ventilation system is used to provide more complete scavenging of crankcase vapors. Fresh air from the air cleaner is supplied to the crankcase, mixed with blowby gases, then passed through a PCV valve into the intake manifold (Figure 11-6). The primary control is through the PCV valve, which meters the flow at a rate depending on the manifold vacuum. In other words, for a PCV to perform prop-

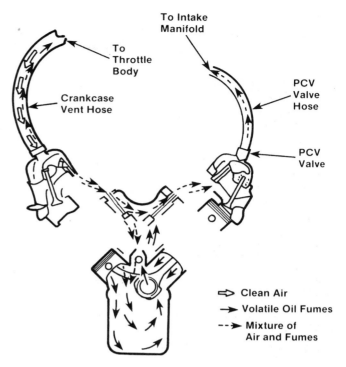

FIGURE 11-6   PCV system flow

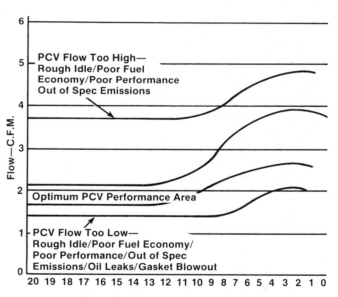

FIGURE 11-7   PCV valve performance

erly, the PCV valve must meter the system of each engine to within specific flow curves (Figure 11-7).

The specific flow curves vary from engine to engine and are critical. For this reason, PCV valves are not interchangeable. A PCV valve that allows excess crankcase vapor and airflow will create an air/fuel mixture that has too much air, causing a rough idle and poor performance. If the carburetor is adjusted to smooth out the idle, poor fuel economy will result. On the other hand, if the PCV valve is clogged and restricts the vapor and airflow to below the desired rate, the air/fuel mixture will contain too much fuel, resulting in reduced fuel economy and poor performance. When this happens, some of the crankcase vapors are drawn back through the air filter, causing the filter element to become loaded with oil.

To maintain idle quality, the PCV valve restricts the flow when the intake manifold vacuum is high. If abnormal operating conditions occur, the system is designed to allow excessive amounts of blowby gases to flow back through the crankcase vent tube into the air inlet duct to be consumed by normal combustion. When vacuum is moderate, during part-throttle acceleration for example, the plunger is pulled to a center position and meters a maximum amount of blowby into the engine. As vacuum increases during idle or constant-speed driving, the plunger is pulled farther into the bore, restricting the amount of blowby that can enter the engine. This helps keep the air/fuel mixture from being upset, which could lead to rough idling or stalling. Finally, at wide-open throttle, low vacuum allows the valve to partially or fully close. In this case, excess blowby is routed to the air cleaner and drawn into the engine through the carburetor.

PCV SYSTEM DIAGNOSIS AND SERVICE PROCEDURES

No adjustments can be made to the PCV system. Service of the system involves a careful inspection, operation, and replacement of faulty parts. When replacing a PCV valve, match the part number on the valve with the vehicle maker's specifications for the proper valve. If the vehicle cannot be identified, refer to the part number listed in the manufacturer's service manual.

PCV System Inspection

The first step in PCV servicing is a visual inspection. As shown in Figure 11-8, the PCV valve can be located in several places. The most common location is in a rubber grommet in the valve or rocker arm cover (Figure 11-8A). Or it can be installed in the middle of the hose connections (Figure 11-8B), as well as installed directly in the intake manifold (Figure 11-8C).

Once the PCV valve is located, make the system inspection by using the following procedure:

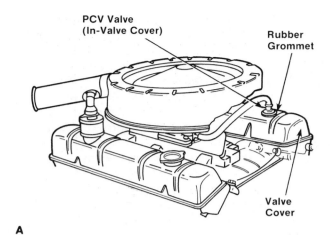

A

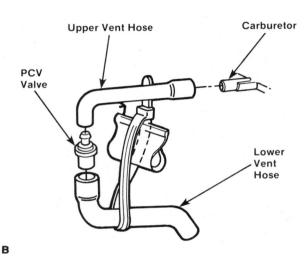

B

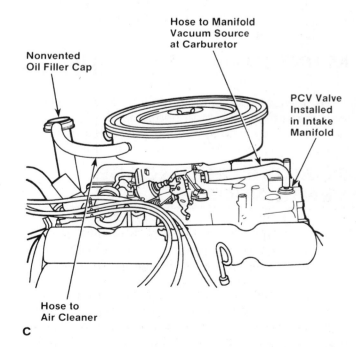

C

FIGURE 11-8   Various locations of the PCV valve

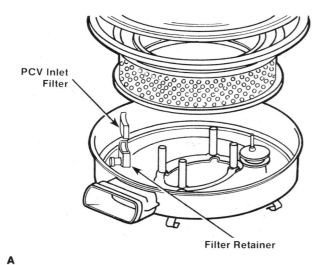

PCV Inlet
Filter

Filter Retainer

A

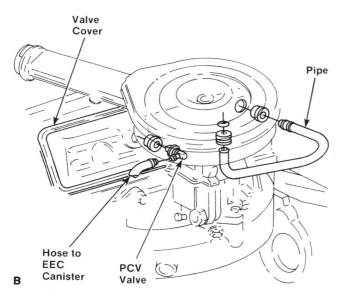

Valve
Cover

Pipe

Hose to
EEC
Canister

PCV
Valve

B

FIGURE 11-9 When inspecting the PCV system, check the air filter in the (A) air cleaner or in the (B) air hose.

1. Make sure all the PCV system hoses are properly connected and that they have no breaks or cracks.
2. Remove the air cleaner and inspect the carburetor or fuel injector air filter. Crankcase blowby can clog these with oil. Clean or replace such filters as described in Chapter 3.
3. Check the crankcase inlet air filter located in the air cleaner (Figure 11-9A), in the air hose (Figure 11-9B), or in the crankcase inlet air cleaner in the valve corner. As the filter does its job, it gets dirty. If it is oil

soaked, it is a good indication that the PCV valve is not working the way it should. When the filter becomes so dirty that it restricts the flow of clean air to the crankcase, it can cause the same problems as a clogged PCV valve. So be sure to check this filter when checking the PCV system and replace it as necessary.

4. Inspect for dirt deposits that could clog the passages in the manifold or carburetor base. These deposits can prevent the system from functioning properly, even when the PCV valve, valve filter, and hoses are not clogged.

Functional Checks of PCV Valve

If an engine is idling roughly, check for a clogged PCV valve or plugged hose. Check the PCV part number to be sure the correct PCV is installed. Replace as required using the following procedure:

1. Disconnect the PCV valve from the valve cover, intake manifold, or hose.
2. Start the engine and let it run at idle. If the PCV valve is not clogged, a hissing will be heard as air passes through the valve.
3. Place a finger over the end of the valve to check for vacuum (Figure 11-10). If there is little or no vacuum at the valve, check for a plugged or restricted hose. Replace any plugged or deteriorated hoses.
4. Turn off the engine and remove the PCV valve. Shake the valve and listen for the rattle of the check needle inside the valve. If the valve does not rattle, replace it.

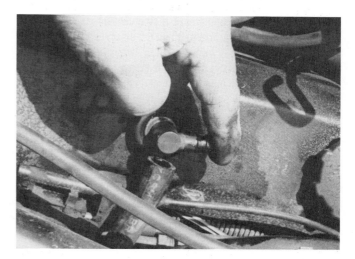

FIGURE 11-10 To check for vacuum, place a finger over the end of the PCV valve.

Proper operation of the PCV system depends on a sealed engine. If oil sludging or dilution is noted and the PCV system is functioning properly, check the engine for a possible cause and correct it to ensure that the system will function as intended.

 SHOP TALK

When diagnosing poor idle condition, check out the PCV system very carefully because it can affect the vehicle's idling. Service technicians must remember that an engine operated without any crankcase ventilation can be damaged. Therefore, it is important to replace the PCV valve and air cleaner breather at proper intervals (Table 11-1). Periodically inspect the hoses and clamps and replace any that show signs of deterioration. An engine that runs too richly during warm-up but runs well if the PCV valve is removed should be checked for crankcase lube oil contamination and the cause of the contamination should be investigated (for example, short trips, fouled plugs, thermostat, or leaking injectors).

Replacing the PVC Valve

If the PVC valve must be replaced, follow a procedure such as this:

1. Disconnect the PCV system hose from the PCV valve by simply pulling the hose from the valve or by removing a hose clamp with pliers or a screwdriver and pulling the hose free of the valve.
2. If the valve is in-line, use pliers to remove it or just pull off the hoses, depending on the hose connection. If the valve is located in a rubber grommet in the engine valve cover, remove it with pliers. Wiggle it back and forth while pulling it from the grommet.

 SHOP TALK

If the grommet comes out with the PCV valve, it can be difficult to replace. Coating the grommet with a silicone lubricant will make it easier to reinstall.

3. Install the new PCV valve by reversing steps 1 and 2. When replacing the valve, be sure to check the part number with the manufacturer's specifications that appear on the emission decal (Figure 11-11).

Replacing PCV Hoses

To replace a PCV hose—both the inlet hose to the crankcase and the valve hose to the carburetor or injectors—proceed as follows:

1. Disconnect the hoses at both ends using either a screwdriver or pliers if the hose is secured with a hose clamp. If no clamp is used, simply pull the hose away from its connection. Then remove the hose.
2. Use only special hose made for PCV and fuel system service as a replacement. Federal regulations prohibit the use of any type not specifically approved for use in PCV systems. Ordinary heater hose or water hose will not withstand the blowby vapors.
3. Using a sharp knife, cut the new hose to the same length as the old one.
4. Install the new PCV system hose, reversing the order of step 1.

Replacing the Crankcase Ventilation Filter Element

To replace this breather filter element in a typical PCV system, proceed as follows:

1. Remove the air filter housing cover.
2. Unsnap the hose from the filter in the air cleaner.
3. Slide the retainer clip from the filter nipple and remove the filter from the air cleaner. This type of filter cannot be cleaned.
4. Insert a new polyurethane filter. Use a new retainer clip supplied with the replacement filter pack or clean and reuse the old retainer clip.

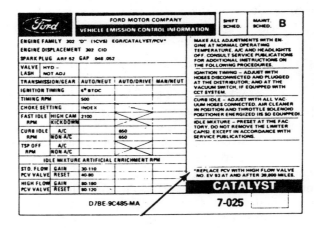

FIGURE 11-11 Replace PCV valves according to manufacturer's specifications.

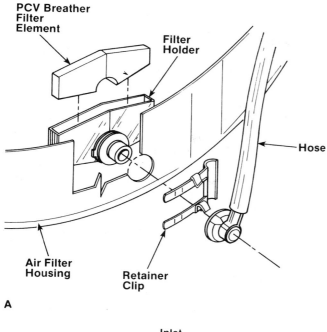

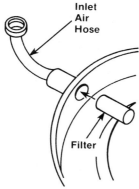

FIGURE 11-12   (A) Polyurethane filter; (B) wire gauze PCV filter

5. Secure the filter in the air cleaner with the retainer clip.
6. Snap the hose on the filter.
7. Install the air cleaner cover.

As shown in Figure 11-12, other PCV air inlet filter arrangements are used in some vehicles. In the system shown in Figure 11-12A, a polyurethane filter is used; it cannot be cleaned and reused. The wire gauge PCV filter used in the unit shown in Figure 11-12B should be cleaned in a proper solvent recommended by the OEM. The recommended replacement for either design is given in the vehicle's service manual.

The PCV system diagnosis given in Table 11-2 is a summary of PCV problems, their possible causes, and remedies for the conditions.

EXHAUST GAS RECIRCULATING (EGR) SYSTEMS

Exhaust gas recirculating (EGR) systems as described in Chapter 10 are used to control emissions of nitrogen oxides (NO_x). The EGR system dilutes the air/fuel mixture with controlled amounts of exhaust gas. Since exhaust gas does not burn, this reduces the peak combustion temperatures. At lower combustion temperatures, very little of the nitrogen in the air combines with oxygen to form NO_x. Most of the nitrogen is simply carried out with the exhaust gases. As shown in Figure 11-13, it is desirable for driveability/performance to have the EGR valve opening (and the amount of gas flow) proportional to the throttle opening. Driveability is also improved on most applications by shutting off the EGR when the engine is started cold, at idle, and at full throttle. Since the NO_x control requirements vary on different engines, there are several different systems with various controls to provide these functions.

EGR SYSTEM DIAGNOSIS AND SERVICE PROCEDURES

The amount of EGR gas flow is carefully calibrated for every engine. If there is too much or too little, it can cause performance problems by changing the engine breathing characteristics. Also, with little EGR flow, the engine can overheat and detonate. Typical problems that show up in ported EGR systems are

- *Rough Idle.* This can be caused by an EGR valve stuck open, a PVS failed open, dirt on the valve seat, or loose mounting bolts. Loose mounting will cause a vacuum leak and a hissing noise. A stuck valve can be diagnosed by a functional test and visual inspection.
- *Surge, Stall, or Will Not Start.* Can be caused by the valve stuck open.
- *Detonation (Spark Knock).* Any condition that prevents proper EGR gas flow can cause detonation. This includes a valve stuck closed, leaking valve diaphragm, restrictions in flow passages, EGR disconnected, or a problem in the vacuum source. Detonation on engines with high spark advance is serious enough to destroy them.
- *Lead Poisoning.* If leaded gas is used improperly, it can cause deposits on the seat

TABLE 11-2: PCV SYSTEM PROBLEMS AND THEIR REMEDY		
Condition	**Possible Cause**	**Remedy**

Reduced airflow
  Rough or unstable idle
  Poor economy
  Power loss at all speeds
  Stalling or slow idle speed
  Reduced idle speed
  Slow starting (with normal
   cranking)
  Oil leaks or increased oil
   consumption
  Oil in the throttle body or plenum
	Plugged hose
	PCV valve stuck closed
	Dirty filter pad in air cleaner
		Check hose, filter pad, or PCV valve.
		Replace as needed.

Increased pressure in crankcase
  Oil sludging
  Oil in air cleaner
  Blue smoke in exhaust
		Replace PCV valve.

Crankcase vapors to atmosphere
reduce airflow
  Rough or unstable idle
  Poor economy
  Power loss
  Stalling
  High idle speed
  Lean running engine if equipped
   with MAF
	Hose broken or disconnected ahead
	of valve or orifice
		Replace hose or connect properly.

Vacuum leak; reduced airflow and
reduced scavenging of crankcase
  Rough or unstable idle
  Poor performance
  Hissing noises
  Oil sludging
	Hose broken or disconnected
	beyond PCV valve
		Replace hose or connect properly.

and valve. This will restrict flow, causing detonation and possibly overheating.

- *Poor Fuel Economy.* This is an EGR condition only if it is related to detonation or other symptoms of restricted or zero EGR flow.

EGR Valve Testing Equipment

The most used pieces of testing equipment for checking EGR valves and systems are:

- Vacuum gauge
- Hand vacuum pump
- Off-car bench tester
- Engine analyzer

Vacuum Gauge. A vacuum gauge can be used to check for excessive exhaust back pressure by using the following procedure:

1. Disconnect a vacuum line connected to an intake manifold port.

2. Put a vacuum gauge between the disconnected vacuum line and the intake manifold port.
3. Connect a tachometer.
4. Start the engine and gradually increase speed to 2000 rpm with the transmission in NEUTRAL.
5. The reading from the manifold vacuum gauge (Figure 11-14) should be above 16 inches mercury. If not, there could be an excessive back pressure in the exhaust system. To verify an excessive back pressure problem or a vacuum leak, perform the following:

- Turn off the engine.
- Disconnect the exhaust system at the exhaust manifold.
- Repeat steps 4 and 5.
- If 16 inches of mercury is still not obtained on the gauge, the exhaust mani-

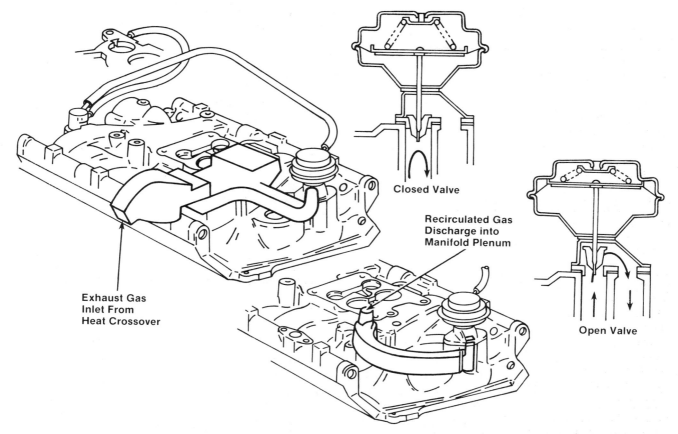

Closed Valve

Recirculated Gas
Discharge into
Manifold Plenum

Exhaust Gas
Inlet From
Heat Crossover

Open Valve

FIGURE 11-13 EGR system allows a metered amount of exhaust gas into the intake manifold

FIGURE 11-14 Reading from the manifold vacuum gauge should be above 16 inches mercury

fold is probably blocked. Visually inspect the exhaust manifold for blockage.

• If a reading of 16 inches of mercury is obtained, the blockage is either in the

muffler, exhaust pipe(s), or catalytic converter. If the problem is a damaged converter, inspect the muffler to be sure that converter debris has not entered the muffler, causing an obstruction.

Hand Vacuum Pump. A hand vacuum pump can be used to check the operation of the EGR valve by using the following procedure:

1. Check all vacuum lines for correct routing. Ensure that they are attached securely. Replace cracked, crimped, or broken lines.
2. With the engine at normal operating temperature, be certain there is no vacuum in the EGR valve at idle.
3. Install a tachometer.
4. On EFI engines (multi-point injection), disconnect the throttle air bypass valve solenoid.
5. Remove the vacuum supply hose from the EGR valve port and plug the hose.
6. Start the engine and idle with the transmission in NEUTRAL, then observe the engine

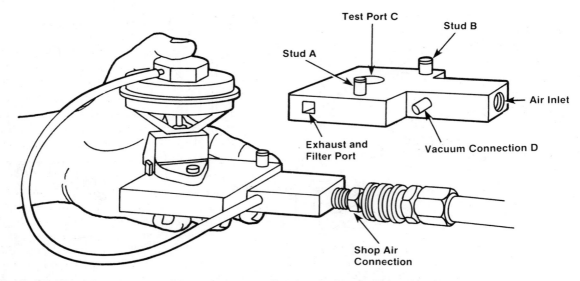

FIGURE 11-15 Bench tester provides vacuum while simulating exhaust back pressure

idle speed. If necessary, adjust idle speed to the emission decal specification.

7. Slowly apply 5 to 10 inches of mercury vacuum to the EGR valve vacuum port, using a hand vacuum pump.

8. If any of the following occurs when vacuum is applied to the EGR valve, replace the valve.
   - Engine does not stall.
   - Idle speed does not drop more than 100 rpm.
   - Idle speed does not return to normal (±25 rpm) after the vacuum is removed.

9. Reconnect the throttle air bypass valve solenoid, if removed.

10. Unplug and reconnect the EGR valve vacuum supply hose. The EGR valve is operating properly.

Off-Car Bench Tester. With a bench tester (Figure 11-15), the EGR valve is hooked up to the vacuum port on the tester while shop air at 60 psi minimum is blown through the test fitting. With both ported and positive back pressure EGR valves, the valve stem will move up when the EGR valve is put on the tester. With negative back pressure EGR valves, the valve stem will go up when the vacuum is applied, but it will go back down (close) when it is put on the tester and exposed to simulated back pressure.

Four-Gas Infrared Exhaust Emission Analyzer. An engine analyzer can be used to check the emission systems for correct operation. By inserting the exhaust analyzer probes into the exhaust system, a reading can be obtained to check that the emission standards for the vehicle are being met. Details on the use of an emission analyzer are given later in this chapter.

EGR VALVES AND SYSTEMS TESTING

As described in Chapter 10, the type of EGR valve that opens whenever vacuum is applied to it and lacks the back pressure feature is usually referred to as a *ported vacuum EGR valve.* Those that open only when there is the right combination of supply vacuum and exhaust back pressure are called *back pressure EGR valves.*

There are two types of back pressure EGR valves. The most common type is the positive back pressure EGR valve. This design uses positive (increasing) exhaust back pressure to increase EGR flow. As exhaust back pressure builds, it pushes up the back pressure diaphragm, which closes small air bleeds around the diaphragm. This allows full vacuum to reach the EGR valve causing the valve to open. As back pressure drops, the diaphragm moves down, uncovers the air bleeds, and cuts the vacuum to the EGR valve. The valve now goes shut. This process continues in a sort of cyclic motion as exhaust back pressure is balanced against EGR flow.

The other type of back pressure EGR valve is that which uses negative back pressure. Instead of reacting to rising back pressure, it reacts to decreasing back pressure. This type of valve is typically used on applications that have less natural back pressure, such as high-performance automobiles with free-flowing mufflers and large-diameter exhaust tubing.

It is important to know the difference between positive and negative back pressure EGR valves because they work differently and are tested differently. Never substitute an OE positive back pressure EGR valve for a negative one or vice versa—but some aftermarket positive EGR valves can be used to replace OE negative types if they are properly calibrated for the application.

To tell the difference between the types, as a rule, back pressure EGR valves (both positive and negative) have a slightly larger diameter diaphragm housing than ported vacuum EGR valves. In addition, a pair of tiny vent holes can be found in the end of the valve stem that allow the exhaust to flow up through the hollow valve stem to reach the back pressure transducer to make the system operate like a positive or negative back pressure EGR system.

One way to tell positive from negative back pressure EGR valves is to turn the valve upside down and note the pattern on the diaphragm plate. Positive valves have a slightly raised X-shaped rib whereas negative EGR valves are raised considerably higher. But this is not always true. On some General Motors engines the only way to tell a positive from a negative is to look for a letter next to the date code below the part number. "N" means negative and "P" means positive.

There are also some dual diaphragm EGR valves that are not back pressure valves. GM has some valves with two vacuum connections. One is attached to ported vacuum while the other goes directly to intake vacuum. With this design, manifold vacuum is used like exhaust back pressure to indicate engine load. By balancing ported vacuum against intake vacuum, the valve is able to adjust its flow rate to suit the engine load.

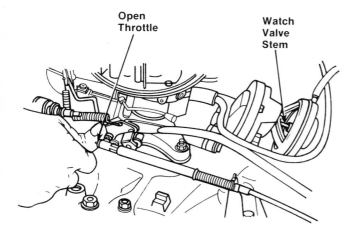

FIGURE 11-16 EGR valve operation test

 SHOP TALK _____

A word of caution about back pressure EGR valves—their operation can be affected by any changes in exhaust back pressure. When an OE back pressure EGR valve is specified for a certain catalytic converter, it is calibrated to the exhaust back pressure of the OE exhaust system. Most quality aftermarket exhaust systems are compatible with OE back pressure specs, but beware of the one-size-fits-all mufflers. Using this type of muffler can increase or decrease exhaust back pressure enough to upset the operation of the EGR valve in many instances, and the result can be rough running, hesitating, or stalling if the exhaust back pressure is increased (too much EGR) or detonation if it is reduced excessively (too little EGR or none at all).

Ported EGR Valve Function Test

To prepare for any EGR valve function test, operate the engine at a hot curb idle so that the PVS or TVS will open the vacuum to the valve. It is possible to tell that the engine is hot by feeling the upper radiator hose for coolant flow. Be careful when making this test because hot engine or radiator fluid can cause a serious burn.

To make an operational check on a ported EGR valve system, proceed as follows:

 1. Open the throttle just momentarily to increase engine speed to about 1500 rpm (Figure 11-16).
2. Watch the EGR valve stem as engine speed is increased (Figure 11-16).
   - If the stem moves out (and in when the throttle is released), the vacuum source is good, the diaphragm is good, and the valve is not sticking. Then perform the *EGR gas flow test.*
   - If the stem does not move, either the valve is stuck, the diaphragm is leaking, or there is a problem with the vacuum source. Inspect the vacuum hoses. If they are operating properly, perform the *diaphragm leak test* and *EGR source vacuum test.*

 SHOP TALK _____

In some vehicles, it is often difficult to determine if the EGR stem is moving. Use a mirror to watch for stem movement.

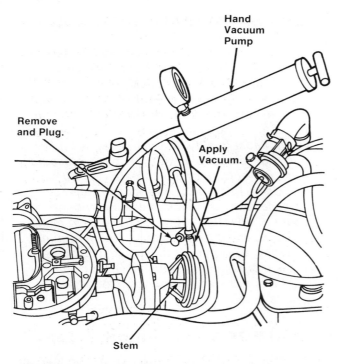

FIGURE 11-17 EGR gas flow test

EGR System Gas Flow Test. The exhaust gas flow test must be made with the engine idling and hot. This test tells the service technician if the gas flow passages are open. (This test cannot be used for EGR systems with integral back pressure to transducers.)

The gas flow test is made in the following manner:

 1. Remove the vacuum line from the valve diaphragm. Plug the line to prevent a vacuum leak or dirt entry.
2. Attach a hand vacuum pump to the valve diaphragm (Figure 11-17).
3. Apply 15 inches mercury of vacuum to the valve, a little at a time. Watch the valve stem for movement.

If the gas flow is correct, the engine will begin to idle roughly at some point, or it might stall. Rough idle or stall means the gas flow is good. If the stem moves but the idle does not change, there is a restricted passage in the valve or spacer plate. Remove the valve for inspection. If the stem does not move or the valve diaphragm fails to hold vacuum, install a new EGR valve.

Diaphragm Leak Test. The diaphragm leak test (Figure 11-18) can be performed with the engine off. The procedure for the test is as follows:

 1. Attach a hand vacuum pump to the vacuum nipple of the EGR valve.

2. Apply 8 inches mercury of vacuum and trap (hold) it.

The diaphragm must hold 7 to 8 inches mercury of vacuum for at least 30 seconds. If it holds vacuum, the diaphragm is good. Perform the source vacuum test. If it will not hold vacuum, replace the EGR valve.

EGR Valve Leakage. Perform this test if the engine stalls, idles roughly, or runs poorly:

 1. Remove and cap the EGR valve vacuum hose. Start the engine. If the engine idle quality has improved noticeably, double-check the vacuum hose routing because the valve might have a vacuum supply at idle.
2. If the engine idle quality does not improve, remove the EGR valve from the spacer. Block the EGR passages with a plate or install a good EGR valve. Start the engine. If the idle quality is still bad, the problem is elsewhere. Reinstall the EGR valve. If the idle quality improves noticeably, the EGR valve has excessive leakage and should be replaced.

Testing for EGR Source Vacuum. Occasionally the EGR system might fail to operate because the vacuum port on the carburetor is plugged or because of a vacuum leak. To test for source vacuum (Figure 11-19), proceed as follows:

1. Remove the EGR supply vacuum hose from the connection at the carburetor. To locate the right hose, trace the vacuum circuit from the valve, through the PVS or TVS, and to the carburetor (Figure 11-20). Late-model vehicles have a vacuum schematic on the emission decal.
2. Attach a vacuum gauge to the source.
3. Start the engine and watch the gauge as you momentarily accelerate to half throttle (3000 rpm maximum). The vacuum should

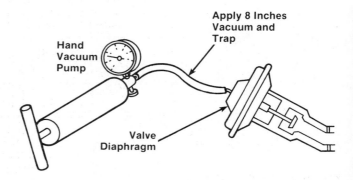

FIGURE 11-18 EGR valve diaphragm leak test

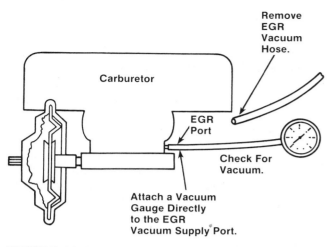

FIGURE 11-19   Testing for EGR source vacuum

rise at off idle, decrease at half throttle, and go to zero at closed throttle.

- If source vacuum is good, test the PVS or TVS and inspect the hoses.
- If vacuum is not good, check for an obstruction in the carburetor or a vacuum leak caused by loose mounting.

Testing a Two-Port PVS Valve (Figure 11-21). The two-port PVS valve is good if it blocks a vacuum from the EGR valve with a cold engine and allows vacuum to get to the valve with the engine hot. Refer to Table 11-3 for valve opening temperature.

Let the engine cool and release cooling system pressure before removing the valve. It also might be necessary to drain out some of the coolant. To make the test, proceed as follows:

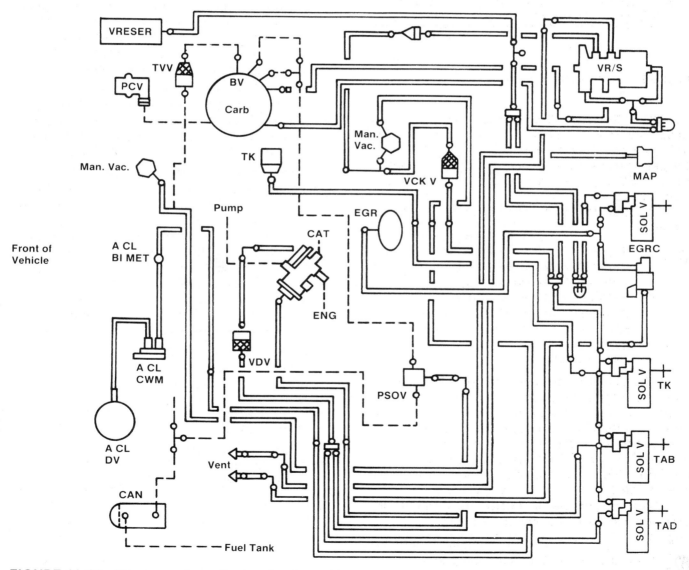

FIGURE 11-20   Vacuum schematic drawing

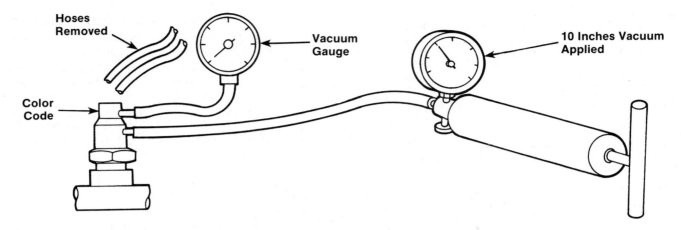

FIGURE 11–21 Testing a two-port PVS valve

TABLE 11-3:	TYPICAL TWO-PORT VALVE OPENING TEMPERATURES*
Color Code	**Coolant Above Temperature**
Green	68° F
Black	100° F
Blue	133° F
Natural	90° F
Gray	155° F

*Check vehicle's service manual for exact temperatures.

1. Remove both hoses from the PVS valve.
2. Connect a vacuum gauge to one port and remove the vacuum supply to the other outlet.
3. Apply 10 inches mercury of vacuum to the PVS valve. If the valve is good there should be no reading on the gauge.
4. Operate the engine until the coolant warms above the valve setting. The vacuum gauge must immediately read the applied vacuum.
5. Replace the PVS valve if it fails either the hot or cold test.

 SHOP TALK

On a four-port PVS valve, test both the upper and lower valves.

Testing a Three-Port PVS Valve. The three-port PVS is operating correctly if the vacuum at the center port is switched to

- Upper port with the engine temperature below the valve setting (see Table 11–4).

- Lower port with the engine temperature above the valve setting.

As shown in Figure 11–22, proceed as follows:

1. Apply vacuum to the center port.
2. Vacuum will be equal to the supply below the temperature setting in Table 11–4.
3. Vacuum should be equal to the supply above the temperature setting. If it fails the test, let the engine cool, release the radiator pressure, and install a new PVS valve.

Testing Air Cleaner TVS Valve. The air cleaner TVS operates by the same rules as the two-port PVS: it must block EGR vacuum with a cold engine and allow vacuum to pass with the engine warm. Since it is easily removed from the air cleaner, a bench test is recommended. To make the cold check (Figure 11–23A), proceed as follows:

1. Cool the TVS valve with liquid refrigerant to 40 degrees or less.

WARNING: When using an air conditioning refrigerant to cool the TVS, be sure to observe appropriate handling precautions. Wear goggles and protect the skin from contact with liquid refrigerant 12. Also, do not spill R-12 on a running engine because a toxic gas is produced when this refrigerant is burned.

2. Apply 16 inches mercury of vacuum. The TVS must hold a minimum of 5 to 16 inches vacuum for at least 30 seconds. If it does not, replace the TVS valve.

To perform the warm check, proceed as follows:

TABLE 11-4: TYPICAL THREE-PORT VALVE OPENING TEMPERATURES*	
Color Code	Coolant Above Temperature
Brown	100° F
Green	133° F
Yellow	165° F

*Check vehicle's service manual for exact temperatures.

1. Warm the TVS valve to 76 degrees Fahrenheit or more with a heat gun (Figure 11–23B).
2. Apply vacuum, but the valve must not hold the vacuum when warm. If it does, replace the TVS valve.

Testing the EGR Shutoff Solenoid. To check the EGR shutoff solenoid, proceed in the following manner:

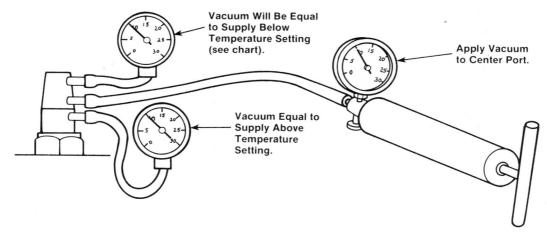

FIGURE 11-22   Testing a three-port PVS valve

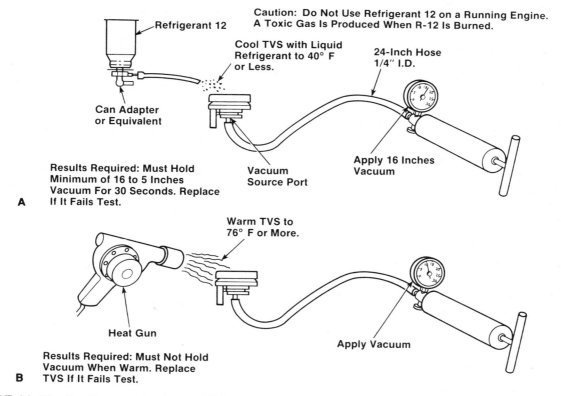

FIGURE 11-23   Testing an air cleaner TVS valve: (A) cold check; (B) warm check

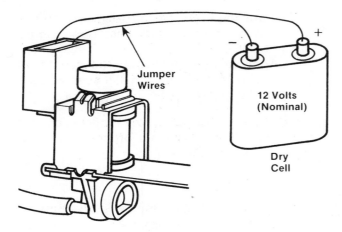

FIGURE 11-24   Testing an EGR shutoff solenoid

 1. Blow through the EGR shutoff solenoid valve. It must not pass the air de-energized.
2. Energize the solenoid with a 12-volt dry-cell battery as shown in Figure 11-24.

CAUTION: Do not allow the jumper to cross if using the vehicle's battery. This could cause a fire or severe burns. A dry-cell battery is recommended for this test.

---

3. If the solenoid is operating properly, a click must be heard and ports must pass air.

EGR Valve Bench Test.   An EGR valve that malfunctions on the vehicle should be bench tested before it is replaced with a new valve. It could be stuck simply because of some foreign material or buildup of deposits. Figure 11-25 illustrates the bench test. The bench test is conducted as follows:

 1. Tap the valve, turn the stem, and open and close the valve a few times to clean out any foreign particles.

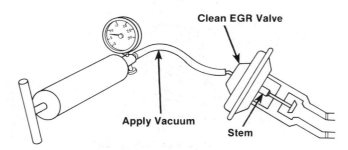

Clean EGR Valve

Apply Vacuum

Stem

FIGURE 11-25   EGR valve bench test

2. Apply the vacuum smoothly from 2 to 10 inches mercury.
3. Carefully watch the stem. If it moves out smoothly as vacuum increases and in smoothly as vacuum decreases, the valve is good. If the valve sticks or chatters, replace it.

The off-car bench tester mentioned earlier in this chapter can be used to make this test. It should be used following the manufacturer's instructions.

EGR SYSTEM INSPECTION AND CLEANING

The procedures that follow are to be used for the inspection and cleaning of the EGR valve and passages.

EGR Passage Cleaning

If the inspection of the EGR passages in the inlet manifold or adapter indicates an excessive buildup of deposits, the passages should be cleaned. Care should be taken to ensure that all loose particles are completely removed to prevent clogging the engine or the EGR valve.

 1. Remove the vacuum hose from the EGR valve.
2. Remove the EGR attaching bolts, or nuts, and the EGR valve assembly from the intake manifold. Discard the gasket.
3. Remove the deposits from the EGR ports by hand using a small drill bit and a screwdriver to clean the holes and chamber. Do not use an electric drill.
4. Brush any small particles down the port into the EGR passage and blow compressed air through the ports. Do not use any solvents for cleaning.
5. Clean and install the EGR valve (see EGR Valve Replacement later in this chapter) using a new gasket.

CAUTION: Do not wash the assembly in solvents or degreaser; permanent damage to the valve diaphragm can result. Also, sandblasting of the valve is not recommended; this could affect the operation of the valve.

EGR Valve Inspection

The inspection procedure of the EGR system is as follows:

1. Remove the EGR valve; discard the gasket.
2. Look for deposits on the valve pintle.
3. Depress the valve diaphragm and inspect for deposits around the valve seating area through the valve outlet.
4. The valve requires cleaning if deposits exist.

EGR Valve Cleaning

Some vehicle manufacturers recommend a cleaning procedure if exhaust deposits have jammed an EGR valve. Other manufacturers recommend replacement of the valve if it does not operate. Refer to the manufacturer's service manual. If cleaning is approved by the OEM, proceed as follows:

1. Hold the valve assembly in hand. Using a light snapping action with a plastic hammer, tap on the end of the round pintle to remove exhaust deposits from the valve seat. Empty loose particles.
2. Clean the mounting surface of the EGR valve with a wire wheel or wire brush, and the pintle with a wire brush. Compressed air can also be used to clean the valve seat and pintle areas.

 SHOP TALK _____

Some service manuals suggest the use of heat control valve solvent on the valve seat and pintle areas. However, when using such a solvent to loosen deposits, extreme care must be taken not to let any of this solvent get into the valve or reach the diaphragm.

The cause of the deposits that collect on the valve poppet (Figure 11–26) or stem and in the ports should be corrected. Black deposits are carbon and indicate a rich exhaust condition. The carburetion should be checked. Brown or beige and sometimes gray deposits come from leaded gas. If leaded gas is being used improperly, the customer should be advised to use unleaded gas as specified for the vehicle.

Negative Back Pressure EGR Valves

There are two types of back pressure EGR valves: negative transducer and positive transducer (Figure 11–27). A vacuum-modulated EGR valve is shown in Figure 11–28.

Since a back pressure transducer is simply a device to modulate EGR vacuum, the problem con-

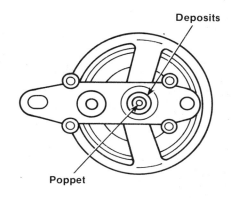

FIGURE 11–26 Deposits in the EGR valve port

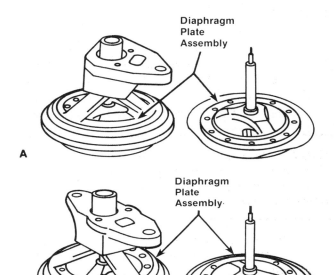

FIGURE 11–27 Back pressure EGR valves: (A) negative and (B) positive

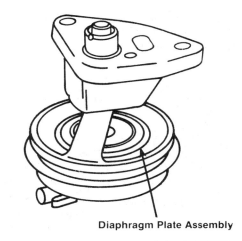

FIGURE 11–28 Vacuum modulated EGR valve

ditions can be the same as on any ported EGR valve system. In fact, with high exhaust back pressure, the transducer does not modulate, and the system functions the same as a ported valve. If the exhaust is obstructed, it will obviously change the EGR valve operation. Thus, excess EGR gas will likely compound the performance problems that result from a restricted exhaust.

If the transducer bleed hole is obstructed or if the vent is plugged or stuck closed, there will be no modulation of vacuum. The result will be too much EGR flow at part throttle and poor performance.

If the back pressure signal does not get to the transducer, it bleeds off all the vacuum. Back pressure EGR systems will not flow any exhaust gas to the intake without a back pressure signal. The results will be the same as any inoperative EGR valve problem. Likewise, if the transducer diaphragm is broken or if the seal does not close (or partially closes) the bleed hole, the system will be completely inoperative.

Vacuum Modulated and Negative Back Pressure EGR Valve.
To check the operation of these valves, proceed as follows:

1. Check the hose routing (refer to the vehicle emission control information decal).
2. Check the EGR valve signal tube orifice for obstructions.
3. Hook up a vacuum gauge between the EGR valve and the carburetor and check the vacuum. The engine must be operating at approximately 195 degrees and 3000 rpm. There should be at least 5 inches mercury of vacuum.
4. Depress the valve diaphragm.
5. With the diaphragm still depressed, hold a finger over the source tube and release the diaphragm.

WARNING: Since the valve is hot, be sure to wear heavy work gloves to avoid burning your fingers.

---

6. Check the diaphragm and the seat for movement. The valve is good if it takes more than 20 seconds for the diaphragm to move to the seated position (valve closed).
7. Replace the EGR valve if it takes less than 20 seconds to move to the seated position.
8. If the EGR valve is working properly, check the EGR-PVS or the TVS or the solenoid (if used).

Positive Back Pressure EGR Valves.
To check this type of EGR valve, proceed as follows:

1. Check the hose routing (refer to vehicle emission control information label).
2. Check the EGR. Check the signal tube orifice for obstructions. On engines with an EGR-TVS or EGR solenoid, there should be no EGR valve movement below the calibration temperature. If there is, test the EGR-TVS, PVS, or EGR solenoid for proper operation.
3. Check the EGR-TVS, TVS, or EGR solenoid for correct operation.
4. Valve check:
   - Remove the EGR valve from the vehicle.
   - Apply a constant (not hand pump) external vacuum supply (10 inches mercury or more) to the EGR vacuum signal tube.
   - The valve should not come open. If it does, the transducer control valve is stuck closed, and the EGR valve must be replaced.
   - With the vacuum supply still applied, direct a stream of air from the low pressure source 16 inches of mercury.
   - The valve should open completely. If it does not open at all, the transducer control valve is stuck open and the EGR valve must be replaced.
   - If the EGR valve and transducer control valve are both operating, clean the EGR mounting surfaces and reinstall the valve.

Testing Variable Transducer (BVT) EGR Systems.
To perform a function test of a BVT-EGR system, proceed as follows:

1. Check all the vacuum hoses for correct routing and secure attachment.
2. Run the engine to hot idle. Check that there is no vacuum to the EGR valve diaphragm at hot curb idle. If there is check the hose routing.
3. On EFI engines only, disconnect the electrical connector to the idle air bypass valve.
4. Install a tachometer. Adjust the curb idle if necessary.
5. Perform an EGR valve gas flow test as follows:
   - Disconnect the hose to the source and transducer (Figure 11–29).
   - With the engine at hot curb idle, slowly apply 8 to 10 inches mercury of vacuum:

—Idle must drop 100 rpm or more.

—Engine might stall.

—Idle speed must return to normal with vacuum removed.

6. Check the transducer for a plugged air bleed. To do this, try to apply 6 inches Hg of vacuum with the engine off. The valve stem and vacuum must not move. Replace the valve (or transducer) if it fails the test. Repeat the test with the engine at hot curb idle. If the vacuum holds, the exhaust system is restricted.

Testing Remote Back Pressure Transducer Systems. Some back pressure EGR systems use a separate back pressure transducer (Figure 11–30). To prepare for the function test of a remote transducer, proceed as follows:

1. Move the air cleaner out of the way so you can see the EGR valve stem. Do not disconnect any vacuum hoses.
2. Check all the hoses for proper routing and good connections.
3. Check the transducer and EGR valve for proper installation and tightness.
4. Before running the functional test, check the transducer for a plugged bleed as described below:

• Disconnect the source vacuum hose from the transducer and attach a hand vacuum pump (Figure 11–31).

• With the engine off, attempt to apply 6 inches mercury of vacuum to the EGR valve through the transducer.

• The transducer should bleed off vacuum and not open the EGR valve. If it holds vacuum, install a new transducer.

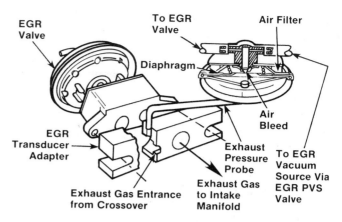

FIGURE 11–30 Some back pressure EGR systems use a separate back pressure transducer

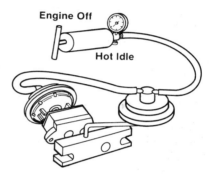

FIGURE 11–31 Air bleed test on remote transducer

5. With the engine at normal operating temperature, slowly open and close the throttle. Do not exceed half throttle or 3000 rpm.
6. The EGR valve stem must move up or down, or oscillate up and down.

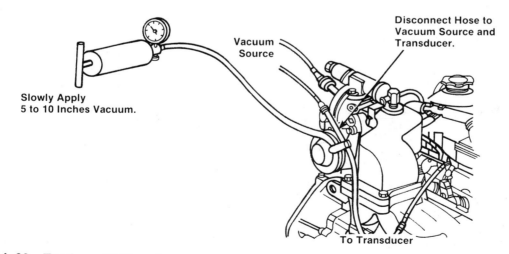

FIGURE 11–29 Testing a BVT system

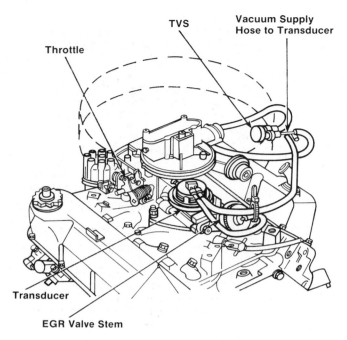

FIGURE 11-32 Functional test of an EGR system with remote back pressure transducer

- If the valve stem moves, the system is in good working order.
- If the stem does not move, proceed to step 7.

7. Disconnect the vacuum supply hose to the transducer and connect the vacuum gauge

to the TVS switch. Slowly open and close the throttle, observing the gauge.

- If the vacuum is less than 4 inches mercury, check the TVS switch (gauge) operation or check for vacuum at the EGR port.
- If the vacuum is greater than 4 inches mercury, proceed to step 8.

8. Recheck, on the engine, the operation of the EGR valve (Figure 11-32). If the valve check is favorable, remove the valve and transducer and check the passages for deposits or loose particles (spacer included).

- If inspection shows that the components are clean, replace the transducer and verify correction.
- If cleaning is required, use a small wire brush or scraper. Do not use solvents or compressed air on the transducer. Recheck EGR valve operation and if not satisfactory, replace the transducer and verify correction.

Basic EGR Control Systems

As described in Chapter 10, there are several controls that operate in conjunction with the EGR valve. Some of the more common EGR controls are:

EGR/CSC Systems. EGR/Coolant Spark Control (Figure 11-33) or CC/EGR/Coolant Con-

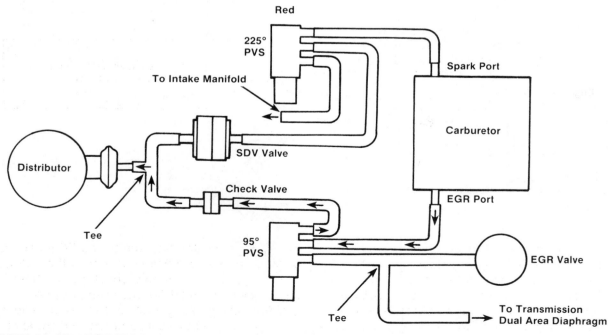

FIGURE 11-33 EGR/CSC system components

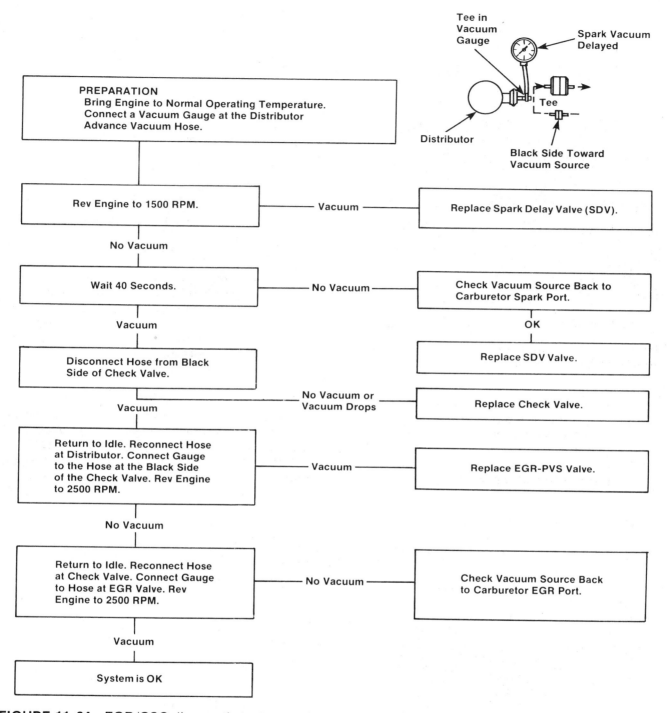

FIGURE 11-34   EGR/CSC diagnostic test

trol/EGR is a special vacuum switching system that regulates both the EGR vacuum flow and the vacuum to the distributor spark advance, depending on the engine temperature. Problems in the EGR/CSC system related to EGR operation are essentially the same as those for a system with a simple two-port PVS vacuum valve. In addition, failure of the three-

port PVS to switch vacuum to the EGR can cause erratic shifts or a false shift feel in the automatic transmission. A defective check valve should not affect the EGR, but it can cause engine performance problems, such as stumble or hesitation. The chart in Figure 11-34 covers a complete performance test of the entire EGR/CSC system.

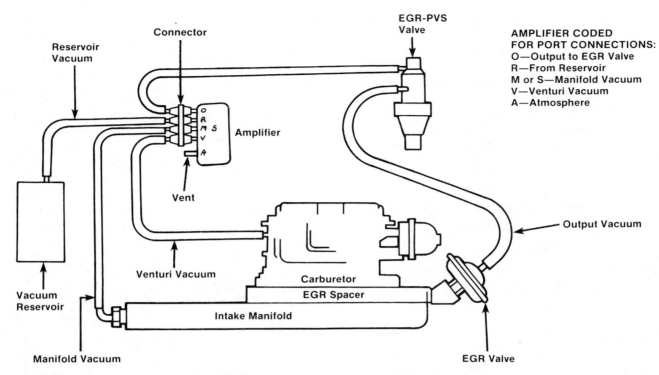

FIGURE 11–35 Typical single connector venturi vacuum amplifier system

 SHOP TALK —————

If this EGR system is disconnected, it can cause erratic transmission operation and/or damage to the transmission internal parts.

EGR/VVA Systems. A venturi vacuum amplifier (Figure 11–35) is used with some EGR systems so that carburetor venturi vacuum can control the EGR valve operation. Venturi vacuum is more desirable because it is proportioned to the airflow through the carburetor. Since it is a relatively weak vacuum signal, the VVA is used to convert it to a strong enough signal to operate the EGR valve. Problems with the VVA system would generally show up in a zero-EGR condition caused by vacuum not getting to the EGR valve. Possible causes of this could be

- Inoperative amplifier
- Leaking vacuum reservoir
- Leaking check valve
- Misrouted, damaged, or improperly connected vacuum hose.

To make a venturi vacuum amplifier test (Figure 11–36), proceed as follows:

 1. Run the engine until it reaches normal operating temperature.

2. Check all the hoses and system components for good condition, proper routing, and tight connections.
3. Disconnect the hose at the EGR valve and install a vacuum gauge.
4. Disconnect the venturi vacuum hose at the carburetor.
5. Check for specified vacuum at curb idle (see service manual). It should be the "bias" setting of the amplifier. If the amplifier

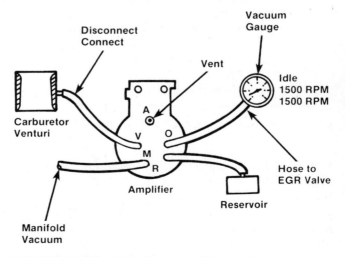

FIGURE 11–36 Vacuum amplifier test

is set for zero bias, it can be no more than 1/2 inch of vacuum.

6. Watch the gauge as the engine accelerates to about 1500 to 2000 rpm. The vacuum should not change with the venturi hose disconnected.

7. Connect the hose to the carburetor venturi port and check for a vacuum increase. If it increases more than 1/2 inch, check the curb idle against decal specifications.

8. Again watch the gauge as the engine accelerates. This time the vacuum must build to 4 inches or more during acceleration and return to bias vacuum at idle.

9. If the vacuum is not as specified in steps 5, 6, or 7, check for a correct manifold vacuum supply. If it is okay, replace the amplifier and repeat the test.

EGR Delay Timer and Solenoid. Some EGR systems, such as the Chrysler venturi vacuum-control shown in Figure 11-37, have a delay timer and solenoid. To test the delay solenoid, proceed as follows:

1. Check the timer delay wiring for proper connections (Figure 11-38).

2. Check for electrical operation by unplugging the connection, jumping one terminal of the solenoid to ground, and jumping the other to the positive terminal of the battery. If the solenoid is working, a click will be heard.

3. Check all delay circuit vacuum hoses and connections. Replace any cracked, broken, or leaking hoses.

4. Check the solenoid's passages. On most cars, unplug the hoses at the solenoid and bypass the valve with a short length of 1/8-inch tubing. On these engines with the idle enrichment system, unplug the hoses from the solenoid and bypass it with a length of 3/16-inch tubing. If normal movement of the EGR valve is restored, then the solenoid is plugged and should be replaced.

5. To check the delay timer, stop the engine, disconnect and reconnect the hose to the EGR valve, then restart. Immediately rev the engine 2000 to 3000 rpm and watch the stem of the EGR valve. If it moves during the first 35, 60, or 90 seconds, according to the system specifications in the service manual, the timer is defective. The timer can also be checked with a test lamp. Connect it across the connector at the sole-

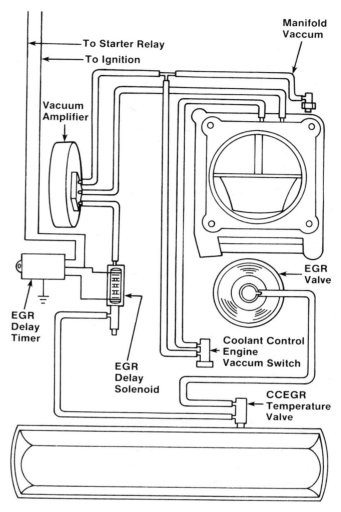

FIGURE 11-37 Chrysler venturi vacuum-control EGR system

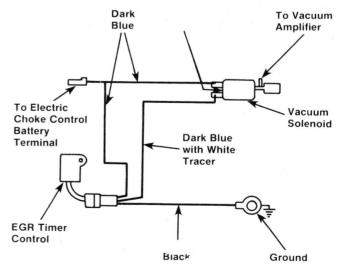

FIGURE 11-38 Wiring diagram of Chrysler's EGR time delay

noid, start the engine, and see that the light glows for the correct number of seconds.

Remember that some late-model engines have a charge temperature sensor wired into the timer circuit, so the timer will not start to run until the mixture in the intake runners is about 60 degrees Fahrenheit. To check the charge temperature sensor itself, begin with the engine cold (under 60 degrees). Remove both wires from the sensor and connect an ohmmeter across its terminals. If resistance is about 10 ohms, replace the CTS. In addition, check the continuity of the switch's ground circuit wiring, which connects to the side terminal.

Vacuum Reservoir and Check Valve Test. The check valve is operating correctly if it allows vacuum to flow freely in one direction and traps vacuum applied to the other port. To test the check valve, connect a vacuum source and gauge to the ports alternately as shown in Figure 11–39. Replace the valve if it fails either test.

Vacuum reservoirs used without an external check valve have only one vacuum port. To test this reservoir:

1. Remove the hose that leads to the amplifier from the reservoir nipple.
2. Attach a vacuum pump to the reservoir.
3. Apply 14 inches of vacuum and trap. If the vacuum holds, the reservoir is good.

If the reservoir has an in-line check valve, test both the reservoir and valve at the tee from the manifold vacuum source. Disconnect the manifold vacuum hose from the tee and proceed as follows:

1. Apply 14 inches mercury vacuum to the amplifier port and trap. If vacuum holds, the reservoir and check valve are in proper

working condition. The maximum leakage allowed is 1 inch in 1 minute.
2. Apply 14 inches of vacuum to the line and trap. If the vacuum reading is 13 inches or higher after 1 minute, the check valve is fine.
3. If the vacuum does not hold in step 1 or 2, replace the leaky component and repeat the steps.

EGR High-Speed Cutoff Systems. In some EGR applications, it is desirable to cut off EGR flow at wide-open throttle, where a richer fuel mixture is required for performance. One device used for this purpose is the wide-open throttle (WOT) valve. The WOT valve is installed so that it can vent the EGR vacuum when it receives a full throttle signal from the carburetor. Its port connections are:

- To the venturi vacuum (control port)
- To the EGR source vacuum
- To the EGR diaphragm

At, or near, wide-open throttle, venturi vacuum is high enough to overcome spring pressure pushing downward on the WOT valve diaphragm. This is called the WOT valve venting position (Figure 11–40A). Venturi vacuum lifts the diaphragm up, and the vent port is uncovered. Vacuum from the EGR port is purged and atmospheric pressure rather than vacuum is applied to the EGR valve diaphragm. The EGR valve remains closed. Exhaust gas recirculation is interrupted. The engine delivers full power because fuel mixture dilution is cut off almost instantaneously.

When the load on the engine is reduced from wide-open throttle, venturi vacuum applied to the WOT valve diaphragm is weakened. WOT valve spring pressure overcomes venturi vacuum. This is

FREE-FLOW CONDITION

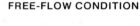
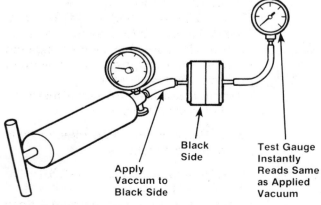

NO-FLOW CONDITION

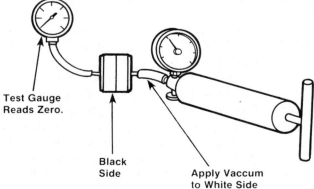

FIGURE 11-39 Vacuum check valve test

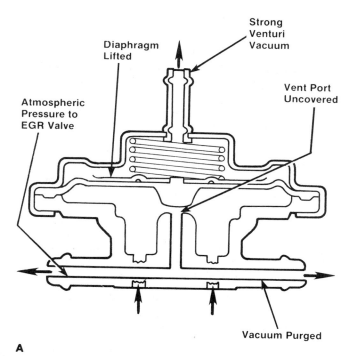

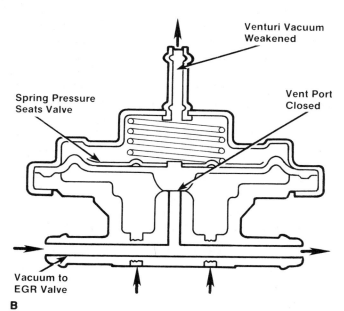

A

B

FIGURE 11-40 WOT valve: (A) in venting position;
(B) in nonventing position

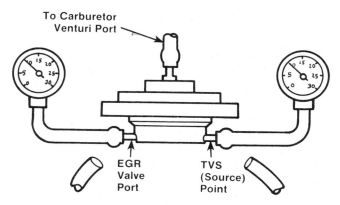

FIGURE 11-41 WOT valve nonventing test

- Check for proper hose routing and connections.
- Warm the engine to the normal operating temperature.
- Perform an EGR valve functional test as described earlier in this chapter.

To make the nonventing test (Figure 11-41), proceed as follows:

1. Remove the hose from the EGR vacuum source port of the WOT valve.
2. Install a vacuum gauge on the valve. Observe the gauge at 1200 to 1500 prm.
   - If vacuum is 4 inches mercury or greater, remove the gauge, reinstall the vacuum hose, and proceed to Step 3.
   - If vacuum is less than 4 inches mercury, check all vacuum source hoses and test the TVS switch operation. Insure a 4-inch vacuum supply, or greater, to the WOT valve before proceeding to Step 3.
3. Remove the hose from the EGR vacuum source port.
4. Install a vacuum gauge and observe the gauge at 1200 to 1500 rpm.
   - If vacuum is 4 inches mercury or more, the WOT valve is functioning correctly in the OPEN position. Leave the gauge installed and proceed to the venting and closing tests.
   - If vacuum is less than 4 inches mercury, replace the WOT valve.

To make the WOT venting and closing tests (Figure 11-42), proceed as follows:

1. Connect a vacuum gauge to the EGR valve port of the WOT valve.
2. Remove the venturi vacuum signal hose and install a hand vacuum tester pump.

called the WOT valve nonventing position (Figure 11-40B). The diaphragm is pushed down, the valve is seated, and the air vent is closed. Vacuum from the EGR port in the carburetor is applied to the EGR valve and the valve resumes normal operation.

To prepare the WOT tests, proceed as follows:

- Remove the air cleaner without disturbing the vacuum hose connections.

TABLE 11-5: EXHAUST GAS RECIRCULATION SYSTEM PROBLEMS

Condition	Possible Cause	Remedy
Engine idles abnormally roughly and/or stalls	EGR valve vacuum hoses misrouted.	Check EGR valve vacuum hose routing. Correct as required.
	Leaking EGR valve	Check EGR valve for correct operation.
	Failed EGR valve gasket or loose EGR attaching bolts	Check EGR attaching bolts for tightness. Tighten as required. If not loose, remove EGR valve and inspect gasket. Replace as required.
	Improper vacuum to EGR valve at idle	Check vacuum from throttle body EGR port with engine at stabilized operating temperature and at curb idle speed. If vacuum is more than 1.0 in. Hg refer to throttle body idle diagnosis and EGR check chart for computer command control system.
Engine runs roughly on light throttle acceleration and has poor part load performance.	EGR valve vacuum hose misrouted.	Check EGR valve vacuum hose routing. Correct as required.
	Check for loose valve.	Torque valve.
	Sticky or binding EGR valve	Clean EGR passage of all deposits. Remove EGR valve and inspect. Replace as required.
	Wrong or no EGR gasket(s) and/or spacer	Check and correct as required. Install new gasket(s), install spacer (if used), torque attaching parts.
Engine stalls on decelerations.	Control valve blocked or airflow restricted.	Check internal control valve function per service procedure.
	Restriction in EGR vacuum line	Check EGR vacuum lines for kinks, bends, etc. Remove or replace hoses as required. Check EGR valve for excessive deposits causing sticky or binding operation. Replace valve.
	Sticking or binding EGR valve	Remove EGR valve and replace.
Part-throttle engine detonation[1]	Insufficient exhaust gas recirculation flow during part-throttle accelerations	Check EGR valve hose routing. Check EGR valve operation. Repair or replace as required. Check EGR passages and valve for excessive deposit. Clean as required. Check EGR per service procedure.
Engine starts but immediately stalls when cold.[2]	EGR valve hoses misrouted.	Check EGR valve hose routings.

[1]Nonfunctioning EGR valve could contribute to part-throttle detonation. Detonation can be caused by several other engine variables. Perform ignition and throttle body related diagnosis.
[2]Stalls after start can also be caused by throttle body problems.

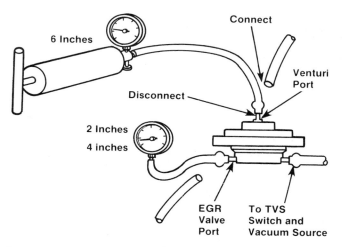

FIGURE 11-42 WOT valve venting and closing tests

3. Apply 6 inches mercury of vacuum to the venturi port.
4. Observe the vacuum gauge at 1200 to 1500 rpm. It should drop to 2 inches or less if the WOT valve is venting properly.
5. Remove the tester to simulate zero venturi vacuum.
6. Check the gauge for a return to 4 inches mercury of vacuum. If vacuum increases to 4 inches mercury, the valve is closing properly.
7. Replace the WOT valve if it fails to vent (Step 4) or to close properly (Step 6).

Table 11-5 shows problems that can occur in an exhaust gas recirculation system, their possible causes, and suggested remedies.

EVAPORATIVE EMISSION CONTROL SYSTEM

As described in Chapter 10, the charcoal canister storage method is the basic evaporative emission control system (EECS) used. This method transfers fuel vapor from the fuel tank to an activated carbon (charcoal) storage device (canister) and holds vapors when the vehicle is not operating (Figure 11-43A). When the engine is running, the fuel vapor is purged from the carbon element by intake airflow and consumed in the normal combustion process. That is, gasoline vapors from the fuel tank flow into the canister. Any liquid fuel goes into a reservoir in the bottom of the canister to protect the integrity of the carbon bed above. These vapors are absorbed into the carbon. The canister is purged when the engine is running above idle speed (Figure 11-43B). Ambient air is allowed into the canister through the air tube in the top. The air mixes with the vapor and the mixture is drawn into the intake manifold.

CANISTER TESTING

Check the canister to make sure that it is not cracked or otherwise damaged (Figure 11-44). Also make sure that the canister filter is not completely saturated. Remember that a saturated charcoal can cause symptoms that can be mistaken for carburetor problems. Rough idle, flooding, and other conditions associated with poor bowl venting can indicate a canister problem. A canister filled with liquid or water causes back pressure in the fuel tank. It can also cause richness or flooding symptoms during purge or start-up. (Some trucks have intentionally

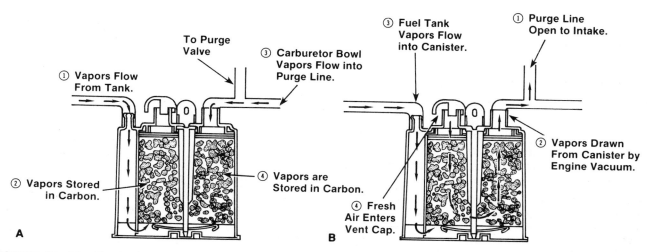

FIGURE 11-43 Typical vapor and airflow: (A) engine off-vapor storage; (B) purging

FIGURE 11–44 Checking a canister

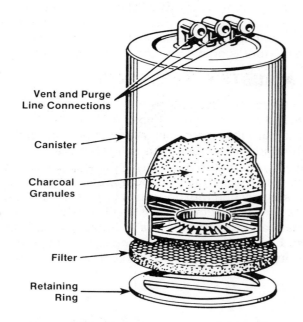

Vent and Purge Line Connections

Canister

Charcoal Granules

Filter

Retaining Ring

FIGURE 11–45 Most carbon canisters have replaceable filters.

pressurized fuel tank systems. Check the calibration and engine decal before diagnosing.)

To test for saturation, unplug the canister momentarily during a diagnosis procedure and observe the engine's operation. If the canister is saturated, either it or the filter must be replaced depending on its design. That is, some models have a replaceable filter, others do not.

In some older models, the fiberglass filter can be replaced without changing the entire canister. To change a canister filter (Figure 11–45), follow the steps in this procedure:

- Disconnect all hoses from the canister. Be sure to mark the hose lines.
- Loosen the hold-down clamps and remove the canister.
- Pull the old fiberglass filter from the bottom of the canister. Most filters are held in place by a retaining ring or bar.
- Install the new fiberglass filter pad under the retainer.
- Replace the canister, tighten the hold-down clamps, and reconnect the canister hoses.

Most late-model canisters have replaceable filters. The entire canister must be replaced if defective. Canisters are held in place by sheet metal straps and screws. Loosen the screws, mark vapor and vacuum lines for replacement, and remove the canister. Install a new canister and reconnect the vapor and vacuum lines correctly.

 SHOP TALK ⸺⸺⸺⸺

Sometimes a partially saturated canister can be evacuated by removing it from its mounting area (leaving the hoses connected), inverting it, and running the engine at high idle for 3 to 5 minutes. Be sure to replace the canister filter if it can be separated from the canister.

PURGE CONTROL SYSTEMS

Four types of purge control systems are currently in use on vehicles.

1. *Air Cleaner Purge.* This system has no control valve or solenoid (Figure 11–46A) and is used primarily on EPI turbo and larger V-8 engines.
2. *Direct Vacuum Purge.* This system's vapor flow depends directly on the strength of the vacuum (Figure 11–46B).
3. *Vacuum-Operated Purge Valve.* This is the most common of the standard purge control system. The purge valve is vacuum controlled and operates as follows:
   - Closed with no control vacuum.
   - Closed with engine off.
   - Open when control vacuum is applied.
   - Vapors flow to the intake when the control vacuum is on.

   In addition to the standard purge valve system shown in Figure 11–46C, some engines use a canister purge shutoff valve with reverse logic to the purge valve.
4. *Canister Purge Solenoid.* This sytem is used with electronic engine control (EEC)

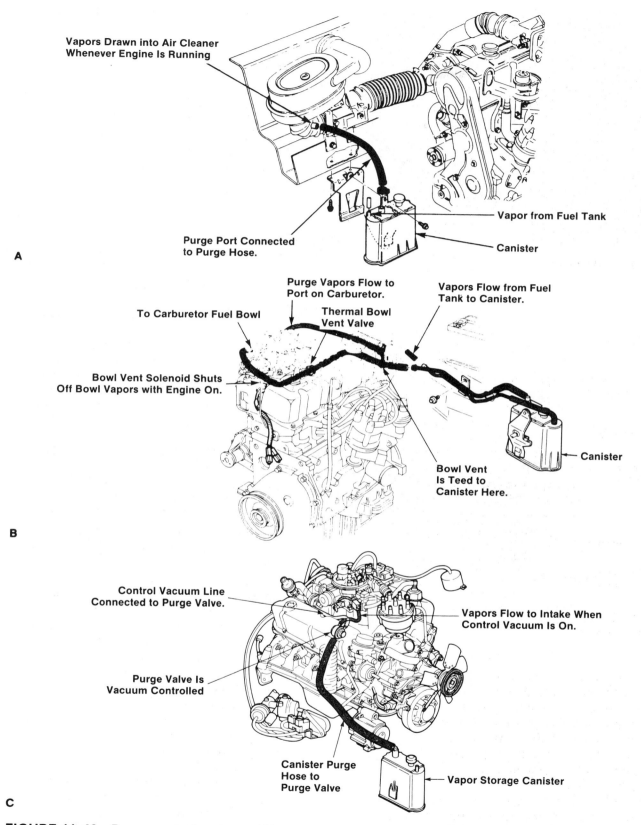

FIGURE 11-46 Purge control systems: (A) purge hose to air cleaner; (B) direct vacuum purge; (C) vacuum-operated purge valve; (D) canister purge solenoid

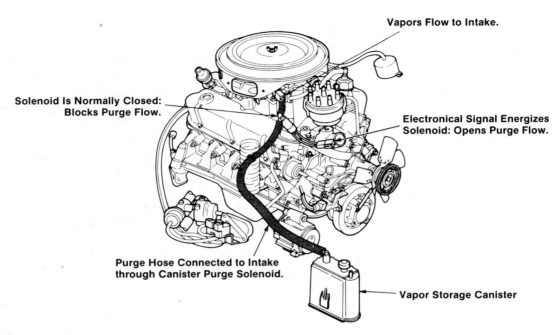

Vapors Flow to Intake.

Solenoid Is Normally Closed: Blocks Purge Flow.

Electronical Signal Energizes Solenoid: Opens Purge Flow.

Purge Hose Connected to Intake through Canister Purge Solenoid.

Vapor Storage Canister

D

FIGURE 11-46 (Continued)

or electronic control module (ECM). The EEC or ECM operates the solenoid control valve, which controls vacuum to the purge valve in the charcoal canister (Figure 11-46D). Under cold engine or idle conditions, the solenoid valve is turned on by the EEC or ECM. This closes the valve and blocks vacuum to the canister purge valve. The EEC or ECM turns off the solenoid valve and allows purge when

• Engine is warm.
• Engine has been running a specified time.
• A specified road speed is reached.
• A specified throttle opening is achieved.

The solenoid control valve opens with vacuum applied and closes with no vacuum. This prevents purge to the intake manifold under conditions of low ported vacuum, such as deceleration. If the solenoid valve is open or is not receiving power, the canister can purge to the intake manifold at all times (Figure 11-47). This allows extra fuel at idle or during warm-up. The result could cause rough or unstable idle or a too rich operation during warm-up. If the solenoid is open or not receiving power, the canister can purge to the intake manifold at all times. This can allow extra fuel to enter the intake manifold at idle or during warm-up, which can cause rough or unstable idle or a too rich operation.

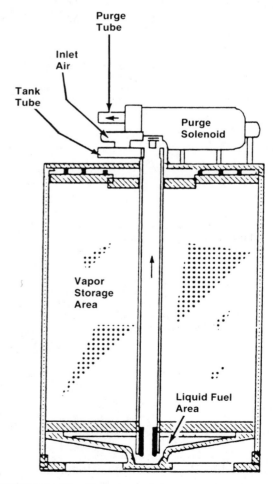

Purge Tube

Inlet Air

Tank Tube

Purge Solenoid

Vapor Storage Area

Liquid Fuel Area

FIGURE 11-47 Purge canister

Testing Canister Purge Valve

A vacuum leak in any of the evaporative emission components or hoses can cause starting and performance problems as can any engine vacuum leak. It can also cause complaints of fuel odor. Incorrect connection of the components can cause rich stumble or lack of purging (resulting in fuel odor). To conduct a vacuum-on, valve-open test on a typical canister purge valve, proceed as follows:

1. Apply and trap 5 inches mercury of vacuum to signal the vacuum port X (Figure 11–48A). Vacuum bleed-off must be less than 1 inch per second.
2. With the vacuum on hold, blow into port Y. It must be easy to blow air into port Y.
3. For a vacuum-off, valve-closed test, disconnect the vacuum source and leave it open to the air. Apply 16 inches mercury of vacuum to port Z (Figure 11–48B). The valve must hold vacuum. The bleed-off should not exceed 1 inch per second.

Replace the canister purge valve if any of the above steps fail. A valve with an internal bleed purge

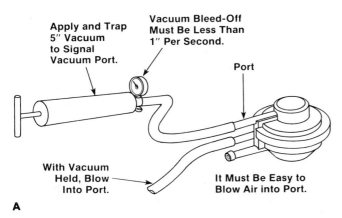

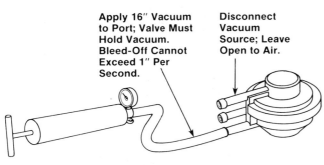

FIGURE 11–48   Testing the canister purge valve; (A) valve open test; vacuum on; (B) valve closed test; vacuum off

might have some flow in Step 3. This is normal. Test these valves as follows:

- Start the engine and allow it to warm up. Then try to blow through the hose while raising the engine speed to about 2000 rpm. It should be possible to blow through the hose.
- Return the engine to idle and try again to blow through the hose. The valve should be closed and block the airflow. If the valve fails either test, replace it.

Testing Canister Purge Solenoid (CPRV)

To test a CPRV, proceed as follows:

1. Apply 5 inches mercury of vacuum and trap (solenoid off).
2. The canister purge valve must not pass air.
3. Energize the solenoid with a jumper wire (Figure 11–49). The valve must open if it is operating properly.

FUEL BOWL VENT CONTROL

If the fuel bowl does not vent properly, there can be flooding on start-up, especially if the vent fails to open with the carburetor hot. Flooding can also occur if vapors back up into the carburetor from the canister due to a failed TVV valve. A failed open carburetor bowl vent solenoid or vacuum/thermal bowl vent can cause suction in the fuel bowl during canister purging, resulting in lean driveability problems.

Testing Fuel Bowl Solenoid (SV-CBV)

To test a SV-CBV, proceed as follows:

1. Blow through the valve with no power to the solenoid. There should be a free airflow.
2. Energize the solenoid as shown in Figure 11–49.

 SHOP TALK ———————

Short-circuiting of the bowl vent solenoid will not hurt the valve but will cause a blown fuse. In most systems, this fuse also supplies power to the ignition circuit so the engine will not run. Any diagnosis of "no start" should include checking this fuse.

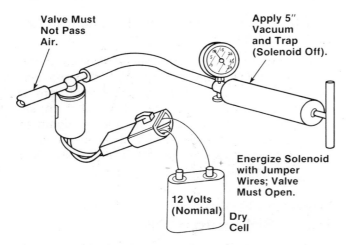

FIGURE 11-49 Testing the cansiter purge solenoid

Testing Vacuum Bowl Vent Valve (VBVV) or Canister Purge Shutoff Valve (CSOV)

To test either the VBVV or CSOV, proceed as follows:

1. Blow into Port A (Figure 11-50). The valve should allow free airflow.
2. Connect a 15- to 40-inch-long hose (3/16-inch diameter) to Port B. Then apply 5 inches mercury of vacuum to Port B and

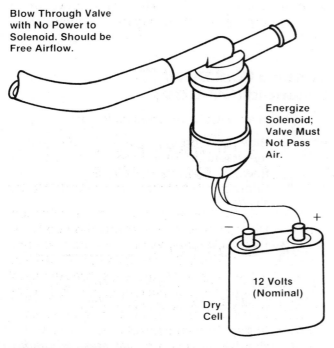

FIGURE 11-50 Testing the fuel bowl vent solenoid

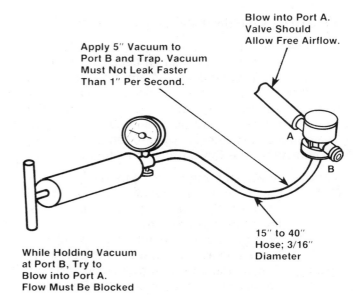

FIGURE 11-51 Testing the vacuum bowl vent valve or canister purge shutoff valve

trap. Vacuum must not leak faster than 1 inch per second.
3. While holding the vacuum at Port B, try to blow into Port A. Airflow must be blocked.
4. Replace the valve if it fails the test of Step 1 or 3.

Testing Vacuum/Thermal Bowl Vent Valve (VTBVV)

To test the vacuum/thermal bowl vent valve, use the same setup as shown in Figure 11-51. Proceed in the following manner:

1. With the valve cold, blow into Port A. The valve should not allow airflow.
2. Warm the valve or allow it to warm to above 120 degrees Fahrenheit. Again, blow into Port A.

   - Vacuum must hold.
   - With continued blowing, the valve will cool and close.
3. With the valve still warm, apply 5 inches mercury of vacuum into Port B and trap.
   - Vacuum must hold.
   - Valve must not permit the flow of air when blowing into Port A.

FUEL TANK SYSTEM DIAGNOSIS

If you smell fuel, inspect the fuel tank system as shown in Figure 11-52. To check the fuel flow through a typical system, proceed as follows:

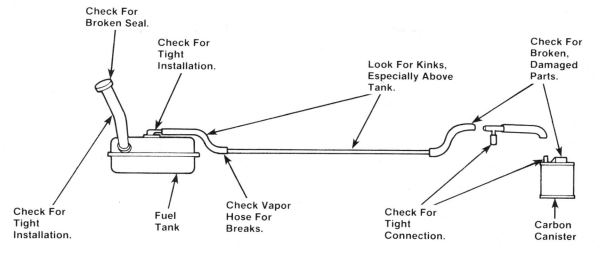

FIGURE 11-52 Inspecting for fuel odors

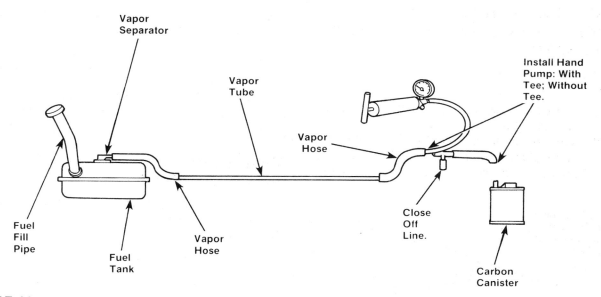

FIGURE 11-53 Functional flow test

1. Install a hand pump with or without a tee as shown in Figure 11-53.
2. Apply 2.5 psi pressure with the hand pump.

CAUTION: Apply no more than 2.5 psi pressure.

---

3. Remove the fuel cap. The pressure should drop to zero immediately. If not, locate the blockage in the fuel flow system.

Table 11-6 is a summary of evaporative emission control system problems and remedies.

AIR TEMPERATURE EMISSION CONTROLS

Warm combustion air improves the vaporization of the fuel in the carburetor, fuel injection body, or intake manifold. This makes more of the fuel burn and reduces HC and CO emissions in the exhaust. As described in Chapter 10, there are three different systems used on various gasoline engines to heat the inlet air and/or the air/fuel mixture.

1. Heated air inlet controls
2. Manifold heat control valves
3. Early fuel evaporation (EFE) heaters

TABLE 11-6:   EEC SYSTEM PROBLEMS AND THEIR REMEDY

Condition	Possible Cause	Remedy
Gasoline odor	Carburetor bowl vent solenoid or vacuum valve (VBVV or V/TBVV) failed in closed or restricted condition. Disconnected/damaged hoses to carburetor fuel bowl vent or canister. Fill cap gasket damaged or relief valve leaks. Vapor separator or hose leaks. Canister damaged or filled with liquid (gas or water). Kinked hose from vapor separator to canister. Carburetor bowl vent line blocked or restricted.	Inspect system very carefully. Correct faulty situation or replace faulty components.
Gasoline odor Possible flooding in some cases.	Purge line blocked. Purge valve failed to close. Disconnected or leaking vacuum signal port on purge valve. Purge solenoid failed to close or is not energized. Purge valve damaged by belts or pulleys. Canister loaded with liquid fuel.	Inspect system very carefully. Correct faulty situation or replace faulty components.
Hard starting Poor idle Poor performance Purging whenever engine runs/cranks	Purge valve or solenoid failed to open. Leaking signal port on purge shutoff valve.	Check purge valve or solenoid and replace if necessary.
Engine vacuum leak, no purging; canister loaded with liquid fuel	Purge line leak downstream of valve.	Repair leak.
Lean hesitation Vacuum in bowl during purge Poor cruise performance	Bowl vent valving failed to open.	Replace valve.

HEATED AIR INLET CONTROL

As described in Chapter 10, a heated air system (Figure 11-54) provides warm air from around the exhaust manifold to the carburetor when underhood air temperatures are less than 100 degrees Fahrenheit. The snorkel of the air cleaner contains a vacuum-operated damper. While the engine is cold, the damper stays closed to divert heated air to the snorkel (Figure 11-55). Control is usually either by a thermostatic coil spring or a vacuum motor on the snorkel. A temperature-operated vacuum sensor (TVS) controls vacuum motor operation.

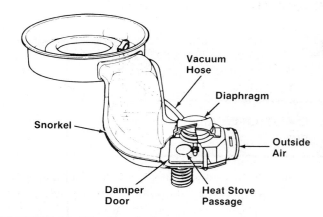

FIGURE 11-54   Snorkel passage open

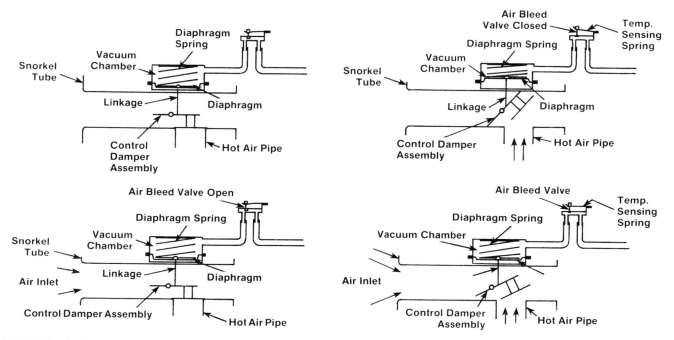

FIGURE 11-55 Inspection and performance test on a heated air inlet system

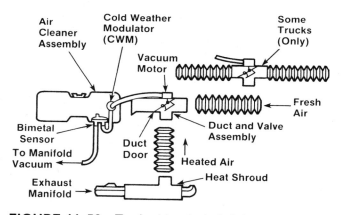

FIGURE 11-56 Typical heated air inlet system

Heated Air Inlet System Inspection and Performance Test

To perform an inspection and performance test on a typical heated air inlet system (Figure 11-56), proceed as follows:

 1. Apply the parking brake and block the wheels.

2. Remove the air cleaner cover and element. Inspect the heated air duct for proper installation and/or damage. Service all of the parts as required.

3. Remove components as necessary to ensure that the duct door is in the OPEN TO FRESH AIR position. If the door is in the CLOSED TO FRESH AIR position, check for binding and sticking. Service or replace as required.

4. Check the vacuum source and integrity of the vacuum hoses to the bimetal sensor, cold weather modulator (CWM), and vacuum motor.

5. Start the engine. If the duct door has moved to the HEAT ON position (closed to fresh air) proceed to Step 7. If the door stays in the HEAT OFF position (closed to warm air), place a finger over the bleed of the bimetal sensor. The air duct door must move rapidly to the HEAT ON position. If the door does not fully move to this position, stop the engine and test the vacuum motor. Repeat this step with a new vacuum motor. The ambient air must be at least 60 degrees Fahrenheit during this test.

6. With the engine off, cool the bimetal sensor and CWM by spraying with liquid from a small can of refrigerant R-12 with a proper adapter for 20 seconds after the liquid contacts the sensor and CWM.

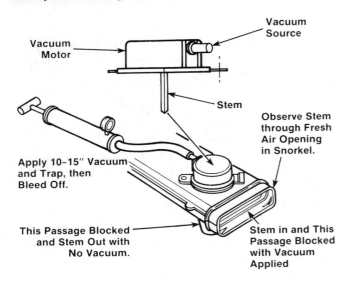

FIGURE 11-57 Testing the air door vacuum motor

Labels in figure:
- Vacuum Motor
- Vacuum Source
- Stem
- Observe Stem through Fresh Air Opening in Snorkel.
- Apply 10-15" Vacuum and Trap, then Bleed Off.
- This Passage Blocked and Stem Out with No Vacuum.
- Stem in and This Passage Blocked with Vacuum Applied

WARNING: Do not cool the bimetal sensor while the engine is running. If refrigerant R-12 is drawn into the intake system while the engine is running, poisonous phosgene gas will be expelled into the test area. Perform this test only in a well-ventilated area. Wear goggles when using refrigerant.

---

7. Start and run the engine briefly (less than 15 seconds). The duct door should move to the HEAT ON position. If the door does not move or moves only partially, replace the sensor. Cool the CWM and bimetal sensor.

8. Shut off the engine and observe the duct door:
   - Vehicles with retard-delay valve: Valve will return slowly to the HEAT OFF position (10 to 30 seconds).
   - Vehicles with CWM: Valve will stay in the HEAT ON position for at least 2 minutes. If less than 2 minutes, replace the CWM and repeat this step after cooling the CWM and bimetal sensor.

Testing the Air Door Vacuum Motor

To check the air door, proceed as follows:

1. Check all hoses for proper hookup. Check for kinked, plugged, or damaged hoses.
2. With the engine off, observe damper door position through snorkel opening. If the position of the snorkel makes observation difficult, use a mirror.

3. At this point, the damper door should be in such a position that the heat stove passage (Figure 11-57) is covered (snorkel passage open). On some engines in cold temperatures, the damper door might be closed due to the use of a vacuum check or delay valve. Momentarily disconnect the vacuum hose at the thermostatic vacuum motor and observe the damper position. If the stove passage is not covered, check for binds in the linkage.

4. Apply at least 7 inches mercury of vacuum to the diaphragm assembly through the hose disconnected at the sensor unit. The damper door should completely close the snorkel passage when vacuum is applied. If not, check to see if the linkage is hooked up correctly and not binding. Check for a vacuum leak in the diaphragm assembly. Replace the assembly, if necessary.

To remove a defective air door vacuum motor, proceed as follows:

1. Remove the air cleaner.
2. Disconnect the vacuum hose from the motor.
3. Drill out the two spot welds with a 1/16-inch drill; then enlarge as required to remove the retaining strap. Do not damage the snorkel tube (Figure 11-58).
4. Remove the motor retaining strap.
5. Lift up the motor, cocking it to one side to unhook the motor linkage at the control damper assembly.

To install a new air door vacuum motor, proceed as follows:

1. Drill a 7/64-inch hole in the snorkel tube at point A (Figure 11-59).
2. Insert the vacuum motor linkage into the control damper assembly.
3. Use the motor retaining strap and sheet metal screw or pop rivet provided in the motor service package to secure the retaining strap and motor to the snorkel tube.
4. If a screw is used, make sure that it does not interfere with the operation of the damper assembly. Shorten the screw if required.
5. Connect the vacuum hose to the motor and install the air cleaner.

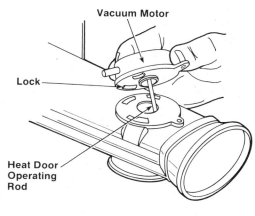

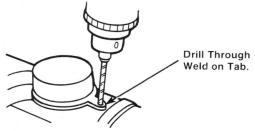

FIGURE 11-58   Apply vacuum at the sensor unit

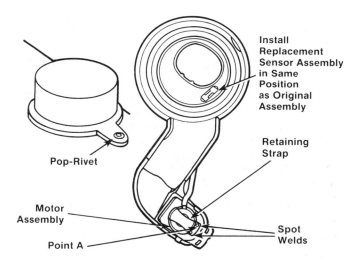

FIGURE 11-59   Removing THERMAC motor

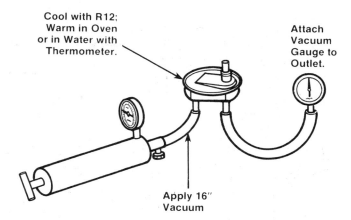

FIGURE 11-60   Testing the air cleaner bimetal sensor

Testing the Air Cleaner Bimetal Sensor

The bimetal sensor tests should be made out of the air cleaner tray. It must have a specified vacuum output at both ends as given in the vehicle's service manual. Table 11-7 shows a chart indicating typical bimetal sensor checks.

To test the bimetal sensor, proceed in the following manner (Figure 11-60):

1. Attach a vacuum to one outlet of the sensor.
2. Cool the sensor with R-12 refrigerant.

WARNING: Wear goggles or safety glasses when using the refrigerant. The engine must be off before cooling with refrigerant.

3. Warm sensor in an oven or in water with a thermometer (Figure 11-61) to the specified temperature.
4. Apply 16 inches mercury to the other sensor outlet.

TABLE 11-7: TYPICAL BIMETAL SENSOR CHECKS				
Color Code	Temp. Setting	Vacuum Output with 16 In. Hg. at Source	Bleed Setting	Vacuum Output With 16 In. Hg. Source
Brown	75° F	1 to 5 In. Hg.	40° F	11 In. Hg.
Purple, Black	90° F	1 to 5 In. Hg.	55° F	11 In. Hg.
Green, Yellow	105° F	1 to 5 In. Hg.	70° F	11 In. Hg.
Pink	90° F	4 to 9 In. Hg.	65° F	11 In. Hg.
Blue	105° F	4 to 9 In. Hg.	80° F	11 In. Hg.

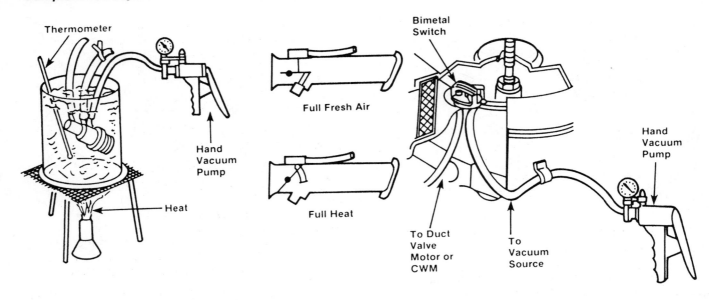

FIGURE 11-61 Checking a TVS valve and vacuum motor

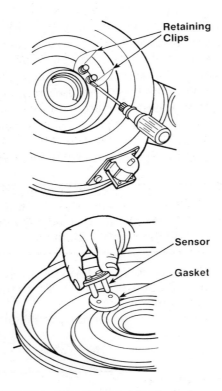

FIGURE 11-62 Removing a fault sensor

5. Replace the sensor if it fails either test mentioned in Table 11-7.

To remove a faulty sensor, proceed as shown in Figure 11-62:

1. Remove the air cleaner.
2. Detach the hoses at the sensor.

3. Pry up the tabs on the sensor retaining clip. Remove the clip and sensor from the air cleaner. Note the position of the sensor for installation.

To install a new sensor, proceed in the following manner:

1. Install the sensor and gasket assembly in their original positions.
2. Press the retainer clip on the hose connectors.
3. Connect the vacuum hoses and install the air cleaner on the engine.

Testing the Cold Weather Modulator (CWM)

This test can be conducted in or out of the air cleaner housing. The engine should be off if conducted in the cleaner. To make the CWM test, proceed as follows:

1. Attach the vacuum source to the larger (center) port (Figure 11-63).
2. Apply 16 inches mercury of vacuum and trap.
3. Cool with R-12; then heat the unit. The modulator must hold the vacuum cold and release it.
4. If the modulator fails the test, replace.

Testing the Retard-Delay Valve

To check a typical retard-delay valve found in some air cleaner systems, proceed as follows (Figure 11-64):

1. Attach the vacuum pump to the small port of the retard-delay valve.
2. Attach a vacuum gauge to the larger gauge.
3. Apply 10 inches mercury of vacuum and trap.
4. Check that vacuum gauge reads 10 inches.
5. Open the bleed valve on the vacuum pump and count the seconds it takes to drop from 8 inches mercury of vacuum to zero vacuum. Compare the drop times against those given in the service manual. If the valve does not meet delay time specifications, replace it.

Troubleshooting the Heated Inlet Control System

In a modern gasoline engine, emission controls that have any effect on the mixture necessarily have an effect on driveability, since all the systems are calibrated to operate as one integral system. Problems with the heated air inlet fall into the following groups:

- Obstructions to airflow, causing hard starting, performance problems, and poor fuel economy
- Vacuum loss, causing cold driveability problems since the system goes to full fresh air
- Air leaks that bypass the control and usually cause cold driveability problems.
- Failure of the vacuum trap (CWM or retard-delay valve) can cause stumble on cold acceleration because the engine suddenly gulps in cold air
- Failure to switch to fresh air can overheat the mixture and cause warm driveability prob-

lems and detonation. A summary of possible problems and their remedies in a heater air inlet system is given in Table 11–8.

MANIFOLD HEAT CONTROL VALVES

As described in Chapter 10, there are two basic types of manifold heat control valves.

1. Vacuum-operated valve
2. Thermostat-operated valve

Vacuum Exhaust Heat Control Valve

The typical exhaust heat control valve such as that shown in Figure 11–65 operates as follows:

1. When a cold engine starts, the valve is closed by the intake manifold vacuum acting on the vacuum motor diaphragm.
2. The valve stays closed until the coolant PVS warms up and vents the vacuum or the vacuum drops significantly due to the throttle opening in acceleration.

Both electronic and nonelectronic systems are in use: this chapter covers nonelectronic systems; electronic control systems are mentioned in Chapter 12. Figure 11–66 shows the basic nonelectronic system with the symbols that appear on the emission decal schematics but without the actual hose routing. The major components are:

- Manifold vacuum, which is connected to the top port of a three-port PVS (ported vacuum switch). The PVS is labeled VCV (vacuum control valve) on the schematic.

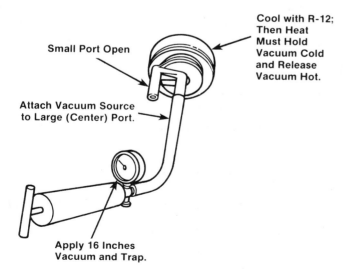

FIGURE 11–63 Testing a cold weather modulator

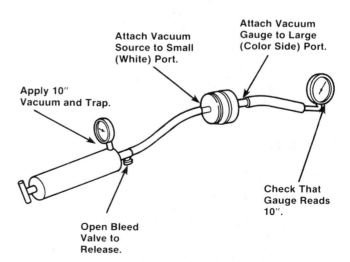

FIGURE 11–64 Testing a retard-delay valve

TABLE 11-8: HEATER AIR INLET SYSTEM PROBLEMS AND THEIR REMEDY

Condition	Possible Cause	Remedy
Cranks normally but starts hard Hesitates or stalls on acceleration Lack of power Poor fuel economy	Debris in snorkel or air door Collapsed air duct Dirty air filter	Check for restricted airflow and correct the problem.
Hesitates on cold acceleration Lack of power Poor fuel economy	Vacuum leak Bimetal stuck in bleed-off Vacuum motor diaphragm leak Air door stuck open.	Correct to be sure system has heated air.
Warm driveability problems in cold ambient air conditions.	Heated air duct leaking or not installed Air cleaner not sealed tight Broken or leaking duct or air cleaner	Correct all fresh air leaks in the system.
Lack of power in warm engine Might overheat Might detonate	Air door stuck open Sensor fails to bleed off when it warms	Check air door motor and sensor. Repair or replace as necessary.
Poor acceleration cold Hesitates or stumbles on acceleration	No vacuum trap to air door motor (at full throttle) Check valve leaks in cold weather modulator Retard-delay valve leaks.	Check air door motor, CWM, and retard-delay valve; replace if faulty.

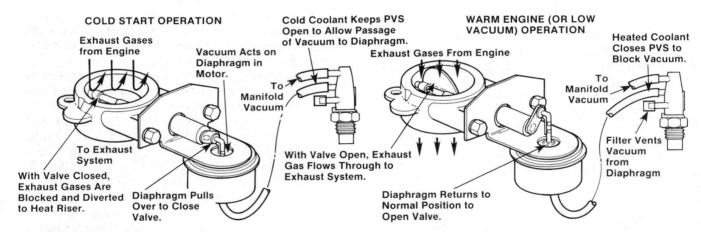

FIGURE 11-65 Operation of an exhaust heat control valve

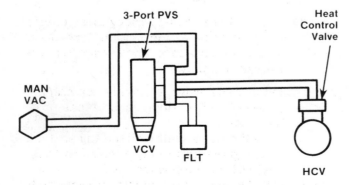

FIGURE 11-66 Basic nonelectronic system with emission decal schematic symbols

• The top port, which is open to the center port and the heat control valve (HCV) diaphragm when the PVS is cold. When the engine coolant warms, the PVS switches and opens the center port to the bottom port. Manifold vacuum is blocked at the PVS. Vacuum in the HCV motor is vented through the filter (FLT) on the PCV lower port.

Testing and Maintenance of a Vacuum-Operated Valve

To check a vacuum-operated valve, proceed as follows (Figure 11-67):

1. Inspect the valve for any abnormal condition. Repair or replace it as necessary.
2. Disconnect the hose from the PVS.
3. Apply 15 inches mercury vacuum to the vacuum motor diaphragm. Trap for 60 seconds. The valve must leak no more than 2 inches vacuum in 60 seconds.
4. Watch the position of the vacuum motor stem. It must go to full closed with the vacuum on and to full open when the vacuum is released.
5. If necessary, lubricate the shaft with a graphite lube.

Functional Testing a Vacuum-Operated Valve System

To check the function of a vacuum-operated valve, proceed as follows:

 1. With the engine off, note the position of the valve motor stem. It should be fully extended (Figure 11-68).
2. Start the engine cold. The stem should pull into the vacuum motor housing.
3. If the stem does not move, stop the engine. Attach a vacuum pump to the vacuum motor diaphragm and apply 15 inches vacuum. The valve stem must pull in and stay in with the vacuum trapped. It may not leak more than 2 inches in 60 seconds.

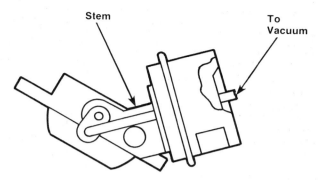

FIGURE 11-68   Valve motor stem should be fully extended

4. Lubricate the valve shaft if the stem sticks when pulling in or releasing.
5. If the valve and motor are all right, check the vacuum system. Look for a leaking or restricted hose; repair as necessary. Check the hoses for proper routing.
6. If the valve closes with the engine started, let it warm. When the coolant warms, it should close.

If the valve operates with a separate vacuum source, but not from the PVS, test the PVS as described earlier in this chapter.

Because of its location, the exhaust heat control valve can stick if the shaft is not serviced regularly with graphite lubricant or heat control solvent. Sticking closed can cause overheating, detonation, and hot performance problems. Sticking cold can cause poor idle or poor performance cold. A vacuum loss usually causes the valve to fail closed, resulting in cold performance conditions. Table 11-9 is a summary of possible vacuum valve problems and their remedies.

Thermostat-Operated Heat Control Valve

Although a thermostat-operated heat control valve is not as common as the vacuum type, some vehicles have them. To inspect and service this type of valve, proceed as follows:

 1. Inspect the valve assembly for damage (Figure 11-69). Replace or repair it as necessary.
2. Turn the valve shaft by hand to see if it is free. It must turn freely and return to the CLOSED position when cold.
3. Cool or heat the thermostat as required to check for proper opening or closing.
4. Lubricate the valve with a graphite lube.

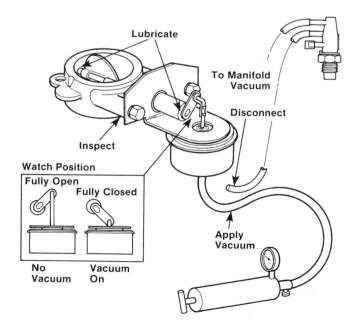

FIGURE 11-67   Testing a vacuum-operated valve

TABLE 11-9: VACUUM HEAT-CONTROL VALVE PROBLEMS

Condition	Possible Cause	Remedy
Poor idle when cold Uneven acceleration when cold Poor performance Hesitation on cold acceleration	Valve stays open; no heat to riser	Check for vacuum leak, PVS failed in warm position, or motor diaphragm leaks; correct.
Overheating Hard starting hot Poor idling Hot stalling Detonation	Heat to the riser when engine is hot caused by mechanical sticking closed or PVS failed in cold position.	Check possible cause and correct.

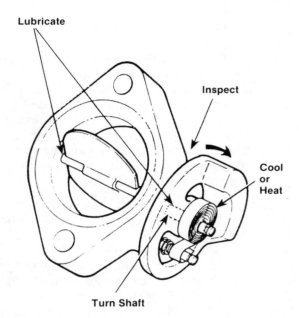

FIGURE 11-69 Inspect the valve assembly for damage

EARLY FUEL EVAPORATION (EFE) HEATER

The EFE heater system operation is best understood and described by looking at the electrical diagram such as the one shown in Figure 11-70.

An open circuit in EFE will cause no heating of the mixture on a cold engine resulting in performance problems the first minute or two of cold operation. If the heater is powered continually for some reason, it can cause warm engine driveability problems and possibly overheating and detonation.

To perform a typical EFE electrical system check, use a 12-volt test light and jumper wires as needed for open-circuit testing. A service manual should also be available to provide the circuit check-

point; a typical manual electrical circuit (Figure 11-70) is used as an example of how to check the coolant temperature switch or solenoid and heater relay. Start the procedure as follows:

 1. If the heater relay does not click when the engine is cold:
   - Test for power with the key in the ON position. If there is none, check the fuse link.
   - Remove the wire at point 45 in Figure 11-70 and jump to ground. Replace the switch if it is closing when the engine is cold.

2. If the relay clicks but the heater does not, check the following:
   - Test for power at the heater relay. If there is none, repair the fuse link.
   - Test for power out of the relay. Replace the relay if it is defective.
   - Test for power to the heater. Replace the heater if it does not work.

3. If the heater is on constantly, check the following:
   - Test to be sure that the coolant switch fails to close.
   - Test to be sure that there is no short to ground in the electrical circuit.

Vacuum EFE Controls

Although most EFE systems are similar to the one just described, some use vacuum in place of an electrical arrangement. The method of testing a vacuum system is basically the same as the heated air inlet controls described earlier in the chapter.

Before testing the coolant temperature vacuum switches, carefully inspect the vacuum lines. A cracked, pinched, or burned hose can cause an EFE malfunction. Then, with the engine running cold (coolant at room temperature), apply vacuum from a pump to the inlet port of the thermal vacuum switch.

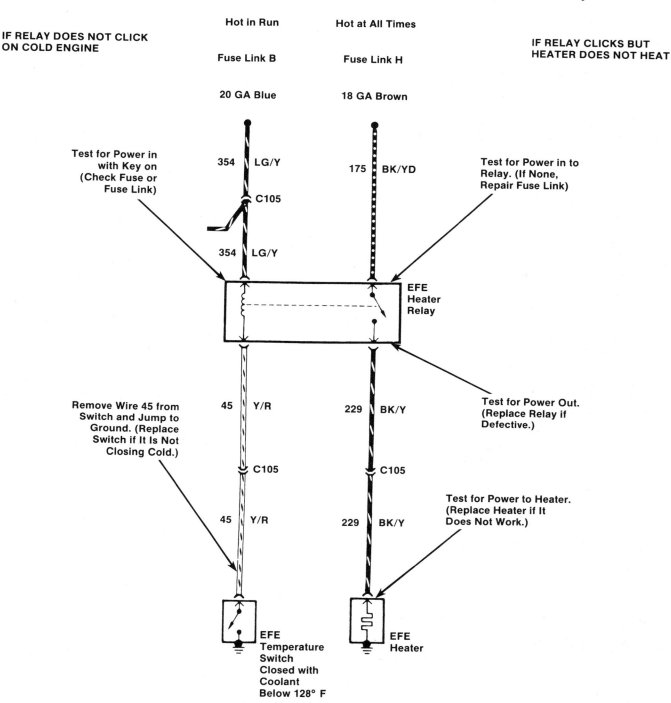

FIGURE 11-70   Testing an electric circuit

Check for vacuum at the outlet port. If there is none, replace the thermal vacuum switch.

With the engine warmed up, the thermal vacuum switch must be closed. Remember that on engines that use an oil temperature sensitive switch, it might take up to 10 miles of cold weather driving to reach the temperature at which the switch will close.

To check an EFE relay or valve for proper operation, inspect the valve and linkage for obvious damage. Apply at least 10 inches mercury of vacuum from a vacuum source of the EFE valve actuator motor. The valve should move freely to the CLOSED position, and the vacuum diaphragm should hold vacuum for at least 1 minute.

If the valve binds, lubricate it with special heat valve lubricant following the directions on the can. If the lubricant will not free the valve, it must be replaced. If the diaphragm lets the vacuum leak down in less than a minute, replace the vacuum actuator.

AIR INJECTION SYSTEM

As described in chapters 9 and 10, fresh air is supplied to the exhaust system by either a pump-type air injection system or a pulse air injection system. In the hot exhaust system, the oxygen in the air combines with some of the HC and CO gases to change them to water vapor and carbon dioxide. It is important to keep in mind that neither system will cause an engine operation problem if it fails. There will be no poor idling or poor driveability, unless backfiring in the exhaust system could be considered a driveability problem. Instead, there might be noise (rumbles, chirps, knocks, squeaks), and there will be an increase in the HC and CO levels.

PUMP-TYPE AIR INJECTION SYSTEM

The typical system (Figure 11–71) usually consists of the following:

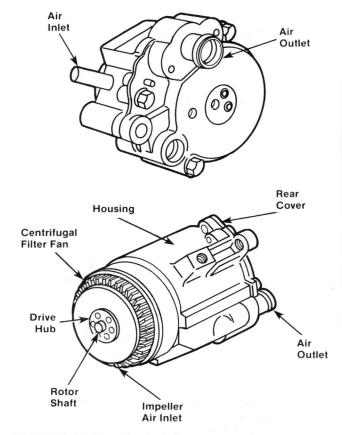

FIGURE 11–72 Typical thermactor air pump

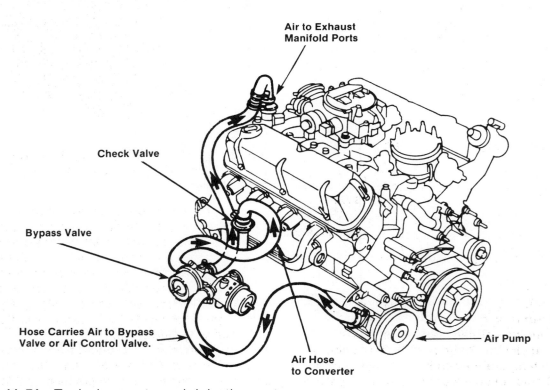

FIGURE 11–71 Typical pump-type air injection system

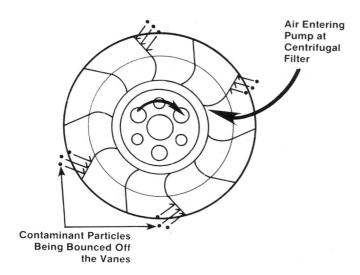

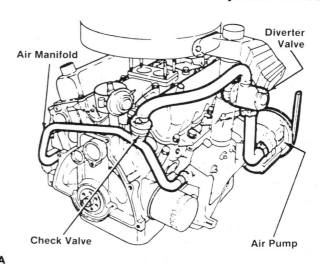

A

FIGURE 11-73 Centrifugal filter operation

- A positive displacement, vane-type air pump motor (Figure 11-72) that is driven by a belt on the front of the engine and supplies the air to the system. Intake air passes through a centrifugal filter fan (Figure 11-73) at the front of the pump where foreign materials are separated from the air by centrifugal force.
- A control diverter or bypass valve that directs air through a check valve to the exhaust ports or the atmosphere.
- A check valve that prevents backflow of exhaust into the pump in the event of an exhaust backfire or pump drive belt failure.
- The necessary plumbing, which can be either internal (Figure 11-74A) or external (Figure 11-74B)

Operation of an Air Injection System

As mentioned in Chapter 10, various auto manufacturers have different names for their pump-type air injection systems. Each system can vary slightly, but basic operation is the same. However, it is always wise to check the vehicle service manual for the exact procedures for testing the air injection system.

In any system, fresh air is supplied when the engine is running but is dumped or bypassed to the atmosphere in modes where the exhaust is rich. There are two typical types:

1. *Conventional Air System.* This system supplies fresh air to the exhaust manifold (upstream air only); it uses an air bypass or diverter valve. It can use a conventional

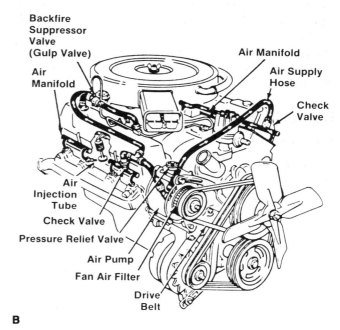

B

FIGURE 11-74 Air injection system with (A) internal plumbing; (B) with external plumbing

catalytic converter arrangement. The operating schematic is given in Figure 11-75.

2. *Managed Air System.* This system supplies fresh air to the manifold or the catalytic converter (upstream or downstream air); it uses a combination or switching air valve. It is used with a three-way catalyst (dual bed converter) and controls airflow in the following ways:
   - To the atmosphere when the exhaust is rich. This prevents backfire and catalyst damage and removes pump load from the engine.

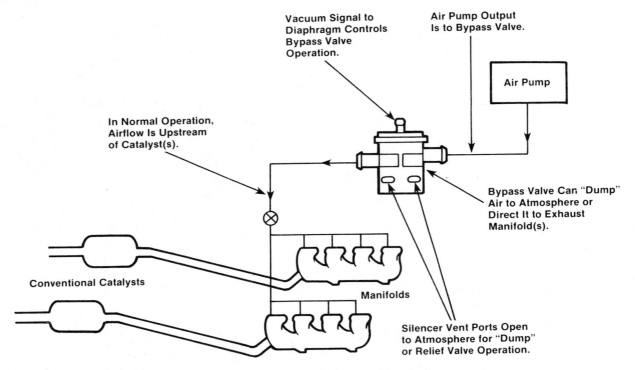

FIGURE 11-75 Operating schematic for conventional air pump system

- Upstream to the exhaust manifold when the engine is first started and the catalyst is still cold.
- Downstream to the mid-bed port of the catalyst when the engine/catalyst warms up.

Operating schematics for the two principal managed system are shown in Figure 11-76.

Air Pump

The air pump rarely causes a problem, only noise. The system is never completely free of noise. In a positive displacement pump, the noise increases in pitch with this frequency; that is, with any increase in engine rpm. A slightly noisy new pump might become less noisy with break-in mileage; 500 miles is sufficient to break in the pump. If the noise persists after break-in or is excessive, replacement of the pump might be necessary. Other causes of noise at the pump pointed out in Table 11-10 are a loose pulley, loose mounting, incorrectly adjusted belt, or kinked or restricted hose.

The modern air pump motor requires no periodic maintenance and with most types should never be oiled. Since service parts are not readily available, most shops install a new or rebuilt air pump rather than try to repair it.

To conduct an air pump output test, proceed as follows:

1. Start the engine, set the parking brake, place the transmission in PARK or NEUTRAL, and warm to normal operating temperature. Shut off the engine.
2. Disconnect the air outlet hose from the diverter or bypass valve.
3. Attach a pressure gauge to the air outlet hose (Figure 11-77).

CAUTION: Make certain that the gauge is securely fastened in place so that it cannot be blown off the hose when pump pressure is applied. Also be sure that the gauge is arranged so that output air is directed away from the technician.

4. Connect a tachometer and restart the engine.
5. Increase engine speed slowly up to 1000 rpm and check the reading on the gauge. If the output pressure is 1 psi or more, the air pump is operating properly. If the output pressure is below 1 psi or is not steady, the pump should be replaced.

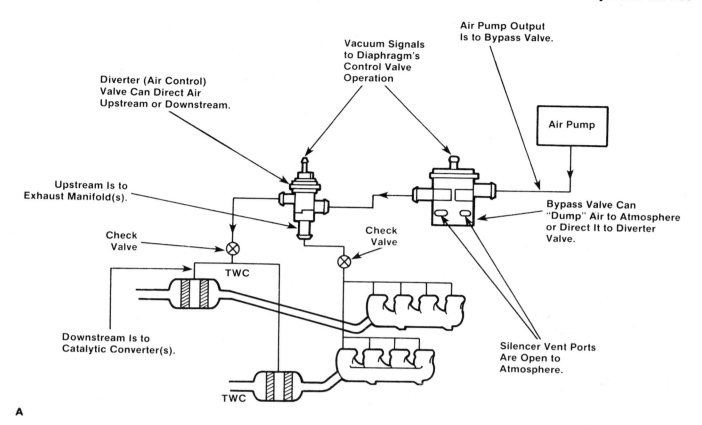

Diverter (Air Control) Valve Can Direct Air Upstream or Downstream.

Vacuum Signals to Diaphragm's Control Valve Operation

Air Pump Output Is to Bypass Valve.

Air Pump

Upstream Is to Exhaust Manifold(s).

Check Valve

TWC

Check Valve

Bypass Valve Can "Dump" Air to Atmosphere or Direct It to Diverter Valve.

Downstream Is to Catalytic Converter(s).

TWC

Silencer Vent Ports Are Open to Atmosphere.

A

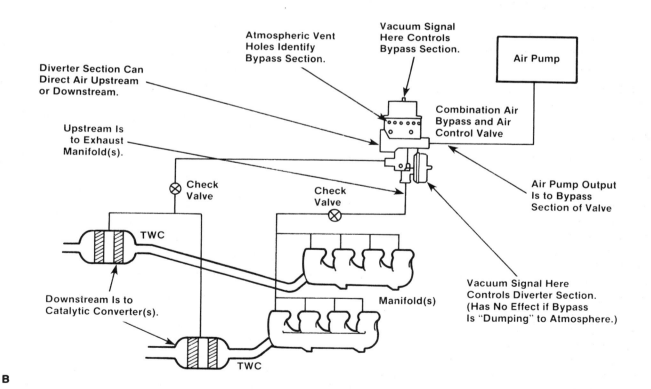

Diverter Section Can Direct Air Upstream or Downstream.

Atmospheric Vent Holes Identify Bypass Section.

Vacuum Signal Here Controls Bypass Section.

Air Pump

Upstream Is to Exhaust Manifold(s).

Combination Air Bypass and Air Control Valve

Check Valve

Check Valve

TWC

Air Pump Output Is to Bypass Section of Valve

Downstream Is to Catalytic Converter(s).

Manifold(s)

TWC

Vacuum Signal Here Controls Diverter Section. (Has No Effect if Bypass Is "Dumping" to Atmosphere.)

TWC

B

FIGURE 11-76 (A) System with air bypass and diverter valves; (B) with combination bypass/diverter valve

TABLE 11-10: AIR PUMP NOISE DIAGNOSIS

Condition	Possible Cause	Remedy
Excessive belt noise	Loose belt	Torque to manufacturer's specification. CAUTION: Do not use a pry bar to move an air pump.
	Seized pump	Replace pump.
	Loose pulley	Replace pulley and/or pump if damaged. Torque bolts to manufacturer's specifications.
	Loose or broken mounting brackets or bolts	Replace parts as required and torque bolts to specification.
Excessive mechanical clicking	Overtightened mounting bolt	Torque to manufacturer's specification.
	Overtightened drive belt	Same as loose belt.
	Excessive flash on the air pump adjusting arm boss	Remove flash from the boss.
	Distorted adjusting arm	Replace adjusting arm.
Excessive air injector system noise (putt-putt, whirling, or hissing)	Leak in hose	Locate source of leak using soap solution and replace hoses as necessary.
	Loose, pinched or kinked hose	Reassemble, straighten, or replace hose and clamps as required.
	Hose touching other engine parts	Adjust hose to prevent contact with other engine parts.
	Bypass valve inoperative	Test the valve.
	Check valve inoperative	Test the valve.
	Pump or pulley mounting fasteners loose	Tighten fasteners to specification.
	Restricted or bent pump outlet fitting	Inspect fitting and remove any flash blocking the air passageway. Replace bent fittings.
	Air dumping through bypass valve (at idle only)	On many vehicles, the thermactor system has been designed to dump air at idle to prevent overheating the catalyst. This condition is normal. Determine that the noise persists at higher speeds before proceeding.
	Air dump through bypass valve (decel and idle dump)	On many vehicles, the thermactor air is dumped in air cleaner or remote silencer. Make sure hoses are connected and not cracked.

TABLE 11-10: AIR PUMP NOISE DIAGNOSIS (CONTINUED)		
Condition	**Possible Cause**	**Remedy**
Excessive pump noise (chirps, squeaks, and ticks)	Loose pulley or mounting bolts or worn or damaged pump	Check the air injector system for wear, loose pulley mounting bolts, or damage and make necessary corrections.
	Loose or leaking hose	Locate leak with soap solution. Replace hose or tighten connection.
	Hose touching other engine parts	Reposition and secure hose.

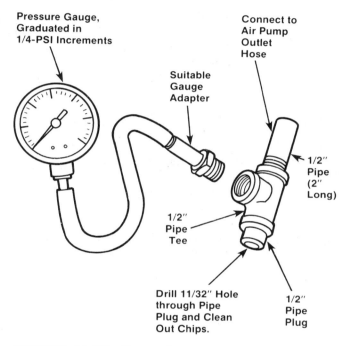

FIGURE 11-77   Type of adapter to be used with a pressure gauge on the air pump

Pump Removal.   To remove the air pump from the system, proceed as follows:

1. Disconnect all hoses attached to the air pump.
2. Relieve drive belt pressure by loosening the bolts holding the adjusting arm and mounting brackets (Figure 11-78). Then remove the pump's drive belt. On some vehicles, it might be necessary to loosen the air conditioning to get at the air pump belt.
3. Remove the air pump pulley and mounting bolts.
4. Lift the pump away from the mounting bracket and out of the engine compart-

ment. If any valves are fastened to the pump, it should be removed from the motor.

Pump Reinstallation.   To reinstall the pump, proceed in the following manner:

1. If any valves were removed from the air pump, refasten them on the new unit.
2. Set the air pump back on the engine and loosely install the mounting bolts. If a gasket was used on the rear of the pump, install a new one in the same location on the new unit.
3. Install the air pump pulley and torque the mounting bolts to approximately 10 foot-pounds. Replace the air conditioning compressor and drive belt if they were removed.

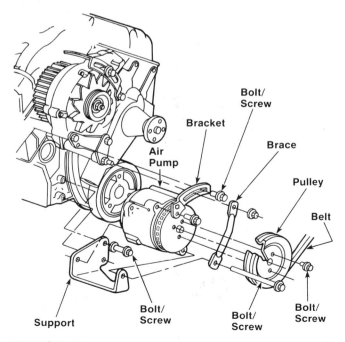

FIGURE 11-78   Air pump mounting

4. Adjust the drive belt to the proper tension as described later in this chapter.

5. Torque the mounting bolts to the manufacturer's specifications.

6. Using new clamps, connect all hoses. Make sure all are securely fastened.

Drive Belt Adjustment. Before attempting to adjust the drive belt, inspect it carefully for wear, cracks, or other damage. If necessary, install a new belt, but be sure it is installed as shown in the service manual (Figure 11–79). When making a belt adjustment, be sure the engine is hot, then proceed as follows (Figure 11–80):

1. Check the pulley mounting nuts. Torque to the manufacturer's specifications given in the service manual.

2. Check the belt tension with a tension gauge. (See service or shop manual for the specifications.)

3. To adjust the belt, loosen the mounting and adjusting arm bolts.

4. Move the pump and hold it at the proper tension. Most vehicle manufacturers provide a special belt tension tool for this purpose.

CAUTION: Do not use a pry bar against the pump housing. The pump housing could become dented or creased, which could cause noise or a seized pump.

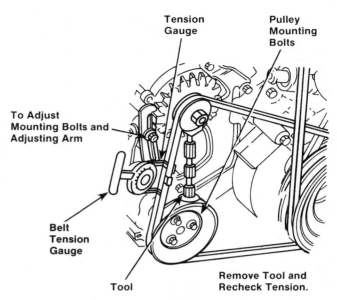

FIGURE 11–80 Air pump drive belt adjustment

5. Remove the tension tool and recheck the tension.

Table 11–10 gives a summary of pump noises, their causes, and how to solve them. However, do not service the air pump unless the car has traveled at least 500 miles.

Testing the Air Injection Valves

Although many early air injection systems used several valves, modern systems use only two or three different types.

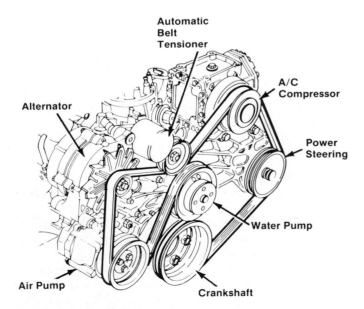

FIGURE 11–79 Drive belt installation

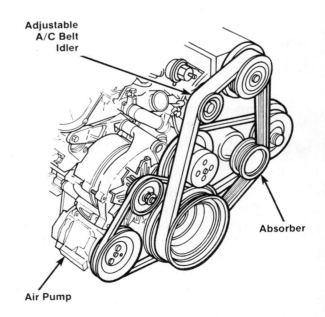

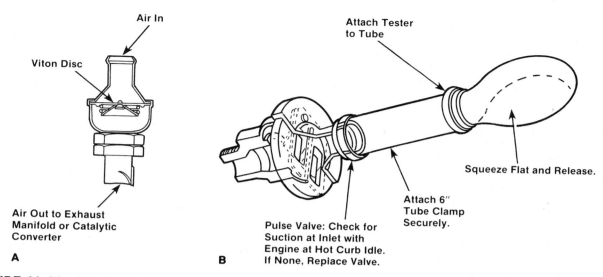

FIGURE 11-81 (A) Air check valve; (B) testing air and pulse check valves

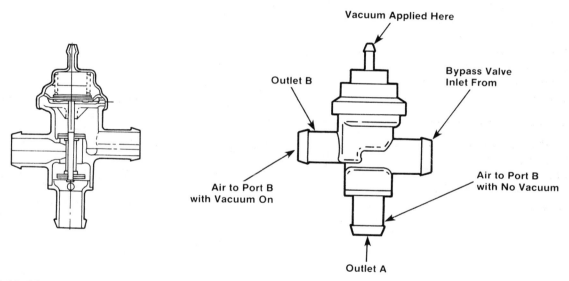

FIGURE 11-82 Air diverter valve

Air Check Valve Test. If a check valve (Figure 11-81A) fails in an air pump system, exhaust gas will escape, contaminating the pump and hoses. This can also cause back pressure loss and offset the EGR system operation. With the engine at a normal operating temperature and turned off, proceed as follows:

 1. Attach the 6-inch tube and clamp securely. Then attach a bulb tester similar to the one shown in Figure 11-81B.
2. Squeeze the bulb flat and release. It must stay creased for 8 seconds and return to normal in less than 15 seconds. Replace the valve if it leaks faster.

Diverter Valve Test. To check a diverter valve, proceed as follows:

 1. With the engine at fast idle (about 1500 rpm), disconnect the air hoses or squeeze the hose to verify the airflow.
2. Remove the air hose and connect the vacuum hose.
3. Outlet A (Figure 11-82) must have airflow with the vacuum hose off; there should be little or no airflow with the vacuum on.
4. Outlet B must have airflow with vacuum on and little or none with vacuum off.
5. Test with the external valve if the valve fails to divert to outlet A with the vacuum on.

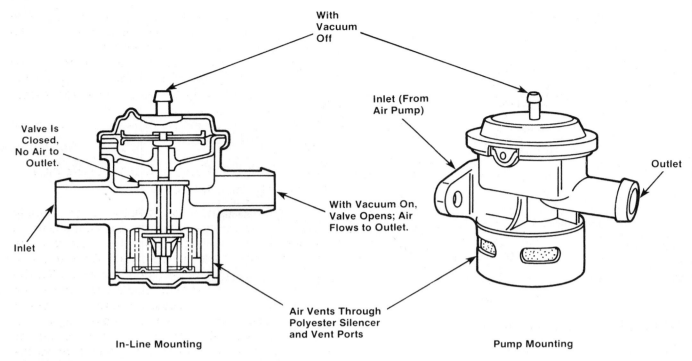

FIGURE 11–83 Normally closed air bypass valve operation

6. Check the control system if the valve is performing properly in Step 5. If not, replace it.

Testing a Bypass Valve. Bypass valves and combination bypass/diverter valves can cause a loss of air to the exhaust if they fail in the bypass mode. If they fail open, they can put air in the exhaust when it is rich. This often causes backfire and can melt the catalyst substrate. To check a normally closed bypass valve (Figure 11–83), proceed as follows:

1. With the engine at a fast idle, remove the vacuum hose (Point A in Figure 11–84); then connect. Air must come out of the vents that are usually located at the bottom of the bypass valve.
2. Connect the vacuum hose. Air must come out of Port B. The vent airflow must be reduced significantly. If there is no airflow, apply the vacuum with an external source.
3. If the bypass valve fails the test, replace it. If the valve works properly only with an external vacuum source, check the vacuum control system.

Combination Air Injector Bypass/Diverter Valve Test. To check the typical combination air injector bypass/diverter valves, proceed as follows:

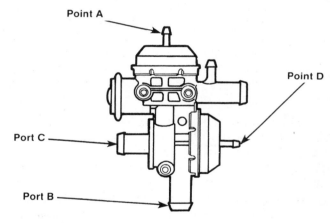

FIGURE 11–84 Testing the combination air bypass/diverter valve

1. Run the engine at fast idle (1500 rpm). Squeeze the hoses to feel the airflow; disconnect if necessary.
2. Apply vacuum at Point A (Figure 11–84), then release. With the vacuum off, air must come out of the dump ports. With vacuum on, air must come out of Port B.
3. Trap the bypass vacuum so the air is from Port B.
4. Apply the vacuum and trap. Air must divert to Port C with vacuum at Point D. When the

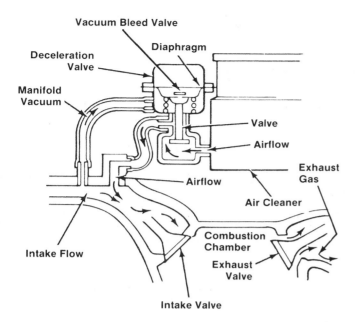

FIGURE 11-85   Deceleration control valve

vacuum is released, the air must switch back to Port B.

5. Test the controls by connecting vacuum hoses one at a time. With the bypass vacuum on, air must come out of Port B. It should shift to Port C with the diverter at 3000 rpm and the engine warm. Check the vacuum sources if they are not operating properly.

Deceleration or Gulp Valve Test. To help prevent backfiring during high vacuum conditions (deceleration), the deceleration or gulp valve (Figure 11-85) allows air to flow into the intake manifold. This air enters the air/fuel mixture to lean the rich condition created when the throttle valve closes during deceleration. When the throttle plate closes, manifold vacuum becomes high. This draws the deceleration valve diaphragm down and opens the valve, allowing air from the air cleaner to flow into the intake manifold. The deceleration valve is used primarily on manual transmission engines.

To check a deceleration valve, proceed as follows:

1. Connect the tachometer, start the engine, and allow the idle to stabilize.
2. With the engine running at specified idle speed, remove the small deceleration valve signal hose from the manifold vacuum source.
3. Reconnect the signal hose and listen for airflow through the ventilation pipe and

into the deceleration valve. There should also be a noticeable speed drop when the signal hose is reconnected.

4. If the airflow does not continue for at least 1 second or engine speed does not drop noticeably, check valve hoses for restrictions or leaks.
5. If no restrictions or leaks are found, replace the valve.

Troubleshooting a Conventional Air Injector System

Remember that the air injector system usually does not affect engine performance. Most problems other than noise will show up only during tail pipe probe testing of the exhaust gas. Anything that causes the air to bypass continually will increase the HC and CO in the exhaust. Usually the cause is a loss of vacuum to the bypass diaphragm or vent. A normally open bypass valve must have a tight connection to the manifold vacuum in order to dump the air. If the vacuum is lost, it will divert air to the manifold all the time. This could cause backfire in rich exhaust conditions or even catalyst damage (if used on a vehicle with a catalyst).

CAUTION: Any time a restricted catalytic converter has to be replaced because of too much exhaust back pressure, be sure to check the air injector system functions.

Table 11-11 is a summary of problems that can occur in a conventional air injector system.

Troubleshooting a Managed Injector Air System (Table 11-12)

Problems in the managed injector air control system result when:

*   Anything that causes the air to bypass continually will increase the HC and CO in the exhaust. Usually the cause is a loss of vacuum to the bypass diaphragm.
*   If the air is diverted upstream too early or with the exhaust too rich, it can cause backfiring in a rich exhaust. This can be caused by a vacuum loss if a normally open air bypass valve solenoid is used.
*   If the air goes downstream when it should be upstream, it can cause catalyst damage in a rich exhaust condition by overheating the catalytic converter.

TABLE 11-11: CONVENTIONAL CONTROL SYSTEMS DIAGNOSIS

Condition	Possible Cause	Remedy
No air to exhaust High tail pipe emissions	Vacuum source leaked or plugged ahead of bypass diaphragm Bypass valve fails closed. WOT valve fails closed. TVS fails closed. Venturi port on WOT valve to wrong source Airflow restricted Timing orifice plugged in bypass valve	Check possible cause and correct the problem.
Air to rich exhaust • Induction backfire • Poor performance if catalyst is damaged	Bypass valve fails open TVS fails open Vacuum hose misrouting	Check possible causes and correct problem.

TABLE 11-12: MANAGED INJECTOR AIR SYSTEM PROBLEMS AND THEIR REMEDY

Condition	Possible Cause	Remedy
No air to exhaust High tail pipe emission	Vacuum source leaked or plugged ahead of air bypass valve diaphragm. Cold weather modulator (or TVS) fails closed. Air bypass valve solenoid stuck closed.	Check possible causes and correct the problem.
High NOx emission Two-way catalytic converter overheated Air upstream only; no downstream	PVS fails closed. Air bypass valve solenoid stuck closed.	Check either possible cause and correct.
Air downstream too soon High HC and CO Three-way catalytic converter	PVS fails open. Air bypass valve solenoid stuck open.	Check either possible cause and correct.
Air upstream with cold engine Backfire Catalytic converter damage	Air bypass valve stuck open. Vacuum connected directly to air bypass valve diaphragm.	Check either possible cause and correct.

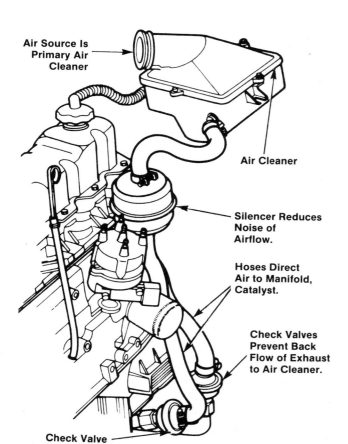

Air Source Is Primary Air Cleaner

Air Cleaner

Silencer Reduces Noise of Airflow.

Hoses Direct Air to Manifold, Catalyst.

Check Valves Prevent Back Flow of Exhaust to Air Cleaner.

Check Valve

FIGURE 11-86 Pulse air system

PULSE AIR INJECTION SYSTEM

The pulse air injection system (Figure 11-86) uses natural pulses in the exhaust system to pull air in. Air flows when the air cleaner pressure is higher than exhaust back pressure. It can deliver fresh air to the exhaust manifold or to the manifold and the catalyst (catalytic converter).

The pulse air system is considered the simplest of all air injection systems. It consists of one or two air check valves (Figure 11-87A) mounted in the line between the air cleaner and the exhaust system. As long as the hoses, fittings, and connections are good and it is possible to feel suction pulsations (with no subsequent exhaust blow back) from the inlet side of the valve, the pulse air system is working correctly.

To check the pulse air check valve, follow the same procedure for the pump-type injector system, except check for suction at the inlet of the valve with the engine at hot curb idle. If there is none, replace the pulse check valve.

Some pulse air systems have a silencer assembly (Figure 11-87B) that filters the air and clamps out pulse noises. If this assembly should become clogged, it should be replaced. A silencer assembly is used on an air pump that is not equipped with a filter fan.

Aspirator Air System

About the only thing that goes wrong with the aspirator air system is failure of the aspirator valve itself. Valve failure usually results in loud exhaust noises under the hood, especially at idle. Also, the rubber hose between the air cleaner and the valve will become brittle and cracked from the escaping exhaust gases that flow back through the hose.

Before condemning an aspirator valve however, check to see if exhaust noises are coming from a leak in the exhaust manifold or in the junction where the aspirator air system connects to the exhaust system.

To determine if the valve has failed, remove the inlet hose (air cleaner side) from the valve. With the

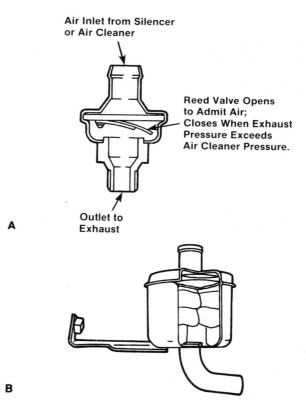

Air Inlet from Silencer or Air Cleaner

Reed Valve Opens to Admit Air; Closes When Exhaust Pressure Exceeds Air Cleaner Pressure.

Outlet to Exhaust

A

B

FIGURE 11-87 Pulse air check valve (A) not interchangeable with check valves in air pump systems; (B) silencer assembly used with pulse air or with air pumps not equipped with a filter fan.

engine at idle, the technician should be able to feel the fresh air being drawn into the valve during the negative pulses in the exhaust system. If the hot gases are escaping from the inlet, the aspiration valve should be replaced.

CATALYTIC CONVERTER INSPECTION

A catalytic converter does not require regular service or maintenance. Its operation should be checked during tune-ups and at state motor vehicle inspections.

 SHOP TALK

Regular tune-up service is essential because any presence of overly rich mixture can cause converter damage.

A converter can fail for any one of a number of reasons. Clogging is a common mode of failure that can cause a severe case of CC "constipation." Free breathing is choked off, backing the exhaust up into the engine and creating excessive back pressure. Clogging can be caused by an accumulation of carbon deposits in the converter, by contamination, by melting, or by breakdown of the ceramic substrate.

In the older style GM converters, for example, the ceramic pellets can fracture and crack as a result of thermal cycling and jostling. The smaller pieces tend to settle and clog the spaces between the pellets, causing a restriction.

As mentioned many times in this book, leaded gasoline quickly poisons the converter and renders the catalyst useless after a couple of tankfuls (there is no way to clean out a poisoned converter). But it is also important to keep in mind that contaminants such as phosphorus and zinc that are found in motor oil can also poison the CC if sufficient oil is drawn past worn valve guides or rings.

Converter meltdown, which also causes plugging, can happen when the converter overheats. The converter is an afterburner of pollutants, so the more pollutants it has to burn, the hotter it gets. Anything that makes the fuel mixture run rich or prevents fuel from being completely burned dumps an extra load on the converter. A single fouled spark plug, for example, can increase the amount of unburned hydrocarbons in the exhaust from 100 parts per million to 1600 to 2000 ppm.

An OE converter is required by law to go 5 years or 50,000 miles. That assumes, of course, that the motorist does not try to save a few cents per gallon by punching out the filler neck on the gas tank to burn leaded gas, and that the engine is regularly serviced and does not suffer any ignition, fuel, or mechanical problems that would dump unburned fuel into the exhaust.

Catalytic converters can become damaged from being bumped or scraped. External physical damage frequently indicates internal damage, too, and the CC might require replacement.

The following are some precautions that will help customers avoid converter damage and should be mentioned to them:

- Avoid pushing any car equipped with a CC and a manual transmission for the purpose of starting the engine. Delayed firing of the power plant can allow unburned fuel to reach the CC. If and when the engine does fire, the accumulated fuel could explode from the extreme heat and damage the CC. Use jumper cables or a booster system.

- With prolonged use, some gasoline additives can be harmful chemically to either type of CC. These additives would include the usual carburetor cleaners and carbon solvents. This type of chemical damage to the CC substrate will not be evident, of course, but the catalytic process would be affected. Check with the manufacturer.

- Avoid running a CC-equipped car for a long period in a closed area. Lack of normal air circulation around the CC can cause it to run even hotter than normal (800 degrees Fahrenheit) and result in the scorching of adjacent sheet metal or body insulation. The sulfuric acid vapor smell is not pleasant, either.

- When jacking or lifting any CC-equipped car built by GM or AM, make certain that the lift pads do not contact the CC unit, which is mounted quite close to the frame. Ford and Chrysler's tubular CC unit is positioned a good distance away from the chassis.

- Under all circumstances—booster-cable starting, starter servicing, compression testing and so on—do not crank the engine for any length of time without having it fire. Raw fuel might collect in the CC and then explode when ignition finally occurs. This precaution also applies to the practice of shutting off the ignition before the vehicle has stopped moving.

- Fix obvious problems like dieseling, power surge, backfiring, sticking choke, or any other problems immediately.

- Do not disconnect spark plugs to test the ignition. If this is unavoidable, do not run the engine for more than 30 seconds when a plug is not firing.

It must be remembered that sulfuric acid or rotten egg odor is not necessarily an indication of a bad converter. Besides controlling HC and CO emissions, a converter can also produce small quantities of its own emissions. Most gasoline contains a small amount of sulfur and other compounds that are not completely removed during refining. The sulfur can react with water vapor that is produced in the converter to make hydrogen sulfide. This toxic substance causes a rotten egg odor, which is usually noticed while the engine is warming up or during deceleration.

This odor can also indicate that the engine is not operating properly, especially when it happens at normal operating temperatures. This is often caused by the engine being out of tune or running too rich. Adjusting the idle screw will not necessarily eliminate the odor since it is probably not caused by an incorrect mixture adjustment but by some other problem in the engine ignition or carburetion.

The amount of sulfur in gasoline blends can vary, so switching brands might eliminate the odor.

CHECKING CATALYTIC CONVERTERS

Assuming all goes well and nothing catastrophic happens to damage the catalytic converter, a new CC will eventually be needed because the converter's ability to clean the exhaust gradually diminishes over time. The catalyst itself never wears out, but its ability to precipitate the chemical reactions that clean up the exhaust gradually deteriorates as soot and other contaminants build up inside the converter.

The EPA is cracking down on what it calls the unnecessary replacement of catalytic converters as well as the use of junk converters that fail to meet minimum test standards. A used converter, for example, will not receive the EPA's approval unless it first passes a test that proves it can still clean up 50 percent of the unburned hydrocarbon (HC) and carbon monoxide (CO) emissions within 2 minutes and 75 percent of the HC and CO emissions within 200 seconds. Aftermarket replacement converters likewise must be certified to show they will continue to work for at least 25,000 miles and not fail due to construction defects for 5 years or 50,000 miles.

Caught in the middle is the installer, who is now required by law to do the following:

1. Verify that the catalytic converter is not missing.
2. Certify in writing that a vehicle's converter must be replaced and why.
3. Get the customer's approval in writing prior to changing the converter (Figure 11–88).
4. Install only a replacement converter that is EPA accepted, meaning it is an OE equivalent, an approved aftermarket converter, or a tested and certified recycled converter.
5. Make sure the existing converter is no longer covered by the vehicle manufacturer's 5-year, 50,000-mile emissions warranty. This means that if the converter fails before the vehicle manufacturer's 5-year, 50,000-mile emissions warranty has expired, the customer is supposed to return to the dealer for free repairs. Under the new regulations, independent repair facilities are prohibited from replacing converters still under warranty except under limited circumstances.

In addition, a copy of this signed statement must be retained for six months and the old converter for a period of fifteen days. Replaced converters must be marked in such a way that they can be identified with particular customer invoices and statements, and be available for EPA inspection.

Of equal importance, the EPA wants to make sure the converters installed are good ones, that the CC is not contaminated or clogged, and that it does a reasonable job of cleaning up the exhaust. The days of selling reconditioned converters that often turn out to be lead poisoned or clogged are over. All approved replacement converters are required to carry a permanent label that identifies what type they are ("N" for new, "U" for used), along with a code number issued to the manufacturer by the EPA, an application part number, and a manufacturing date. If a converter does not have a label, do not install it. Failure to meet any of the above requirements can make you liable for a $2,500 fine for each converter replaced.

Because of EPA regulations, accurate diagnosis of catalytic converter problems has become an absolute must. For example, clogging will produce the characteristic symptoms of an exhaust restriction: loss of power at high speeds, stalling after starting (if totally blocked), a drop in engine vacuum as engine rpms increase, or sometimes popping or backfiring at the carburetor.

There are several ways to test for excessive back pressure. The easiest way is to hook up a vacuum

Catalytic converter are emission control devices which are designed to last the life of the vehicle and do not normally require replacement. Furthermore, if the vehicle is properly used and maintained, original converters are covered by the emissions control warranty for five years or 50,000 miles. Federal law prohibits repair businesses from replacing these devices except under certain limited circumstances.

☐ The vehicle is over five years old or has more than 50,000 miles on it and the catalytic converter required replacement because:

_____

_____

(For example: "The catalytic converter was plugged" or "Unrepairable leaks")

—OR—

☐ The vehicle's catalytic converter was missing when the vehicle was brought in.

_____   _____
Date   Vehicle Owner's Signature

_____
Facility Representative's Signature

FIGURE 11–88 Customer agreement for changing a catalytic converter

gauge to a source of manifold vacuum and then increase the engine speed to 2500 rpm. The vacuum reading should drop momentarily, then return to about the same level (14 to 22 inches mercury) and hold steady for at least 15 seconds if there is no restriction. If the vacuum readings drop off, this might signal an exhaust restriction. The vacuum hose on the EGR valve must be disconnected and plugged prior to performing this test. Otherwise the EGR valve will open, causing a "false" vacuum leak. The one drawback to this test is that a vacuum leak can act like an exhaust restriction.

WARNING: Allow all exhaust system components to cool completely before attempting work on catalytic converters. Heat in excess of 1800 degrees Fahrenheit is generated by the catalytic converter. Severe burns can result from touching or brushing against a hot catalytic converter.

An alternate method, therefore, is to read exhaust back pressure directly. This can be done by disconnecting the air pump check valve from the air injection plumbing and inserting a tapered fitting connected to a fuel pressure gauge into the air injection plumbing. Exhaust back pressure will back up through the plumbing and register on the gauge. At

2500 rpm, the gauge should show less than 2-1/2 pounds of back pressure. If there are 3 or more pounds of back pressure, something is clogging the exhaust. This is one of the method's disadvantages: The technician cannot tell if the converter, a crushed exhaust pipe, or a bad muffler is causing the restriction. If a visual inspection under the car fails to reveal any obvious damage, try disconnecting the exhaust system aft of the converter to see if the readings change (this connection will have to be broken anyway if the converter is going to be replaced). If the back pressure readings are still high, then CC is bad; if the readings suddenly drop, the restriction is before the converter.

Converter efficiency is harder to test because it requires special test equipment. One way is to use a hand-held digital pyrometer, an electronic device that measures heat. By touching the pyrometer probe to the exhaust pipe just ahead of the converter and just aft of the converter, it is possible to read an increase of at least 100 degrees as the exhaust gases pass through the converter. If the outlet temperature is the same or lower, nothing is happening inside the converter. But do not be quick to condemn the converter, however. To do its job efficiently, the converter needs a steady supply of oxygen from the air pump. A bad pump, faulty diverter valve or control valve (three-way CC), leaky air connections, or faulty computer control over the air injection system could be preventing the needed oxygen from reaching the converter.

One new piece of equipment that tests for both internal clogging and temperature efficiency consists of two liquid crystal heat sensors and two pressure gauges and test fittings for tapping into the exhaust system fore and aft of the converter. By noting both temperature and pressure readings, the technician can tell whether or not the converter is working or clogged.

As described later in this chapter, the best way to check the operation of a catalytic converter is with a four-gas infrared exhaust emission analyzer. Its complete operational procedure is given there.

REPLACING A CATALYTIC CONVERTER

Once a CC is determined faulty, replacement is usually best. There are two types of replacement:

1. Installation kits (Figure 11-89). Such kits include all necessary components for the installation.
2. Direct fit catalytic converters (Figure 11-90).

The exception to the replacement (although questioned by many experts in the field) is the pellet converter designed by GM in which the pellets can be replaced without having to remove the converter from the vehicle. To do this, proceed as follows:

1. The plug in CC is removed from the pellet converter. A vibrator is attached to the catalytic converter to shake the pellets into a container (Figure 11-91A).
2. To refill the converter, new beads are placed in the can (Figure 11-91B). The air

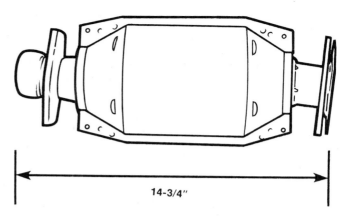

FIGURE 11-90   Direct fit catalytic converter

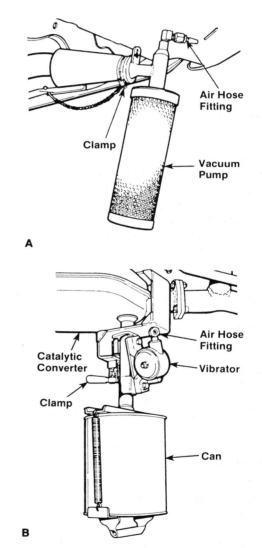

FIGURE 11-91   (A) Aspirator attached to tail pipe; (B) attach vibrator and can of catalyst to the converter fill hole.

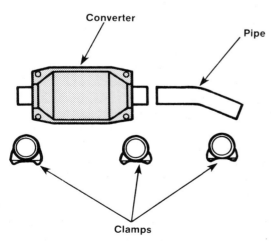

FIGURE 11-89   Installation kit

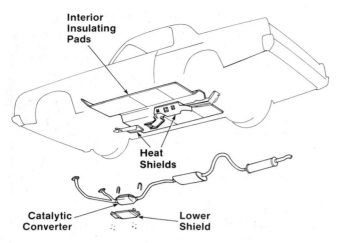

FIGURE 11-92 Location of heat shields

hose is then attached to the vibrator unit, and a vacuum pump is attached to the tail pipe. New beads will be drawn into the converter. When full, the tools are removed and the converter plug is coated with nickel-based antiseize compound and replaced.

The procedure for removing and replacing a CC is basically the same as that for removing or replacing a vehicle's muffler. To remove a converter (after it has had a chance to cool down), proceed as follows:

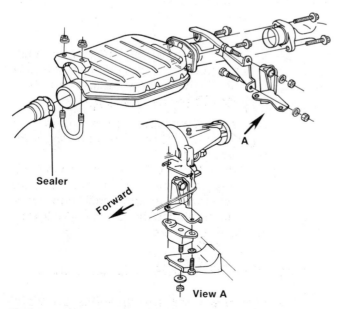

FIGURE 11-93 Fasteners that hold the catalytic converter in position

1. Raise the vehicle to the height desired.
2. Remove any heat shield that surrounds the CC (Figure 11-92).
3. Remove the fasteners that hold the CC in position on the exhaust system tubing (Figure 11-93). Separate the converter from the tubing or pipe and remove it from the vehicle.

To replace the catalytic converter, proceed as follows:

1. Position the upper heat shield (if used) and connect the converter to the inlet pipe. Always install new gaskets.
2. Connect the CC flanges to the exhaust tubing and loosely attach the bolts.
3. Align all parts (including the lower heat shields) and tighten all nuts and bolts to the specifications given in the service manual.
4. To complete the installation, lower the car, start the engine, and check the system for leakage.

 SHOP TALK _____

Remember that each CC installation and replacement might vary slightly from one model to another. Always check the service manual for the proper procedure.

EXHAUST EMISSION ANALYZER

As vehicle emission standards become stricter and emission control devices become more sophisticated, it becomes necessary for the service technician to understand the theory and testing procedures of four-gas engine emissions analysis. In addition to monitoring emissions for inspection and maintenance (I/M), an infrared exhaust gas analyzer doubles as one of the best diagnostic tools available.

The infrared exhaust analyzer (Figure 11-94) operates on a single beam of infrared light passing through the gas sample. The beam is chopped by a rotating wheel containing optical filters, which derive a signal proportional to the percentage of carbon monoxide and the hydrocarbon level in the sample. The signal is electronically processed to separate the carbon monoxide signal and the hydrocarbon signal and apply them to direct reading meters on the analyzer face. The sample is taken from a location in the exhaust system specified on the Emission Control Information Label. The probe is connected to the analyzer through a sampling hose and a water trap to remove moisture.

If the percent of O_2 exceeds the percent of CO, the engine is said to be running to the lean side of stoichiometric (Figure 11-95). On converter/reactor vehicles, if O_2 exceeds CO and the CO reading is .5 percent or higher, chances are the catalyst is damaged or ineffective. It is getting enough oxygen but not reducing CO. Hydrocarbon readings are also usually higher in this condition.

USING CO_2 and O_2 READINGS

Referring to Figure 11-95, it is apparent that the stoichiometric point occurs when CO_2 is on the lean side of 14.7/1 air/fuel ratio (approximate). At the same time, CO and O_2 readings are approximately the same. Notice that stoichiometric occurs just as O_2 begins to increase and CO_2 begins to drop. The idea is to check the vehicle's CO_2, CO, and O_2 readings to ensure proper functioning of the carburetor.

CAUTION: If any adjustment procedures are to be performed, they must be done in accordance with manufacturer's specifications per federal guidelines.

ANALYZER TEST PROCEDURES

The basic tests with an infrared analyzer are performed at curb idle speed. Check the individual

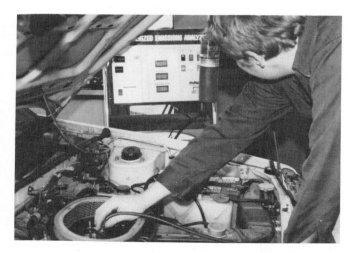

FIGURE 11-94 Check oxygen reading accuracy

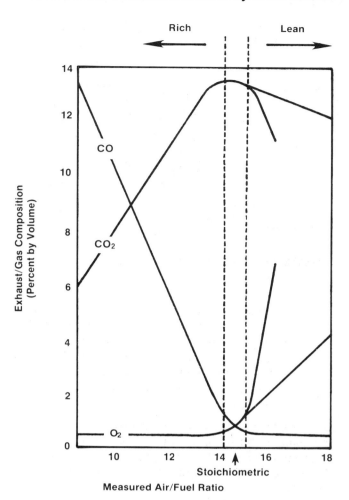

FIGURE 11-95 Stoichiometric point at which fuel mix burns most efficiently

---

manufacturer's procedure on the underhood label, service manual, or technical bulletins for instructions for making the idle speed adjustment (see Chapter 4). Then proceed as follows:

1. Connect a tachometer to the engine. Run the engine until it reaches operating temperature.
2. Read the HC and CO levels on the meters. If the engine is operating at specified curb idle speed and is in good condition, readings comparable to those shown (or lower) should be obtained.

 SHOP TALK _____

Always check the specifications on the Emission Control Information Label in the engine compartment.

Many tests and checks can be performed with the emission analyzer to quickly diagnose and pinpoint the causes of excessive emissions.

Catalytic Converter Check

To check the catalytic converter with the analyzer, proceed as follows:

1. Insert an exhaust gas sample probe into the tail pipe at least 6 inches. The engine must be at normal operating temperature and at curb idle.
2. The CO reading on the analyzer meter should rear near zero. It should not exceed 0.3 percent CO maximum.
3. If the reading is over 0.3 percent CO, advise the owner that repairs or adjustments might be needed to correct excessive emissions because the CC is not oxidizing fuel from the engine. The converter might be faulty.

Steady State Speed Test

Perform this test after completing the basic test procedures at curb idle speed.

1. Accelerate the engine slowly while watching the HC and CO meters.
2. Increase engine speed to about 2500 rpm and allow the engine to stabilize.

When the engine stabilizes, note the HC and CO readings. Both readings should have dropped slowly and steadily to below curb idle speed levels. If the engine reacts as described, efficient fuel system and emission control system operation is indicated. If it does not, the need for further testing is indicated to locate the cause.

 SHOP TALK _____

HC and CO levels are already low at curb idle speed in a properly functioning engine. Because of this, the drop in HC and CO levels might not be great during this test.

Blowout Procedure

Excessive idle time during adjustment can adversely affect HC and CO readings. To prevent this, accelerate the engine to 2000 to 2500 rpm for 10 seconds between measurements. Meter readings should stabilize between 30 seconds and one minute after return to idle speed. If they do not, repeat the blowout procedure. If adjustments require more

than 2 or 3 minutes, repeat this procedure to maintain a clean engine.

Fast Idle Speed Test

The fast idle speed test is a variation of the steady state speed test and is conducted in the following manner:

1. Operate the engine with the fast idle speed screw on the second step of the fast idle cam.
2. Check the tachometer to be sure that the idle speed is as specified on the Emission Control Information Label.
3. Adjust to specifications if necessary. Be sure the engine is at normal operating temperature. Read the HC and CO levels.

Exhaust gas analysis is obviously the most sophisticated way to measure converter efficiency. On CC-equipped cars, HC readings should usually be less than 100 ppm and CO levels under 1.0 percent if the CC is doing its job and the engine is in good condition. If both HC and CO levels are elevated, momentarily disconnect the air pump to see if the readings change. If the CC is bad, the readings will not change. If they go up and then down when the air pump is reconnected, however, the problem is ignition or fuel related.

When using a four-gas exhaust analyzer, look at the oxygen readings in relation to the HC and CO readings. If the oxygen reading is higher than CO, it tells the technician there should be sufficient oxygen in the exhaust for the converter to do its job. If HC and CO levels remain high in spite of plentiful oxygen, however, it is a good indication that the converter is not oxidizing any longer. The engine and converter must be at normal operating temperature before the readings can be relied upon. Some converters cool off enough at idle to "flame out." Running the engine at fast idle for 15 seconds or so should relight the converter and get it oxidizing HC and CO again.

Before operating an analyzer, check the following points:

- Always perform the exhaust analysis in a well-ventilated area. The presence of exhaust gas in the air can affect test accuracy by adding impurities to the test gas sample.
- Make sure the exhaust system of the vehicle being tested is free of leaks. A leaking exhaust can allow outside air to dilute the test sample, resulting in inaccurate readings. Also, make sure the manifold heat valve moves freely and is in the OPEN position.

FIGURE 11-96 Getting a sample of exhaust gas

- On cars with dual exhaust, insert the probe in the side opposite the manifold heat control valve.
- To obtain an accurate sample of exhaust gas for analysis, make sure the infrared's probe is placed at least 18 inches into the tail pipe (Figure 11-96). After the probe has been placed, make sure you secure it to prevent it from falling out during testing.
- Always follow the car manufacturer's instructions (on the emission decal) before disabling any emission control devices including the air pump and emission canister. Normally, when performing diagnostic tests on the engine, the air pump should be blocked off to make all four gases more apparent. However, if following a programmed test sequence found on the more sophisticated computerized models (where the machine tells you what to do), some units require the air pump to remain operative. If you are not sure what to do, read the operating manual supplied with the test equipment.
- Keep a spare analyzer filter on hand. Trying to obtain a test sample with clogged filters can affect the accuracy of the readings. The low flow light is there for a reason, so do not ignore it. Remember to take the probe out of the tail pipe as soon as the diagnosis is complete. The longer it is in, the harder the exhaust's emissions will be on the analyzer's filters, traps, plumbing, and the like.
- Before using the analyzer, allow it to warm up, along with the car. Then calibrate the machine as required. On some models, man-

ual calibration is necessary to insure accurate readings.

At the center of the emission control techniques is the control of the air/fuel ratio. By closely controlling the ratio, the engine can burn the fuel mixture efficiently. The point at which the fuel mixture burns most efficiently is called stoichiometric (see Figure 11-95), and low HC/CO readings are present in the exhaust. As the figure shows, the stoichiometric point is at approximately 14.7/1 air/fuel ratio. This can vary slightly because of the differences in fuel blending.

HC and Air/Fuel Ratio

Figure 11-97 shows the relationship between the air/fuel ratio and HC. The actual point of misfire will vary depending on the vehicle. HC's are low during normal operating air/fuel ratios. At approximately 17/1 and greater, a lean misfire condition begins and HC readings increase.

On preconverter/reactor vehicles, the HC increase was used to find the lean limit when adjusting the idle mixture screws. On vehicles with converter/reactors, O_2 is used to check carburetor adjustment (see Chapter 4).

CO and Air/Fuel Ratio

Figure 11-98 shows the relationship between CO and the air/fuel ratio. CO is a good indicator of a rich running engine between 10/1 and 14.7/1. Above

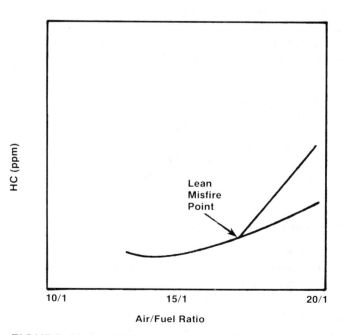

FIGURE 11-97 HC and air/fuel ratio

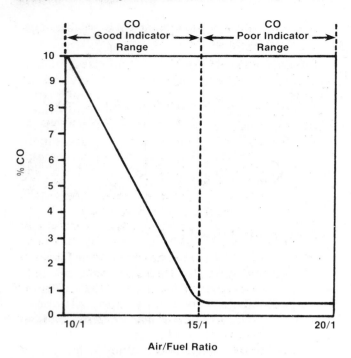

FIGURE 11-98   CO and air/fuel ratio

15/1, CO does not change appreciably. Therefore, CO alone is a poor indicator when air/fuel ratios are lean.

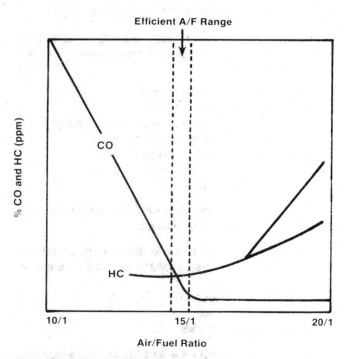

FIGURE 11-99   HC/CO on nonconverter-equipped cars

HC/CO on Nonconverter Vehicles

When checking the carburetor adjustment on nonconverter/reactor vehicles, the idea is to set the carburetor to a point where HC/CO readings are minimal. Figure 11-99 shows air/fuel ratio limits and the relationship of HC/CO readings on nonconverter/reactor vehicles. Richer than approximately 15/1 means CO increases. Leaner than approximately 14.4/1 means HC increases. Somewhere between 14.4/1 and 15/1 is the maximum efficiency range.

HC/CO on Converter/ Reactor Vehicles

Catalytic converters and thermal reactors, when working properly, are so effective they virtually eliminate HC/CO readings from the exhaust. This reduction can sometimes cause confusion during exhaust gas diagnosis. Figure 11-100 shows the effects of tail pipe HC/CO levels on converter/reactor-equipped vehicles. There is a marked lack of HC increase at the lean misfire point. CO readings are very low, and HC is no longer effective for indicating misfire on a lean running engine. Since converter/reactor vehicles must run lean to be effective, some other kind of leanness indicator is needed to check the carburetor's air/fuel ratio.

Oxygen-Lean Indicator

It has already been stated that O_2 is a good indicator of a lean-running engine. By using CO and

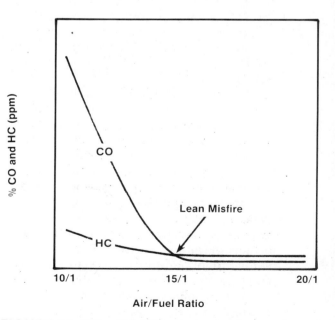

FIGURE 11-100   HC/CO on converter-equipped cars

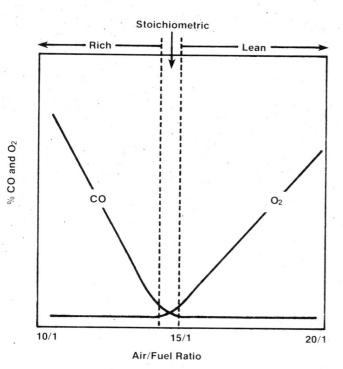

FIGURE 11-101 Pitch and lean indications with CO and O_2

O_2 readings, the carburetor can be checked for efficiency.

If the percent of CO exceeds the percent of O_2, the vehicle is said to be running on the rich side of stoichiometric (Figure 11-101). If this condition occurs on a converter/reactor vehicle, the catalyst will be oxygen-starved and ineffective in reducing HC/CO. This does not necessarily mean the converter/reactor is defective. It might simply mean more O_2 must be added for it to work properly.

High levels of hydrocarbon can be caused by one of the following malfunctions (Figure 11-102):

1. Poor compression
   - Bad valves, guides, or lifters
   - Bad piston rings
   - Bad pistons or cylinders
   - Head gasket leaks or is blown.
   - Valve timing off, loose or slipped timing chain, or valves timed wrong.
   - Wrong pistons or crankshaft
2. Problems usually related to the ignition system such as:
   - Bad spark plugs
   - Bad plug wires
   - Bad dirt cap or rotor
   - Bad coil or condenser
   - Wrong dwell
   - Incorrect timing

3. Extremely lean fuel mixture or overly rich mixture
4. Vacuum leaks
5. PCV system inoperative

A high carbon monoxide level (Figure 11-103) is usually caused by overly rich fuel mixture because of

1. Choke problems such as choke stuck or binding, set too rich, vacuum break set wrong or inoperative, choke thermostat bad, no current to electric choke.
2. Carburetor problems such as float level set incorrectly, leaky float needle and seat, wrong jets, metering rods out of adjustment, mixture adjusted too rich, air bleeds plugged, carburetor air filter dirty or plugged.
3. Heated air intake system not working properly.
4. Heat riser problems
5. PCV valve plugged
6. Crankcase oil contaminated with fuel.
7. Wrong timing
8. Excessive crankcase dilution with fuel caused by fuel pump diaphragm rupture, power valve stuck open, choke stuck closed.
9. Carbon canister system not purging adequately or has not had time to purge.

A higher than normal reading of HC and CO can be caused by one or more of the following conditions:

1. Inoperative PCV system
2. Heat riser stuck open
3. Air pump inoperative or disconnected
4. Air cleaner preheat door stuck
5. Converter not working.
6. Air pump not working.
7. Rich fuel mixture

A lower than normal CO_2 reading can be caused by one or more of the following:

1. Exhaust gas sample dilution because of leaky exhaust system or leaky sample hose/probe
2. Rich carburetor
3. Dirty air filter
4. Defective choke

A lower than normal O_2 and higher than normal CO reading can be caused by one or more than one of the following malfunctions:

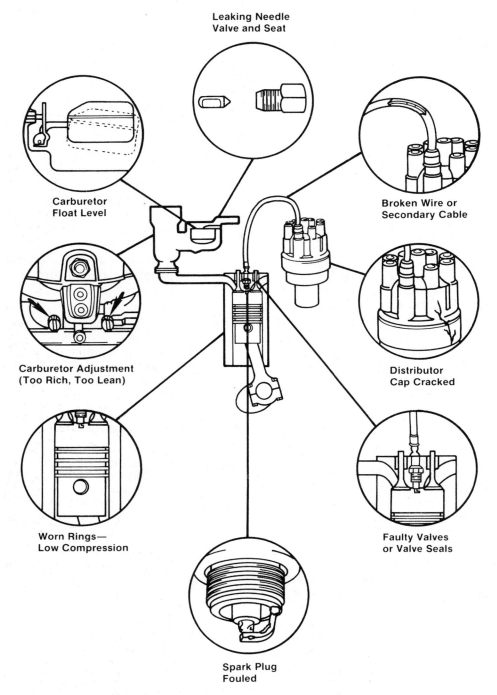

FIGURE 11-102 Causes of high levels of hydrocarbon

1. Rich carburetor
2. Dirty air filter
3. Defective choke
4. Idle mixture adjustment improperly set
5. Restricted air bleeds
6. Restricted PCV system
7. Improper idle speed
8. Carbon canister system not purging properly
9. Carbon canister recently subjected to high quantities of fuel vapors and has not had time to purge properly

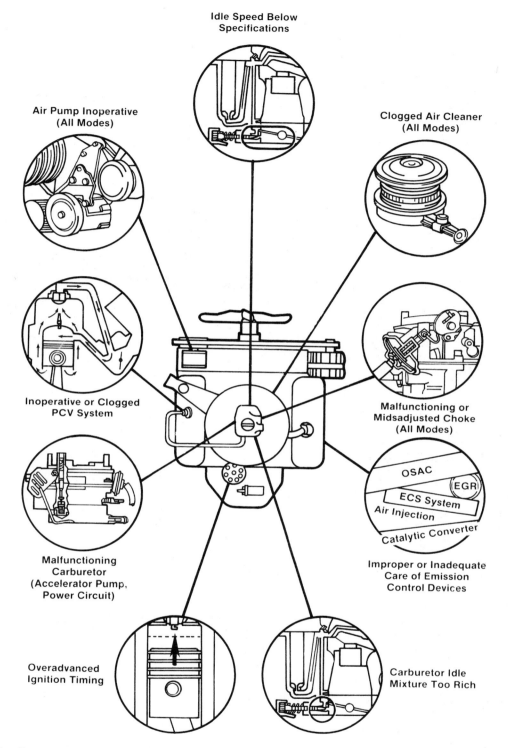

FIGURE 11-103   Causes of high carbon monoxide levels

A higher than normal O_2 and lower than normal CO reading can be caused by one or more of the following malfunctions:

1. Lean carburetor
2. Vacuum leak
3. Idle mixture improperly adjusted

TABLE 11-13: GO/NO GO SPECIFICATIONS FOR VEHICLES WITH EMISSION CONTROL DEVICES

Vehicle Year	HC (ppm) Less Than	CO_2 (%) Greater Than	CO (%) Less Than	O_2 (%) in the Range of
1971–75	400	8	3	0.1–7
1976–79	200	8	2	0.1–7
1980–88	100	8	0.5	0.1–7

4. Improper idle speed
5. Meter rods improperly positioned
6. Internal air/vacuum leak in carburetor
7. Restricted idle system
8. Restricted main metering system
9. Air pump connected during air pump disconnected testing

To pinpoint specific compression and ignition system deficiencies, conduct detailed tests of these areas. After servicing, recheck emission levels using the analyzer. Newer vehicles, when properly tuned, can have CO levels below 0.5 percent and HC levels below 100 ppm (Table 11–13).

Accelerator Pump Test

To check the accelerator pump, proceed as follows:

1. Operate the engine at curb idle speed. Be sure it is at proper operating temperature.
2. Pump the accelerator pedal momentarily then release it while watching the meters on the analyzer.
3. HC and CO levels should rise steadily and rapidly to readings that are at least triple those of the curb idle speed levels.
4. Causes of failures:
   - When the accelerator pump is working properly a steady stream of fuel will be visible at the pump discharge nozzles as the throttle linkage is operated slowly from idle. The accelerator pump in the carburetor is malfunctioning. If the CO level drops off sharply before starting to rise or if there is excessive play in the accelerator pump, the cup is not seating or the check valves are leaking.
   - Certain types of fuels can cause the accelerator pump cup to swell. The swelling causes the accelerator pump to react more slowly than normal, which

results in hesitation, stalling, hard starting, or poor driveability.
   - Rebuild the accelerator pump with a repair kit and perform the accelerator pump test again to verify that proper pump operation has been restored.

Positive Crankcase Ventilation Test

The analyzer provides a quick, positive check on the operation of the positive crankcase ventilation system when used in the following manner:

1. Operate the engine at specified curb idle speed. Be sure the engine is at normal operating temperature. Note the reading on the CO meter.
2. Remove the PCV valve from the cylinder head cover and allow it to hang free. Note the CO reading. It should drop below the CO level at curb idle speed (Figure 11–104). A drop of more than 5 points usually indicates excessive oil dilution with raw fuel. Oil should be changed or a dry-out procedure effected.
3. With the engine operating at specified curb idle speed, place your thumb over the open end of the PCV valve. Watch the CO meter. The CO reading should rise to a level higher than noted at curb idle with the PCV valve installed in the cylinder head cover.
4. If the CO levels do not react as described while performing the positive crankcase

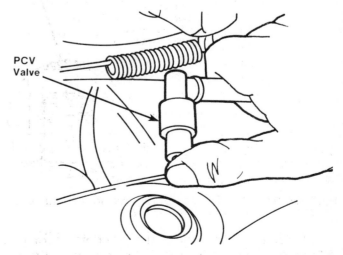

FIGURE 11-104 PCV valve removed from cylinder head cover

ventilation system test, clean or replace the PCV valve and retest the system.

5. Remove the PCV valve from the cylinder head cover and let it draw in fresh air. If this produces a normal or leaner than normal idle mixture, the crankcase lubricant probably has excessive fuel dilution. Under these circumstances, the vehicle should be operated at high speeds for sufficient time to boil off the fuel vapor (15 to 30 minutes) or drain the crankcase and fill with fresh oil (it is not necessary to change the oil filter), then adjust the idle mixture to specifications.

Fuel Dilution

Cold starts combined with short-trip driving or shipment of vehicles in the nosedown position can result in fuel dilution. When excessive fuel dilution of the crankcase lubricant is present, fuel vapors fed to the intake manifold through the crankcase ventilation system can produce overrich mixtures, which might fail HC and CO requirements when tested at idle for emission compliance (in some cases even with the mixture screws completely seated).

Carburetor Power Valve Test

The analyzer can also be used to check operation of the carburetor power valve under load. This check is made in two steps—at no load and under load.

1. Operate the engine at curb idle speed. Note the CO reading. It should be between 0.2 and 0.3 percent.
2. Accelerate the engine to about 2000 rpm and allow it to stabilize. As the engine accelerates, the CO reading should increase sharply.
3. When the engine stabilizes at 2000 rpm, recheck the CO reading. It should decrease the level previously shown on the meter.

If it is richer, the power enrichment system could be faulty or the float could be stuck. Allow the engine to return to curb idle and recover. Correct any discrepancies before proceeding (see Chapter 4).

Power Valve Load Test

Prepare the vehicle (with automatic transmission only) and observe all safety precautions. Block both front wheels and apply the parking brake. Keep the service brakes applied as well while performing

the test. Be sure that the automatic transmission fluid is at the specified level and that the engine is at operating temperature, since this test is similar to the converter stall test on cars with automatic transmission.

WARNING: Keep everyone away from the car during this test.

---

1. Put the transmisison into DRIVE (automatic transmission); momentarily accelerate the engine to wide-open throttle.
2. Hold the throttle open no longer than necessary and never longer than 5 seconds at a time.

As in the no-load test, the CO level should rise sharply and then fall to the level achieved during no-load acceleration. The meter readings should be virtually identical in both the no-load test and the load test. If they are not and the no-load test results were satisfactory, the carburetor power enrichment system is the most likely cause. Check the power enrichment system on the carburetor following procedures in the appropriate manufacturer's service manual under fuel systems.

 SHOP TALK _____

A whining or siren-like noise is normal with some automatic transmissions during this test. However, loud metallic noises from loose parts or interference within the transmission indicate a defective torque converter.

Inoperative Cylinder

The analyzer can be used to pinpoint an inoperative cylinder quickly and easily by following a procedure such as this:

1. Before starting the engine, loosen each secondary cable from the distributor cap, then reconnect each cable to the cap. This will facilitate easy removal of the cables during the test (Figure 11–105).
2. Operate the engine at specified curb idle speed and note the HC and CO readings.
3. Disconnect the secondary cables from the distributor cap, one at a time, with insulated pliers while watching the HC and CO meters.
4. Note the readings after no more than 10 seconds in each case. Follow manufactur-

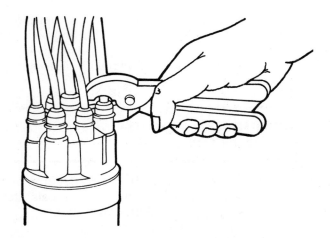

FIGURE 11-105 Disabling a cylinder by removing a spark plug wire

er's guidelines as to the length of time a cylinder can safely be rendered inoperative due to the catalytic converter. When an operating cylinder is disabled by disconnecting the spark plug, the HC level will rise sharply. The CO level might also rise, but not necessarily. If there is no change in the HC reading when the spark plug is disconnected, a mechanical defect in that cylinder is indicated.

5. Inspect the inoperative cylinder carefully and correct all discrepancies.
6. After repairs are complete, repeat the inoperative cylinder test.

Testing for Leaks

To test for various leaks such as air, cylinder, oil, and fuel, proceed as follows:

 1. Remove the air cleaner and plug all disconnected hoses.
2. With the engine running, place your hand lightly over the air horn opening to reduce air intake. Watch the analyzer meters closely.
   - If the engine speed increases and the CO level rises, an air leak is indicated.
   - If the HC level rises at the same time, the carburetor is probably adjusted much too lean. Mixture screws will have to be adjusted until the carburetor is running rich and then readjusted to the proper setting.
   - If the HC level rises sharply and the engine labors and threatens to stall, excessive blowby is indicated. Perform a

compression or a cylinder leakage test, depending on available equipment to locate the source of the problem.
   - If an air leak is indicated, begin by making a careful visual inspection of all vacuum hoses and lines. Replace all segments of burned or brittle vacuum hose. Check for leaks with a vacuum gauge. Small leaks can be detected with a mechanic's stethoscope with the probe removed. It will amplify the hissing sound caused by a leak.
   - If vacuum leaks are found and corrected, replace the air cleaner and attempt to adjust the mixture screws. If no leak is found in the vacuum lines or hoses, check the carburetor.
3. Use engine oil to check the gasket flange area for oil or fuel leaks. With the engine running, carefully spread oil over the carburetor exterior with a small brush (Figure 11–106).
4. Check a small area at a time while watching the HC and CO meters. If the oil seals leak, the CO level will rise. If the leak is large, the HC level will rise also. If the carburetor is leaking, it must be replaced.
5. If no leak is found in the carburetor or intake manifold and the CO level remains low, disconnect the vacuum accessories one at a time and plug their hoses or insert a vacuum gauge. If the CO level rises when an accessory is disconnected and the hose plugged, the leak is in the accessory unit.

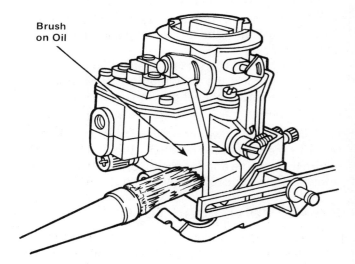

Brush on Oil

FIGURE 11-106 Brush on oil to seal leaks momentarily

Critical Check Points

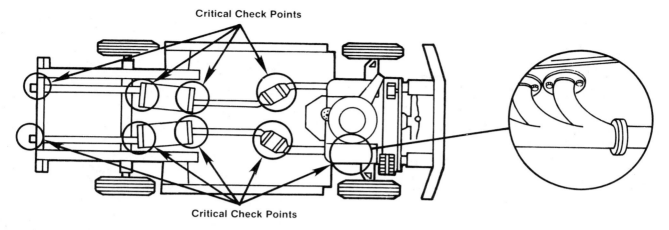

Critical Check Points

FIGURE 11-107 An exhaust system showing points of possible leakage

Repair or replace the components and/or hoses as required.

When all leaks have been found and corrected, replace the air cleaner and connect all hoses. Set the ignition timing, then adjust the mixture screws until both the HC and CO levels meet specifications and recheck engine idle speed.

Exhaust System Fume Test

The analyzer can be used as a quick and accurate check on the exhaust system condition. It will locate any leaks in the exhaust manifold, piping, mufflers, or catalytic converters (Figure 11-107).

Remove the long pickup hose and probe from the tail pipe or catalytic converter. Check for cracks in the exhaust manifold by moving the probe around the junction of the manifold and exhaust pipe. Check the surfaces of all exhaust system components. Pay special attention to junction points. Even a small leak will register with an increased HC reading.

Luggage Compartment Fume Test

An excessive concentration of CO in the passenger compartment or trunk can be deadly. The analyzer can be used to detect the presence of both HC and CO in the passenger compartment better than any other method.

Insert the probe on the long hose into the passenger compartment through a window and close the window on the hose. If necessary, the window opening can be plugged with shop towels.

Start the engine and allow it to operate for 15 to 20 minutes at idle. Watch the HC and CO meters. Any leakage into the passenger compartment will register on the meters. If leakage is present, carefully check the exhaust system components with the analyzer and replace those that are leaky.

Combustion Leak Test

The analyzer will detect combustion leaks much more quickly and with less trouble than pulling the upper radiator hose and watching for bubbles at the thermostat housing. Place the analyzer probe above the coolant in either the radiator filler neck or the coolant recovery tank (Figure 11-108).

CAUTION: Do not immerse the probe in the coolant.

Analyzer Probe Is Over, Not In, Radiator Filler Neck.

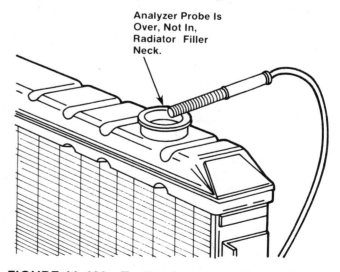

FIGURE 11-108 Testing for combustion leaks

Then load the engine by blocking the front wheels, setting the parking brake and service brakes, and accelerating the engine in gear for no longer than 5 seconds. Combustion gas leakage through the head gasket or block into the coolant will cause a reading on the HC meter.

Idle Adjustment Using Propane Assist

Because of the combination of catalytic converters and very lean mixtures used on today's engines, it is very difficult to properly perform curb idle speed and mixture adjustments. These adjustments must be properly made to insure that emission control devices limit CO, HC, and NO_x to specified levels.

The solution as described in Chapter 4 by Ford, Chrysler, and General Motors is to use propane gas to assist in achieving correct idle settings.

SPARK ADVANCE SYSTEMS

Spark advance systems have been in use since the earliest gasoline engines. It was discovered that the proper timing of the ignition spark helped the car engine to reduce exhaust emission and develop more power output. Throughout the years each car manufacturer developed slightly different spark timing controls according to engine requirements and emission standards for each model year, but the systems and devices all operate on the same principles.

As mentioned earlier a spark from the ignition system ignites the compressed air/fuel mixture from the carburetor in the engine's combustion chamber. The burning process spreads through the combustion chamber as a continuous and progressive action and not as an explosion. It takes approximately 3 milliseconds from the instant the air/fuel mixture from the carburetor ignites until its combustion cycle is complete.

The ignition spark must occur early enough so that combustion pressure will reach its maximum level just after top dead center (ATDC), when the piston is beginning its downward power stroke. Ideally, the combustion process should be completed by about 10 degrees ATDC (Figure 11-109). If the spark occurs too soon before top dead center, the rising piston will have to push against the developed combustion pressure. If the spark occurs too late, the force on the piston will be reduced. In either situation, power is lost and harmful emissions are produced. In extreme cases engine damage is possible. Ignition must start at the proper instant for maximum power and efficiency.

As engine speed increases, piston speed increases. If the air/fuel ratio from the carburetor remains relatively constant, the fuel burning time will remain constant. However, at greater engine speeds, the pistons travel farther during the burning period. The ignition timing must be changed to ensure that the highest combustion pressure occurs at the proper piston position. A change in timing is called spark advance and is typically a job for the distributor's mechanical and vacuum advance devices. Two

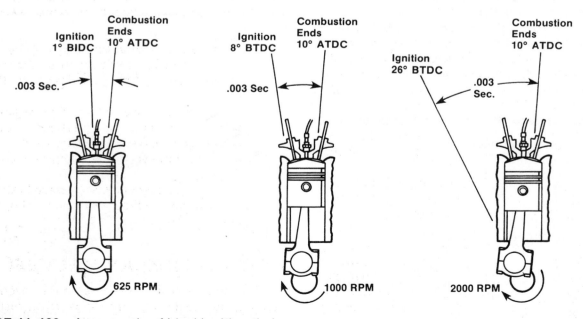

FIGURE 11-109 An example of ideal ignition timing

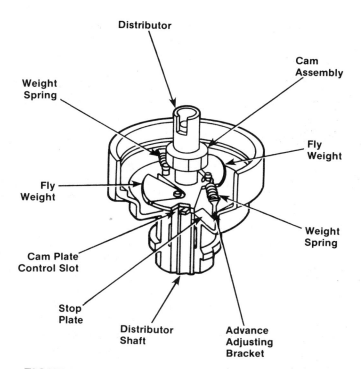

FIGURE 11-110   Mechanical spark advance—centrifugal advance system

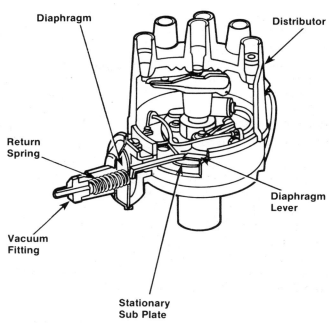

FIGURE 11-111   Mechanical spark advance—vacuum advance system

such systems have been in use with conventional engine control systems: centrifugal advance and vacuum advance systems. These systems operate in much the same way they did on the conventional distributor systems, compensation for engine speed and load. This was done so that maximum pressure was exerted on the top of the piston as soon as the piston rod passed top dead center.

The centrifugal advance system utilized a set of flyweights attached to the lower portion of the distributor's cam assembly. The weights moved against spring tension as the engine speed increased. This motion of the weights turned the timer core assembly in the distributor (Figure 11-110) so that the timer core rotated in the direction of distributor shaft rotation. The teeth on the timer core aligned with the pole piece teeth sooner, signaling the ignition system to fire sooner.

The vacuum advance uses engine vacuum to move a spring-loaded diaphragm, which is connected by a connecting rod to the bottom plate of the magnetic pickup assembly. The magnetic pickup assembly is mounted over the main bearing on the distributor housing so it is able to rotate. When a vacuum signal is applied to the diaphragm, it moves against spring pressure pulling the connecting rod with it. This movement causes the magnetic pickup assembly (pole piece) to rotate in the opposite direc-

tion of the distributor shaft (timer core) rotation (Figure 11-111). This causes the pole piece teeth to align sooner with the timer core teeth signaling the ignition system to fire sooner.

About the time IMCO (Improved Combustion) refinements were made to the engine to reduce exhaust emissions, various controls began to be used only in the vacuum advance for better emission controls.

- In general, retarding (or delaying) the spark timing improves combustion by increasing temperature and allowing leaner fuel mixtures.
- Some controls have bettered the overall emissions level by causing a faster warm-up of a cold engine, allowing reduction in the choke come-off time so that the engine runs lean faster.
- Retarding the spark during deceleration or curb idle also reduces richness and improves emission levels.

VACUUM DELAY VALVES

The vacuum delay valve filtered the carburetor vacuum, making it take longer to get to the distributor advance mechanism. Generally, vacuum must be

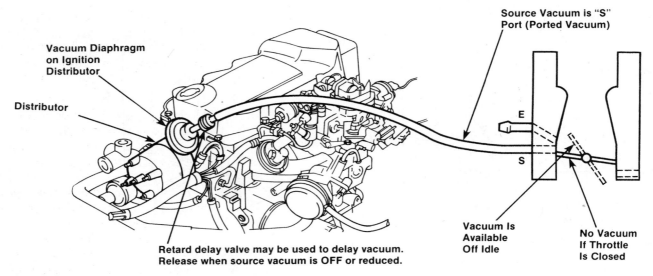

FIGURE 11-112   Ported vacuum spark advance

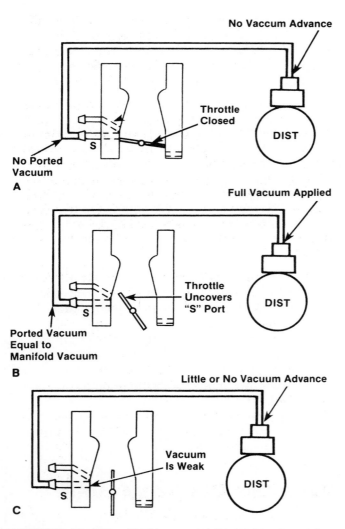

FIGURE 11-113   (A) Hot curb idle or deceleration;
(B) off-idle and cruise; (C) full throttle

in the system for 15 to 30 seconds before it is allowed to affect the advance mechanism. The typical ported vacuum spark advance shown in Figure 11-112 has the following operating modes:

1. Hot curb idle or deceleration (Figure 11-113A):
   - No vacuum advance
   - Retarded spark causes leaner burning; better HC-CO control
2. Off-idle and cruise mode (Figure 11-113B):
   - Full vacuum advance for performance
   - Mechanical advance adds to the total advance
3. Full throttle mode (Figure 11-113C):
   - Little of no vacuum advance
   - Mechanical advance provides the total advance

In some spark advance systems a modulate ported vacuum spark control is used (Figure 11-114). This device is a combination vent valve and spark delay valve. It vents (dumps) vacuum from the diaphragm in the idle or WOT operation and delays vacuum advance during light acceleration to help improve exhaust emissions. It also prevents fuel migration that could damage distributor or other valves. The vacuum spark control must be mounted upright to prevent fuel from accumulating in the valve (Figure 11-115).

Frequently, with a vacuum vent valve, a cold start spark hold (CSSH) system is employed (Figure 11-116A). This system delays the spark retard when a cold engine decelerates; that is, it prevents cold engine stall in deceleration. It provides instant retard

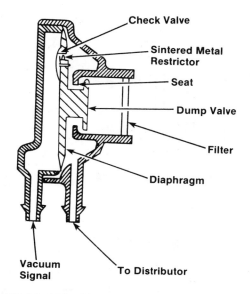

FIGURE 11-114 Vacuum vent valve

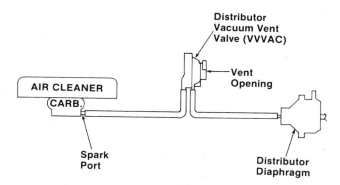

FIGURE 11-115 Vacuum vent valve mounting position

for better emission control. A regulated manifold vacuum spark advance system shown in Figure 11-116B also controls exhaust emissions and improves idle or deceleration conditions.

The following valves are also found in some spark advance systems:

- *Spark Delay Valve (VDV) and Retard Delay Valve (VRDV).* Both valves delay application or release of vacuum in one direction; free flow is available in the opposite direction. Construction of both valves is the same (Figure 11-117) except that the filter is located on the source vacuum side. These valves can be checked as shown in Figure 11-39.
- *Vacuum Regulator Valve (VRV).* This valve provides partial advance at closed throttle for better emission control. Full spark vacuum is on when the throttle opens (Figure 11-118).
- *Ported Vacuum Switch (PVS).* This switch senses the coolant temperature (Figure 11-119). It switches vacuum around a restrictor or delay valve when the engine warms, or can block vacuum cold and open warm.

Another method of vacuum delay was the orifice spark advance control (OSAC) system (Figure 11-120). The OSAC system was an early attempt to limit NO_x by tailoring the spark advance curve to driving conditions (Figure 11-121). It accomplished this by delaying spark advance during acceleration. The OSAC valve was placed in the vacuum line to the distributor advance. A metering orifice in the valve allows full vacuum to reach the distributor only after

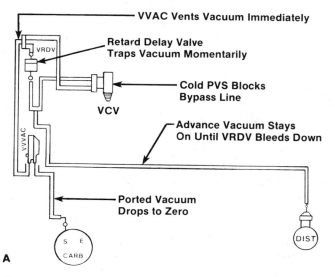

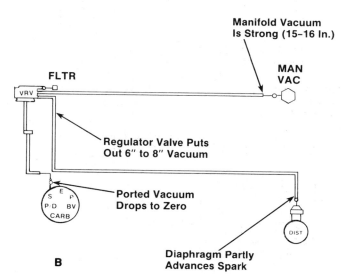

FIGURE 11-116 (A) Cold start spark hold system with vacuum vent valve; (B) regulated manifold vacuum spark advance at idle

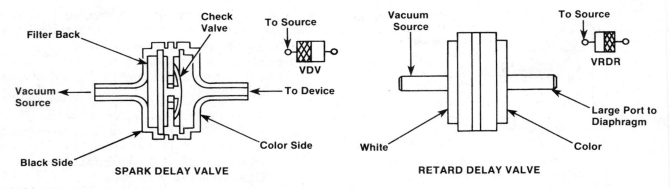

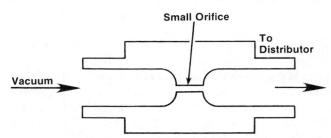

FIGURE 11-117 Spark delay valve and retard delay valve

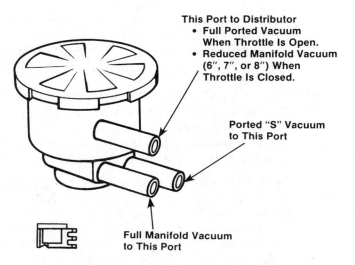

FIGURE 11-118 Vacuum regulator valve

FIGURE 11-120 Cross-sectional view of an OSAC valve

a delay of 10, 17, or 27 seconds, depending on engine calibration. However, if vacuum suddenly dropped—when the throttle is opened wide, for example—a one-way valve in the OSAC allowed the distributor to instantly retard timing. Vacuum also

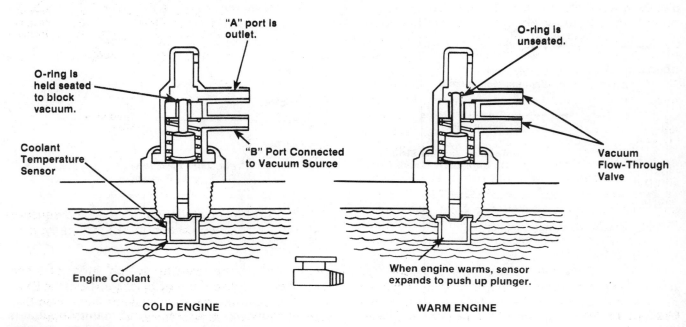

FIGURE 11-119 Ported vacuum switch

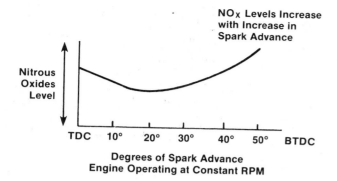

FIGURE 11-121 Spark advance versus nitrous oxide emissions

drops at idle, since a ported vacuum source is used for vacuum advance control.

The basic function of the OSAC valve is like that of any vacuum delay valve. But many OSAC valves also contain a thermal valve as well, which improves the response of a cold engine. Since NO_x is not a problem when the engine is cold, the OSAC valve allows full vacuum to flow either way when the valve is below 58 degrees Fahrenheit. Quicker advance means better cold driveability.

SPEED- AND TRANSMISSION-CONTROLLED TIMING

These systems prevent any distributor vacuum advance when the car is in a low gear or is traveling slowly. A solenoid controls the application of carburetor vacuum to the advance mechanism (Figure 11-122). Current flow through the solenoid winding is controlled by a switch that reacts to the car's operating conditions.

A control switch used with a manual transmission can sense lever position. A control switch used with an automatic transmission will usually work from hydraulic fluid pressure. Both systems prevent any vacuum advance when the car is in a low or intermediate gear.

A speed-sensing switch can be connected to the vehicle speedometer cable (Figure 11-123). The switch signals an electronic control module when the vehicle speed is below a certain level. The module triggers a solenoid that controls engine vacuum at the distributor.

Both vacuum-delay systems and speed- and transmission-controlled systems usually have an engine temperature bypass. This allows normal vacuum advance at high and low engine temperatures. Before March 1973, some systems had an am-

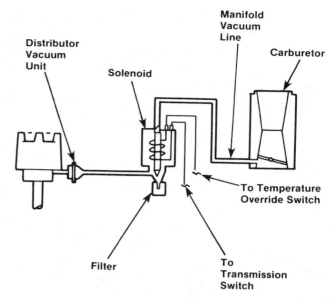

FIGURE 11-122 Simplified circuit diagram of a transmission-controlled spark advance system

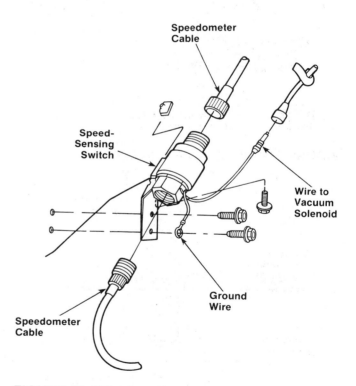

FIGURE 11-123 Installation of a speed-sensory switch on a speed-controlled spark advance system

bient temperature override switch. Most of these switches were discontinued by orders from the EPA. Later temperature override systems were used that sensed coolant temperatures or under-the-hood temperature.

TABLE 11-14:   SPARK ADVANCE PROBLEMS AND THEIR REMEDIES

Condition	Possible Cause	Remedy
Poor performance	Leaky advance diaphragm	Replace diaphragm.
Missing under load	Spark port plugged	Unplug port.
Surge at steady speed	Vent valve stuck in vent position	Check vent and take necessary action.
Poor fuel economy	Vacuum line leak Vacuum regulator valve fails closed.	Repair line leak or replace regulator valve.
Advance delay on acceleration; no retard delay on deceleration; stall on deceleration or quick stop.	Retard delay valve backward	Correct relay valve location.
Poor acceleration	PVS fails to open in a CSSH system	Replace PVs in system.
Poor idle (too fast); high emission in exhaust; overheating; or detonation (knock).	Wrong vacuum routing	Correct vacuum routing.

SPARK ADVANCE PROBLEMS

Incorrect spark timing causes conditions such as misfiring under load, rough idle, detonation, overheating (especially after prolonged idle), and poor engine performance.

If the vacuum advance diaphragm leaks or the vacuum signal does not get to it, there is no vacuum advance. The spark stays at base timing until the centrifugal advance operates. If the spark is not retarded (vacuum advance off) during heavy acceleration, the engine can detonate. Loss of diaphragm vacuum in a system with some advance at idle can cause overheating and rough idle plus detonation. Other problems and their remedies are given in Table 11-14.

REVIEW QUESTIONS

1. Which of the following is not a requirement of the Federal Clean Air Act?
   a. Damaged or faulty parts must be replaced with good parts.
   b. Repairs to the emission control system must be made to manufacturer's specifications.
   c. Emission control systems must be updated every two years.
   d. All replacement parts for emission control systems must satisfy the original design requirements of the manufacturer.

2. Which of the following is not included on a Vehicle Emission Control Information Label?
   a. engine size
   b. vehicle model and year
   c. emission component and vacuum hose schematic
   d. label part number

3. Which of the following is not included in the Vehicle Identification Number code?
   a. catalyst description
   b. country of origin
   c. model year
   d. body type

4. Which of the following is an emission control device that should be inspected and checked?
   a. PVC systems
   b. EGR systems
   c. air injector systems
   d. all of the above

5. An engine is idling roughly and a faulty PCV valve is pinpointed. Technician A attempts to repair the valve. Technician B replaces it. Who is right?
   a. Technician A
   b. Technician B
   c. Both A and B
   d. Neither A nor B

6. Which emission control system controls emissions of oxides of nitrogen?
   a. PCV system
   b. EMC system
   c. EGR system
   d. air injector system

7. To make an operational check on a ported EGR valve system, Technician A operates the engine at a hot curb idle. Technician B makes the test on a cold engine. Who is right?
   a. Technician A
   b. Technician B
   c. Both A and B
   d. Neither A nor B

8. What problem in an EGR system could cause the engine to stall on decelerations?
   a. control valve blocked or airflow restricted
   b. restriction in EGR vacuum line
   c. sticking or binding EGR valve
   d. all of the above

9. What material is located in the filter in the canister of the evaporative emission control system?
   a. charcoal
   b. plastic
   c. rubber
   d. all of the above

10. Which of the following purge control systems are currently in use on vehicles?
   a. air cleaner purge
   b. canister purge solenoid
   c. direct vacuum purge
   d. all of the above

11. Which of the following systems are used on gasoline engines to heat the inlet air and/or the air/fuel mixture?
   a. heated air inlet controls
   b. manifold heat control valves
   c. early fuel evaporation (EFE) heaters
   d. all of the above

12. Which of the following systems supplies fresh air to the exhaust system?
   a. pump-type air injection system
   b. pulse air injection system
   c. both A and B
   d. neither A nor B

13. Excessive mechanical clicking in the air pump could be caused by _____.
   a. overtightened mounting bolt
   b. distorted adjusting arm
   c. overtightened drive belt
   d. all of the above

14. Which of the following is not a requirement of the installer when replacing a catalytic converter?
   a. Only new catalytic converters can be installed.
   b. The installer must get the customer's approval in writing prior to changing the converter.
   c. Only an EPA accepted converter can be installed.
   d. The installer must certify in writing that a vehicle's converter must be replaced and why.

15. Which of the following statements concerning exhaust emission analyzers is not true?
   a. The analyzer can monitor emissions for inspection and maintenance.
   b. The basic tests with an infrared analyzer are performed at fast engine speed.
   c. An infrared exhaust gas analyzer is a useful diagnostic tool.
   d. The infrared analyzer operates on a single beam of infrared light passing through a gas sample.

16. Which of the following is not used to open an EGR valve?
   a. boost pressure
   b. manifold vacuum
   c. venturi vacuum
   d. exhaust back pressure

17. A vehicle is experiencing flooding. Technician A says that the bowl vent valve has failed in the *open* position. Technician B says that the charcoal filter is saturated. Who is probably correct?
   a. Technician A
   b. Technician B
   c. Both A and B
   d. Neither A nor T

18. In which of the following systems would you find a cold weather modulator?
    a. air injection system
    b. EGR system
    c. heated air inlet system
    d. EEC system

19. Which of the following advance systems makes use of flyweights to change ignition timing?
    a. centrifugal advance system
    b. vacuum advance system
    c. electronic advance system
    d. all of the above

20. Which of the following is not an advantage of retarding the spark?
    a. Allows the use of leaner air/fuel mixture.
    b. Causes a faster warm-up period.

    c. Improves emissions during acceleration.
    d. None of the above

21. Which of the following vacuum delay valves is used to control ignition advance based on vehicle coolant temperature speed?
    a. ported vacuum switch
    b. vacuum regulator valve
    c. spark delay valve
    d. all of the above

22. During which of the following operating modes does the centrifugal advance mechanism provide most of the ignition advance?
    a. idle
    b. off-idle cruising
    c. full throttle
    d. deceleration

CHAPTER TWELVE

ELECTRONIC EMISSION CONTROLS

Objectives

After reading this chapter, you should be able to:
- Describe the major electronic emission control systems in use today.
- Describe the operation of the exhaust gas recirculation (EGR) system.
- Define and identify the major components/devices of the EGR system.
- Diagnose and service EGR-related malfunctions.
- Describe the operation of the secondary air system.
- Define and identify the major components/devices of the secondary air system.
- Diagnose and service secondary air system malfunctions.
- Describe the operation of an oxygen sensor.
- Diagnose the operation of an electronic spark advance system.

Modern electronic engine control systems are efficient, economical, innovative, and complex. Understanding electronic control systems requires an understanding of the function of numerous electronic circuits and components. By understanding the basic operating principles behind an electronic control system, the technician can better understand the similarities between different engine types and will be prepared for new technical developments as they are introduced.

The major electronic control systems (Figure 12–1) include such systems as air/fuel control, ignition timing, engine idle speed, and emission control. Of these, the first three systems primarily control engine operating conditions and have been discussed in some detail in preceding chapters of this book. The EGR secondary air and spark advance systems are primarily concerned with controlling the emissions of an engine and will be the focus of discussion in this chapter.

As stated in Chapter 10, there has been a great amount of technical research and development, as well as binding legislation, to control engine emissions. For many years the catalytic converter, a part of the exhaust system, had been at the heart of emission control systems. Electronic controls and the advent of on-board computers have allowed car manufacturers to improve the effectiveness of the catalytic converter by incorporating the secondary air system. These technological advances have also seen the incorporation of the electronic EGR system and the fuel evaporative emission system to further improve the engine's emissions.

ELECTRONIC ENGINE CONTROL OF EMISSION SYSTEMS

In 1976 Chrysler began utilizing the Lean Burn System, which controlled ignition timing and used a carburetor with an air/fuel ratio of 18 to 1. To correct driveability problems associated with this very lean mixture, Chrsyler used timing controls with a wide range. Also in 1976, Cadillac began using a computer to control the fuel injection system on the Seville.

On these early electronic systems, one circuit controlled spark and the other controlled fuel. It was not until 1981 that systems controlling both spark

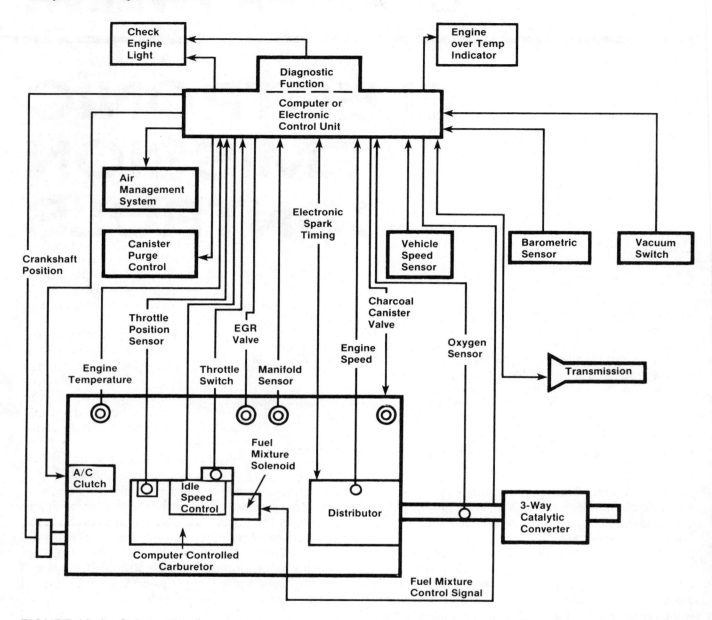

FIGURE 12-1   Schematic of computer system that controls air/fuel mixture ignition timing, engine idle, and exhaust emissions

and fuel became common on federal spec cars. Early systems, however, still had limited functions, but additional functions were introduced with each new model year. Car manufacturers used only those systems that were necessary to meet the emission and CAFE requirements of that model year. During this period, vehicle manufacturers also suggested that they were purposely delaying the introduction of more sophisticated computerized engine controls until there were trained mechanics available to repair them. Despite their claims, with each new model year more controls were added.

Figure 12-2 lists common inputs and outputs used by most computer systems. Not all sensors and functions are used on all cars, and some might have additional sensors or functions that are not listed.

The electronic control module (ECM) or electronic engine control (EEC) is the central processing unit (CPU) where all sensor information is analyzed and compared to the specifications stored in the programmable read only memory (PROM) section of the computer. If the sensor data is not correct, the ECM sends instructions to one or more of the output devices to correct the error.

There are three types of memory in automotive computers: read only, programmable read only, and random access.

1. The read only memory (ROM) is programmed information that can be read only by the ECM. The ROM program cannot be changed. If battery voltage to the computer is lost, ROM information will be retained.

2. The PROM section of memory is factory programmed calibration data containing information for each model. This data varies depending on the car's engine, transmission, axle ratio, weight, and other factors. The data information is contained in chip form (plug-in cartridge). PROM information is not lost if battery voltage is interrupted.

3. The random access memory (RAM) section is a short-term storage area used to store information such as trouble codes and other temporary information. Some vehicles might also be equipped with an option called block learn interrogator, which changes or modifies specifications from the PROM as the vehicle changes. All information stored in the RAM is lost if the battery voltage to the computer is interrupted.

The ECM also contains a power supply that provides the various voltages required by the CPU and an internal clock that controls the rate at which sensor readings and output changes are made. Also contained are protection circuits that safeguard the CPU from interference caused by other systems in the vehicle and diagnostic circuits that monitor all inputs and outputs and signal a warning light (check engine) if any values go outside of the specified parameters.

Engine operating information comes from input sensors. It does not use air cleaner or coolant sensing vacuum switches. The typical closed-loop electronic control system contains a sensing device that feeds back a signal to the exhaust gas recirculation valve to tell it what the output results are. In closed-loop EGR control, the signal can be a position signal, which indicates how far the valve is open, or it can be a pressure signal, which indicates the flow rate of EGR gas.

The ECM compares the results signal with the desired valve position or control pressure condition for the operating mode of the engine. It then decides whether to increase, decrease, or maintain the flow rate. The output control, then, is through a dithering solenoid vacuum valve assembly or an electronic vacuum regulator (EVR) to change or hold the vacuum on the EGR diaphragm.

Closed-loop fuel control regulates the air/fuel mixture in a 14.7-to-1 ratio in cruise operation for

OPERATING CONDITION SENSED (INPUT)   **SYSTEMS CONTROLLED (OUTPUT)**

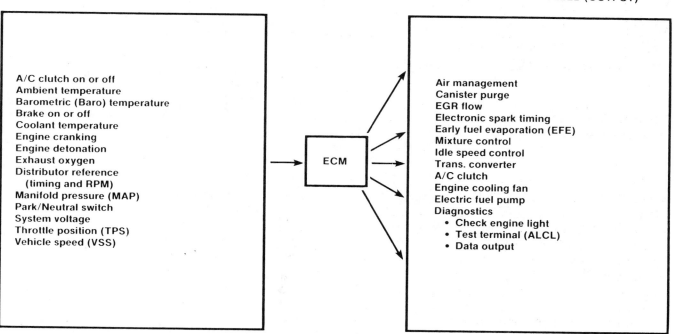

A/C clutch on or off
Ambient temperature
Barometric (Baro) temperature
Brake on or off
Coolant temperature
Engine cranking
Engine detonation
Exhaust oxygen
Distributor reference
   (timing and RPM)
Manifold pressure (MAP)
Park/Neutral switch
System voltage
Throttle position (TPS)
Vehicle speed (VSS)

ECM

Air management
Canister purge
EGR flow
Electronic spark timing
Early fuel evaporation (EFE)
Mixture control
Idle speed control
Trans. converter
A/C clutch
Engine cooling fan
Electric fuel pump
Diagnostics
 • Check engine light
 • Test terminal (ALCL)
 • Data output

FIGURE 12-2   Electronic control module (ECM) input and output conditions.

complete burning. This reduces HC-CO and improves fuel economy. A typical fuel control system operates as follows:

- Principal inputs for fuel control and emission control information are manifold pressure, throttle position, engine coolant temperature, and inlet air charge temperature.
- The ECA operates a feedback control solenoid, stepper motor, or injection solenoid(s) to control fuel flow.
- The exhaust gas oxygen sensor feeds back the richness or leanness of the exhaust so the ECA can adjust fuel delivery. It is important to keep in mind that the closed-loop operation is for the cruising mode of a warm engine. The ECM provides varying enrichment in other modes.

An ECM control of an emission device system depends on the same engine input signals as fuel control. It blocks out or operates air management, the EGR, the evaporative fuel system purge, and the manifold heat valve, depending on the engine calibration and equipment. Remember that the controls shown in Figure 12-3 are considered typical; they are not on all EFC engines. The controls function as follows:

- Air management solenoids control secondary air to the exhaust, catalyst, or atmosphere, depending on the engine mode.
- EGR solenoids (EGRV, EGRC) control the strength (or on/off) of the vacuum to the EGR valve to control the flow volume or blockout, depending on the engine mode. Some EEC engines use an electronic vacuum regulator (EVR) instead of two solenoids.
- ECM also controls timing of the spark signal to the emission module for improved combustion and fast warm-up.
- Most EEC engines have a closed-loop EGR valve control. An EGR valve position sensor (or a pressure sensor) tells the ECM how much EGR gas is flowing. The ECM can then adjust the flow to the operating mode through the EGRV, EGRC solenoids, or EVR.

COMPUTER INPUTS

Computer inputs are signals or measurements of the operating conditions of the engine. Inputs are compared to specifications stored in the PROM. If variations exist, the computer determines what changes should be made. Each input has one or more functions to control and might share its control

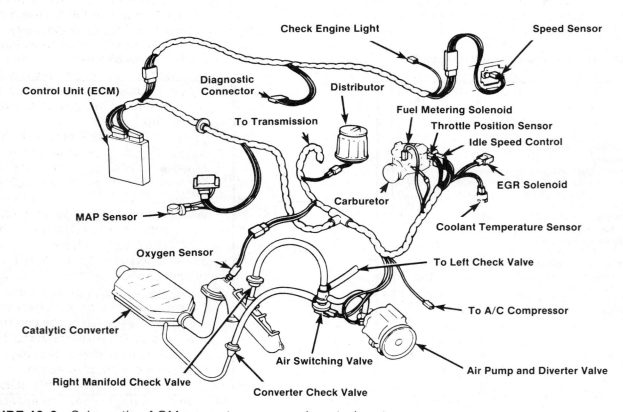

FIGURE 12-3 Schematic of GM computer command control system

with other inputs. The following list gives commonly used inputs. Remember that not all cars use them in the same manner, and some do not use them at all.

- *A/C Clutch On and Off.* This input controls the engine's idle speed control to prevent stalling. The timing on some cars can also be altered to prevent hesitation.
- *Ambient Air Temperature Sensor.* This sensor controls the preheated air and early fuel evaporation systems. In some instances, it might govern a car's automatic climate controls, such as the air conditioning. The AATS's become coated by contaminants drawn into the air intake system. This part should be replaced every 30,000 miles.
- *Brakes On or Off.* This input signals the ECM to disengage the converter clutch. It can also lower idle speed.
- *Barometric Pressure Sensor.* This sensor alters the air/fuel mixture and timing controls, depending upon the altitude in which the car is being operated.
- *Coolant Temperature Sensor.* This sensor controls the air/fuel mixture until the engine is warm enough for the oxygen sensor to take over the mixture control. This sensor can also alter idle speed and timing if the engine is cold or overheated. The operation of the coolant temperature sensor can be equated to the operation of the choke on some cars. It can also prevent heater turn-on until the engine is warm enough to provide heat on cars equipped with automatic climate control. The coolant temperature sensor requires the most frequent replacement of all the computer input systems because of the harsh environment in which it operates. This part is subject to coolant leaks, changes in calibration of the thermister, and resistance buildup on electrical connections. This sensor should be replaced every 30,000 miles.
- *Distributor Reference.* This input has two functions: It tells the ECM the speed of the engine in rpm's and when the piston in each cylinder reaches top dead center (TDC).

This sensor can be a distributor pickup coil, crankshaft position sensor, or camshaft position sensor. It also provides basic information for ignition timing and injector opening in fuel injected cars. On fuel injection equipped cars, loss of the distributor reference will cause a no-start condition.

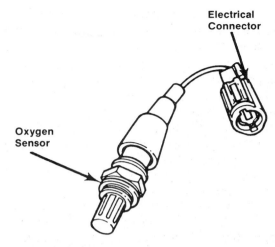

FIGURE 12-4   Exhaust oxygen sensor

- *EGR Position Sensor.* This sensor tells the ECM how far the EGR valve is open. It is used only on Ford vehicles.
- *Exhaust Cranking.* This input retards the timing for easier starting, and on fuel injected cars provides an extra amount of fuel for improved starting.
- *Exhaust Oxygen Sensor.* This sensor (Figure 12-4) controls the mixture to maintain an air to fuel ratio of 14.7 to 1 so the dual-bed catalytic converter can function properly. The oxygen sensor operation and servicing are covered later in this chapter.
- *High Gear Switch.* This input tells the ECM when the car's automatic transmission is in high gear and allows the torque converter clutch to lock up.
- *Knock Sensor.* The knock sensor tells the ECM that the engine is pinging, and in turn the ECM retards the timing. Some of the latest systems retard timing on an individual cylinder basis.
- *MAP Sensor.* This sensor measures engine load and controls spark advance. It might also provide extra fuel if additional power is required. The vacuum hoses leading the MAP sensor should be checked at every tune-up. During winter months, the inside of the vacuum hoses should be checked for moisture buildup because hoses can freeze and prevent the MAP from functioning. However, the MAP sensor might become damaged if the engine is steam cleaned.
- *Park/Neutral Switch.* This switch tells the ECM if the car is not in gear and lowers the idle speed.

- *Power Steering Switch.* This switch causes the ECM to raise the idle speed when the power steering is turned to lock. This prevents stalling when the automobile is engaged in tight turns or is being parked.
- *System Battery Voltage.* Provides power for the system. Poor grounds can cause all or any part of the computer control system to malfunction.
- *Throttle Position Sensor.* This sensor tells the ECM of changes in the position of the throttle. If the throttle has closed, the fuel supply might be cut off to lower HC emissions. If the throttle is open, more fuel and ignition timing will be added to provide additional power. Operation of this sensor should be checked at every tune-up and adjusted per tune-up specs. The TPS might exhibit a wear pattern that will require periodic replacement. This part can be damaged by engine steam cleaning.
- *Vehicle Speed Sensor.* This sensor tells the ECM the vehicle's speed in miles per hour. This input controls when the torque converter clutch locks up and also can be used to control EGR flow and canister purge.

COMPUTER OUTPUTS

Computer outputs are signals to control devices to charge the operating conditions of the car. The major output components are the following:

- *Air Management.* This function controls when and where the output of the air pump goes. When the system is in open loop, the air goes to the exhaust manifold. When the system is in closed loop, the air is sent to the catalytic converter.
- *Canister Purge.* This function controls when stored vapors in the canister should be drawn into the engine and burned. The canister will not purge until the engine is warm and above idle speed. Some cars will not purge until the car is accelerating more than 5 miles per hour.
- *Computer-Controlled Coil Ignition.* This system consists of the ECM, ignition module, camshaft position sensor, and connecting wiring. Each cylinder is paired with the one opposite it, in firing order. Two cylinders, one from the compression stroke and one from the exhaust stroke, are fired at the same time. The cylinder on the exhaust stroke requires little available voltage to arc. The remaining voltage is used by the cylinder on the compression stroke. There are three coils on a V-6 engine and two coils on a 4-cylinder engine. The ECM is used to control spark advance and basic timing. The base reference is supplied by the crankshaft position sensor. The ignition module takes the output signal from the ECM and uses it to control when the spark is fired.
- *EGR Flow.* This function controls when and how much EGR gas is supplied to the engine. Most cars do not have EGR flow at idle. EGR flow is controlled by a pair of electronically controlled vacuum solenoids. One supplies manifold vacuum to the EGR valve when EGR is required, and the second vents the vacuum when EGR is not required.
- *Electronic Spark Control (ESC).* This function controls the amount of spark advance the engine is given. It receives its inputs from the MAP sensor, throttle position sensor, distributor reference, coolant temperture sensor, and knock sensor.
- *Early Fuel Evaporation (EFE).* This function controls the heating of the base of the carburetor or TBI unit. It is achieved by solenoid control of a vacuum chamber that operates a shut-off valve in the exhaust manifold or by control of the electrical circuit that supplies the grid heater mounted under the carburetor or TBI unit.
- *Mixture Control (Carbureted).* This function controls the mixture with an electrical solenoid used to operate the metering rods and idle air bleed valve in the carburetor. The actual mixing still relies on the pressure differential that makes the carburetor work. This system controls approximately 25 percent of the mixture; the remaining 75 percent is still controlled by the carburetor design.
- *Mixture Control (Fuel Injected).* This function uses a solenoid valve to control the amount of fuel supplied to the engine. This system does not rely on the pressure differential that is needed on a carbureted car. The fuel injection system has 100 percent control of the mixture.
- *Idle Speed Control.* This function controls idle speed on carbureted vehicles by an electric motor mounted on the throttle linkage. On fuel injected cars, idle speed is controlled by an air bypass motor that steps in and out, changing the airflow that bypasses the throttle plate. When installing the ISA the pintle must be completely retracted to prevent damage to the throttle body unit.

- *Torque Converter Clutch.* This function controls when the torque converter clutch locks up. Lockup torque converter clutches are used only on cars with automatic transmissions. The transmission must be in high gear, and the car must be accelerating faster than 35 miles per hour.

Whenever engine operation deviates from design specifications, information programmed into the computer is used to provide a dependable backup system. If the sensors fail, the engine develops a malfunction. In some cases, the computer decides the sensor's information is unreliable and substitutes a preprogrammed value so the vehicle can be driven in for repair. Today's computer control systems are complicated. It would take endless amounts of time to diagnose these systems using precomputer control methods. Most manufacturers of these systems have incorporated self-diagnostics into the computer to facilitate troubleshooting. Some systems require the use of special scan testers; others can be diagnosed with simple shop tools. On most systems, the technician activates the computer self-diagnostic mode and monitors a light on the dashboard. The light flashes codes that must be recorded for later reference. These codes represent specific troubles that can be diagnosed using trouble code charts from service repair manuals (Figure 12-5).

Prior to condemning the computer control system, complaints regarding excessive fuel consumption, loss of power, or poor performance should be diagnosed as they would be on an engine without computerized controls.

ELECTRONIC EGR OPERATION AND CONTROL

In vehicles equipped with electronic engine control (EEC) the EGR valve within this system resembles and is operated in a manner similar to the conventional EGR valves. This system uses sensors, solenoids, and an electronic control assembly to modulate and control EGR system components as shown in Figure 12-6. The EEC/EGR valve uses a pintle valve to better control the flow rate of exhaust gases. A sensor mounted on the valve stem (Figure 12-7) sends an electronic signal to the on-board computer, which in turn tells how far the EGR valve is opened. At this time, the EGR control solenoids (EGRV, EGRC) will either maintain or alter the EGR flow, depending on engine operating conditions. Source vacuum is manifold vacuum and is applied or vented, depending on the computer commands.

Code[1]	Circuit Affected
13	Open oxygen sensor circuit
14	Coolant sensor circuit shorted
15	Coolant sensor circuit open
21	TPS signal voltage high
22	TPS signal voltage low
23	M/C solenoid circuit open or grounded (carb. models)
	MAT voltage high (fuel injection)
24	VSS circuit
24B	PARK/NEUTRAL switch
25	MAT sensor signal voltage low
31	Waste gate solenoid
32	BARO sensor circuit
	EGR vacuum control (3.0L & 3.8L turbo)
33	MAP sensor voltage too high
	MAF sensor frequency high (fuel injection)
34	MAP sensor voltage too low
	MAF sensor frequency low (fuel injection)
35	ISC switch circuit shorted
36	MAF sensor burn-off (5.0L & 5.07L)
41	No distributor reference circuit C-3 I ignition (3.8L turbo)
42	EST circuit C-3 I ignition—cam sensor loss (3.8L turbo)
43	ESC retard signal too low
44	Lean oxygen sensor value
45	Rich oxygen sensor value
46	Antitheft fault (5.7L)
51	Faulty PROM, PROM installation, or ECM
52	Faulty CALPAC
53	EGR vacuum control (carb. models) Faulty alternator (fuel injection)
54	M/C solenoid high (carb. models) Fuel pump low voltage (fuel injection)
55	Faulty ECM
61	Degraded O_2 sensor (2.8L)
63	MAP sensor voltage high (2.8L)
64	MAP sensor voltage low (2.8L)

[1] "12" will display only if no reference pulses are received by the ECM. It will never be stored as a malfunction.

NOTE: On EFI models only, "check engine"/"service engine soon" light indicates operational mode of engine. The "check engine"/"service engine soon" light will flash at a rate of one flash per second in closed loop. In open loop, the "check engine"/"service engine soon" light will flash at a rate of 2.5 flashes per second.

FIGURE 12-5 Typical ECM trouble code identification

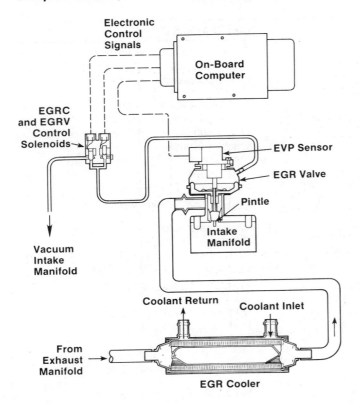

FIGURE 12-6 Typical EGR system used with an EEC.

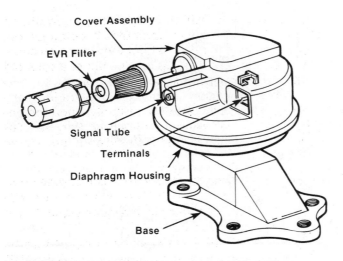

FIGURE 12-7 Normal diverter valve function

Figure 12-8 illustrates a functional schematic of a typical electronic EGR system. A cooler is frequently used to reduce exhaust gas temperatures, which enables the exhaust gas to flow better and in turn reduce the amount of detonation (Figure 12-9).

Early EGR systems used an in-line cooler; later systems use a cooler sandwiched between the EGR

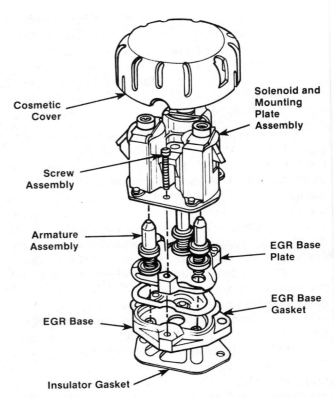

FIGURE 12-8 Schematic of electronic EGR system with EVR vacuum control

valve and carburetor spacer. Like the conventional EGR valves, the EEC/EGR valve is usually closed at idle and during wide-open throttle conditions.

The EGRV/EGRC system is controlled by two solenoids (Figure 12-10). The solenoids respond to voltage signals from the on-board computer and operate in the following manner:

1. An EGR vent solenoid, or EGRV, which is normally an open vent solenoid valve that closes when it is energized (Figure 12-11A).
2. An EGR control solenoid, or EGRC, which is normally a closed solenoid valve that opens when it is energized (Figure 12-11B).

Voltage signals from the on-board computer can trigger the solenoids to:

- Increase EGR flow by applying vacuum to the EGR valve (Figure 12-12).
- Maintain EGR flow by trapping vacuum in the system (Figure 12-13).
- Decrease EGR flow by venting EGR vacuum (Figure 12-14).

In actual operation, both solenoids constantly shift between the three operating conditions men-

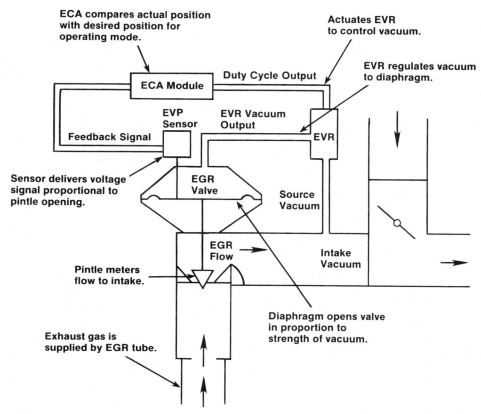

FIGURE 12-9 A coder reduces exhaust gas temperatures

Normally Open EGRV Solenoid Operation

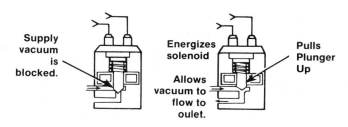

Normally Closed EGRC Solenoid Operation

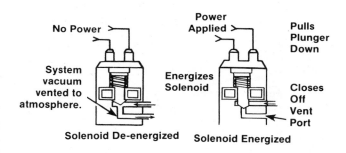

FIGURE 12-10 EGRV/EGRC system is controlled by two solenoids.

tioned above as engine operating conditions change. If the on-board computer is doing its job, you will be able to feel and hear the solenoids clicking on and off. If not, there may be either a solenoid problem or an electronic control problem.

The EGR system is activated only during part-throttle modes of operation and only when the engine is warm and running smoothly. For the EGR system to be enabled, the following conditions must be met:

- The engine must be warm. The engine coolant temperature (ECT) sensor must be sending a moderate- to low-voltage signal to the on-board computer.
- The throttle plate must be at a part-throttle position as sensed by the throttle plate (TP) sensor.
- Manifold absolute pressure (MAP) must be moderate as determined by the sensor and on-board computer.
- A calibrated length of time must have elapsed since start-up or cranking.

On the other end of the operating cycle, the EGR system is disabled during wide-open throttle conditions or closed throttle conditions. During wide-open throttle conditions the engine is allowed

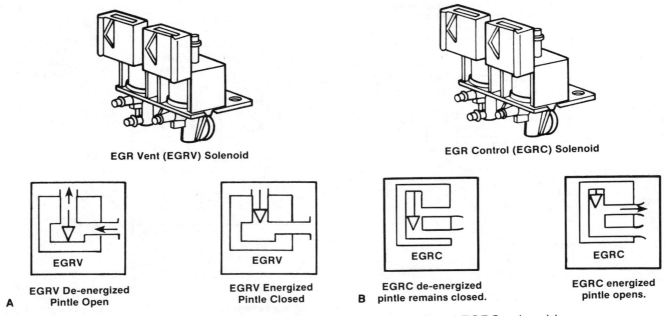

EGR Vent (EGRV) Solenoid

EGR Control (EGRC) Solenoid

EGRV De-energized Pintle Open

EGRV Energized Pintle Closed

EGRC de-energized pintle remains closed.

EGRC energized pintle opens.

A

B

FIGURE 12-11 (A) Normally open EGRV solenoid; (B) normally closed EGRC solenoid

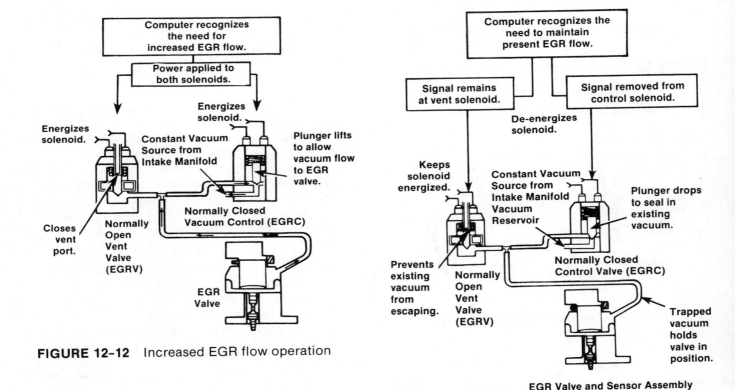

FIGURE 12-12 Increased EGR flow operation

FIGURE 12-13 EGRV/EGRC solenoid operation to maintain EGR flow

to develop maximum power. An enabled EGR system at this time would affect the engine's power output. In closed throttle modes, such as cold starts, warm-ups, idle, and deceleration, the engine can be subjected to hesitation or stalling. Reducing the EGR flow at this time will reduce the chances of the engine stalling.

Figure 12–15 shows the typical locations of various components in an electronic valve in an EVR vacuum control system. Figure 12–16 illustrates the

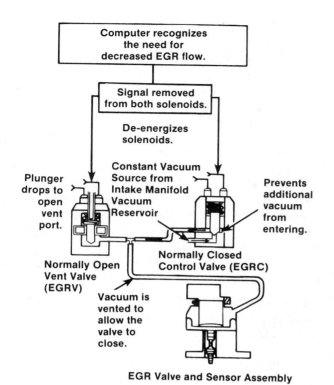

Computer recognizes the need for decreased EGR flow.

Signal removed from both solenoids.

De-energizes solenoids.

Plunger drops to open vent port.

Constant Vacuum Source from Intake Manifold Vacuum Reservoir

Prevents additional vacuum from entering.

Normally Open Vent Valve (EGRV)

Normally Closed Control Valve (EGRC)

Vacuum is vented to allow the valve to close.

EGR Valve and Sensor Assembly

FIGURE 12-14   Decreased EGR flow operation

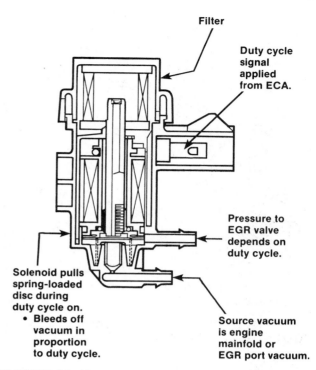

Filter

Duty cycle signal applied from ECA.

Pressure to EGR valve depends on duty cycle.

Solenoid pulls spring-loaded disc during duty cycle on.
• Bleeds off vacuum in proportion to duty cycle.

Source vacuum is engine mainfold or EGR port vacuum.

FIGURE 12-16   Electronic vacuum regulator (EVR) operation

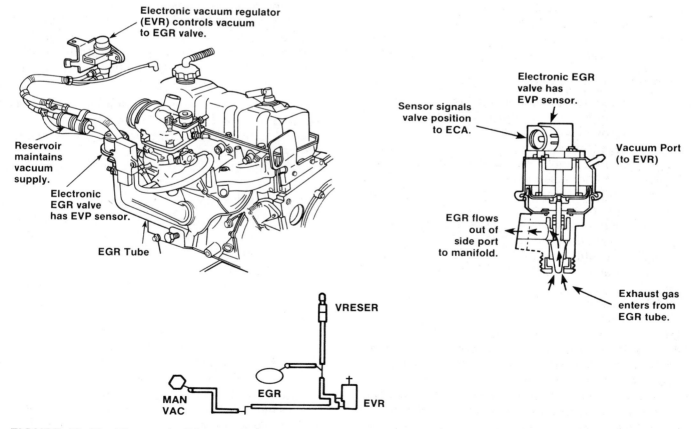

Electronic vacuum regulator (EVR) controls vacuum to EGR valve.

Reservoir maintains vacuum supply.

Electronic EGR valve has EVP sensor.

EGR Tube

Electronic EGR valve has EVP sensor.

Sensor signals valve position to ECA.

Vacuum Port (to EVR)

EGR flows out of side port to manifold.

Exhaust gas enters from EGR tube.

VRESER

MAN VAC

EGR

EVR

FIGURE 12-15   Electronic EGR valve with EVR vacuum control

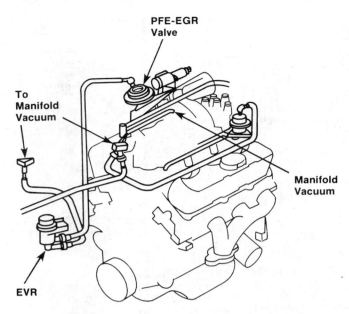

To Manifold Vacuum

PFE-EGR Valve

Manifold Vacuum

EVR

FIGURE 12-17 Typical vacuum schematic for PFE-EVR system

steps of operation of the electronic EGR system with an EVR vacuum control. The EGR valve is functionally a ported valve with a position sensor added; it operates only at part-throttle. The ECM does not supply vacuum in cranking, closed throttle, or wide-open throttle operation. In the controlling modes, the EVR operates on a duty cycle output from the ECM.

Recent developments in some manufacturer internal combustion engine designs have had an effect on the operation of the EGR system. These design improvements are based on a fast burn technology, which allows the air/fuel mixture to burn quickly and thoroughly. With these improvements, it is possible to increase the EGR flow rates.

The on-board computer uses airflow calculations to know exactly how much oxygen is available for combustion. If exhaust gases are recirculated into the intake manifold, a certain volume of air will be displaced with inert EGR gases. This will reduce the measurable amount of oxygen that is available for combustion. Every cubic foot of air in the manifold will now contain EGR gases, thus less oxygen. On vehicles equipped with fuel injection, unless the injector pulse width is adjusted or modified accordingly, the air/fuel mixture will become too rich as the oxygen level is reduced. To adjust the fuel injector pulse width the on-board computer determines the proportion of EGR gases in the intake manifold. The EVP sensor provides the computer with information concerning the amount of inert exhaust gases flowing into the intake manifold. The computer then ac-

curately determines the EGR flow rates and adjusts the fuel injector pulse width accordingly to provide the correct air/fuel ratio.

One of the more recent EGR emission control systems is the pressure feedback electronic (PFE) type that has electronic vacuum regulator (EVR) control. This control (Figure 12-17) applies traps or bleeds off vacuum to the EGR valve. That is, the vacuum to the EGR valve is supplied and controlled by the EVA instead of the dithering solenoid valves. Feedback to the ECA is a controlled pressure between the metering orifice and the EGR valve pintle (Figure 12-18). The EGR flow rate is proportional to the pressure drop across the metering orifice. The operation of the PFE-EVR system is illustrated in the functional schematic (Figure 12-19).

EGR SYSTEM TROUBLESHOOTING

Before attempting to troubleshoot or repair a suspected EGR system on a vehicle, the following conditions should be checked and be within specifications:

- Engine is mechanically sound.
- Injection system is in tune and operating properly.
- Mechanical-vacuum advance is properly adjusted (or the electronic advance system is operating properly).

If one or more of the above conditions is faulty or operating incorrectly, perform the necessary tests and services to correct the problem before servicing the EGR system.

In all the closed-loop electronic control EGR systems, the valve (by itself) functions the same as a ported EGR valve.

- Apart from the electronic control, the system can have all of the problems of any EGR system. Sticking valves, obstructions, and loss of vacuum will produce the same symptoms as on non-EEC systems (see Chapter 11).
- If an electronic control component is not functioning, the condition will usually be recognized by the ECM. The EEC quick test will show a service code to indicate a need to test the components and external circuits, following the EEC pinpoint tests.
- The EGRV and EGRC solenoids, or the EVR, should normally cycle on and off very frequently when EGR flow is being controlled

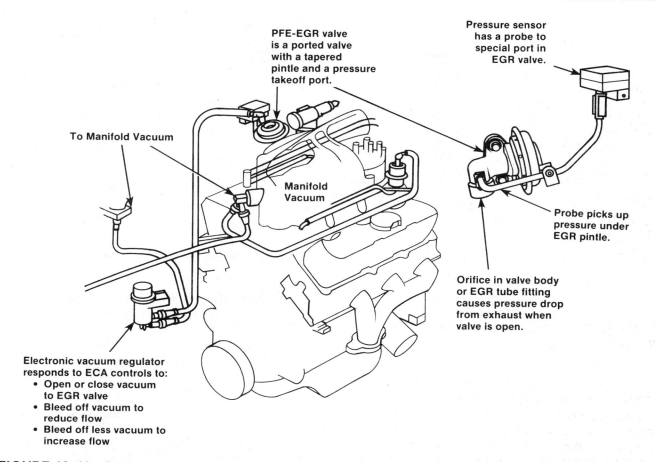

PFE-EGR valve
is a ported valve
with a tapered
pintle and a pressure
takeoff port.

Pressure sensor
has a probe to
special port in
EGR valve.

To Manifold Vacuum

**Manifold
Vacuum**

Probe picks up
pressure under
EGR pintle.

Orifice in valve body
or EGR tube fitting
causes pressure drop
from exhaust when
valve is open.

Electronic vacuum regulator
responds to ECA controls to:
• Open or close vacuum
  to EGR valve
• Bleed off vacuum to
  reduce flow
• Bleed off less vacuum to
  increase flow

FIGURE 12–18 Pressure feedback electronic (PFE) EGR system with electronic vacuum regulator (EVR) control

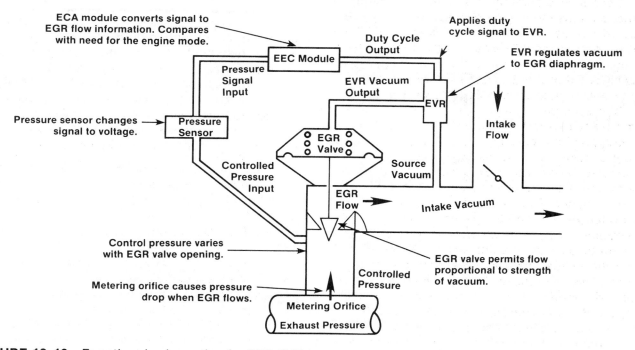

ECA module converts signal to
EGR flow information. Compares
with need for the engine mode.

Applies duty
cycle signal to EVR.

EVR regulates vacuum
to EGR diaphragm.

EEC Module

Duty Cycle
Output

Pressure
Signal
Input

EVR Vacuum
Output

EVR

Pressure sensor changes
signal to voltage.

**Pressure
Sensor**

Intake
Flow

**EGR
Valve**

Controlled
Pressure
Input

Source
Vacuum

EGR
Flow

Intake Vacuum

Control pressure varies
with EGR valve opening.

EGR valve permits flow
proportional to strength
of vacuum.

Metering orifice causes pressure
drop when EGR flows.

Controlled
Pressure

Metering Orifice

Exhaust Pressure

FIGURE 12–19 Functional schematic of a PFE-EVR system

TABLE 12-1: ELECTRONIC EGR SYSTEM DIAGNOSTIC CHART

Condition	Possible Cause	Remedy
Detonation (spark knock) Engine overheat Fails emission NO$_x$ test Rough idle Low power Poor fuel economy	Vacuum leak or loss Valve sticks closed. Leaking valve diaphragm Vacuum reservoir leak No power to solenoid(s) or EVR EGR valve gasket leak Blocked flow passages	Check possible causes and make necessary correction.
Engine stalls after cold start. Engine stalls on deceleration. Vehicle surges during cruise. Rough idle Sluggish; poor performance Poor fuel economy	Valve opens too far; sticking open. TVS or PVS fails open. EGRV solenoid fails closed. EVR vent blocked.	Inspect various components suggested in possible causes and take necessary corrective actions.

(warm engine and cruise rpm). If they do not, it indicates a problem in the electronic control system or the solenoids.

- Generally, an electronic control failure will result in low or zero EGR flow and might cause symptoms like overheating, detonation, and power loss.

Table 12-1 is a summary of problems that can occur in an electronic EGR system, their possible causes, and suggested remedies. But before attempting any testing of the EGR system, visually inspect the condition of all vacuum hoses for kinks, bends, cracks, and flexibility. Replace defective hoses as required. Check vacuum hose routing. (See the underhood decal or the manufacturer's service manual for correct routing.) Correct any misrouted hoses.

FUNCTIONAL TEST OF ELECTRONIC CONTROLLED EGR SYSTEM

A typical functional test of an electronic controlled EGR system is performed as follows:

1. Check all vacuum hoses for correct routing and secure attachment.
2. Run the engine to hot idle. Check that there is no vacuum to the EGR valve diaphragm at hot curb idle. If there is, check the hose routing.
3. On EFI engines only, disconnect the electrical connector to the idle air bypass valve.
4. Install a tachometer. Adjust the curb idle if necessary.
5. Perform an EGR valve gas flow test.

6. Accelerate to 2500 rpm and check for cycling of the EVR or EGRV/EGRC solenoids.

Checking EGR Valve Mounting

1. With the engine off, check the EGR valve mounting bolts for tightness.

CAUTION: If the engine has been run recently, the exhaust manifold and EGR valve could be hot enough to cause severe burns. Use caution and protect the hands by wearing gloves.

---

2. If the mounting bolts are loose, torque to manufacturer's specifications and repeat the vacuum test given above.
3. If the bolts are tight, remove the EGR valve. Inspect the gasket, manifold passages, EGR valve ports, and so on. Clean all surfaces, replace gasket, then clean or replace EGR valve as required.

The methods of checking the EGR system at idle, at high engine speeds, or cold engine operation as well as off vehicle EGR valve tests are basically the same for nonelectronic units. Several components in an electronic EGR system require special care.

EGR VALVE GAS FLOW TEST ON EVR SYSTEM

With the engine at hot curb idle, a gas flow test on an EVR system is conducted according to the following procedure (Figure 12-20):

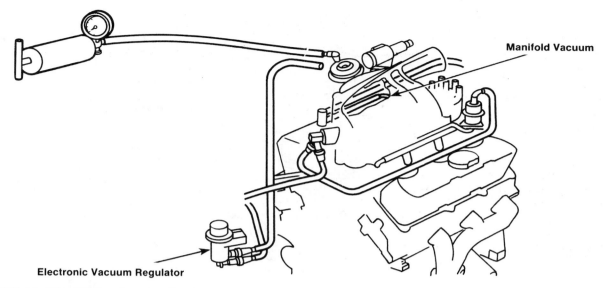

FIGURE 12-20   EGR valve gas flow test on EVR system

 1. Disconnect the hose from the electronic vacuum regulator at the EGR valve diaphragm.
2. Attach a vacuum pump at the EGR valve diaphragm.
3. Slowly apply 5 to 10 inches of vacuum.
   - Idle must drop 100 rpm or more
   - Engine might stall
   - Idle speed must return to normal with the vacuum removed
4. Remove the EGR valve for the bench test/inspection if it fails this test.

TESTING A VACUUM RESERVOIR

To test the vacuum reservoir in an electronic EGR system, proceed as follows:

 1. Remove the manifold vacuum source hose from the reservoir and attach a vacuum pump (Figure 12-21).

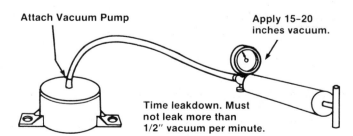

FIGURE 12-21   Testing a vacuum reservoir

2. Apply 15 to 20 inches mercury vacuum to the reservoir.
3. Allow for a time leakdown. The reservoir must not leak more than 1/2 inch mercury vacuum per minute.
4. Replace the reservoir if it fails the test.

TESTING AN EGRC SOLENOID VALVE

To test a normally closed (EGRC) solenoid valve, the procedure is as follows (Figure 12-22):

 1. After removing the hose from the solenoid, blow through the EGRC solenoid valve. It must not pass air and must remain de-energized.
2. Energize the solenoid with a 12-volt source. When energized, the solenoid, if operating properly, must check in and the ports must pass air.
3. Replace the EGRC solenoid valve if it fails the test.

TESTING AN EGRV SOLENOID VALVE

To test a normally open (EGRV) solenoid valve, the procedure is as follows (Figure 12-23):

 1. After removing the hose from the solenoid, blow air into the lower port. The valve must pass air when in a de-energized condition.
2. Energize the solenoid with a 12-volt source. When energized, the solenoid, if

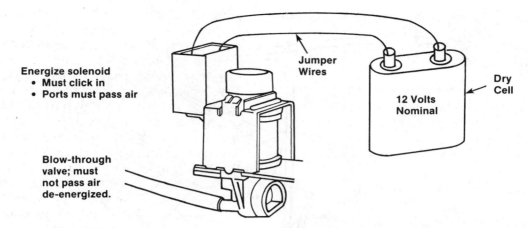

FIGURE 12–22   Testing a normally closed (EGRC) solenoid valve

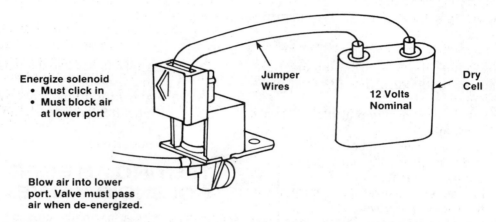

FIGURE 12–23   Testing a normally open (EGRV) solenoid valve

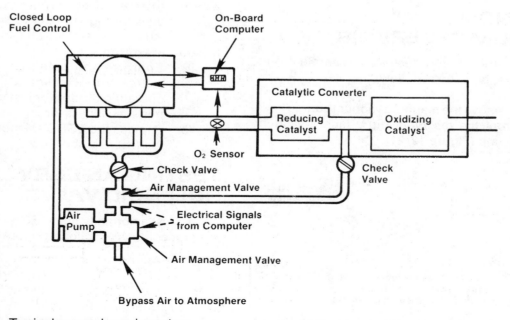

FIGURE 12–24   Typical secondary air system

operating properly, must check in and block air at the lower port.

3. Replace the EGRV solenoid valve if fails the tests.

CAUTION: If using the vehicle's battery when testing EGRC and EGRV solenoid valves, do not allow the jumpers to cross. This could cause fire or severe burns. A dry cell battery is recommended.

SECONDARY AIR SYSTEM

Another principal emissions control system controlled by the on-board computer is the electronic secondary or air injector system. This system (Figure 12-24) is used to provide additional oxygen to continue the combustion process after the exhaust gases leave the combustion chamber. The secondary air system uses an engine-driven pump to inject air into the exhaust part of the cylinder head, exhaust manifold, or the catalytic converter. The system operates continuously and will bypass air dur-

ing high engine speeds and loads on computer command. An air management valve performs the bypass or diverting function, and a check valve protects the air pump from damage by preventing a backflow of exhaust gas. This air system does not have any effect on the air/fuel mixture. A failure that prevents secondary air from being delivered to the exhaust manifold or catalytic converter will not affect engine performance. The failure will, however, reduce the oxidation of HC and CO and can cause the vehicle to fail an emissions test.

The typical electronic secondary air system (Figure 12-25), like the conventional air injection system described in Chapter 10, consists of an air pump, an air control or diverter valve, an air bypass valve, a check valve, and the necessary plumbing. The major difference between the two systems is, of course, the on-board computer and diverter valve (Figure 12-26).

The air pump (Figure 12-27) is driven by a belt on the front of the engine and supplies air to the system. Intake air passes through a centrifugal filter fan at the front of the pump where foreign materials are separated from the air by centrifugal force. Air flows from the pump through either a secondary air bypass (SAB) valve or a secondary air diverter

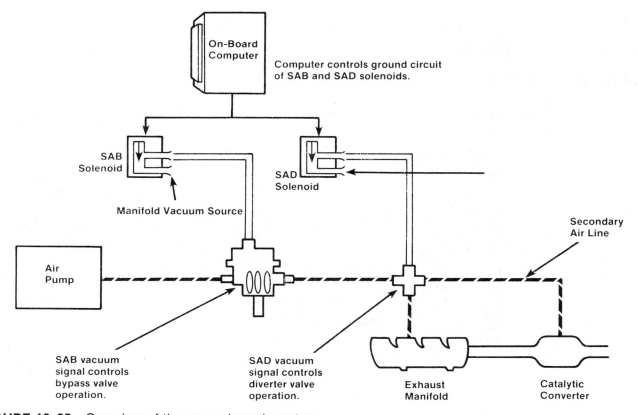

FIGURE 12-25 Overview of the secondary air system

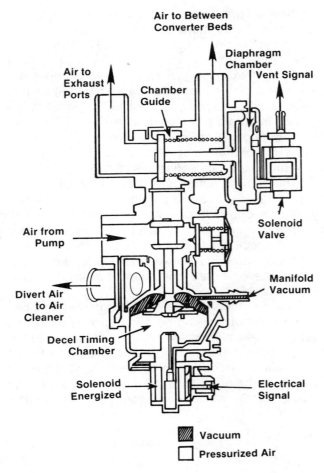

FIGURE 12-26 Diverter valve function

(SAD) valve. Both valves are controlled by the on-board computer. Secondary air also flows through a check valve into the exhaust manifold. The check valve prevents the backflow of exhaust gases into the pump in the event of an exhaust backfire or if the pump drive belt fails. The SAB and SAD valves are usually contained in one housing. The SAB valve provides air directly to the car's air filter for catalytic converter protection or directs air to the SAD valve. The SAD valve in turn directs air to the exhaust manifold (open loop operation) or to the catalytic converter (closed loop operation). Therefore, secondary airflow can be directed to three points:

1. Downstream to the catalytic converter
2. Upstream to the exhaust manifold
3. Vented (or bypassed) to the atmosphere via the air filter. The relationship of the secondary air system to the on-board computer and its associated sensor inputs is shown in Figure 12-28.

SECONDARY AIR SYSTEM COMPONENTS

The main components of the secondary air system are the air bypass (SAB) valve and the air diverter (SAD) valve. The SAB solenoid valve (Figure 12-29) is controlled directly by the on-board computer. The solenoid in turn controls the application of vacuum to the air bypass valve. The SAB is a normal-

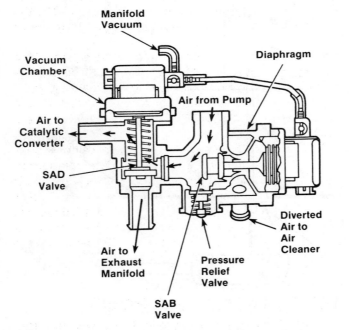

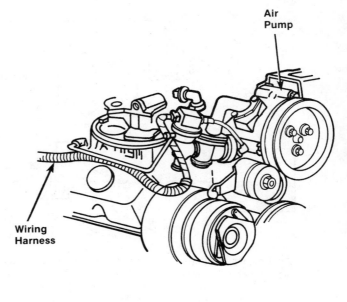

FIGURE 12-27 Air pump location and SAB/SAD valves cutaway view

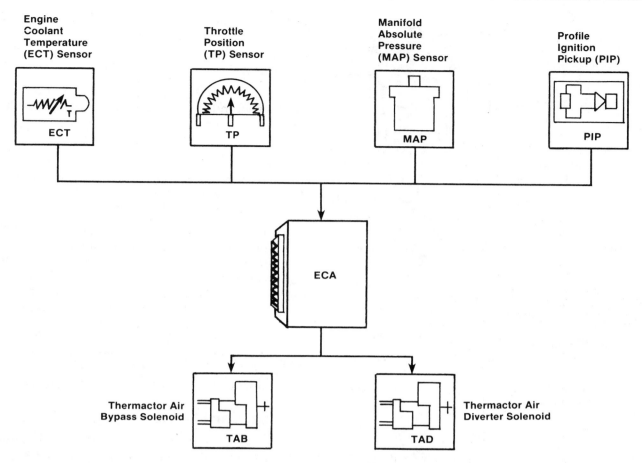

FIGURE 12-28 Relationship of secondary air system to computer sensor inputs

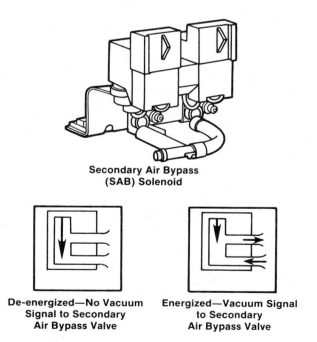

Secondary Air Bypass
(SAB) Solenoid

De-energized—No Vacuum
Signal to Secondary
Air Bypass Valve

Energized—Vacuum Signal
to Secondary
Air Bypass Valve

FIGURE 12-29 Secondary air bypass solenoid operation

ly closed solenoid valve and no vacuum is applied to the SAB valve. When the solenoid is energized by the on-board computer, vacuum is applied to the SAB valve.

The SAB valve is controlled by the action of the solenoid (Figure 12-30) and in turn controls the airflow. With no vacuum applied to the valve, secondary air is vented into the atmosphere. When vacuum is applied to the valve, the airflow is directed to the SAD valve.

The SAD solenoid valve is also directly controlled by the on-board computer. The solenoid in turn controls the application of vacuum to the air diverter valve. The SAD is normally a closed solenoid valve and no vacuum is applied to it. Vacuum is applied to the diverter valve when the on-board computer energizes the solenoid. Operation of the SAD valve is shown in Figure 12-31.

SECONDARY AIR SYSTEM OPERATION

The secondary air system has three principal modes of operation:

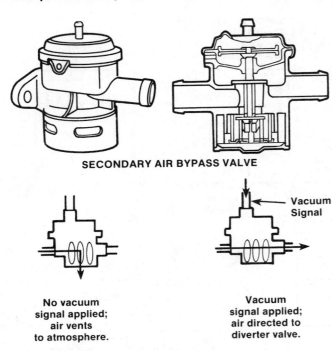

SECONDARY AIR BYPASS VALVE

No vacuum
signal applied;
air vents
to atmosphere.

Vacuum
Signal

Vacuum
signal applied;
air directed to
diverter valve.

FIGURE 12-30 Secondary air bypass valve operation

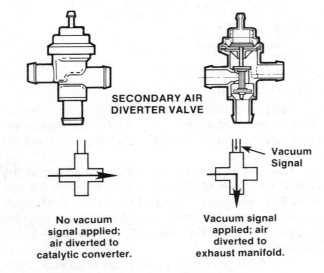

SECONDARY AIR DIVERTER VALVE

No vacuum
signal applied;
air diverted to
catalytic converter.

Vacuum
Signal

Vacuum signal
applied; air
diverted to
exhaust manifold.

FIGURE 12-31 SAD valve operation

1. *Bypass Mode.* Secondary air is vented or bypassed to the atmosphere and does not mix with exhaust gases.
2. *Upstream Mode.* Secondary air is directed to the exhaust manifold near the cylinders. Air mixes with exhaust gases thereby assisting the combustion of any excess HC.
3. *Downstream Mode.* Secondary air is directed to the catalytic converter. This

mode is activated throughout most engine operations.

Bypass Mode

Figure 12-32 shows secondary air entering the SAB valve and then being vented to the atmosphere. Secondary air may be vented or bypassed for several reasons, such as the following:

- A fuel-rich condition
- When the on-board computer recognizes a problem in the system
- During deceleration

Secondary air is typically bypassed during cold engine cranking and cold idle conditions.

When engine coolant temperature is below 55 degrees Fahrenheit at start-up, secondary air is automatically bypassed. The system will maintain a bypass mode of operation until the coolant temperature reaches 170 degrees Fahrenheit. The on-board computer has an internal electronic time that keeps track of the length of time since the engine was started. Secondary air is also maintained in the bypass mode until a pre-set length of time has elapsed.

Upstream Mode

Figure 12-33 shows secondary air being routed through the SAB valve and the SAD valve to the exhaust manifold. The upstream mode is actuated when the on-board computer senses a warm crank/start-up condition. The secondary airflow remains upstream for 1 to 3 minutes after start-up to help control emissions.

The air/fuel mixture at start-up is typically very rich. This rich mixture results in unburned HC and CO to remain in the exhaust after combustion. By switching to the upstream mode, the hot HC and CO mix with the incoming secondary air and are burned up. This reburning of HC and CO compounds causes the exhaust gases to get hotter, which in turn heats up the oxygen sensor (Figure 12-34). Therefore, switching to the upstream mode allows the electronic engine control system to switch to the closed loop operation sooner.

The warm (and activated) oxygen sensor sends exhaust gas oxygen information to the on-board computer. It should be noted that the upstream mode of operation increases the oxygen level in the exhaust gases. The voltage signal from the oxygen sensor to the computer maintains a continuous low level. The on-board computer interprets this signal as a continuous lean condition; that is, too much oxygen in the air/fuel mixture. It can readily be seen,

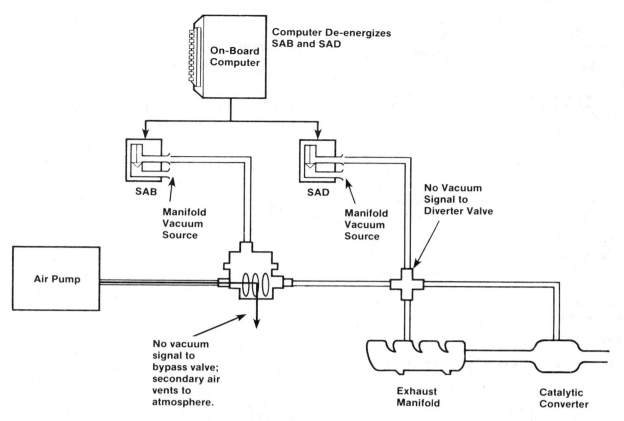

FIGURE 12–32 Secondary air system bypass mode of operation

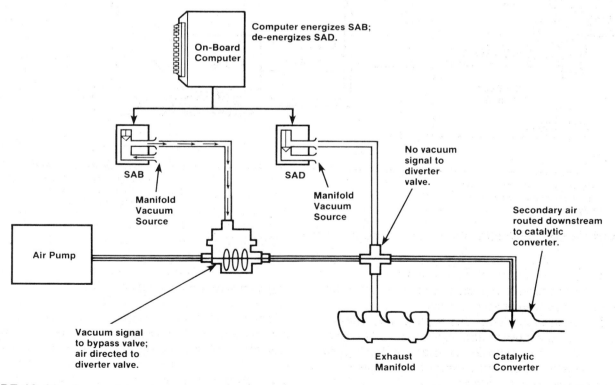

FIGURE 12–33 Secondary air system upstream mode of operation

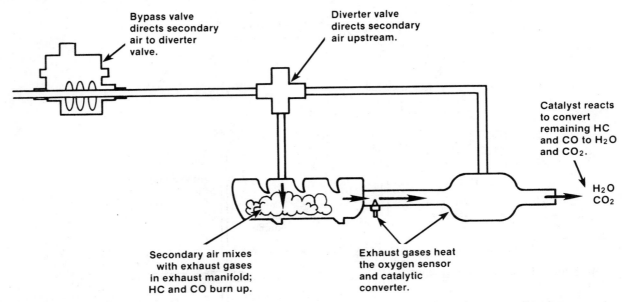

Bypass valve directs secondary air to diverter valve.

Diverter valve directs secondary air upstream.

Catalyst reacts to convert remaining HC and CO to H_2O and CO_2.

H_2O
CO_2

Secondary air mixes with exhaust gases in exhaust manifold; HC and CO burn up.

Exhaust gases heat the oxygen sensor and catalytic converter.

FIGURE 12-34 Upstream secondary air heats the oxygen sensor and catalytic converter.

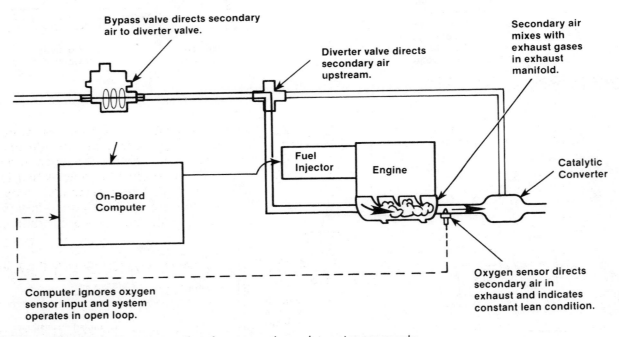

Bypass valve directs secondary air to diverter valve.

Diverter valve directs secondary air upstream.

Secondary air mixes with exhaust gases in exhaust manifold.

Fuel Injector

Engine

Catalytic Converter

On-Board Computer

Computer ignores oxygen sensor input and system operates in open loop.

Oxygen sensor directs secondary air in exhaust and indicates constant lean condition.

FIGURE 12-35 Open loop operation for secondary air upstream mode

then, that the upstream mode results in inaccurate exhaust gas oxygen measurements.

To solve this dilemma, the on-board computer automatically switches to the open loop fuel control whenever the upstream mode is activated. Figure 12-35 illustrates open loop operation for the secondary air upstream mode of operation.

Downstream Mode

The secondary air system operates in the downstream mode during a majority of engine conditions. After the engine is started and runs for a few minutes and the secondary air system is activated in the downstream mode, the catalytic converter attains a

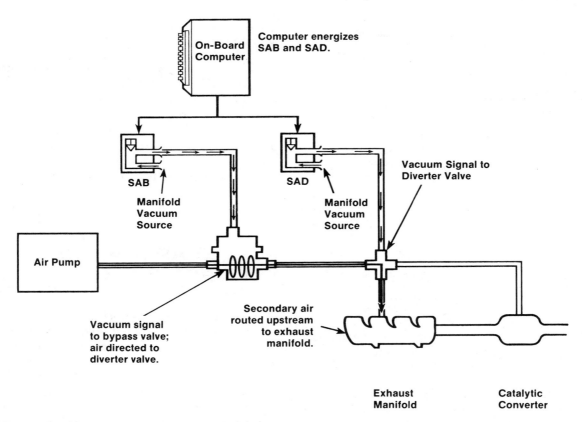

FIGURE 12-36 Secondary air downstream mode of operation

high operating temperature. At the same time, the oxygen sensor will also be warmed and activated. At this time the electronic engine control system can operate in the closed loop fuel control mode. Figure 12-36 illustrates secondary air being routed through the SAB valve and the SAD valve to the catalytic converter.

After the engine has warmed up sufficiently, the air/fuel mixture tends to run lean, leaving fewer excess hydrocarbons remaining after combustion. Therefore, it is not necessary to run the secondary air system in the upstream mode of operation. The on-board computer automatically switches the system to the downstream mode to allow the secondary air to mix with the exhaust gases inside the catalytic converter. The catalytic converter is most efficient at reducing NO_x when the exhaust gases entering it are near stoichiometry (Figure 12-37). If the secondary air is upstream, the high oxygen level interferes with this operation. However, with secondary air diverted downstream, fresh air assists the converter in reducing the NO_x emissions.

The electronic engine control system can operate in the closed loop fuel control mode only when

the secondary air system is diverted downstream. In this mode, the oxygen sensor can provide accurate information about the control of oxygen in the exhaust gases to the on-board computer, which in turn maintains the air/fuel mixture at stoichiometry. Figure 12-38 illustrates closed loop operation for the secondary air downstream mode of operation.

SECONDARY AIR SYSTEM TROUBLESHOOTING

Both the SAB valve and the SAD valve are operated by solenoids that are controlled by the on-board computer. If no air (oxygen) enters the exhaust stream at the exhaust manifold or catalytic converter, HC and CO emission levels will be too high. Air flowing to the exhaust manifold at all times can increase the temperature of the catalytic converter and cause damage to the catalysts. On the other hand, if air is flowing continuously to the catalytic converter, it can cause the converter to overheat during fuel rich operations. An electrical feature

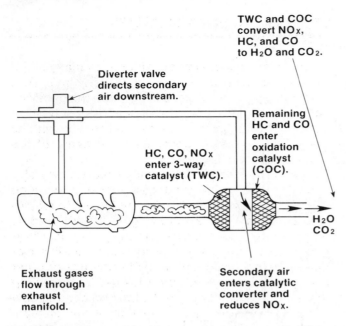

FIGURE 12-37 Downstream secondary air helps the catalytic converter to reduce NO_x emissions.

(open circuit) of the SAB valve will divert air to the atmosphere for all engine operations. Secondary air will continuously flow to the catalytic converter if an open circuit occurs in the SAD valve. Mechanical failures in either or both of the valves can cause

incorrect airflow to the exhaust manifold or the catalytic converter.

SYSTEM EFFICIENCY TEST

Run the engine at idle with the secondary air system on (enabled). Using an exhaust gas analyzer, measure and record the oxygen (O_2) levels. Next, disable the secondary air system and continue to allow the engine to idle. Again measure and record the oxygen level in the exhaust gases. The secondary air system should be supplying 2 to 5 percent more oxygen when it is operational (enabled).

FUEL EVAPORATIVE EMISSION SYSTEM

In carburetor vehicles a fuel evaporative system is used to control exhaust emissions. The system uses a carbon canister to trap evaporating fuel vapors from the carburetor bowl vent and the fuel tank. During warm engine cruise conditions, these fuel vapors are purged or vented from the canister and directed into the cylinders. The on-board computer enables a purge solenoid to initiate the purge cycle.

The carbon canister is filled with activated carbon or charcoal elements. Vapors flow from the fuel

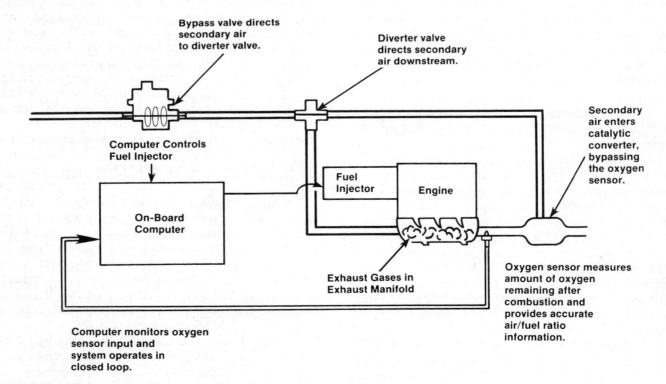

FIGURE 12-38 Closed loop operation for secondary air downstream mode

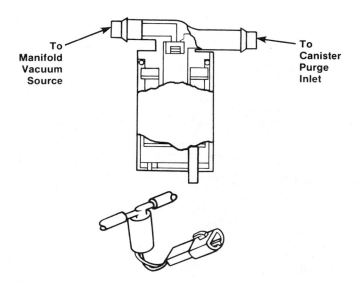

FIGURE 12-39 Canister purge solenoid

tank or carburetor fuel bowl into the canister. A purge line is connected to the canister purge solenoid valve. A vent cap on top of the canister allows fresh air to enter the canister during the purge cycle.

The canister purge solenoid (Figure 12-39) is mounted in-line between the canister and intake manifold. The purge solenoid is a normally closed solenoid. The on-board computer cycles the solenoid while the solenoid itself controls the release of fuel vapors trapped in the charcoal canister. When the solenoid is energized by the computer, the purge valve opens and allows the intake manifold vacuum to draw the trapped fuel vapors from the canister.

Fuel vapors are purged only under the following conditions:

- After a predetermined length of time has elapsed following engine start-up.
- The engine coolant temperature sensor must indicate that engine temperature is above a cold threshold and below an overheat threshold.
- The engine must be operating at stabilized, part-open throttle condition.
- The engine must be operating above a predetermined high rpm limit.

The canister is purged only during a stable, part-open throttle condition. The fuel evaporative system is shut off at idle and closed throttle conditions.

Normal service of the fuel evaporative system consists of replacing the canister or canister filter at the manufacturer's suggested interval and replacing any worn or damaged hoses. Failure of the system to purge on command indicates a failed canister purge solenoid valve.

OXYGEN SENSORS

Oxygen sensors are considered by many emissions technicians to be the kingpin of the input sensors. Although no one sensor, component, or device can be said to have total responsibility for solving all performance problems, the oxygen sensor, with its unique contribution, deserves a fair share of the credit.

An oxygen sensor is actually a galvonic battery capable of producing a voltage signal that can range from 0.1 to 0.9 volts (100 to 900 millivolts). It resembles an 18 mm spark plug (Figure 12-40), and its operation is similar to a car's battery. A typical oxygen sensor consists of a porous ceramic-like element made of zirconium dioxide (the use of other materials such as titonium dioxide is increasing) that acts as the electrolyte. A platinum coating covering both the inside and outside of this element forms the electrodes. The inside electrode is connected to a wire leading from the sensor to the electronic control unit. The outside electrode is grounded to the sensor's metal shell. Openings in the sensor's protective rubber or metal cover allow outside air to enter the zirconium element.

The strength of the voltage signal is related to the amount of oxygen in the exhaust gas compared to the oxygen content of the outside air. From an electrochemical reaction inside the sensor, electricity is produced when oxygen ions (charged particles) from the outside air move from the positively charged inside electrode through the zirconium dioxide (electrolyte) to the negatively charged outside electrode. As in a battery, the charged particle movement can occur and produce electricity only when there is a difference in electrical pressure or potential between the two electrodes.

In the case of the O_2 sensor, the greatest potential occurs when the exhaust is rich. The oxygen content of a rich exhaust is much lower than the oxygen content in the outside air. This greater difference in electrical pressure causes the oxygen ions to move from the inside electrode (+) to the outside electrode (−), resulting in the production of almost 1 volt. On the other extreme, the leaner the mixture, the more oxygen in the exhaust. When comparing this higher content of exhaust oxygen to the high oxygen content in the outside air, the difference in potential is low. This low potential produces a low voltage of approximately 0.1 volt.

In a normally operating O_2 sensor, the switching from high to low voltage occurs many times each second. The voltage signal is then sent to the microprocessor for evaluation and control. Depending on

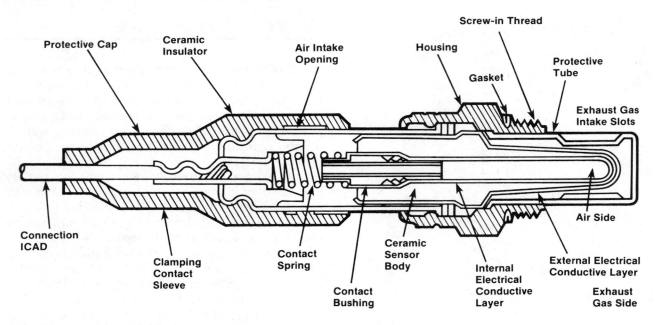

FIGURE 12-40 Oxygen sensor

the strength of the signal, the computer instantly corrects the air/fuel ratio by governing a mixture control device on carbureted engines or the injector pulse width on fuel injected models. The goal is precise control of the air/fuel mixture for maximum performance, economy, and emission control. Only when the air/fuel ratio is at or near the stoichiometric optimum of 14.7:1 can this be accomplished. In terms of oxygen sensor operation, a voltage range of 400 to 600 millivolts indicates the ideal air/fuel ratio is being achieved.

Before an oxygen sensor can function, however, it must reach a minimum operating temperature of approximately 600 degrees Fahrenheit. Prior to reaching that temperature and under other conditions such as low coolant temperature (cold start), wide-open throttle, and rapid deceleration, the computer is said to be operating in open loop. In this mode, the computer ignores signals from the O_2 sensor. Once the computer starts controlling the engine's operation based on signals generated by the oxygen sensor (feedback), the system is in closed loop.

Whenever the engine is in open loop, precise emission control is lost. In an effort to utilize oxygen sensor data more effectively and for longer periods of time, some manufacturers have designed an electric heating element into the sensor to bring it up to operating temperature more quickly. A typical heated sensor consists of a ground wire, sensor output wire (to the computer), and a third wire supplying battery voltage to a heating unit inside the sensor (when the key is in the RUN position only).

TROUBLESHOOTING THE OXYGEN SENSOR

To develop a troubleshooting procedure for tracking down a suspect oxygen sensor, a thorough knowledge of sensor operation and its relationship to system performance is essential, but do not overlook basic care. Before plunging into any course of action, consider the following service suggestions:

- All sensors, regardless of manufacturer, operate and are serviced in the same manner. Take note of special service requirements that call for periodic cleaning (making sure the air flutes are not plugged) or replacement of the sensor at regular intervals. Failure to heed these intervals could result in sluggish sensor operation and cause a surging or similar driveability problem.
- Keep in mind that oxygen sensors are covered by the federal 5-year/50,000-mile emission warranty. If an oxygen sensor needs replacement and it qualifies, have the customer return it to an appropriate dealer for free replacement.
- Be careful not to obstruct the vent hole on the outside of the sensor that allows outside air to enter the inner element. Any obstruction of this opening, whether from dirt or improper placement of the protective boot, can render the sensor useless. This includes using electrical tape or RTV sealant to repair a split or damaged boot.

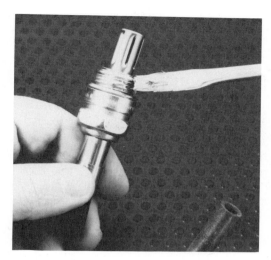

FIGURE 12–41 Coating the sensor's threads with an antiseize compound

- The use of anything but a high-impedance voltmeter (10 megohms) could cause damage to the sensor's internal calibration. This includes the use of ohmmeters, analog voltmeters, and test lights. If you think this is being too cautious, consider that some sources say simply holding the sensor by the output lead and housing at the same time can result in sensor damage. It seems this "human circuit" with the same low impedance as an analog meter could short out the sensor and destroy it.
- Always make sure the sensor's threads are clean and coated with a proper antiseize compound (Figure 12–41) and that torque specifications (never exceed 30 foot-pounds) are followed when it becomes necessary to replace or service an existing oxygen sensor.
- If the sensor lead becomes broken or damaged, a poor quality repair could cause unwanted resistance and result in a false electrical signal. If the lead is damaged, the best advice is to replace the sensor. Trying to save the customers a few dollars with a short-term repair will only lead to long-term headaches and a blemished reputation when they come back.
- Do not use high volatility silicone sealants (RTV) or sprays where they could come into contact with the sensor's element (either directly or indirectly). Clogged or contaminated sensors will cause major mixture control problems. Should a white powdery substance coating be observed on the sensor, it is a sure sign of silicone contamination. Also,

take note that methanol, carbon buildup, leaded gasoline, and silicates (found in antifreeze and many fuel additives) can form deposits that block and prevent proper sensor operation. In these cases, the signals coming from the sensor are delayed, but the computer still reacts with lightning speed. This time lag throws everything out of whack, and the air/fuel mixture starts reacting to an erroneous oxygen signal, which can cause the engine to start surging.

Most computer-controlled engines provide some type of self-diagnostic capability, but before jumping into this mode, the first thing you should do is make quick visual inspection of the sensor (without removing it). Make sure the sensor's outside air passage is clear (the sensor will not work if the outside air passage is blocked by dirt, grease, oil, and so on) and the related wiring and connectors are intact. If everything seems to check out after this initial inspection, then one or more of the following conditions exists:

- The sensor is worn, clogged, or malfunctioning.
- The voltage signals from the sensor are not reaching the computer.
- Something other than the sensor such as the computer, computer wiring, carburetor (or injectors), or ignition system is malfunctioning.

To find out which problem exists, further testing is required. After performing a preliminary inspection, find out if the computer has stored any trouble codes. For example, on many Chrysler feedback systems a trouble code of 21 (oxygen sensor circuit) indicates an oxygen sensor problem. If a code exists, go to the appropriate diagnostic chart. These charts were designed to isolate a specific malfunction with a minimum of time and monetary investment.

Sometimes a trouble code appears and the problem can be solved using the diagnostic chart. But many times trouble codes are not set and the technician must rely on past experience and a working knowlege of the system to solve a problem.

A quick way to check sensor operation is to monitor its voltage output. Several methods can be used here, the most accurate of which is the specialized analyzer. If a monitoring device is unavailable, a DVOM (Figure 12–42) will work (keep in mind this typical example is valid for most cars equipped with O_2 sensors, but not for all of them).

When using the DVOM, make sure there is a closed loop (if in rpm for about 30 to 45 seconds to

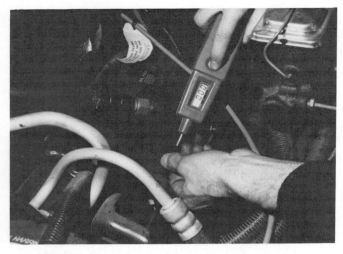

FIGURE 12-42 Monitoring the output and reaction time of a sensor with a DVOM

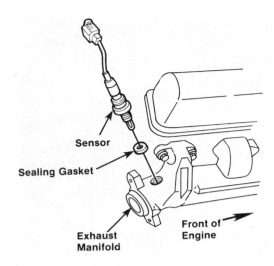

FIGURE 12-44 Sealing gasket

heat up the sensor), then ground the meters black lead and connect the red lead to the sensor output. As various mixture conditons are simulated, such as closing the choke to enrich the mixture or pulling a vacuum hose to lean the mixture, note if there is a rapid change in voltage from high to low. If this change in voltage corresponds to the simulated conditions, the O_2 sensor is good.

If the results do not seem right, try cleaning (decarbonizing) the sensor on the car. To accomplish this task (Figure 12-43), ground the computer input lead or M/C solenoid for 2 to 3 minutes (longer than this can damage the sensor) and hold the engine at approximately 2500 rpm. Under these condi-

FIGURE 12-43 Grounding the computer input lead will cause the computer to ask for a full lean condition.

tions, the computer places the system in full lean and the hot exhaust generated by the lean mixture acts to incinerate any foreign particles contaminating the sensor.

To avoid inaccurate sensor readings and/or exhaust gas leaks into the engine compartment, make sure the sealing gasket (Figure 12-44) is not left lying on the workbench.

When a specific procedure is not given to lean the mixture, try creating a massive vacuum leak. If the car still runs, this should force a lean enough mixture that running under this condition at 2500 rpm for a couple of minutes will increase the chances of cleaning the sensor.

A quick check to determine if a problem in the computer and/or related wiring (Figure 12-45) exists is to simulate an oxygen sensor using the natural resistance of your body. This can be done by unplugging the sensor, holding the computer input lead in one hand, and touching the positive battery terminal with the other. The system should be in closed loop and read a high voltage (DVOM connected to the O_2 output lead).

If the positive battery lead is moved to the negative lead, the voltage should drop substantially. By performing the above test while monitoring other information such as mixture control dwell variation (solenoid on or off time) or injector pulse width, it is possible to tell if the system is reacting to the oxygen sensor input.

SPARK ADVANCE

An engine with an EEC does not have a vacuum or mechanical (centrifugal) advance mechanism in

FIGURE 12-45 When performing a visual inspection before servicing or diagnosing an oxygen sensor-related problem, always check the sensor's wiring.

the distributor as described in Chapter 11. Instead, the ECM receives signals from several inputs and computes the best timing for the engine's operating conditions (Figure 12-46). The significant operating conditions are manifold pressure, throttle position, coolant temperature, inlet air temperature, and exhaust gas richness.

The ECM also receives a basic timing signal from the pickup in the distributor and the ignition module. The computed timing signal (advance or retard) goes back to the ignition (TFI) module to control spark timing. (If there is a failure in the ECM control or wiring to the ignition module, the module automatically sets a fixed advance so the engine can run.)

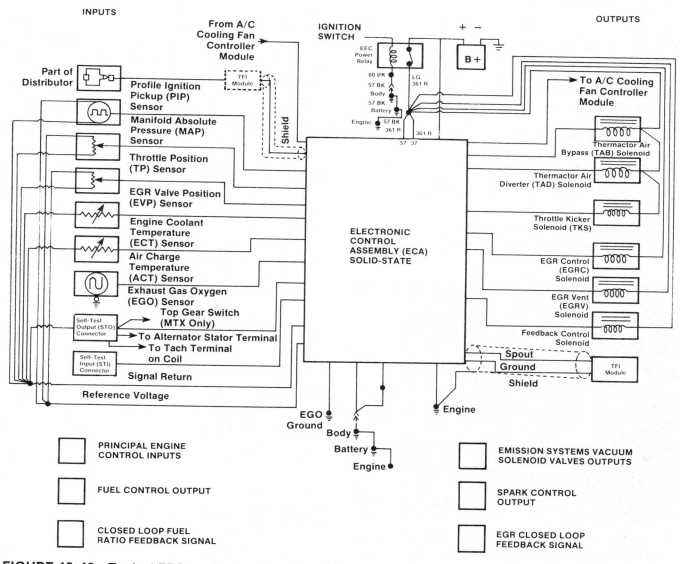

FIGURE 12-46   Typical EEC engine functions for closed-loop fuel control and emissions control.

REVIEW QUESTIONS

1. Which of the following memory circuits is used to store trouble codes and other temporary information?
   a. read only memory
   b. programmed read only memory
   c. random access memory
   d. all of the above

2. Which of the following is not controlled by an electronic control module in an electronic emission control system?
   a. PCV system
   b. EGR system
   c. air injection system
   d. EFE system

3. A car is receiving a 30,000 mile tune-up. Technician A says that the coolant temperature sensor should be replaced as a preventive maintenance measure. Technician B says that the exhaust oxygen sensor should also be replaced every 30,000 miles. Who is correct?
   a. Technician A
   b. Technician B
   c. Both A and B
   d. Neither A nor B

4. Which of the following inputs tells the ECM the engine rmp?
   a. MAP sensor
   b. high gear switch
   c. distributor
   d. throttle position sensor

5. Input from which of the following components would not cause the ECM to alter the idle speed?
   a. coolant temperature sensor
   b. Park/Neutral switch
   c. power steering switch
   d. vehicle speed sensor

6. During which of the following operation modes is the EGR valve normally open?
   a. wide-open throttle
   b. part throttle cruising
   c. idle
   d. none of the above

7. Which of the following is not used to bleed off vacuum from the EGR valve?

   a. EGR vent valve
   b. EGR control valve
   c. electronic vacuum regulator
   d. all of the above

8. Adding too much exhaust gas to the intake charge will create a _____ .
   a. rich mixture
   b. lean mixture
   c. high NO_x emissions
   d. high compression ratios

9. A vehicle is experiencing a sluggish performance when idling or under heavy load. Technician A says the vent valve might be stuck open. Technician B says that the EGR valve is probably failed in the *closed* position. Who is correct?
   a. Technician A
   b. Technician B
   c. Both A and B
   d. Neither A nor B

10. Which of the following provides additional oxygen to the catalytic converter to aid in oxidizing emissions?
    a. exhaust gas recirculation system
    b. air injection system
    c. feedback carburetor
    d. thermostatic air cleaner

11. Which of the following components in a secondary air system is controlled by an electronic control module?
    a. diverter valve
    b. air pump
    c. check valve
    d. all of the above

12. During which condition does the air injection system operate in the bypass mode?
    a. warm engine operation
    b. warm crank/start-up condition
    c. deceleration
    d. all of the above

13. During which of the following conditions is the EEC canister purged?
    a. idle
    b. part throttle
    c. wide-open throttle
    d. deceleration

GLOSSARY OF TERMS

A

Actuator   A device that delivers motion in response to an electrical signal.

Additive   As used with reference to automotive oils, a material added to the oil to give it certain properties. For example, a material added to engine oil to lessen its tendency to congeal or thicken at low temperature.

Advance   Spark or ignition advance. Causing the ignition spark to occur earlier to compensate for faster engine operation or slower fuel combustion.

Air   A gas containing approximately 4/5 nitrogen, 1/5 oxygen, and some carbonic gas.

Air Bypass Valve   A valve in the air pump system. At high engine vacuum, it vents pressurized air from the air pump to the atmosphere in order to prevent backfiring. At other times it sends air to the exhaust manifold; on cars with a three-way catalyst, it sends air only to the oxidation catalyst when the engine warms up. Also known as the **anti-backfire valve** or **diverter valve.**

Air Charge Temperature Sensor   A thermistor sensor responsible for inputting to the processor the temperature of an air stream in the air filter or intake manifold. If air is cold, signals choke to let off slowly. Alters engine speed after choke is off, and below a certain temperature dumps thermactor air for catalyst protection.

Air Charge Temperature Sensor   A thermistor sensor responsible for inputting to the processor the temperature of an air stream in the air filter or intake manifold. If air is cold, signals choke to let off slowly. Alters engine speed after choke is off, and below a certain temperature dumps thermactor air for catalyst protection.

Air Cleaner   A device for filtering, cleaning, and removing dust from the air admitted to a unit, such as an engine or air compressor.

Air Cleaner Bimetal Sensor   Senses the temperature of incoming fresh air. Bleeds off vacuum when air cleaner is warm.

Air Cleaner Duct and Valve Vacuum Motor   Opens and closes the air cleaner duct to provide heated or unheated air to the engine dependent upon the temperature of the incoming air.

Air Control Valve   A vacuum-controlled valve in the thermactor system that diverts air pump air to either the upstream (exhaust manifold) or downstream (underbody catalyst) air injection points as required; a diverter valve or a combination bypass/diverter valve.

Air/Fuel Mixture   A ratio of the amount of air that is mixed with fuel before it is burned in the combustion chamber.

Air Gap   The space between spark plug electrodes, motor and generator armatures, field shoes, etc.

Air Lock   A bubble of air trapped in a fluid circuit that interferes with normal circulation of the fluid.

Air Pump   A device to produce a flow of air at higher than atmospheric pressure. Normally referred to as a thermactor air supply pump.

Alloy   A mixture of different metals such as solder, which is an alloy consisting of lead and tin.

Altitude Compensation System   An altitude barometric switch and solenoid used to provide better driveability at more than 4,000 feet above sea level.

Aluminum   A metal noted for its lightness and often alloyed with small quantities of other metals for automotive use.

Ambient Temperature   Temperature of air surrounding an object.

Ammeter   An electrical meter used to measure current flow in amperes.

Ampere   The unit of measure of current flowing in an electrical circuit.

Amplifier   A circuit or device used to increase the voltage or current of a signal.

Annealing   A process of softening metal. For example, the heating and slow cooling of a piece of iron.

Antibackfire Valve   Often called a gulp valve, the antibackfire valve is located downstream from the air bypass valve. Its purpose is to divert a portion of the thermactor air to the intake manifold when it is triggered by intake manifold vacuum on deceleration. The valve operates only during periods of sudden decrease in intake manifold pressure.

Antifreeze A material such as alcohol, glycerin, etc., added to water to lower its freezing point.

Arcing Electrical energy jumping across a gap. Arcing across ignition points causes pitting and erosion.

Atmospheric Pressure The weight of the air at sea level; about 14.7 pounds per square inch; less at higher altitudes.

Automotive Emissions Gaseous and particulate compounds that are emitted from a car's crankcase, exhaust, carburetor, and fuel tank (hydrocarbons, nitrogen oxide, and carbon monoxide).

B

Backfire Suppressor Valve A device used in conjunction with the early design thermactor exhaust emission system. Its primary function is to lean out the excessively rich fuel mixture, which follows closing of the throttle after acceleration. Allows additional air into the induction system whenever intake manifold vacuum increases.

Barometric Pressure A sensor or its signal circuit that sends a varying frequency signal to the processor relating actual barometric pressure.

Battery A cell or group of cells connected together to deliver electrical energy (in the automobile battery, by chemical action).

Boiling Point The temperature at atmospheric pressure at which bubbles or vapors rise to the surface and escape.

Bore The diameter of a hole, such as a cylinder; also to enlarge a hole as distinguished from making a hole with a drill.

Boring Bar A stiff bar equipped with multiple cutting bits, which is used to bore a series of bearings or journals in proper alignment with each other.

Boss An extension or strengthened section, such as the projections within a piston, which support the piston pin or piston pin bushings.

Bowl Vent Port Port in a carburetor that vents fumes and excess pressure from the bowl to maintain atmospheric pressure.

Breaker Arm The movable member of a pair of breaker points.

Breaker Cam The multilobed cam rotating in the ignition distributor, which interrupts the primary circuit to induce a high-tension spark for ignition.

Breaker Plate The plate in the distributor that supports the breaker points.

Breaker Points Contact points in the distributor that close and open the ignition primary circuit. (Also called ignition points, distributor points, or points.)

British Thermal Unit A measurement of the amount of heat required to raise the temperature of 1 pound of water 1 degree Fahrenheit.

Burnish To smooth or polish by the use of a sliding tool under pressure.

Bushing A removable liner for a bearing.

Bypass An alternate path for a flowing substance.

Bypass Valve Valve that opens under certain conditions to permit a flow of liquid or gas by some alternate to its normal route. See **Air Bypass Valve.**

C

Calibrate To determine or adjust the graduation or scale of any instrument giving quantitative measurements.

Caliper (Inside and Outside) An adjustable tool for determining the inside or outside diameter by contact and retaining the dimension for measurement or comparison.

Cam The rotating member of the distributor with lobes that open and close the breaker points as many times per distributor-shaft revolution as there are cylinders in the engine.

Cam Angle Also known as dwell period. Referring to an ignition distributor, the number of degrees of rotation of the ignition distributor shaft during which the contact points are closed.

Cam Ground Piston A piston ground to a slightly oval shape which, under the heat of operation, becomes round.

Camshaft The shaft containing lobes or cams to operate the engine valves.

Canister A container in an evaporative emission control system that contains charcoal to trap vapors from the fuel system.

Canister Purge Shutoff Valve A vacuum-operated valve that shuts off canister purge when the thermactor air diverter valve dumps air downstream.

Canister Purge Solenoid Electrical solenoid or its control line. Solenoid opens valve from fuel vapor canister line to intake manifold when energized. Controls flow of vapors between carburetor bowl vent and carbon canister.

Canister Purge Valve Valve used to regulate the flow of vapor from the charcoal canister to the engine.

Carbon A common nonmetallic element that is an excellent conductor of electricity. It also forms in the combustion chamber of an engine during the burning of fuel and lubricating oil.

Carbon Dioxide Compressed into solid form, this material is known as dry ice and remains at a temperature of 109 degrees. It goes directly from a solid to a vapor state.

Carbonize The process of carbon formation within an engine, such as on the spark plugs and within the combustion chamber.

Carbon Monoxide Gas formed by incomplete combustion; colorless, odorless, very poisonous.

Carburetor A device for automatically mixing fuel in the proper proportion with air to produce a combustible gas.

Case Harden To harden the surface of steel.

Catalyst A compound or substance that can speed up or slow down the reaction of other substances without being consumed itself. In an automatic catalytic converter, special metals (i.e., platinum, palladium) are used to promote more complete combustion of unburned hydrocarbons and a reduction of carbon monoxide.

Catalytic Converter A muffler-like component in the exhaust system that promotes a chemical reaction that converts certain air pollutants in the exhaust gases into harmless substances.

Cathode Ray Tube An electronic tube with an electron gun at the rear that projects a stream of electrons to a spot on the screen end of the tube. The height of the spot on the screen is controlled by the height of the voltage applied to the tube.

Centigrade A measurement of temperature used principally in foreign countries, zero on the centigrade scale being 32 degrees on the Fahrenheit scale.

Centrifugal Force A force that tends to move a body away from its center of rotation. An example is supplied by whirling a weight attached to a string.

Chamfer A bevel or taper at the edge of a hole.

Charging Circuit The alternator (or generator) and associated circuit used to keep the battery charged and to furnish power to the automobile's electrical systems when the engine is running.

Chase To straighten up or repair damaged threads.

Chassis Dynamometer A machine for measuring the power delivered to the drive wheels of a vehicle.

Check Valve A gate or valve that allows passage of gas or fluid in one direction only.

Chilled Iron Cast iron on which the surface has been hardened.

Chromium Steel An alloy of steel with a small amount of chromium to produce a metal which is highly resistant to oxidation and corrosion.

Circuit The complete path of an electrical current, including the generating device. When the path is unbroken, the circuit is closed and current flows. When the circuit continuity is broken, the circuit is open and current flow stops.

Clearance The space allowed between two parts, such as between a journal and bearing.

Clockwise Rotation Rotation in the same direction as the hands of a clock.

Closed Circuitry A circuit that is uninterrupted from the current source and back to the current source.

Closed Loop A system that feeds back its output to the input side of the processor, which monitors the output and makes corrections as necessary.

Closed Loop Mode Mode in which the ECA operates with EGO sensor feedback.

Closed System Crankcase ventilation system that vents crankcase pressure and vapors back into the engine where they are burned during combustion rather than venting to the atmosphere.

Coefficient of Friction A measurement of the amount of friction developed between two surfaces pressed together and moved one on the other.

Coil A spiral winding of electrical wire. In the automobile ignition system, a step up transformer (pair of coils) that changes lower voltage and high current in the primary coil winding to high voltage and low current in the secondary coil windings.

Cold Manifold An intake manifold to which the exhaust gas is not applied for heating purposes.

Cold Weather Modulator A vacuum modulator located in the carburetor air cleaner on some models. Prevents the air cleaner duct door from opening to nonheated intake air when fresh air is below 55 degrees Fahrenheit. Looks like a **Temperature Vacuum Switch,** differentiated by color codes.

Combination Thermactor Air Bypass and Air Diverter Valve Combines the function of a normally closed air bypass valve and an air control valve in one integral valve.

Combustion The process of burning.

Combustion Space or Chamber In automobile engines, the volume of the cylinder above the piston with the piston on top center.

Compound A mixture of two or more ingredients.

Compression The reduction in volume or the squeezing of a gas. As applied to metal, such as a coil spring, compression is the opposite of tension.

Compression Ratio The volume of the combustion chamber at the end of the compression stroke as compared to the volume of the cylinder and chamber with the piston on bottom center.

Compression Rings The upper rings on a piston designed to hold the compression in the cylinder.

Computed Timing The relationship of spark plug firing to crankshaft position expressed in crankshaft degrees. On some vehicles, the crankshaft position is determined by the crankshaft position (CP) sensor. On the others, this function is controlled by the profile ignition pickup (PIP) sensor.

Computer A device that takes information, processes it, makes decisions, and outputs those decisions. Computers are generally large and not portable (see **Microcomputer**).

Condensation The process of a vapor becoming a liquid; the reverse of evaporation.

Condenser An electrical device used to store the excess coil energy above spark-plug firing needs and discharge this excess energy back through the primary ignition circuit. By providing a place for primary current to flow while the points are opening, the condenser prevents arcing at the points.

Concentric Two or more circles have a common center.

Conductor Any material that permits electrical current to flow easily. A nonconductive material is called an insulator.

Connecting Rod Rod that connects the piston to the crankshaft.

Contraction A reduction in mass or dimension; the opposite of expansion.

Convection A transfer of heat by circulating heated air.

Conventional Oxidation Catalyst Catalyst that acts on two major pollutants: hydrocarbons and carbon monoxide.

Coolant The liquid that circulates in an engine cooling system.

Corrode To eat away gradually as if by gnawing, especially by chemical action such as rust.

Counterbore To enlarge a hole to a given depth.

Counterclockwise Rotation Rotating the opposite direction of the hands on a clock. The same as anticlockwise rotation.

Countersink To cut or form a depression to allow the head of a screw to go below the surface.

Crankcase The housing within which the crankshaft and many other parts of the engine operate.

Crankcase Breather A port or tube that vents fumes from the crankcase. An inlet breather allows fresh air into the crankcase.

Crankcase Dilution Under certain conditions of operation, unburned portions of the fuel find their way past the piston rings into the crankcase and oil reservoir where they dilute or thin the engine lubricating oil.

Cranking Circuit The starter and its associated circuit, including battery, relay (solenoid), ignition switch, neutral start switch (on vehicles with automatic transmission), and cables and wires.

Crankshaft The main shaft of an engine which in conjunction with the connecting rods changes the reciprocating motion of the pistons into rotary motion.

Crankshaft Counterbalance A series of weights attached to or forged integrally with the crankshaft so placed as to offset the reciprocating weight of each piston and rod assembly.

Crude Oil Liquid oil as it comes from the ground.

Current The movement of electrical energy. Generally considered to be the movement of electrons, or negatively charged particles, within the conductor. Measured in amperes.

Cylinder Head A detachable portion of an engine fastened securely to the cylinder block, which contains all or a portion of the combustion chamber.

Cylinder Sleeve A liner or tube interposed between the piston and the cylinder wall or cylinder block to provide a readily renewable wearing surface for the cylinder.

D

Dead Center The extreme upper or lower position of the crankshaft throw at which the piston is not moving in either direction.

Deceleration A decrease in speed.

De-energized Having the electric current or energy source turned off.

Degree Abbreviated "deg." or indicated by a small "°" placed alongside a figure; can be used to designate temperature readings or to designate angularity, one degree being 1/360 part of a circle.

Delay Valve Vacuum restriction used to retard or delay the application of a vacuum signal.

Delay Valve Two-Way A delay valve that functions in both directions

Density Compactness; relative mass of matter in a given volume.

Detergent A compound of a soap-like nature used in engine oil to remove engine deposits and hold them in suspension in the oil.

Detonation As used in an automobile, indicates a too rapid burning or explosion of the mixture in the engine cylinders. It becomes audible through a vibration of the combustion chamber walls and is sometimes confused with a "ping" or spark knock.

Dial Gauge A type of micrometer wherein the readings are indicated on a dial rather than on a thimble.

Diaphragm A flexible partition or wall separating two cavities.

Die One of a pair of hardened metal blocks for forming metal into a desired shape or a thread die for cutting external threads.

Die Casting An accurate and smooth casting made by pouring molten metal or composition into a metal mold or die under pressure.

Diode An electronic device that permits flow of electricity in only one direction. In automotive application, used at the alternator to change alternating current into direct current.

Direct Current Current flowing in one direction only.

Disable A type of microcomputer decision that results in an automotive system being deactivated and not permitted to operate.

Distortion A warpage or change in form from the original shape.

Distributor The mechanism in an ignition system that opens and closes the primary circuit through the distributor breaker points and directs the secondary high voltage to the spark plugs at the correct time for firing.

Distributor Terminal That terminal of the ignition coil to which primary current flows from the breaker point circuit of the distributor, permitting the coil's magnetic field to saturate and collapse as the points close and open.

Distributor Vacuum Vent Valve Required by some engines to prevent fuel migration to distributor advance diaphragm and to act as a spark advance delay valve.

Diverter Valve Valve that directs secondary air downstream to mid-bed port of three-way catalyst. See **Air Control Valve.**

Dowel Pin A pin inserted in matching holes in two parts to maintain those parts in fixed relation one to the other.

Draw Filing A method of filing wherein the file is drawn across the work while held at a right angle to the length of the file.

Drill A tool for making a hole or to sink a hole with a pointed cutting tool rotated under pressure.

Drive-Fit Also known as a force-fit or press-fit. This term is used when the shaft is slightly larger than the hole and must be forced in place.

Drop Forging A piece of steel shaped between dies while hot.

Dwell The period during which the breaker points are closed. Measured in degrees of distributor shaft rotation.

Dwell Period Also known as cam angle.

Dwell Section That segment of the ignition pattern on a scope from the point at which breaker points close to the point at which they open.

Dynamometer A machine for measuring the actual power produced by an internal combustion engine.

E

Early Fuel Evaporation A device to heat the air fuel mixture entering the intake manifold when the engine is cold.

Eccentric One circle within another circle wherein both circles do not have the same center. An example of this is a cam on a camshaft.

EGR Exhaust Gas Recirculation.

EGR Act EGR solenoid pressure valve assembly.

EGR Control Solenoid Electrical solenoid or its control line. Solenoid switches engine manifold vacuum to operate EGR valve when solenoid is energized.

EGR Cooler Assembly Heat exchanger using engine coolant to reduce exhaust gas temperature.

EGR Valve Position Sensor Potentiometric sensor or its signal line used in electronically controlled EGR systems. Sensor wiper position is proportional to EGR valve pintle position. This allows the electronic control assembly (ECA) to determine actual EGR flow at any point in time.

EGR Vent Solenoid Electrical solenoid or its control line. Solenoid normally vents EGRC vacuum line. When EGRV is energized, EGRC can open the EGR valve.

EGR Venturi Vacuum Amplifier Device that uses relatively weak venturi vacuum to control a manifold vacuum signal to operate the EGR valve. Contains a check valve and relief valve that open whatever venturi vacuum signal is equal to or greater than manifold vacuum.

Electrode   The firing terminals that are found in a spark plug.

Electromechanical   Refers to a device that incorporates both electronic and mechanical principles together in its operation.

Electronic   Pertaining to the control of systems or devices by the use of small electrical signals and various semiconductor devices and circuits.

Electronic Control Assembly   A vehicle computer consisting of a calibration assembly containing the computer memory and thus its control program and processor assembly, which is the computer hardware.

Electronic Engine Control   A computer-directed system of engine control; it can be controlled by one or more sensors. As a rule it controls engine timing and/or fuel system (with the FBC system). The typical electronic engine control system operates on the circuit theory. Once the ignition switch is turned on, EEC is in operation.

Enable   A type of microcomputer decision that results in an automotive system being activated and permitted to operate.

Energized   Having the electrical current or electrical source turned on.

Engine   As used in automobiles, the term applies to the prime source of power generation used to propel the vehicle.

Engine Coolant Temperature Sensor   Refers to thermistor sensor or its signal line. Sensor is immersed in engine coolant. Provides engine coolant temperature information to ECU, which is used to alter spark advance and EGR flow during warm-up or overheat condition. (Replaces cooling and EGR PVS.)

Engine Displacement   The sum of the displacement of all the engine cylinders.

"E" Port   A carburetor source for EGR vacuum.

Evaporative Emission Controls   Emission control system responsible for preventing fuel vapors from entering the atmosphere, primarily from the fuel tank and carburetor.

Evaporative Emission Shed System   A system for containment of fuel vapors (evaporative emissions) introduced in 1978. Annual improvements have modified this system.

Exhaust Gas Check Valve   Allows thermactor air to enter exhaust manifold but prevents reverse flow in the event of improper operation of other components.

Exhaust Gas Oxygen Sensor   Exhaust Gas Oxygen Sensor or its signal line. Sensor changes its output voltage as exhaust gas oxygen content changes when compared to the oxygen content of the atmosphere. Constantly changing voltage signal is sent to the processor for analysis and adjustment to the air/fuel ratio.

Exhaust Gas Recirculation   A procedure in which a small amount of exhaust gas is readmitted to the combustion chamber to reduce peak combustion temperatures and thus reduce nitrogen oxide emissions.

Exhaust Heat Control Valve   A valve that routes hot exhaust gases to the intake manifold heat riser during cold engine operation. The valve can be thermostatically controlled or vacuum operated. In many vehicles, this valve is controlled by the signal from the ECA.

Exhaust Manifold   That part of an engine through which exhausted or spent gases flow to the exhaust system.

Exhaust Pipe   The pipe connecting the engine to the muffler to conduct the exhausted or spent gases away from the engine.

Exhaust Valve   A valve that permits the exhausted or spent gases to escape from a chamber.

Expansion   An increase in size. For example, when a metal rod is heated, it increases in length and perhaps also in diameter; expansion is the opposite of contraction.

F

Fahrenheit   A scale of temperature measurement ordinarily used in English-speaking countries. The boiling point of water is 212 degrees Fahrenheit as compared to 100 degrees Celsius.

Feedback Carburetor Actuator   A computer-controlled stepper motor that varies the carburetor air/fuel mixture.

Feeler Gauge   A metal strip or blade finished accurately with regard to thickness used for measuring the clearance between two parts. Such gauges ordinarily come in a set of different blades graduated in thickness by increments of 0.001 inch.

Ferrous Metal   Metal that contains iron or steel and is therefore subject to rust.

File   To finish or trim with a hardened rasp or file.

Fillet   A rounded filling between two parts joined at an angle.

Filter (Oil, Water, Gasoline, Etc.)   A unit containing an element, such as a screen of varying degrees of fineness. The screen or filtering element is made of various materials depending on the size of the foreign particles to be eliminated from the fluid being filtered.

Firing Section   That segment of the ignition pattern on a scope from the point at which the breaker points open, initiating the firing of the spark plug, to the point at which the spark extinguishes.

Firing Spike The vertical line appearing on the scope as the plug begins to fire, representing the voltage required to jump the spark plug gap.

Fit A kind of contact between two machined surfaces.

Flange A projecting rim or collar on an object for keeping it in place.

Floating Piston Pin A piston pin that is not locked in the connecting rod or the piston but is free to turn or oscillate in both the connecting rod and piston.

Flutter or Bounce As applied to engine valves, refers to a condition wherein the valve is not held tightly on its seat during the time the cam is not lifting it.

Flywheel A heavy wheel in which energy is absorbed and stored by means of momentum.

Foot-Pound This is a measure of the amount of energy or work required to lift 1 pound to 1 foot.

Force-Fit Also known as a press-fit or drive-fit. This term is used when the shaft is slightly larger than the hole and must be forced in place.

Forge To shape metal while hot and plastic by hammering.

Four-Cycle Engine Also known as Otto cycle, wherein an explosion occurs every other revolution of the crankshaft; a cycle being considered as half a revolution of the crankshaft. These strokes are intake, compression, power, and exhaust.

Freewheeling A mechanical device that engages the driving member to impart motion to a driven member in one direction but not the other. Also known as an overrunning clutch.

Fuel Knock Same as **Detonation.**

Fuel (Rich, Lean) A qualitative evaluation of air/fuel ratio based on the voltage from the exhaust gas oxygen sensor.

Fuel Tank Vapor Valve A valve mounted in the top of the fuel tank responsible for venting excess vapor and pressure from the fuel tank into the evaporative emission control system.

Fuel-Vacuum Separator Used to filter waxy hydrocarbons from carburetor ported vacuum to protect the vacuum delay and distributor vacuum controls.

Fulcrum The support on which a lever turns in moving a body.

G

Gap A break in the continuity of a circuit. The distance between the electrodes of a spark plug. The distance between breaker points at the high point on the distributor cam.

Gasket Anything used as a packing, such as a nonmetallic substance, placed between two metal surfaces to act as a seal.

Glaze As used to describe the surface of the cylinder, an extremely smooth or glossy surface such as a cylinder wall highly polished over a long period of time by the friction of the piston rings.

Glaze Breaker A tool for removing the glossy surface finish in an engine cylinder.

Grind To finish or polish a surface by means of an abrasive wheel.

Ground The negatively charged side of a circuit. A ground can be a wire, the negative side of the battery or even the vehicle chassis.

Ground Circuit The return side of an electric circuit.

Gum In automotive fuels, this refers to oxidized petroleum products that accumulate in the fuel system, carburetor, or engine parts.

H

Hall Effect A process in which current passes through a small slice of semiconductor material at the same time as a magnetic field to produce a small voltage in the semiconductor.

Harmonic Balancer A device to reduce the torsional or twisting vibration that occurs along the length of the crankshaft used in multiple cylinder engines. Also known as a vibration damper.

H.C. High compression.

Heated Air Inlet System System that operates during cold weather and cold start. Its purpose is to bring warm, filtered air into the engine to control the volume of air entering the engine, vaporize the fuel better, and reduce hydrocarbon and carbon monoxide emissions.

Heated Exhaust Gas Oxygen Sensor An EGO sensor with a heating element.

Heat Riser The passage in the manifold between the exhaust and intake manifold.

Heat Treatment A combination of heating and cooling operations timed and applied to a metal in a solid state in a way that will produce desired properties.

Helical Shaped like a coil of wire or screw thread.

Hg Chemical symbol for mercury. Also, a reference to amount of vacuum, i.e., "inches of mercury."

High Tension High voltage. In automotive ignition systems, voltages (up to 40 kilovolt) in the secondary circuit of the system as contrasted to the low, primary circuit voltages (nominally 6 or 12 volts).

Hone An abrasive tool for correcting small irregularities or differences in diameter in a cylinder, such as an engine cylinder or brake cylinder.

Hot Spot Refers to a comparatively thin section or area of the wall between the inlet and exhaust manifold of an engine, the purpose being to allow the hot exhaust gases to heat the comparatively cool incoming mixture. Also used to designate local areas of the cooling system which have attained above average temperatures.

HP Horsepower, the energy required to lift 550 pounds 1 foot in 1 second.

Hydrocarbon A chemical composition, made up of hydrogen and carbon, that is a component of exhaust emissions.

Hydrocarbon Engine An engine using petroleum products, such as gas, liquefied gas, gasoline, kerosene, or fuel oil, as a fuel.

Hydrogen Highly flammable elemental gas. Chemical symbol is H.

I

I.D. Inside Diameter.

Idle Refers to the engine operating at its slowest speed with a vehicle not in motion.

Idle Vacuum Valve This device can be used in conjunction with other vacuum controls to dump thermactor air during extended periods of idle. Provides protection to catalyst.

Ignition In internal-combustion gasoline engines, the process of igniting the air/fuel mixture in the combustion chamber by means of an electrical spark from a spark plug.

Ignition Distributor An electrical unit that provides a means for conveying the secondary or high tension current to the spark plug wires as required.

Ignition Module Signal The signal produced by the ECA that controls the ignition module on and off time and, therefore, the spark plug timing and coil dwell.

Ignition System The means for igniting the fuel in the cylinders.

IHP Indicated horsepower developed by an engine and a measurement of the pressure of the explosion within the cylinder expressed in pounds per square inch.

Induction The transfer of electricity by magnetism rather than direct flow through a conductor. When current flows through a wire or coil, it builds up a magnetic field. If another conductor moves through this field (or the field moves through another conductor), a voltage is induced in the other conductor. This is the principle of the ignition coil and also of the engine analyzer's induction pick-ups.

Inertia A physical law that tends to keep a motionless body at rest or also tends to keep a moving body in motion; effort is thus required to start a mass moving or to retard or stop it once it is in motion.

Infrared Analyzer Instrument used to measure exhaust emissions.

Inhibitor A material to restrain or hinder some unwanted action, such as a rust inhibitor, which is a chemical added to cooling systems to retard the formation of rust.

Injector Electronic fuel injection solenoid (one of two) or its control line. Solenoid, when energized, allows fuel to flow into throttle body and then into one plane of the dual-plane intake manifold.

Inlet Valve or Intake Valve A valve that permits a fluid or gas to enter a chamber and seals against an exit.

Input Information provided to a microcomputer to allow accurate control of a system.

Input Conditioner A device or circuit that conditions or prepares an input signal for use by a microcomputer.

Intake Manifold or Inlet Pipe The tube used to conduct the gasoline and air mixture from the carburetor to the engine cylinders.

Integrated Circuit A small semiconductor device with circuitry that can perform numerous functions.

Intermediate Section That segment of the ignition pattern on a scope from the point at which the spark extinguishes to the point at which the breaker points close.

J

Journal A bearing within that a shaft operates.

K

Keep Alive Memory A series of vehicle battery-powered memory locations in the microcomputer that allow the microcomputer to store information on input failure, identified in normal operations for use in diagnostic routines. Keep Alive Memory adopts some calibration parameters to compensate for changes in the vehicle system.

Key A small block inserted between the shaft and hub to prevent circumferential movement.

Keyway or Key Seat A groove or slot cut to permit the insertion of a key.

Knock A general term used to describe various noises occurring in an engine; can be used to describe noises made by loose or worn mechanical parts, preignition, detonation, etc.

Knock Sensor A device designed to vibrate at approximately the same frequency as the engine knock frequency. The knock sensor provides information on engine knock to an electronic engine control microcomputer.

L

Lapping The process of fitting one surface to another by rubbing them together with an abrasive material between the two surfaces.

Liner Usually a thin section placed between two parts, such as a replaceable cylinder liner in an engine.

Liter A metric measure equal to 2.11 pints.

M

Magnet Any body with the property of attracting iron or steel.

Magnetic Field The area surrounding the poles of a magnet that is affected by its attraction or repulsion forces.

Malleable Casting A casting that has been toughened by annealing.

Manifold A pipe with multiple openings used to connect various cylinders to one inlet or outlet.

Manifold Absolute Pressure Sensor A pressure sensitive disk capacitor sensor used to measure air pressure inside the intake manifold.

Manifold Control Valve A thermostatically operated valve in the exhaust manifold for varying heat to the intake manifold with respect to the engine temperature.

Manifold Vacuum Vacuum generated below the throttle plates of a carburetor. Vacuum present in the intake manifold. Manifold vacuum is high at idle and lowers as the throttle plates open.

Mechanical Efficiency (Engine) The ratio between the indicated horsepower and the brake horsepower of an engine.

Microcomputer A device that takes information, processes it, makes decisions, and outputs those decisions. Microcomputers are generally small and portable and are located inside a processor.

Micrometer A measuring instrument for either external or internal measurement in thousandths and sometimes tenths of thousandths of inches.

Microprocessor An integrated circuit within a microcomputer that controls information flow within the microcomputer. Also called the Central Processing Unit (CPU).

Microprocessor Control Unit Integral part of electronically controlled feedback carburetor using TWC catalyst. Various sensors monitor mode conditions. MCU is widely used on Ford-built vehicles for the control of air/fuel ratios.

Mill To cut or machine with rotating tooth cutters.

Millimeter One millimeter is the metric equivalent of 0.039370 of an inch or one inch being the equivalent of 25.4 mm.

Misfiring Failure of an explosion to occur in one or more cylinders while the engine is running; can be continuous or intermittent failure.

Monolithic Substrate The ceramic honeycomb structure used as a base to be coated with a metallic catalyst material for use in the catalytic converter.

Motor This term should be used in connection with an electric motor and should not generally be used when referring to the engine of an automobile.

Muffler A chamber attached to the end of the exhaust pipe that allows the exhaust gases to expand and cool. It is usually fitted with baffles or porous plates and serves to subdue much of the noise created by the exhaust.

N

Needle Bearing An antifriction bearing using a great number of small-diameter rollers of greater length.

Negative A pole from which electricity flows. Having an excess of electrons (negatively-charged particles). Opposite of **Positive.**

Neutral Drive Switch A sensor that provides information on transmission status to the ECA.

Nitrogen An elemental gas that is inert. Seventy-eight percent of the air is nitrogen.

Nitrogen Oxides A compound formed during the engine's combustion process when oxygen in the air combines with nitrogen in the air to form the nitrogen oxides, which are agents in photochemical smog.

Nonferrous Metals This designation includes practically all metals that contain no iron or very little iron and are therefore not subject to rusting.

Normally Closed Refers to a switch or solenoid that is closed when no control or force is acting on it.

Normally Open Refers to a switch or solenoid that is open when no control force is acting on it.

Normal Pattern A scope pattern that represents ignition system operation of an engine completely and accurately tuned to manufacturer's specifications.

O

Octane Number   A unit of measurement on a scale intended to indicate the tendency of a fuel to detonate or knock.

Octane Selector   A calibrated device for adjusting the timing of the ignition distributor in accordance with the characteristics of the fuel in use.

Ohm   A unit of electrical resistance opposing current flow. Resistance varies in different materials and with temperature.

Ohmmeter   An instrument that measures resistance of a conductor in units called ohms.

Open Circuit   An electrical circuit whose path has been interrupted or broken either accidentally (e.g., a broken wire) or intentionally (e.g., a switch turned off).

Open Loop   An electronic control system in which sensors provide information, the microcomputer gives orders, and the output actuators obey the orders without feedback to the microcomputer.

Open System   Descriptive term for crankcase emissions control system, which vents to the atmosphere.

Oscillation   A flow of electricity rapidly changing direction.

Oscilloscope   A cathode ray tube with controls and associated circuitry to present on its screen wave forms or patterns for observation and measurement of voltage with respect to time.

Output Driver   A transistor in the output control area of the processor that is used to turn various actuators on and off.

P

Particulate   A tiny bit of solid matter found in exhaust gases.

Paraded   A scope display mode that presents cylinder patterns horizontally one after the other in firing order, beginning with No. 1 plug at the left.

Pattern   In automotive ignition system service, designates the wave form on the screen of a scope.

Peen   To stretch or clinch over by pounding with the rounded end of a hammer.

Petcock   A small valve placed at various points in a fluid circuit usually for draining purposes.

Petroleum   A group of liquid and gaseous compounds composed of carbon and hydrogen that are removed from the earth.

Phosphor-Bronze   An alloy consisting of copper, tin, and lead sometimes used in heavy-duty bearings.

Photochemical   The action of light (photo) on air pollutants (chemicals), which creates smog.

Piston   A cylindrical part closed at one end, which is connected to the crankshaft by the connecting rod. The force of the explosion in the cylinder is exerted against the closed end of the piston causing the connecting rod to move the crankshaft.

Piston Collapse   A condition describing a collapse or a reduction in diameter of the piston skirt due to heat or stress.

Piston Displacement   The volume of air moved or displaced by moving the piston from one end of its stroke to the other.

Piston Head   That part of the piston above the rings.

Piston Lands   Those parts of a piston between the piston rings.

Piston Pin   The journal for the bearing in the small end of an engine connecting rod, which also passes through piston walls; also known as a **Pin Boss.**

Piston Ring   An expanding ring placed in the grooves of the piston to provide a seal to prevent the passage of fluid or gas past the piston.

Piston Ring Expander   A spring placed behind the piston ring in the groove to increase the pressure of the ring against the cylinder wall.

Piston Ring Gap   The clearance between the ends of the piston ring.

Piston Ring Groove   The channel or slots in the piston in which the piston rings are placed.

Piston Skirt   That part of the piston below the rings.

Piston Skirt Expander   A spring or other device inserted in the piston skirt to compensate for collapse or decrease in diameter.

Polarity   The particular state, either positive or negative, with reference to the two poles or to electrification.

Poppet Valve   A valve structure consisting of a circular head with an elongated stem attached in the center which is designed to open and close a circular hole or port.

Port   In engines, the openings in the cylinder block for valves, exhaust and inlet pipes, or water connections. In two-cycle engines, the openings for inlet and exhaust purposes.

Ported Vacuum   Vacuum tapped from a passage just above the throttle plate in a carburetor; S-port vacuum.

Ported Vacuum Switch   A temperature-actuated switch that changes vacuum connections when the coolant temperature changes. (Originally used to switch spark port vacuum; now used for any vacuum switching function that requires coolant temperature sensing.)

Porting As generally applied to racing engines, the enlarging, matching, streamlining, and polishing of the inside of the manifolds and valve ports to reduce the friction of the flow of gases.

Positive The positive plate of an electrolytic cell. The terminal of a battery toward which electrons are presumed to flow. Opposite of **Negative.**

Positive Crankcase Ventilation System An emission control system that routes engine crankcase fumes into the intake manifold or air cleaner, where they are drawn into the cylinders and burned along with the air/fuel mixture.

Positive Crankcase Ventilation Valve A valve that controls the flow of vapors from the crankcase into the engine.

Preheating The application of heat as a preliminary step to some further thermal or mechanical treatment.

Preignition Ignition occurring earlier than intended. For example, the explosive mixture being fired in a cylinder as by a flake or incandescent carbon before the electric spark occurs.

Press-Fit Also known as a force-fit or drive-fit. This term is used when the shaft is slightly larger than the hole and must be forced in place.

Primary Pertaining to that circuit of a transformer that induces current. The low-voltage circuit in an ignition system.

Primary Wires The wiring circuit used for conducting the low tension or primary current to the points where it is to be used.

Processor The on-board computer that receives data from a number of sensors and other electronic components. Based on data programmed into the processor's memory, the processor generates output signals to control various engine functions.

Program A set of detailed instructions a microcomputer follows when controlling a system.

PSI A measurement of pressure in pounds per square inch.

Pulse Air System A type of exhaust emission control system that uses the natural exhaust pulses in a turned pipe. A reed-type check valve responds to the negative pressure pulses and permits air to be drawn into the exhaust system.

Purge Control Valve Valve used to control the release of fuel vapors from the charcoal canister into the engine.

Purge Solenoid Device used to control the operation of the purge valve in evaporative control emission systems.

Pushrod A connecting link in an operating mechanism, such as the rod interposed between the valve lifter and rocker arm on an overhead valve engine.

Q

Quenching A process of rapid cooling of hot metal by contact with liquid, gases, or solids.

R

Race As used with reference to bearings, a finished inner and outer surface in which or on which balls or rollers operate.

Race Cam A type of camshaft for race cars that increases the lift of the valve, increases the speed of valve opening and closing, increases the length of time the valve is held open, etc. Also known as full, three-quarter, or semi-race cams, depending upon design.

Radiation The transfer of heat by rays, such as heat from the sun.

Random Access Memory A type of memory which is used to store information temporarily.

Raster Scope display mode; identical to stacked. (See **Stacked.**)

Ratio The relation or proportion that one number bears to another.

Read A microcomputer operation wherein information is retrieved from memory.

Read Only Memory A type of memory used to store information permanently. Information cannot be written to ROM; as the name implies, information can only be read from ROM.

Ream To finish a hole accurately with a rotating fluted tool.

Reciprocating A back-and-forth movement, such as the action of a piston in a cylinder.

Reed Valve Check valve used on some secondary air applications. Prevents reverse flow of air from exhaust manifold to intake air cleaner.

Reference Voltage A voltage provided by a voltage regulator to operate potentiometers and other sensors at a constant level.

Relay A switching device operated by a low-current circuit that controls the opening and closing of another circuit of high current capacity.

Relief The amount one surface is set below or above another surface.

Relieving As applied to racing engines, the removal of some metal from around the valves and between the cylinder and valves to facilitate flow of the gases.

Resistance The opposition offered by a substance or body to the passage of electric current through it.

Resistor A current-consuming piece of metal wire or carbon inserted into an electrical circuit to decrease the flow.

Retard When used with reference to an ignition distributor, it means to cause the spark to occur at a later time in the cycle of engine operation; opposite of **Spark Advance.**

Rocker Arm In an automobile engine, a lever located on a fulcrum or shaft, one end of the lever being on the valve stem, the other being on the pushrod.

Rockwell Hardness A scale for designating the hardness of a substance.

Roller Bearing An inner and outer race upon which hardened steel rollers operate.

Rotary Engine An engine construction wherein the crankshaft remains stationary and the cylinder spins around it as in a pinwheel.

Rotary Valve A valve construction in which ported holes come into and out of register with each other to allow entrance and exit of fluids or gases.

RPM Revolutions per minute.

Rubber An elastic vibration-absorbing material of either natural or synthetic origin.

Running-Fit Where sufficient clearance has been allowed between the shaft and journal to allow free running without overheating.

S

SAE Society of Automotive Engineers.

SAE Steels A numerical index system used to identify composition of SAE steel. Basically the first digit indicating the type to which the steel belongs; thus 1 indicates a carbon steel, 2 a nickel steel, etc. The second digit indicates the approximate percentage of the predominant alloying element. Usually the last two or three digits indicate the approximate average carbon content in points or hundredths of 1 percent. Thus "SAE 2340 steel " indicates a nickel steel of approximately 3 percent nickel and 0.40 percent carbon.

SAE Thread A table of threads set up by the society of Automotive Engineers and determines the number of threads per inch. For example, a 1/4-inch-diameter rod with an SAE thread would have 28 threads per inch.

Safety Relief Valve A spring-loaded valve designed to open and relieve excessive pressure in a vessel when it exceeds a predetermined safe point.

Sampling The act of periodically collecting information, as from a sensor. A microcomputer samples input from various sensors in the process of controlling a system.

Saturation The point reached when current flowing through a coil or wire has built up the maximum magnetic field.

Saybolt Test A method of measuring the viscosity of oil with the use of a viscosimeter.

Scale A flaky deposit occurring on steel or iron. Ordinarily used to describe the accumulation of minerals and metals in an automobile cooling system.

Score A scratch, ridge, or groove marring a finished surface.

Seat A surface, usually machined, upon which another part rests or seats; for example, the surface upon which a valve face rests.

Secondary Pertaining to that side of a transformer in which current is induced (by the primary). The high-voltage (high-tension) part of the ignition system.

Seize When one surface moving upon another scratches, it is said to seize. An example is a piston score or abrasion in a cylinder due to lack of lubrication or overexpansion.

Semiconductor General term for transistors, integrated circuits, and other electronic devices made of material, such as silicon, that are used to moderate the flow of electricity.

Sensor A device that measures an operating condition and provides an input signal to a microcomputer.

Short Circuit A usually unintentional grounding of a circuit such that current is diverted from part of the normal circuit by following a shorter (or easier) path to the current source. Commonly called a short.

Shrink Fit Where the shaft or part is slightly larger than the hole in which it is to be inserted. The outer part is heated above its normal operating temperature or the inner part chilled below its normal operating temperature and assembled in this condition; upon cooling an exceptionally tight fit is obtained.

Signal A voltage condition that transmits specific information in an electronic system.

Sleeve Valve A reciprocating sleeve or sleeves with ported openings placed between the piston and cylinders of an engine to serve as valves.

Sliding Fit Where sufficient clearance has been allowed between the shaft and journal to allow free running without overheating.

Slip-In Bearing A liner made to extremely accurate measurements that can be used for replacement purposes without additional fitting.

Sludge As used in connection with automobile engines, it indicates a composition of oxidized petroleum products along with an emulsion formed by the mixture of oil and water. This forms a pasty substance and clogs oil lines and passages and interferes with engine lubrication.

Smog   Air pollution created by the reaction of nitrogen oxides to sunlight.

Solder   An alloy of lead and tin used to unite two metal parts.

Soldering   To unite two pieces of metal with a material, such as solder having a comparatively low melting point.

Solenoid   A wire coil with a movable core that changes position by means of electromagnetism when current flows through the coil.

Solenoid Vent Valve   Energized by the ignition switch to control fuel vapor flow to the canister. When the ignition is off, the valve is open.

Solid State   Any electronic circuit that does not use tubes can be considered solid state. The term refers to circuits that use transistors, integrated circuits, and/or other semiconductors.

Solvent   A solution that dissolves some other material. For example, water is a solvent for sugar.

Spacer Entry EGR System   Exhaust gases are routed directly from the exhaust manifold through a stainless steel tube to the carburetor base.

Spark   An electrical current possessing sufficient pressure to jump through the air from one conductor to another.

Spark Advance   When used with reference to an ignition distributor, means to cause the spark to occur at an earlier time in the cycle of engine operation; opposite of **Retard.**

Spark Delay Valve   A valve that operates spark vacuum advance during rapid acceleration from idle or from speeds below 15 miles per hour and to cut off spark advance immediately on deceleration. Has an internal sintered orifice to slow air in one direction, a check valve for free airflow in opposite direction, and a filter.

Spark Gap   The space between the electrodes of a spark plug through which the spark jumps. Also a safety device to provide an alternate path for the current when it exceeds a safe value.

Spark Knock   Same as **Preignition.**

Spark Line   The normally horizontal line of a scope pattern that represents the firing of the plug once the plug gap has been bridged by the spark. Also called the firing line.

Spark Output (Spout)   The output signal from the processor that triggers the TFI-IV Module to fire the ignition coil.

Spark Plug   A device inserted into the combustion chamber of an engine containing an insulated central electrode for conducting the high-tension current from the ignition distributor. This insulated electrode is spaced a predetermined distance from the shell or side electrode to control the dimensions of the gap for the spark to jump across.

S-Port   A special carburetor port to take off ported vacuum.

Stacked   A scope display mode that presents cylinder patterns vertically one above another in firing order with No. 1 plug at the bottom. Also called raster display.

Stamping   A piece of sheet metal cut and formed into the desired shape with the use of dies.

Standard Thread   Refers to the U.S.S. table of the number of threads per inch. For example, a 1/4-inch-diameter standard thread would have 20 threads per inch.

Steel Casting   Cast iron to which varying amounts of scrap steel have been added.

Stellite   An alloy of cobalt, chrome and tungsten which is often used for exhaust valve seat inserts. It has a high melting point, good corrosion resistance and unusual hardness when hot.

Stoichiometric   Chemically correct. An air/fuel mixture is considered stoichiometric when it is neither too rich nor too lean; stoichiometric ratio is 14:6 parts of air for every part of fuel.

Strategy   A plan the microcomputer follows to achieve a desired goal.

Stress   The force or strain to which a material is subjected.

Stroke   In an automobile engine, the distance moved by the piston.

Stroking   As applied to racing engines, remachining the crankshaft throws off center to alter the stroke.

Studs   A rod with threads cut on both ends, such as a cylinder stud that screws into the cylinder block on one end and has a nut placed on the other end to hold the cylinder head in place.

Suction   Suction exists in a vessel when the pressure is lower than the atmospheric pressure; also see **Vacuum.**

Superimposed   A scope display mode that presents the patterns of all cylinders superimposed on one another.

Switch   A device used to open, close, or redirect the current in an electrical circuit.

T

Tachometer   A device for measuring and indicating the rotative speed of an engine.

Tap   To cut threads in a hole with a tapered, fluted, threaded tool. In England, term for drain cock.

Tappet   The adjusting screw for varying the clearance between the valve stem and the cam.

Can be built into the valve lifter in an engine or installed in the rocker arm on an overhead valve engine.

Temperature Vacuum Switch Located in the carburetor air cleaner to control vacuum to the EGR valve or other diaphragm. A cold air lockout. Responds to the temperature of the inlet air heated by the exhaust manifold. Looks like a cold weather modulator, differentiated by color.

Test Lead A wire or cable from an engine analyzer connected directly or by induction pickup to specific points in an ignition system to sense and transmit electrical signals to the analyzer.

Thermactor An air injection type of exhaust emission control system.

Thermactor Air Bypass Solenoid Electrical solenoid or its control line. Solenoid switches engine manifold vacuum. Vacuum bypasses thermactor air to atmosphere.

Thermactor Air Control Solenoid Vacuum Valve Assembly Used on thermactor air control systems, the valve assembly consists of two normally open solenoid valves with vents.

Thermactor Air Control Valve Combines a bypass (dump) valve with a diverter (upstream/downstream) valve. Controls the flow of air in response to vacuum signals to its diaphragms.

Thermactor Air Diverter Solenoid Electrical solenoid or its control line. Solenoid switches engine manifold vacuum. Vacuum switches thermactor air from downstream (past EGO sensor in exhaust system) to upstream (before EGO sensor) when solenoid is energized.

Thermal Reactor Exhaust manifold design using heat and air to burn unburned hydrocarbons in the exhaust gases to reduce air pollution. Requires a secondary air source to the exhaust system.

Thermal Vacuum Switch See **Temperature Vacuum Switch.**

Thermal Vent Valve Temperature-sensitive valve assembly located in the canister vent line. Closes when engine is cold and opens when hot. Prevents fuel tank vapors from being vented through the carburetor fuel bowl when fuel tank heats up before engine compartment.

Thermistor A resistor that changes its resistance with temperature.

Thermo Efficiency A gallon of fuel contains a certain amount of potential energy in the form of heat when burned in the combustion chamber. Some of this heat is lost and some is converted into power. The thermal efficiency is the ratio of work accomplished compared to the total quantity of heat contained in the fuel.

Thermostat A heat controlled valve used in the cooling system of an automobile engine to regulate the flow of water between the cylinder block and the radiator or used in the electrical circuit of the car heating system to control the amount of heat supplied to the passengers.

Thick Film Integrated Module An electronic module that is mounted on the distributor and fires the ignition coil.

Three-Way Catalyst A catalyst designed to simultaneously convert hydrocarbons, carbon monoxide, and oxides of nitrogen into relatively inert substances. Only effective when the exhaust gases are chemically correct; that is, when the air/fuel ratio of the mixture burned in the cylinders is 14:6:1. Also see **Stoichiometric.**

Throttle Position Sensor Potentiometric sensor or its signal line. Sensor wiper position is proportional to throttle position. Defines engine operation mode: closed, partly open, or wide open throttle; also throttle angle rate for accelerator pump function (replaced carburetor spark/ERG port.) There are two types of throttle positioners: rotary and linear.

Throw With reference to an automobile engine, usually the distance from the center of the crankshaft main bearing to the center of the connecting rod journal.

Timer A feature of microcomputers that allows them to measure time intervals, such as "time since engine start" or "time in neutral idle."

Timing Chain Chain used to drive camshaft and accessory shafts of an engine.

Timing Gears Any group of gears that are driven from the engine crankshaft to cause the valves, ignition, and other engine-driven apparatus to operate at the desired time during the engine cycle.

Tolerance A permissible variation between the two extremes of a specification of dimensions.

Torque An effort devoted toward twisting or turning.

Torque Wrench A special wrench with a built-in indicator to measure the applied force.

Transformer An electrical device that transforms electrical voltage from high to low or from low to high (as in an ignition coil). Total energy is not changed because current increases proportionately to voltage decrease and vice-versa (see **Watt**).

Transistor A solid semiconducting combination of chemical elements used in place of vacuum (radio) tubes. As used in automotive electronic ignition systems, the transistor is a means of furnishing greater current to the ignition coil, resulting in greater secondary voltage to fire the spark plugs while at the same time requiring less current through the breaker points.

Turbulence A disturbed or disordered irregular motion of fluids or gases.

U

Upper Cylinder Lubrication A method of introducing a lubricant into the fuel or intake manifold to permit lubrication of the upper cylinder, valve guides, etc.

V

Vacuum The lack of air or air pressure. There are three type of vacuums important to engine and component function: manifold, ported, and venturi. The strength of these vacuums depends on throttle opening, engine speed, and load.

Vacuum Advance Advances ignition timing with relation to engine load conditions. This is achieved by using engine intake manifold vacuum to operate the distributor diaphragm.

Vacuum Check Valve A one-way valve used to retain vacuum signal in a line after vacuum source is gone; prevents back flow.

Vacuum Control Valve A ported vacuum switch. Controls vacuum to other emission devices during engine warm-up. The two-port types simply open when engine coolant reaches their pre-determined calibration temperatures. The four-port types open likewise, since they are nothing more than two-port types in one housing. The three-port types switch the vacuum source to the center port from the top or the bottom ports. Most VCV's respond to a sensing bulb immersed in engine coolant by utilizing a wax pellet principle.

Vacuum Differential Valve A valve used in the Thermactor system with a catalyst to cut off vacuum to the air bypass valve during deceleration.

Vacuum Gauge An instrument designed to measure the degree of vacuum existing in a chamber.

Vacuum Regulator Provides constant vacuum output from manifold when vehicle is at idle. Switches to engine vacuum off idle.

Vacuum Regulator Valve Two-Port Vacuum regulator that provides a constant output signal when the input level is greater than a preset level. At a lower input vacuum, the output equals the input.

Vacuum Regulator Valve Three- and Four-Port Used to control the vacuum advance to the distributor.

Vacuum Reservoir Stores vacuum and provides "muscle vacuum." Prevents rapid fluctuations and sudden drops in a vacuum signal, such as during acceleration.

Vacuum Restrictor Used to control the flow rate and/or timing in actions to different emission control components.

Vacuum Retard Delay Valve A valve that delays a decrease in vacuum at the distributor vacuum advance unit when the source vacuum decreases. (Can be used to delay release of vacuum from any diaphragm; a "momentary" trap for vacuum.)

Vacuum Vent Valve Controls the induction of fresh air into a vacuum system to prevent chemical decay of the vacuum diaphragm that can occur on contact with fuel.

Valve A valve is a device for opening and sealing an aperture.

Valve Clearance The air gap allowed between the end of the valve stem and the valve lifter or rocker arm to compensate for expansion due to heat.

Valve Face The part of a valve that mates with and rests on a seating surface.

Valve Grinding A process of lapping or mating the valve seat and valve face usually performed with the aid of an abrasive; also called **Valve Lapping.**

Valve Head The portion of the valve upon which the valve face is machined.

Valve-in-Head Engine Same as overhead valve engine.

Valve Key or Valve Lock The key, washer, or other device that holds the valve spring cup or washer in place on the valve stem.

Valve Lifter A pushrod or plunger placed between the cam and the valve on an engine; often adjustable to vary the length of the unit.

Valve Margin On a poppet valve the space or rim between the surface of the head and the surface of the valve face.

Valve Overlap An interval expressed in degrees where both valves of an automobile engine cylinder are open at the same time.

Valve Seat The matched surface upon which the valve face rests.

Valve Spring A spring attached to a valve to return it to the seat after it has been released from the lifting or opening means.

Valve Stem That portion of a valve that rests within a guide.

Valve Stem Guide A bushing or hole in which the valve stem is placed that allows only lateral motion.

Vane Airflow Meter A sensor with a movable vane connected to a potentiometer calibrated to measure the amount of air flowing through the sensor. The output signal is relayed to the ECA which then translates the information into the amount of air flowing into the engine.

Vane Air Temperature Sensor   Located in the vane airflow meter, it senses the temperature of the air flowing into the engine. Changes of temperature result in changes in a resistive element in the sensor.

Vapor Lock   A condition wherein the fuel boils in the fuel system forming bubbles that retard or stop the flow of fuel to the carburetor.

Vehicle Emission Control Information Decal   Critical specifications for servicing emission systems. Decal located in engine compartment.

Venturi Vacuum Amplifier   Used with some EGR systems so that carburetor venturi vacuum can control EGR valve operation. Venturi vacuum is desirable because it is proportional to the airflow through the carburetor.

Vibration Damper   A device to reduce the torsional or twisting vibration that occurs along the length of the crankshaft used in multiple cylinder engines; also known as a harmonic balancer.

Viscosimeter   An instrument for determining the viscosity of an oil by passing a certain quantity at a definite temperature through a standard size orifice or port. The time required for the oil to pass through expressed in seconds gives the viscosity.

Viscosity   The resistance to flow or adhesiveness characteristics of an oil.

Volatility   The tendency for a fluid to evaporate rapidly or pass off in the form of vapor. For example, gasoline is more volatile than kerosene because it evaporates at a lower temperature.

Volt   A unit of measurement of electromotive force. One volt of electromotive force applied steadily to a conductor of one ohm resistance will produce a current of one ampere.

Voltage   The total electromotive force applied to a given circuit, expressed in volts.

Volume   The measure of space expressed as cubic inches, cubic feet, etc.

Volumetric Efficiency   A combination between the ideal and actual efficiency of an internal combustion engine. If the engine completely filled each cylinder on each induction stroke the volumetric efficiency of the engine would be 100 percent. In actual operation, however, volumetric efficiency is lowered by the inertia of the gases, the friction between the gases and the manifolds, the temperature of the gases, and the pressure of the air entering the carburetor. Volumetric efficiency is ordinarily increased by the use of large valves, ports, and manifolds and can be further increased with the aid of a supercharger.

Vortex   A whirling movement or mass of liquid or air.

W

Watt   A unit of power equal to the rate of work represented by a current of 1 ampere under a pressure of 1 volt. (Formula: Watts = amperes × volts.)

White Metal   An alloy of tin, lead, and antimony having a low melting point and a low coefficient of friction.

Winding   A single complete turn of the wire wound into a coil or transformer.

Wringing-Fit   Wherein the clearance is less than for a running or sliding fit and the shaft will enter the hole by means of twisting and pushing by hand.

Wrist Pin   The journal for the bearing in the small end of an engine connecting rod that also passes through piston walls; also known as a piston pin.

Write   A microcomputer operation wherein information is sent to and stored in memory.

ABBREVIATIONS

The following fuel system and emission control abbreviations are some of the more common ones used today.

A	Ampere		CAFE	Corporate Average Fuel Economy
A/C	Air Conditioning		CANP	Canister Purge Solenoid
A/T	Automatic Transmission		CAT	Catalytic Converter
AAS	Aspirator Air System		CAV	Coasting Air Valve
AAV	Anti-afterburn Valve		CBVV	Carburetor Bowl Vent Valve
AB	Air Bleed		CC	Catalytic Converter
ACC	A/C Compressor Signal Switch		CCC	Computer Command Control
ACCS	A/C Clutch Cycling Switch		CCEGR	Coolant Controlled EGR
ACD	A/C Demand		CCEVS	Coolant Controlled Engine Vacuum Switch
ACT	Air Charge Temperature Sensor			
ACV	Air Control Valve		CCIE	Coolant Controlled Idle Enrichment
AFC	Airflow Controlled Fuel Injector		CCO	Converter Clutch Override
AFR	Air/Fuel Ratio		CEC	Computerized Emission Control
AFS	Airflow Sensor		CFI	Central Fuel Injection
AI	Air Injection		CIS	Continuous Injection System
AICV	Air Injection Check Valve		CKV	Check Valve
AIR	Air Pump		CO	Carbon Monoxide
AIR-BPV	Air Bypass Valve		COC	Conventional Oxidation Catalyst
AIR-CHV	Air Check Valve		COV	Control Valve
AIR-DV	Air Diverter Valve		CP	Crankshaft Position
AIR-IVV	Air Idle Vacuum Valve		CPU	Central Processing Unit
AIR-RV	Air Pump Relief Valve		CPRV	Canister Purge Regulator Valve
AIS	Automatic Idle Speed Motor		CPS	Canister Purge Solenoid
AIV	Air Injection Valve		CPV	Canister Purge Valve
ALDL	Assembly Line Diagnostic Link		CRT	Cathode Ray Tube
AMGV	Air Management Valve		CSSA	Cold Start Spark Advance
ANTBV	Antibackfire Valve		CSSH	Cold Start Spark Hold
APDV	Air Pump Diverter Valve		CT	Conventional Thermactor
AS	Airflow Sensor		CTO	Coolant Temperature Override
ASE	Automotive Service Excellence		CTS	Coolant Temperature Sensor
ASIA	Automotive Service Industry Association		CVSCC	Coolant Vacuum Switch Cold Closed
ASRV	Air Switching Relief Valve		CVSCO	Coolant Vacuum Switch Cold Open
ASS	Air Switching Solenoid		DBC	Dual Bed Catalytic
ASV	Air Switching Valve		DCLV	Deceleration Valve
ATS	Air Temperature Sensor		DCS	Deceleration Control System
ATX	Automatic Transaxle		DCTO	Dual Coolant Temperature Override
AWD	Air Warning Device		DEFI	Digital Fuel Injection
BARO	Barometric Pressure Sensor		DLV	Delay Valve
BAT+	Battery Positive Terminal		DME	Digital Motor Electronics
BAT−	Battery Negative Terminal		DMV	Distributor Modulator Valve
BCDD	Boost Controlled Deceleration Device		DP	Dashpot
BDC	Bottom Dead Center		DRCV	Distributor Retard Control Valve
BHP	Brake Horsepower		DS	Detonation Sensor
BMAPS	Barometric/Manifold Absolute Pressure Sensor		DSAV	Deceleration Spark Advance System
			DSTVS	Distributor Spark Terminal Vacuum Switch
BOOST	Turbo Boost Actuator			
BPEGR	Back Pressure EGR		DV	Deceleration Valve
BPS	Barometric Pressure Sensor		DVCV	Distributor Vacuum Control Valve
BTU	British Thermal Unit		DVDV	Differential Vacuum Delay Valve
C-3	Computer Command Center		DVOM	Digital Volt-Ohmmeter
C-4	Computer Controlled Catalytic Converter		DVTRV	Diverter Valve
C-31	Computer Controlled Ignition Coil		DVVV	Distributor Vacuum Vent Valve

EAC	Electronic Air Control		IAA	Idle Speed Control
ECA	Electronic Control Assembly		IAC	Idle Air Control
ECC	Electronic Controlled Carburetor		IAS	Inlet Air Solenoid
ECCS	Electronic Concentrated Control System		IAT	Intake Air Temperature
ECI	Electronic Controlled Ignition		IC	Integrated Circuit
ECM	Electronic Control Module		ICM	Ignition Control Module
ECS	Evaporation Control System		ID	Inside Diameter
ECT	Engine Coolant Temperature Sensor		IDM	Ignition Diagnostic Monitor
ECU	Electronic Control Unit		I/M	Inspection/Maintenance
EEC	Electronic Engine Control		IM	Ignition Module
EESS	Evaporative Emission Shed System		IMCO	Improved Combustion System
EFC	Electronic Feedback Carburetor		IMS	Inferred Mileage Sensor
EFC	Electronic Fuel Control		IMVC	Intake Manifold Vacuum Control
EFE	Early Fuel Evaporation		ISC	Idle Speed Control
EFE-TVS	EFE Thermal Vacuum Switch		ISC	Idle Speed Compensator
EFI	Electronic Fuel Injection		ISS	Idle Speed Solenoid
EGC	Exhaust Gas Check Valve		ITCS	Ignition Timing Control System
EGI	Electronic Gasoline Injection		ITEC	ITEC Fuel Injection
EGO	Exhaust Gas Oxygen		ITS	Idle Tracking Switch
EGOR	EGO Return Signal		ITVS	Ignition Timing Vacuum Switch
EGR	Exhaust Gas Recirculation		JAS	Jet Air System
EGR-EPV	EGR External Pressure Valve		JCAV	Jet Controlled Air Valve
EGR-EVR	EGR Electronic Vacuum Regulator		KAM	Keep Alive Memory
EGR-FDLV	EGR Forward Delay Valve		KS	Knock Sensor
EGR-RSR	EGR Reservoir		LCV	Load Control Valve
EGR-TVS	EGR Thermal Vacuum Switch		LED	Light-Emitting Diode
EGRV	EGR Vent		LOS	Limited Operational Strategy
EGR-VCV	EGR Vacuum Control Valve		LUS	Locking Upshift Solenoid
EGR-VSOL	EGR Vacuum Solenoid		M/C	Mixture Control
EGR-VVA	EGR Venturi Vacuum Amplifier		M/T	Manual Transmission
EGRC	EGR Controller		MAF	Mass Airflow Sensor
EGRCS	EGR Control Solenoid		MAFTS	Manifold Air/Fuel Temperature Sensor
EGRV	EGR Vent Solenoid		MAJC	Main Air Jet Control
EIS	Electronic Ignition System		MAP	Manifold Absolute Pressure Sensor
ELB	Electronic Lean Burn		MAT	Manifold Absolute Temperature
EPA	Environmental Protection Agency		MCN	Microprocessor Control Unit
EPR-SOL	Exhaust Pressure Regulator Solenoid		MCS	Mixture Control System
EPR-VLV	Exhaust Pressure Regulator Valve		MCU	Microprocessor Control Unit
ESC	Electronic Spark (Knock) Control		MEC	Motronic Engine Control
ESS	Engine Speed Sensor		MFI	Multiport Fuel Injection
ESSM	Engine Speed Switch Module		MHCV	Manifold Heat Control Valve
ETC	Electronic Throttle Control		MON	Motor Octane Number
EVAP	Evaporative Control System		MPH	Miles Per Hour
EVCR	Emission Vacuum Control Regulator		MTA	Managed Thermactor Air
EVP	EGR Valve Position		N	Nitrogen
EVR	EGR Vacuum Regulator		NA	Not Applicable
EZF	Performance Graph Ignition		NDS	Neutral Drive Switch
FBC	Feedback Carburetor System		NGS	Neutral Gas Switch
FBCA	Feedback Carburetor Actuator		NIASE	National Institute for Automotive Service Excellence
FCS	Fuel Control System			
FCS	Feedback Carburetor Solenoid		NO_x	Oxides of Nitrogen
FES	Fuel Evaporation System		NPS	Neutral Pressure Switch
FISR	Fast Idle Solenoid Relay		OC	Oxidation Catalytic
FP	Fuel Pump Relay		OCC	Output Cycling Check
FSS	Fuel Shutoff System		OD	Outside Diameter
FVEC	Fuel Vapor Emission Control		OE	Original Equipment
GND	Ground		OS	Oxygen Sensor
HAC	High Altitude Compensator		PA	Pulse Air
HC	Hydrocarbon		PACV	Pulse Air Check Valve
HCV	Heat Control Valve		PAF	Pulse Air Feeder
HEGO	Heat Exhaust Gas Oxygen		PAI	Pulse Air Injection
HEI	High Energy Ignition		PAS	Pulse Air System
HIC	Hot Idle Compensator		PCOV	Purge Control Valve
HPCA	Housing Pressure Cold Advance		PCV	Positive Crankcase Ventilation
HRV	Heat Riser Valve		PCVS	PCV Solenoid

PECV	Power Enrichment Control Valve	TDS	Time Delay Solenoid
PFE	Pressure Feedback EGR	TES	Thermal Electric Switch
PFI	Port Fuel Injection	TFI	Thick Film Integrated Ignition Module
PFI	Programmed Fuel Injection	TFI-IV	Thick Film Integrated
PIP	Profile Ignition Pickup	TGS	Top Gear Switch
POT	Potentiometer	TIC	Thermal Ignition Control
PPM	Parts Per Million	TIDC	Thermostatic Ignition Distributor Control
PROM	Programmable Read Only Memory	TKA	Throttle Kicker Actuator
PSPS	Power Steering Pressure Switch	TKS	Throttle Kicker Solenoid
PSV	Pulse Air Shutoff Valve	TP	Throttle Position
PVCS	Ported Vacuum Control System	TPI	Tuned Port Injection
PVFFC	Pressure/Vacuum Fuel Filler Cap	TPS	Throttle Position Sensor
PVS	Ported Vacuum Switch	TRC	Throttle Return Control
R	Resistance	TSD	Throttle Solenoid
RAM	Random Access Memory	TSP	Throttle Solenoid Positioner
RC	Rear Catalytic Converter	TVBV	Turbocharger Vacuum Bleed Valve
RDV	Reverse Delay Valve	TVS	Temperature Vacuum Switch
ROM	Read Only Memory	TVV	Thermal Vent Valve
RON	Research Octane Number	TWC	Three-Way Catalyst
RVSV	Rollover/Vapor Separator Valve	UIC	Universal Integrated Circuit
S-V	Sol-Vac	V	Volt
SA-FV	Separator Assembly-Fuel Vacuum	V-RSR	Vacuum Reservoir
SBC	Single Bed Catalytic	V-RST	Vacuum Restrictor
SCC	Spark Control Computer	VAV	Vacuum Control Valve
SCS	Spark Control System	VB	Vacuum Break
SCVAC	Speed Control Vacuum Control	VBAT	Vehicle Battery Voltage
SDV	Spark Delay Valve	VCKV	Vacuum Check Valve
SEC ACT	Secondary Actuator	VCS	Vacuum Control Switch
SFI	Sequential Fuel Injection	VCV	Vacuum Control Valve
SHED	Sealed Housing Evaporative Determination System	VDV	Vacuum Delay Valve
		VDV	Vacuum Differential Valve
SIS	Solenoid Idle Stop	VECI	Vehicle Emission Control Information
SLV	Solevac	VIN	Vehicle Identification Number
SOL	Solenoid	VMV	Vacuum Modulator Valve
SOL V	EGR Solenoid Vacuum Valve Assembly	VOM	Volt-Ohmmeter
SP-CTO	Spark Coolant Temperature Override	VOTM	Vacuum-Operated Throttle Modulator
SPFI	Single-Point Fuel Injection	VP	Vacuum Pump
SPOUT	Spark Output	VR/S	Vacuum Regulator Solenoid
SSD	Subsystem Diagnostic	VRDV	Vacuum Retard Delay Valve
SSI	Solid State Ignition	VREF	Voltage Reference
STI	Self-Test Input	VRES	Vacuum Reservoir
STO	Self-Test Output	VREST	Vacuum Restrictor
STRN	Signal Return Line	VRV	Vacuum Regulator Valve
SUC	Shift-Up Control	VS	Vacuum Switch
SVV	Solenoid Vent Valve	VSA	Vacuum Switch Assembly
TAB	Thermactor Air Bypass Solenoid	VSS	Vacuum Switch Solenoid
TAC	Thermostatic Air Cleaner	VSV	Vacuum Solenoid Valve
TAD	Thermactor Air Diverter	VTP	Vacuum Throttle Positioner
TBI	Throttle Body Injection	VVA	Venturi Vacuum Amplifier
TC	Throttle Closer	VVC	Variable Voltage Choke
TCC	Transmission Controlled Spark	WOT	Wide-Open Throttle
TCP	Temperature-Compensated Accelerator Pump	WOTS	Wide-Open Throttle Switch
		WOTV	Wide-Open Throttle Valve
TCS	Transmission Converter Switch	WST	Waste Gate
TDC	Top Dead Center	WSTC	Waste Gate Control

APPENDIX C

DECIMAL AND METRIC EQUIVALENTS

			DECIMAL AND METRIC EQUIVALENTS		
Fractions	Decimal (in.)	Metric (mm)	Fractions	Decimal (in.)	Metric (mm)
1/64	.015625	.397	33/64	.515625	13.097
1/32	.03125	.794	17/32	.53125	13.494
3/64	.046875	1.191	35/64	.546875	13.891
1/16	.0625	1.588	9/16	.5625	14.288
5/64	.078125	1.984	36/64	.578125	14.684
3/32	.09375	2.381	19/32	.59375	15.081
7/64	.109375	2.778	39/64	.609375	15.478
1/8	.125	3.175	5/8	.625	15.875
9/64	.140625	3.572	41/64	.640625	16.272
5/32	.15625	3.969	21/32	.65625	16.669
11/64	.171875	4.366	43/64	.671875	17.066
3/16	.1875	4.763	11/16	.6875	17.463
13/64	.203125	5.159	45/64	.703125	17.859
7/32	.21875	5.556	23/32	.71875	18.256
15/64	.234275	5.953	47/64	.734375	18.653
1/4	.250	6.35	3/4	.750	19.05
17/64	.265625	6.747	49/64	.765625	19.447
9/32	.28125	7.144	25/32	.78125	19.844
19/64	.296875	7.54	51/64	.796875	20.241
5/16	.3125	7.938	13/16	.8125	20.638
21/64	.328125	8.334	53/64	.828125	21.034
11/32	.34375	8.731	27/32	.84375	21.431
23/64	.359375	9.128	55/64	.859375	21.828
3/8	.375	9.525	7/8	.875	22.225
25/64	.390625	9.922	57/64	.890625	22.622
13/32	.40625	10.319	29/32	.90625	23.019
27/64	.421875	10.716	59/64	.921875	23.416
7/16	.4375	11.113	15/16	.9375	23.813
29/64	.453125	11.509	61/64	.953125	24.209
15/32	.46875	11.906	31/32	.96875	24.606
31/64	.484375	12.303	63/64	.984375	25.003
1/2	.500	12.7	1	1.00	25.4

Complete
Fuel Systems and
Emission Control

Student
Technician's
Shop
Manual

 Delmar Publishers Inc.®

CONTENTS

CHAPTER ONE

INTRODUCTION TO FUEL SYSTEMS AND EMISSION CONTROLS

Objectives

After reading this chapter, you should be able to:
- Explain how a gasoline engine operates, including the four-stroke cycle.
- Describe the three basic design characteristics of an engine.
- Understand the function of the systems that are related to engine operation.
- Name the components of a fuel system and explain their operation.
- Name the components of an exhaust system and explain their operation.
- Name the components of an emission control system and explain their operation.

PRACTICE QUESTIONS

1. What mechanical engine component is indicated by letter A in Figure 1–1?

   a. crankshaft
   b. intake manifold
   c. camshaft
   d. exhaust manifold

2. What mechanical engine component is indicated by letter B in Figure 1–1?

   a. crankshaft
   b. intake manifold
   c. camshaft
   d. exhaust manifold

3. What mechanical engine component is indicated by letter C in Figure 1–1?

   a. crankshaft
   b. intake manifold
   c. camshaft
   d. exhaust manifold

4. What stroke in the operating cycle of the gasoline engine is depicted in Figure 1–2?

   a. intake stroke
   b. compression stroke

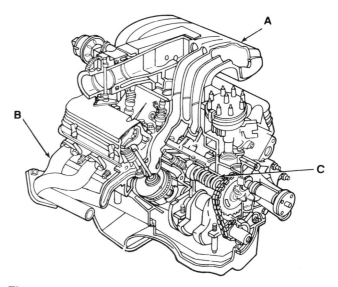

Figure 1–1

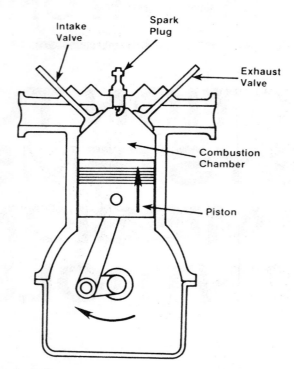

Intake Valve

Spark Plug

Exhaust Valve

Combustion Chamber

Piston

Figure 1-2

c. power stroke
d. exhaust stroke

5. The air system and the fuel system join in what component?
   a. intake manifold
   b. camshaft
   c. carburetor or fuel injector system
   d. fuel pump

6. What reduces NO$_x$ by diluting the air/fuel mixture and lowering combustion temperatures?
   a. EGR
   b. PCV
   c. EEF
   d. ECM

7. A lift is leaking at the packing gland. Technician A uses it anyway, but only for a short time. Technician B refuses to use it. Who is right?
   a. Technician A
   b. Technician B
   c. Both A and B
   d. Neither A nor B

8. Which of the following is not an advantage that pneumatic tools have over electrically powered tools?
   a. flexibility

b. low-cost operation
c. heavier construction
d. safety

9. What is the most common type of fuel pump?
   a. electric fuel pump
   b. mechanical fuel pump
   c. impeller fuel pump
   d. bellows fuel pump

10. A ticking sound can be heard when an engine is running. Technician A says the sound is normal. Technician B checks the gaskets at the exhaust manifold. Who is right?
    a. Technician A
    b. Technician B
    c. Both A and B
    d. Neither A nor B

SHOP ASSIGNMENT 1

Examine the fuel system in a vehicle with carburetion and locate the following components: fuel tank, fuel line, accelerator linkage, fuel pump, filter, emission control vapor canister, canister purge line, fuel bowl vent line, carburetor, vapor line, and emission control vapor separator. Refer to Figure 1–3 as needed.

SHOP ASSIGMENT 2

Examine the exhaust system in a vehicle and locate the following components: exhaust manifold (header), exhaust inlet pipe, catalytic converter, muffler, tail pipe, and resonator. Refer to Figure 1–4 as needed.

SHOP ASSIGMENT 3

Examine the emissions system in a vehicle and locate as many of the following components as are applicable: leaded-fuel restrictor, pressure-vacuum relief filler cap, domed fuel tank, catalytic converter, rollover valve, orifice spark advance control valve, exhaust gas recirculation, charcoal canister, coolant control idle enrichment valve, CCEGR temperature valve, fresh air induction, air pump, exhaust port air injection, closed crankcase ventilation, and electronic spark control. Refer to Figure 1–5 as needed.

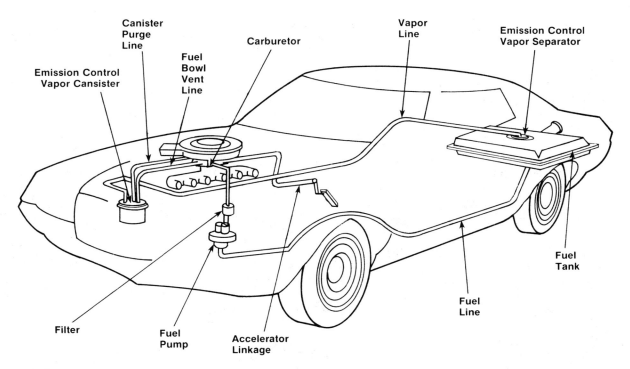

Canister
Purge
Line

Emission Control
Vapor Cansister

Fuel
Bowl
Vent
Line

Carburetor

Vapor
Line

Emission Control
Vapor Separator

Fuel
Tank

Fuel
Line

Filter

Fuel
Pump

Accelerator
Linkage

Figure 1-3

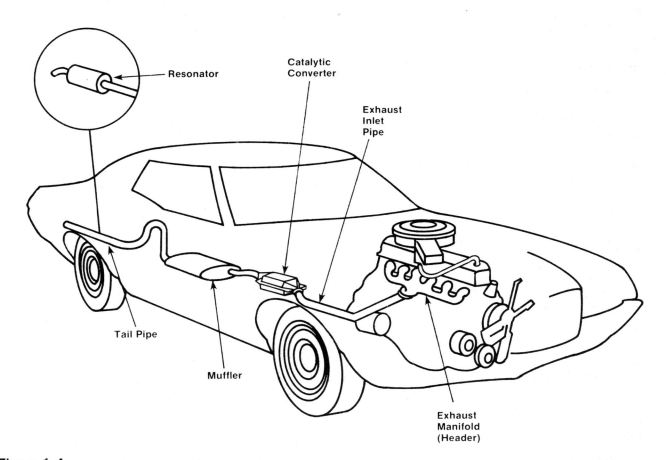

Resonator

Catalytic
Converter

Exhaust
Inlet
Pipe

Tail Pipe

Muffler

Exhaust
Manifold
(Header)

Figure 1-4

Chapter One   Introduction to Fuel Systems and Emission Controls

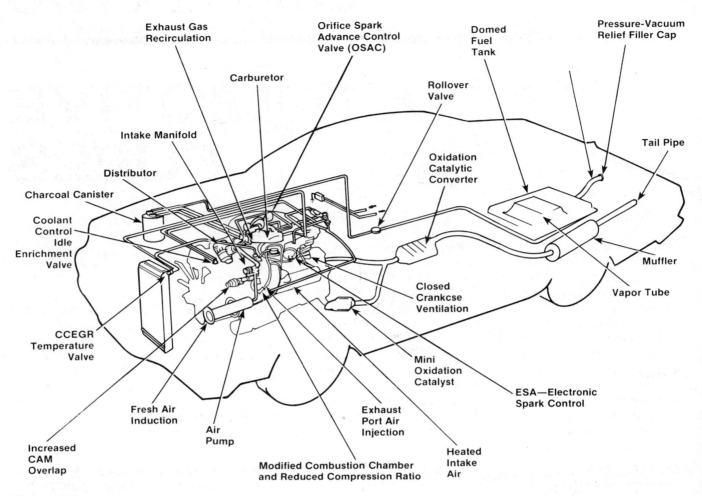

Exhaust Gas
Recirculation

Orifice Spark
Advance Control
Valve (OSAC)

Domed
Fuel
Tank

Pressure-Vacuum
Relief Filler Cap

Carburetor

Rollover
Valve

Tail Pipe

Intake Manifold

Oxidation
Catalytic
Converter

Distributor

Charcoal Canister

Coolant
Control
Idle
Enrichment
Valve

Muffler

Vapor Tube

Closed
Crankcse
Ventilation

CCEGR
Temperature
Valve

Mini
Oxidation
Catalyst

ESA—Electronic
Spark Control

Fresh Air
Induction

Air
Pump

Exhaust
Port Air
Injection

Heated
Intake
Air

Increased
CAM
Overlap

Modified Combustion Chamber
and Reduced Compression Ratio

Figure 1–5

CHAPTER TWO

AUTOMOTIVE FUELS

Objectives

After reading this chapter, you should be able to:
- Identify the services that a good motor fuel must provide.
- Explain how compression ratio and detonation affect a gasoline engine.
- Identify and explain the major factors that affect fuel performance.
- Name the performance characteristics and driving conditions affected by the volatility of gasoline, and describe how each one is affected.
- Explain the differences between leaded and unleaded gasoline.
- Identify commonly used gasoline additives and the function they perform.
- Explain the differences between gasoline and diesel engines.
- Name the characteristics of diesel fuel.
- Explain the principles behind diesel fuel's cetane rating.
- Describe the relationship of carbon residue and sulfur content to diesel engines.
- Explain the use of LP-gas.

PRACTICE QUESTIONS

1. Which of the following factors can affect knock in an engine?
   a. ignition timing overadvance
   b. compression ratio
   c. excessive carbon deposits
   d. all of the above

2. What is the main difference between gasoline and diesel engines?
   a. basic construction
   b. means of initiating combustion
   c. engine cycle
   d. all of the above

3. Which of the following statements concerning diesel engines is incorrect?
   a. In a diesel engine, knock is desirable.
   b. In diesel engines, there is continuous combustion during most of the power stroke
   c. Diesel engines are called spark ignition.
   d. Diesel engines operate on a higher compression ratio than gasoline engines.

4. What component of a compression ignition engine is depicted by the letter A in Figure 2–1?
   a. accelerator pump
   b. carburetor jet
   c. fuel filter
   d. injection nozzle

5. The temperature at which diesel fuel solidifies and no longer flows is called _____ .
   a. pour point
   b. wax appearance point
   c. specific gravity
   d. flash point

6. The ignition quality of diesel fuels is measured by _____ .
   a. octane rating
   b. cetane rating
   c. specific gravity
   d. sulfur content

7. Technician A makes a blended fuel by combining gasoline and diesel fuel. Technician B never blends gasoline and diesel fuel. Who is right?

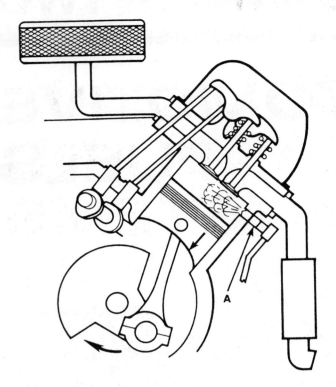

Figure 2-1

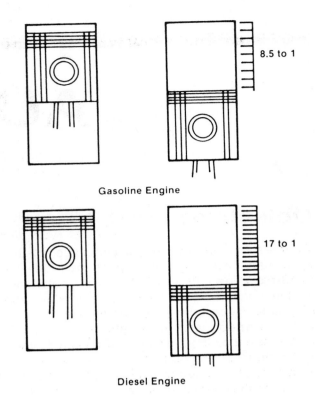

Gasoline Engine

8.5 to 1

Diesel Engine

17 to 1

Figure 2-2

a. Technician A
b. Technician B
c. Both A and B
d. Neither A nor B

8. The ability of a liquid to change into a vapor is called _____ .

   a. viscosity
   b. cetane rating
   c. volatility
   d. pour point

9. Abnormal combustion is known as

_____ .

   a. detonation
   b. compression
   c. oxidation
   d. none of the above

10. Which of the following gasoline additives is not an octane enhancer?

   a. ethanol

b. methanol
c. MTBE
d. isopropyl alcohol

SHOP ASSIGNMENT 4

Examine a gasoline engine and a diesel engine to compare and contrast their operation and fuel usage. See Figure 2-2.

SHOP ASSIGMENT 5

Examine the containers and their labels for as many of the following types of gasoline additives as possible: anti-icing or deicer (isopropyl alcohol), gum or oxidation inhibitors, metal deactivators and rust inhibitors, aromatics (BTX group), ethanol, methanol, and MTBE. Be sure to understand the method of usage, purpose, disadvantages, and cautions for each product.

CHAPTER THREE

VACUUM, AIR INTAKE, AND FUEL SYSTEMS

Objectives

After reading this chapter, you should be able to:
* Define *driveability* and symptoms of driveability problems.
* Define *vacuum* and explain how it affects and assists vehicle operation and driveability.
* Define the air intake system components and their functions, including air intake ductwork, air cleaner/filter assembly, and intake air temperature controls.
* Define the fuel transport system components and their functions, including the fuel tank, fuel lines, filters, and gauges.
* Conduct inspection and servicing procedures for the vacuum system, air intake system, and fuel system.
* Define the components and their functions in an electronic engine control (EEC) fuel delivery system.

PRACTICE QUESTIONS

1. What term refers to a pressure level that is lower than the earth's atmospheric pressure at any given altitude?
   a. driveability
   b. kilopascal
   c. vacuum
   d. none of the above

2. Which of the components identified in Figure 3-1 are vacuum controlled?
   a. altitude compensator
   b. antibackfire valve
   c. thermactor air pump
   d. all of the above

3. What diagnostic tool is depicted in Figure 3-2?
   a. vacuum pump
   b. vacuum gauge
   c. dial indicator
   d. feeler gauge

4. Where do most problems in a vacuum system occur?

   a. in the vacuum pump
   b. at filters
   c. in hose lines
   d. in other systems first

5. When replacing defective vacuum hose lines, Technician A removes all defective hoses and then installs the new ones together. Technician B removes one hose at a time and installs a new one before removing another hose. Who is right?
   a. Technician A
   b. Technician B
   c. Both A and B
   d. Neither A nor B

6. What component of a typical air intake system is indicated by letter A in Figure 3-3?
   a. shroud assembly
   b. preheater
   c. air cleaner assembly
   d. inlet tube

7. Technician A uses a flashlight when working on the fuel system. Technician B uses a trouble

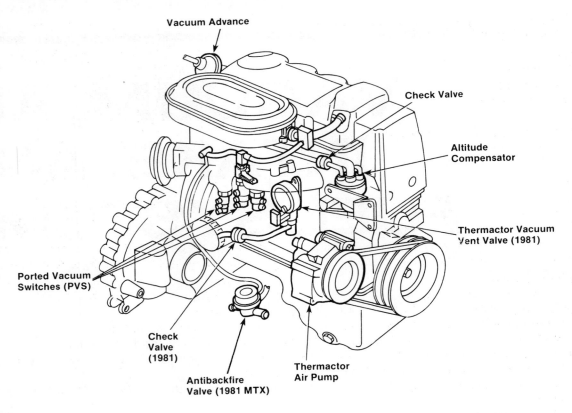

Vacuum Advance

Check Valve

Altitude
Compensator

Thermactor Vacuum
Vent Valve (1981)

Ported Vacuum
Switches (PVS)

Check
Valve
(1981)

Antibackfire
Valve (1981 MTX)

Thermactor
Air Pump

Figure 3-1

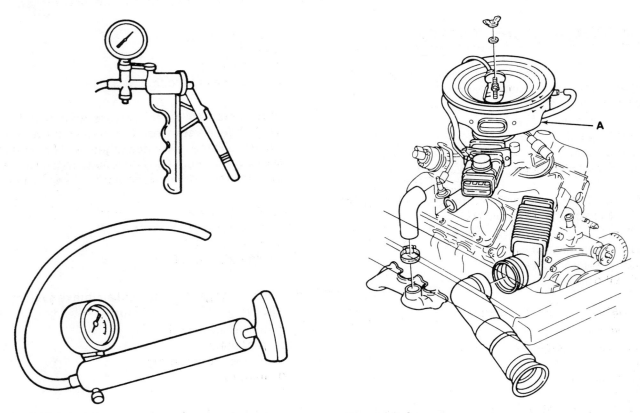

A

Figure 3-2

Figure 3-3

Chapter Three   Vacuum, Air Intake, and Fuel Systems

light or droplight when working on the fuel system. Who is right?

  a. Technician A
  b. Technician B
  c. Both A and B
  d. Neither A nor B

8. What component controls the rate of vapor flow in a fuel vent line?

  a. fuel tank
  b. hose
  c. clamp
  d. restrictor

9. The gasoline filter that is replaceable and might require regular service is located _____ .

  a. in the fuel tank
  b. in the engine compartment
  c. next to the strainer
  d. all of the above

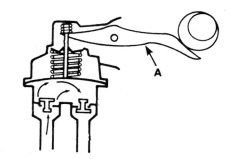

Figure 3-4

10. What component of a mechanical fuel pump assembly is indicated by letter A in Figure 3-4?

  a. rocker arm
  b. shaft
  c. diaphragm
  d. camshaft eccentric

JOB SHEET

SHOP ASSIGNMENT 6
REPLACE A TYPICAL NYLON VACUUM HOSE WITH BONDED NYLON OR RUBBER CONNECTORS

NAME _____ STATION _____ DATE _____

Tools and Materials

New hoses and connectors
Hand tools
Rubber hoses, tees, and elbows, if necessary
Rubber union, if necessary

Protective Clothing

Safety goggles or glasses with side shields

Procedure

1. Is the damaged area on the hose 1/2 inch or more from a connector?

   Yes _____   No _____

   a. If no, go to step 2.

   Not applicable _____   Task completed _____

   b. If yes, cut out the damaged section but not more than 1/2 inch of hose.

   Not applicable _____   Task completed _____

   c. Install a rubber union. (See Figure 3–5a).

   Not applicable _____   Task completed _____

2. If the remaining hose is too short or the damaged portion is longer than 1/2 inch, remove the entire hose.

   Not applicable _____   Task completed _____

   a. Replace with 5/32-inch hose, 7/32-inch hose, and tees. (See Figure 5–5b).

   Not applicable _____   Task completed _____

3. Is a connector damaged?

   Yes _____   No _____

   a. If no, go to step 4.

   Not applicable _____   Task completed _____

   b. If only part of a connector is damaged, cut the connector apart. (See Figure 3–5c).

   Not applicable _____   Task completed _____

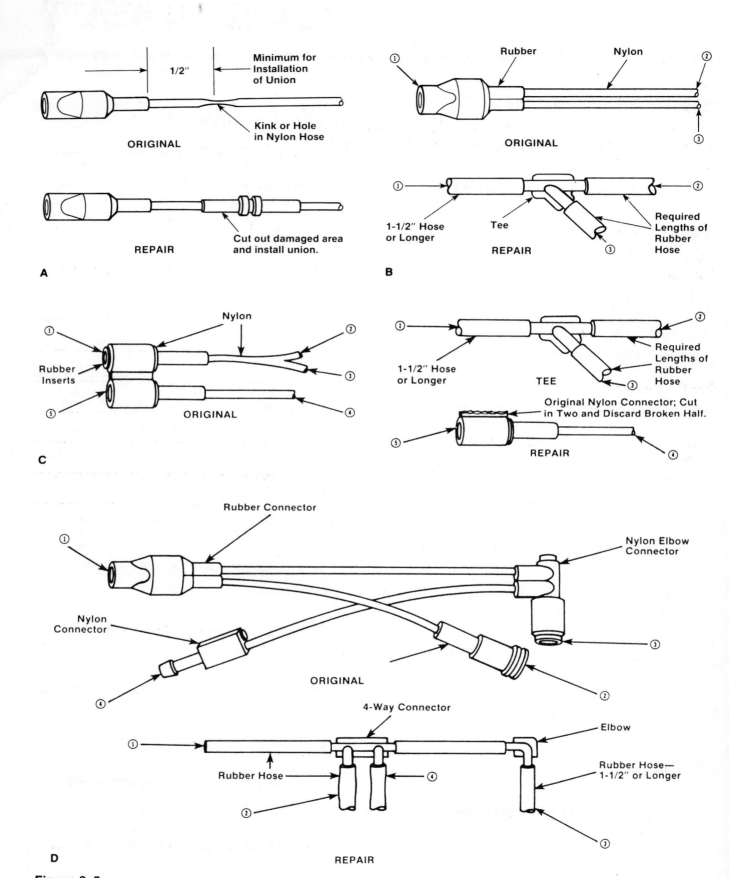

Figure 3-5

Chapter Three   Vacuum, Air Intake, and Fuel Systems

c. Discard the damaged half of the harness.

Not applicable _____   Task completed _____

d. Replace it with rubber vacuum hoses and a tee.

Not applicable _____   Task completed _____

4. If an entire harness is damaged, remove the connectors.

Not applicable _____   Task completed _____

a. Cut replacement hoses (5/32-inch and 7/32-inch), tees, and elbows to required lengths but never less than 1-1/2 inches.

Not applicable _____   Task completed _____

b. Install replacement hoses, tees, and elbows. (See Figure 3–5d).

Not applicable _____   Task completed _____

5. Make sure all hoses are properly routed.

Task completed _____

PROBLEMS ENCOUNTERED: _____

_____

_____

INSTRUCTOR'S COMMENTS: _____

_____

_____

JOB SHEET

SHOP ASSIGNMENT 7
REPLACE A FUEL TANK

NAME _____ **STATION** _____ **DATE** _____

Tools and Materials

Clean safety can for holding gasoline
Masking tape
Siphon hose or similar device
Lift or hoist

Protective Clothing

Safety goggles or glasses with side shields

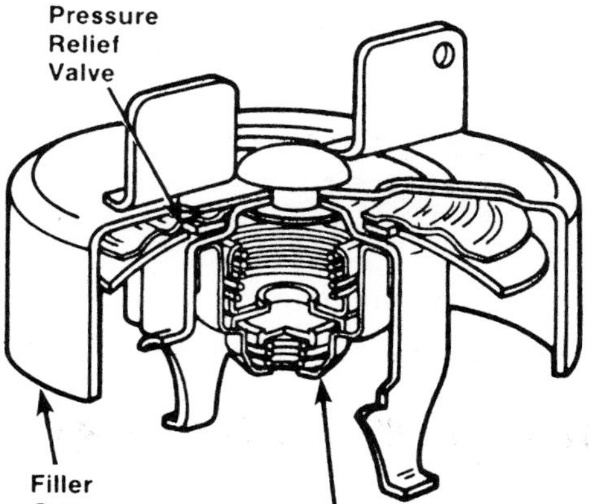

Procedure

1. Disconnect the negative terminal from the battery.

   a. Raise the vehicle.

2. Disconnect the fuel line at the tank that runs to the fuel pump.

3. Connect a siphon hose or similar device to the tank.

   a. Draw the fuel into a clean safety can.

   b. Dispose of contaminated fuel according to local laws.

4. Attach a piece of masking tape as an identifying tag to each tank line to insure correct reinstallation.

   a. Disconnect the vent lines.

   b. Disconnect the plugged wire connected to the sending unit. (The wire is usually mounted on the upper front or top section of the fuel tank.)

5. Unfasten the filler from the tank (See Figure 3-6).

Pressure Relief Valve

Filler Cap

Vacuum Relief Valve

Figure 3-6

Task completed _____

Task completed _____

Task completed _____

Task completed _____

Task completed _____

Task completed _____

Task completed _____

Task completed _____

Task completed _____

Not applicable _____ Task completed _____

a. If it is a rigid one-piece tube, remove the screws around the outside of the filler neck near the filler cap.

Not applicable _____   Task completed _____

b. If it is a three-piece unit, remove the neoprene hoses after the clamp has been loosened.

Not applicable _____   Task completed _____

6. Loosen the bolts holding the fuel tank straps until they are about two threads from the end.

Task completed _____

a. Holding the tank securely against the underchassis with one hand, remove the strap bolts and lower the tank to the ground.

Task completed _____

b. Use a drain pan to catch any spilled fuel.

Not applicable _____   Task completed _____

PROBLEMS ENCOUNTERED: _____

_____

_____

INSTRUCTOR'S COMMENTS: _____

_____

_____

JOB SHEET

SHOP ASSIGNMENT 8
REPLACE AN IN-CARBURETOR GASOLINE FILTER

NAME _____ STATION _____ DATE _____

Tools and Materials

Two wrenches
Gasoline filter, correct size and type for applications
Service manual

Protective Clothing

Safety goggles or glasses with side shields

Procedure

1. Remove the fuel line from the carburetor fuel inlet nut using one wrench to
   hold the inlet nut and another to turn the fuel line fitting.

   Task completed _____

2. Remove the inlet nut from the carburetor.

   Task completed _____

   a. Discard the used filter.

   Task completed _____

   b. If the gasket is separate, keep it with the inlet nut for reinstallation.

   Not applicable _____   Task completed _____

3. Refer to the service manual to choose the correct size and type of gasoline
   filter.

   Task completed _____

4. Install the filter by reversing the steps to remove the old one.

   Task completed _____

5. Turn on the engine.

   Task completed _____

   a. Check fuel line connections for leaks.

   Task completed _____

PROBLEMS ENCOUNTERED: _____

_____

_____

INSTRUCTOR'S COMMENTS: _____

_____

_____

JOB SHEET

SHOP ASSIGNMENT 9
TEST A MECHANICAL FUEL PUMP
FOR PRESSURE AND VOLUME

NAME _____ STATION _____ DATE _____

Tools and Materials

Rubber hose
Hand tools
Service manual
Compressed air or hand air pump
Container with ounces measured on it
Quart container of gasoline
Fuel pump pressure test gauge with restrictor

Protective Clothing

Safety goggles or glasses with side shields

Procedure

1. Remove the air cleaner assembly.

                                                    Task completed _____

2. Remove the in-carburetor fuel filter (if used) or disconnect the fuel line or
   fuel filter at the carburetor or injectors.

                                                    Task completed _____

3. Connect a pressure gauge, a restrictor, and a flexible hose between the
   fuel filter and the carburetor (See Figure 3–7).

                                                    Task completed _____

   a. Place the other end of the rubber hose in a suitable container that has a
      measuring scale, by ounces, on it.

                                                    Task completed _____

4. Operate the engine at normal operating temperature and at idle speed.

                                                    Task completed _____

5. Close the hose restrictor

                                                    Task completed _____

6. Allow the pressure to stabilize and note the reading.

                                                    Task completed _____

7. Is the pump pressure within specifications?

                                          Yes _____   No _____

   a. If no, then replace the pump.

                          Not applicable _____   Task completed _____

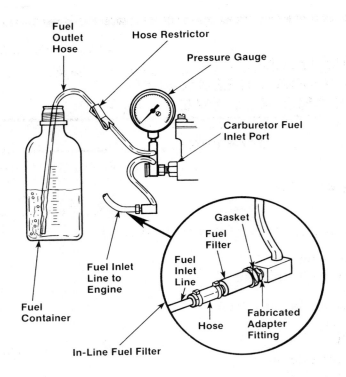

Figure 3–7

b. If yes, then proceed with the volume test in step 8.

Not applicable _____   Task completed _____

8. Operate the engine at the specified idle rpm.

Task completed _____

9. Open the hose restrictor.

Task completed _____

10. Allow the fuel to discharge into the container for a specified time.

Task completed _____

a. Close the restrictor.

Task completed _____

11. Is the pump volume below specifications?

Yes _____   No _____

a. If no, go to the end.

Not applicable _____   Task completed _____

b. If yes, repeat the test using an auxiliary fuel supply and a new fuel filter.

Not applicable _____   Task completed _____

12. Does the pump volume meet specifications?

Yes _____   No _____

a. If yes, go to 12 e.

Not applicable _____   Task completed _____

b. If no, are there restricted or leaking fuel lines, clogged filter hoses, or dirt or sludge at the fuel pickup in the tank?

Yes _____   No _____

c. If yes, replace any damaged parts and blow out any restrictions with compressed air.

Not applicable _____   Task completed _____

d. If no, replace the pump and go to the end.

Not applicable _____   Task completed _____

e. Check for a restriction in the fuel supply from the tank.

Task completed _____

f. Check for improper tank ventilation.

Task completed _____

g. Correct any problems.

Not applicable _____   Task completed _____

PROBLEMS ENCOUNTERED: _____

_____

_____

INSTRUCTOR'S COMMENTS: _____

_____

_____

JOB SHEET

SHOP ASSIGNMENT 10
REPLACE A MECHANICAL FUEL PUMP

NAME _____ STATION _____ DATE _____

Tools and Materials

Socket wrench
Plug
Gasket sealer

Protective Clothing

Safety goggles or glasses with side shields

Procedure

1. Remove the inlet and outlet lines at the pump.

   Task completed _____

2. Use a plug to stop the flow of fuel from the tank.

   Task completed _____

3. Use the correct size socket wrench to remove the mounting bolts.

   Task completed _____

   a. Remove the pump from the engine.

   Task completed _____

4. Clean the old gasket material from the engine block.

   Task completed _____

   a. Apply gasket sealer to the mounting surface on the engine.

   Task completed _____

   b. Apply gasket sealer to the threads of the mounting bolts.

   Task completed _____

5. Install the new gasket.

   Task completed _____

   a. Position the pump by tilting it either toward or away from the block to correctly place the lever against the cam.

   Task completed _____

   b. Is the pump driven by a pushrod?

   Yes _____   No _____

   c. If yes, hold up the rod to permit the rocker arm to go under it.

   Not applicable _____   Task completed _____

6. Is there an internal squeaking noise with each movement of the pump?

Yes _____   No _____

   a. If no, make sure that the pump is properly positioned.

Not applicable _____   Task completed _____

7. Tighten the mounting bolts firmly.

Task completed _____

8. Attach the inlet and outlet lines.

Task completed _____

   a. Start the engine.

Task completed _____

   b. Are any fuel leaks evident?

Yes _____   No _____

   c. If yes, repair.

Not applicable _____   Task completed _____

PROBLEMS ENCOUNTERED: _____

_____

_____

INSTRUCTOR'S COMMENTS: _____

_____

_____

CHAPTER FOUR

CARBURETOR SERVICING AND ADJUSTMENTS

Objectives

After reading this chapter, you should be able to:
- Explain the purpose and functions of a carburetor.
- Identify the main systems and components of a carburetor.
- Recognize carburetor-related performance problems.
- Troubleshoot and adjust a carburetor.

PRACTICE QUESTIONS

1. What component in a typical carburetor float system is indicated by letter A in Figure 4–1?

   a. float
   b. fuel bowl
   c. vent
   d. control switch

2. What component in a typical carburetor float system is indicated by letter B in Figure 4–1?

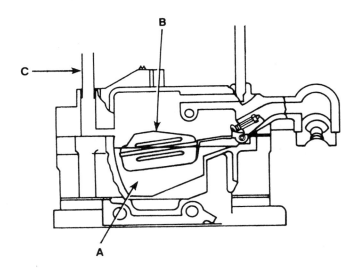

Figure 4–1

   a. float
   b. fuel bowl
   c. vent
   d. control switch

3. What float system component is indicated by letter C in Figure 4–1?

   a. float
   b. fuel bowl
   c. vent
   d. control switch

4. Carburetors that are recommended for use in areas where competition performance carburetors are not allowed are listed under the heading

   _____ .

   a. "Not legal for street use"
   b. "Competition/Off-road"
   c. "Race only"
   d. "Emission Design/Street Performance"

5. Technician A repairs or replaces a carburetor while the engine is at normal operating temperature. Technician B repairs or replaces a carburetor while the engine is cold. Who is right?

   a. Technican A
   b. Technician B
   c. Both A and B
   d. Neither A nor B

6. A vehicle's engine hesitates on acceleration. Technican A and Technician B turn the engine off, remove the air cleaner, and look into the carburetor bores while manually opening the throttle valve. They see a full stream of fuel emit from the pump jet and strike near the center of the venturi area. Technician A says the accelerator pump system is not functioning properly. Technician B says other possibilities, such as dirt, loose bolts, or a leaking air horn, are causing the problem. Who is right?

   a. Technician A
   b. Technician B
   c. Both A and B
   d. Neither A nor B

7. Before removing or disassembling a carburetor, Technician A checks the choke. Technician B checks the hose connections. Who is right?

   a. Technician A
   b. Technician B
   c. Both A and B
   d. Neither A nor B

8. A vehicle's choke valve or linkage is sticking, causing the engine to start hard when it is cold. Technician A realigns the choke valve or linkage. Technician B oils the choke linkage. Who is right?

   a. Technician A
   b. Technician B
   c. Both A and B
   d. Neither A nor B

9. Which of the following is not a possible cause of engine backfire?

   a. defective accelerator pump
   b. old spark plugs or wires
   c. partially clogged fuel filter
   d. float level too low

10. What mechanical device is used to eliminate choke enrichment during cranking of an engine by partially opening the choke at wide-open throttle?

   a. bowl vent valve
   b. choke unloader
   c. vacuum kick
   d. dashpot

JOB SHEET

SHOP ASSIGNMENT 11
REPLACE A CARBURETOR

NAME _____ STATION _____ DATE _____

Tools and Materials

Masking tape
Shop rags or towels
Carburetor repair stand
Soft brush
Carburetor or decarbonizing cleaner
Compressed air
Replacement carburetor parts, if necessary
Hand tools
Vacuum cleaner or magnet, if necessary
New fuel lines, hoses, and clamps, if necessary
New fuel filter
New base gasket
New oil filter

Protective Clothing

Safety goggles or glasses with side shields

Procedure

1. Read the manufacturer's service manual or the instructions enclosed with the replacement carburetor.

   Task completed _____

   a. Is the engine cold?

   Yes _____   No _____

   b. If no, wait until the engine cools down before continuing.

   Not applicable _____   Task completed _____

   c. Disconnect the battery ground cable.

   Task completed _____

2. Tag all fuel lines, vacuum lines, and electrical connections for identification.

   Task completed _____

   a. Disconnect all vacuum lines from the carburetor.

   Task completed _____

   b. Disconnect any electrical connections to components such as the choke, etc.

   Task completed _____

3. Wrap the fuel line connection to the carburetor with a shop rag or towel to ensure that fuel does not spray when the line is disconnected.

Task completed _____

   a. Disconnect the fuel line.

Task completed _____

4. Disengage the throttle and choke linkage by removing the appropriate retaining clips.

Task completed _____

5. Tag the two or three rods and springs that connect to the throttle lever.

Not applicable _____   Task completed _____

   a. Disconnect the rods and springs.

Not applicable _____   Task completed _____

6. Does the vehicle use a hot-water choke system?

Yes _____   No _____

   a. If no, go to step 7.

Not applicable _____   Task completed _____

   b. If yes, loosen the radiator cap to release any pressure.

Not applicable _____   Task completed _____

   c. Place a shop rag or towel around the hot-water choke to soak up escaping coolant.

Task completed _____

   d. Remove the hot-water choke.

Task completed _____

7. Remove the carburetor mounting nuts.

Task completed _____

8. Are all connections disconnected from the carburetor?

Yes _____   No _____

   a. If no, disconnect.

Not applicable _____   Task completed _____

   b. Lift the carburetor off the manifold.

Task completed _____

   c. Drain fuel from the carburetor and dispose of in recommended manner.

Task completed _____

9. If dirt or parts get into the manifold once the carburetor is removed, use a vacuum cleaner or magnet to remove the foreign matter.

Not applicable _____   Task completed _____

10. Inspect all fuel lines, hoses, and clamps.

Task completed _____

a. Replace any that are old, cracked, or brittle.

Not applicable _____   Task completed _____

11. Replace the fuel filter.

Task completed _____

12. Remove the old base gasket on the manifold.

Task completed _____

13. Disassemble the carburetor as recommended in the vehicle's service manual.

Task completed _____

a. Mount the carburetor on a carburetor repair stand.

Task completed _____

14. Clean the carburetor parts as recommended, using carburetor or decarbonizing cleaner, clean cloth, soft brush, hot water, and/or compressed air.

Task completed _____

15. Inspect all parts for reuse or replacement.

Task completed _____

16. Install a new base gasket.

Task completed _____

17. Install the replacement carburetor on the intake manifold.

Task completed _____

a. Install the carburetor mounting bolts, but do not tighten.

Task completed _____

18. Connect the fuel line while the carburetor can still be moved.

Task completed _____

19. Tighten the carburetor mounting bolts in a criss-cross pattern and do not exceed 115 foot-pounds.

Task completed _____

20. Install the remaining vacuum lines and electrical connections.

Task completed _____

21. If the vehicle uses a hot-water choke, install the choke mechanism.

Not applicable _____   Task completed _____

22. Engage the linkage with the appropriate retaining clip(s).

Task completed _____

23. Connect the throttle lever springs and rods.

Task completed _____

a. Make sure the throttle can be closed and opened.

Task completed _____

24. Install a new oil filter and install the filter top.

Task completed _____

25. Install a ground back or battery.

Task completed _____

26. Start the engine. (It will crank longer than usual because the fuel bowl is empty and needs time to fill.)

Task completed _____

27. Set the idle according to specifications.

Task completed _____

PROBLEMS ENCOUNTERED: _____

_____

_____

INSTRUCTOR'S COMMENTS: _____

_____

_____

JOB SHEET

SHOP ASSIGNMENT 12
ADJUST CURB IDLE SPEED WITH SOLENOID

NAME _____ STATION _____ DATE _____

Tools and Materials

Hand tools
Plugs

Protective Clothing

Safety goggles or glasses with side shields

Procedure

1. Make sure engine is at normal operating temperature.

   Task completed _____

   a. Apply parking brake.

   Task completed _____

   b. Make sure air cleaner or filter is in place.

   Task completed _____

   c. Place transmission in NEUTRAL.

   Task completed _____

   d. Turn air conditioning off.

   Not applicable _____ Task completed _____

2. Make sure the throttle stop solenoid is energized and extended.

   Task completed _____

3. To make the adjustments, turn the curb idle speed screw that contacts the throttle stop solenoid. (See Figure 4–2).

   Task completed _____

4. When adjusting the idle, identify and disconnect the vapor storage canister, distributor vacuum, and EGR hoses.

   Task completed _____

   a. Plug the ends of these hoses.

   Task completed _____

5. To prevent variations of fuel and temperature when setting idle mixture, after 3 minutes of idling, increase the engine speed to 2000 rpm for 1 minute.

   Task completed _____

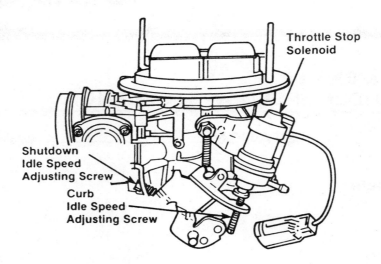

Throttle Stop
Solenoid

Shutdown
Idle Speed
Adjusting Screw

Curb
Idle Speed
Adjusting Screw

Figure 4-2

   a. Continue with the adjustment.

Task completed _____

   b. Do not idle the engine for more than 3 minutes without repeating step 5.

Task completed _____

6. To make a curb idle speed adjustment when no solenoid is used, follow steps 1 through 5.

Not applicable _____   Task completed _____

   a. The carburetor throttle lever tang should be contacting the curb idle screw instead of the solenoid plunger.

Not applicable _____   Task completed _____

7. Adjust the idle adjusting screw to specifications.

Task completed _____

8. After adjustments are completed, unplug and reconnect the hoses to their proper fittings.

Task completed _____

   a. Reconnect all connections.

Task completed _____

PROBLEMS ENCOUNTERED: _____

_____

_____

INSTRUCTOR'S COMMENTS: _____

_____

_____

JOB SHEET

SHOP ASSIGNMENT 13
ADJUST FAST-IDLE

NAME _____ STATION _____ DATE _____

Tools and Materials

Hand tools
Rubber bands, if necessary

Protective Clothing

Safety goggles or glasses with side shields

Procedure

1. Idle the engine at normal operating temperature.

   Task completed _____

2. Place the fast-idle lever on the specified step of the fast-idle cam.

   Task completed _____

   a. In many cases, the EGR must be disconnected and the vacuum line plugged.

   Not applicable _____   Task completed _____

3. Make sure the high cam speed positioner lever is disengaged.

   Task completed _____

4. Turn the fast-idle adjusting screw clockwise to increase speed and counter-clockwise to decrease speed.

   Task completed _____

5. To adjust the fast-idle cam set, remove the choke cap.

   Not applicable _____   Task completed _____

6. Place the fast idle lever in the corner of the specified step of the fast-idle cam (counting the highest step as the first) with the high cam speed positioner retracted.

   Not applicable _____   Task completed _____

7. Hold the throttle lightly closed with a rubber band to maintain the cam position. (This step is not required if the adjustment is done on the vehicle.)

   Not applicable _____   Task completed _____

8. Install the stator cap.

   Not applicable _____   Task completed _____

   a. Rotate it clockwise until the lever contacts the adjusting screw.

   Task completed _____

9. Turn the fast-idle cam adjusting screw until the index mark on the stator cap lines up with the specified notch on the choke casting.

Not applicable _____   Task completed _____

10. Remove the stator cap.

Not applicable _____   Task completed _____

a. Reinstall the choke cap.

Not applicable _____   Task completed _____

PROBLEMS ENCOUNTERED: _____

_____

_____

INSTRUCTOR'S COMMENTS: _____

_____

_____

CHAPTER FIVE

BASICS OF FUEL INJECTION

Objectives

After reading this chapter, you should be able to:

• Compare the operation of a fuel injected system to a carbureted system.
• Explain the two methods of fuel injection: electronic fuel injection (EFI) and continuous injection system (CIS).
• Explain the basic operation of a fuel injection system.
• Name the advantages and limitations of a fuel injection system.
• List the components of a basic fuel injection system.
• Describe the functions of the components of a basic fuel injection system.
• Differentiate between closed and open loop operations.

PRACTICE QUESTIONS

1. In what fuel system do all cylinders receive the same fuel mixture?

  a. fuel injection
  b. carburetion
  c. both A and B
  d. neither A nor B

2. What is the most widely used fuel injection system?

  a. electronic fuel injection
  b. mechanical injection
  c. continuous injection
  d. fuel distributor injection

3. What type of injector is depicted in Figure 5-1?
  a. top-feed injector
  b. throttle-body injector
  c. port-body injector
  d. bottom-feed injector

4. What component of a typical fuel rail assembly is indicated by letter A in Figure 5-2?

  a. pressure regulator
  b. fuel inlet
  c. injector
  d. test point

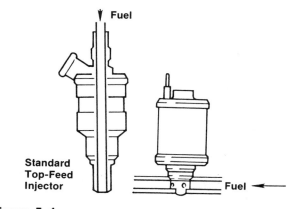

Figure 5-1

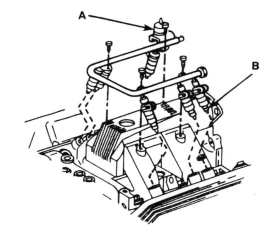

Figure 5-2

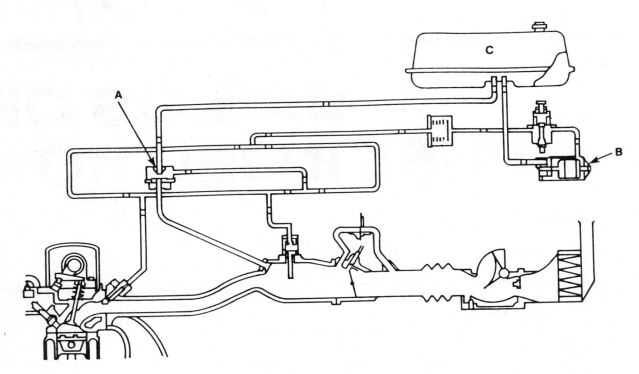

Figure 5-3

5. What component of a typical fuel rail assembly is indicated by letter B in Figure 5-2?

   a. pressure regulator
   b. fuel inlet
   c. injector
   d. test point

6. What component is the principal difference between the fuel systems of electronic fuel injection and continuous fuel injection?

   a. fuel tank
   b. control pressure regulator
   c. fuel pump
   d. accumulator

7. Diesel fuel injection systems deliver fuel directly into the _____ .

   a. precombustion chamber
   b. cylinder
   c. both A and B
   d. neither A nor B

8. What component of a fuel circuit of a gasoline fuel injection system is indicated by letter A in Figure 5-3?

   a. fuel pump
   b. fuel tank
   c. intake valve
   d. pressure regulator

9. What component of a typical diesel mechanical fuel injection system is indicated by letter B in Figure 5-3?

   a. fuel pump
   b. fuel tank
   c. intake valve
   d. pressure regulator

10. What component of a typical diesel mechanical fuel injection system is indicated by letter C in Figure 5-3?

   a. fuel pump
   b. fuel tank
   c. intake valve
   d. pressure regulator

SHOP ASSIGNMENT 14 ELECTRONIC FUEL INJECTION

Examine an electronic fuel injection system on a vehicle. Locate and identify the following components: electronic control unit (ECU), throttle switch, pressure regulator, inlet manifold pressure sensor or airflow sensor, thermo time switch, fuel rail assembly, injectors, fuel tank, fuel pump, and fuel filter. Refer to Figure 5-4 as needed.

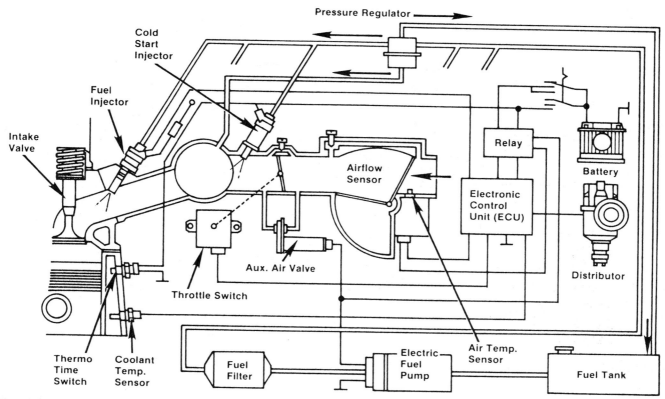

Figure 5-4

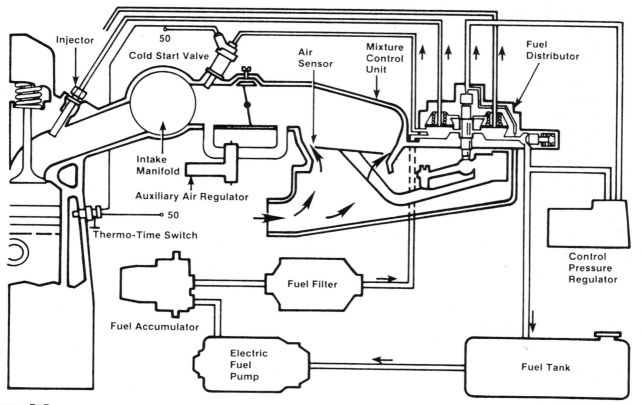

Figure 5-5

SHOP ASSIGNMENT 15
CONTINUOUS FUEL
INJECTION

Examine a gasoline continuous injection system (CIS). Locate and identify the following components: fuel distributor, air sensor, mixture control unit, fuel tank, fuel pump, fuel filter, accumulator, control pressure regulator, injectors, and thermo time switch. Refer to Figure 5–5 as needed.

THROTTLE BODY AND PORT FUEL INJECTION SYSTEMS

Objectives

After reading this chapter, you should be able to:
* Explain the operation of a throttle body fuel injection system.
* Identify the major parts of a throttle body fuel injection system and explain the function of each part.
* Explain the operation of a port fuel injection system.
* Identify the major parts that are unique to port fuel injection and those that are common to both throttle body and port injection systems.
* Identify and describe the physical and operating differences among the fuel injection systems marketed by major domestic car manufacturers.
* Explain the advantages and disadvantages of both throttle body and port injection.

PRACTICE QUESTIONS

1. When the engine is first started and the rpm is well above 600, the throttle-body injection system goes into _____ .
   a. open loop operation
   b. closed loop operation
   c. both A and B
   d. neither A nor B

2. What is the term for a resistor that changes value based on temperature?
   a. sensor
   b. thermister
   c. PFI
   d. MAF

3. Which of the following symptoms can indicate a misadjusted, open, shorted, or loose throttle position sensor?
   a. poor idle
   b. unstable idle speed
   c. constantly closed throttle
   d. all of the above

4. What type of PFI fuel rail assembly is depicted in Figure 6–1?
   a. center base type
   b. crossover type
   c. throttle body type
   d. all of the above

5. What component of a fuel rail assembly is indicated by letter A in Figure 6–1?
   a. injector
   b. fuel pump
   c. pressure regulator
   d. fuel outlet fitting

6. What component of a fuel rail assembly is indicated by letter B in Figure 6–1?

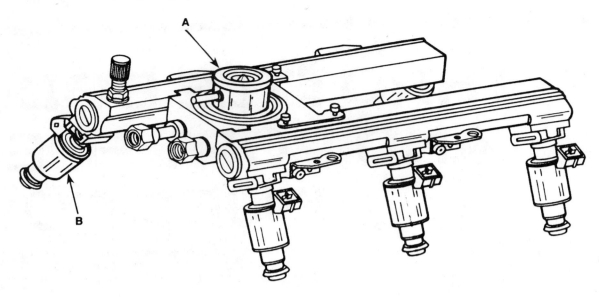

Figure 6-1

a. injector
b. fuel pump
c. pressure regulator
d. fuel outlet fitting

7. The American Motors system that is basically a port fuel injection system is called

   _____ .

   a. metered fuel injection
   b. airflow-controlled fuel injection
   c. central fuel injection
   d. none of the above

8. What major auto manufacturer markets the digital fuel injection (DFI) system?

   a. Chrysler
   b. AMC
   c. GM
   d. Ford

9. Which of the following is a subsystem of the Ford central fuel injection (CFI) system?

   a. air induction
   b. data sensors
   c. fuel delivery
   d. all of the above

10. What component on 1988 Chrysler vehicles with throttle body injection replaces both the power module and the logic module?

    a. single module engine controller (SMEC)
    b. electronic control assembly (ECA)

c. electronic control module (ECM)
d. none of the above

SHOP ASSIGNMENT 16 THROTTLE BODY INJECTION

Examine a throttle body fuel injection system (TBI) and disassemble, if possible. Locate and identify key components such as those indicated in Figure 6-2.

SHOP ASSIGNMENT 17 PORT FUEL INJECTION

Examine a port fuel injection system (PFI) and disassemble, if possible. Locate and identify key components such as those indicated in Figure 6-3.

SHOP ASSIGNMENT 18 AIRFLOW-CONTROLLED INJECTION

Examine the American Motors airflow-controlled fuel injection system. Note the similarities and differences between this system and the TBI and PFI already discussed. Examine the fuel subsystem and the control subsystem. Refer to Figure 6-4 as needed.

Chapter Six   Throttle Body and Port Fuel Injection Systems

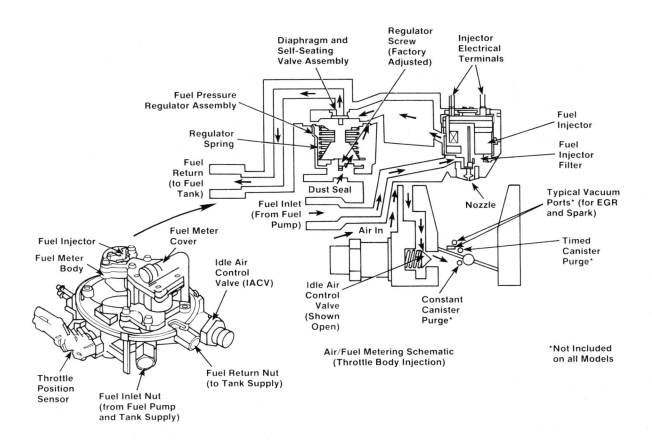

Diaphragm and Self-Seating Valve Assembly

Regulator Screw (Factory Adjusted)

Injector Electrical Terminals

Fuel Pressure Regulator Assembly

Regulator Spring

Fuel Injector

Fuel Injector Filter

Fuel Return (to Fuel Tank)

Fuel Inlet (From Fuel Pump)

Dust Seal

Air In

Nozzle

Typical Vacuum Ports* (for EGR and Spark)

Timed Canister Purge*

Idle Air Control Valve (Shown Open)

Constant Canister Purge*

Air/Fuel Metering Schematic (Throttle Body Injection)

*Not Included on all Models

Fuel Injector

Fuel Meter Cover

Fuel Meter Body

Idle Air Control Valve (IACV)

Throttle Position Sensor

Fuel Return Nut (to Tank Supply)

Fuel Inlet Nut (from Fuel Pump and Tank Supply)

Figure 6-2

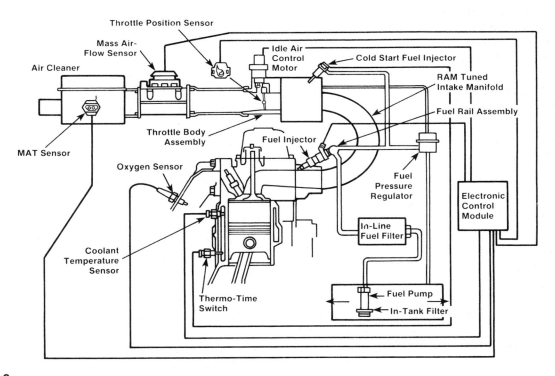

Throttle Position Sensor

Mass Air-Flow Sensor

Idle Air Control Motor

Cold Start Fuel Injector

Air Cleaner

RAM Tuned Intake Manifold

Fuel Rail Assembly

MAT Sensor

Throttle Body Assembly

Fuel Injector

Oxygen Sensor

Fuel Pressure Regulator

Electronic Control Module

Coolant Temperature Sensor

In-Line Fuel Filter

Thermo-Time Switch

Fuel Pump

In-Tank Filter

Figure 6-3

Chapter Six   Throttle Body and Port Fuel Injection Systems

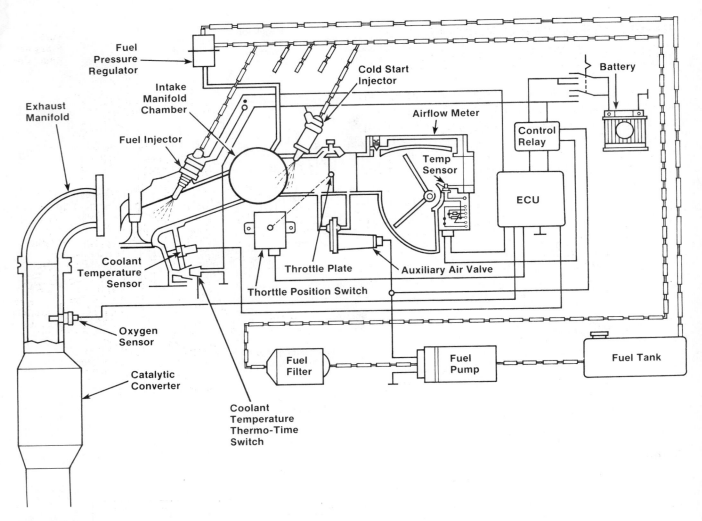

Figure 6-4

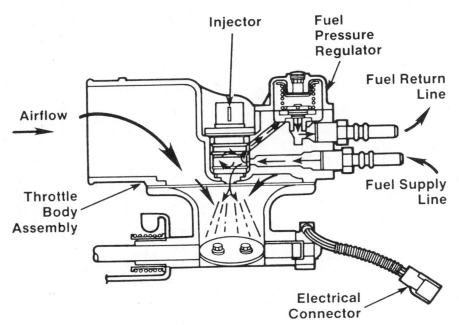

Figure 6-5

Chapter Six   Throttle Body and Port Fuel Injection Systems

SHOP ASSIGNMENT 19
CENTRAL FUEL INJECTION

Examine the Ford Motor Company central fuel injection system. Note the similarities and differences between this system and the TBI and PFI already discussed. Examine the four subsystems: fuel delivery, air induction, engine data sensors, and electronic control assembly (ECA). Refer to figures 6-5 and 6-6 as needed.

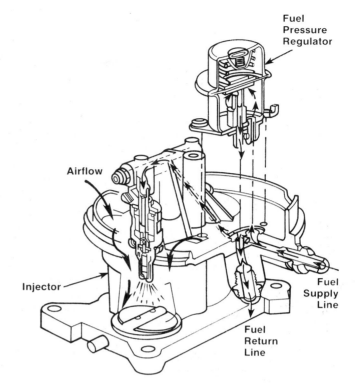

Figure 6-6

CHAPTER SEVEN

FUEL INJECTION SERVICE

Objectives

After reading this chapter, you should be able to:
- Relieve fuel pressure in the system prior to performing maintenance or repairs.
- Perform preventive maintenance checks.
- Use troubleshooting guides to isolate a faulty component.
- Check for fuel delivery on a fuel injection system.
- Describe the diagnostic systems of the three major domestic car manufacturers.
- Know and identify the various trouble codes generated by diagnostic systems.
- Use testers with the applicable diagnostic systems.

PRACTICE QUESTIONS

1. What fuel injection system is used on most import cars?
   a. Bosch CIS
   b. GM PFI
   c. MPI
   d. None of the above

2. What component of a Bosch CIS, not found on domestic systems, maintains the system pressure or prevents it from dropping when the engine is off?
   a. fuel distributor
   b. solenoid
   c. fuel accumulator
   d. diaphragm

3. Technician A relieves the fuel pressure in a CIS by energizing the cold start valve for about 10 seconds. Technician B wraps the fuel line connection on the pressure regulator with a clean shop rag and then cracks loose the connection. Who is right?
   a. Technician A
   b. Technician B
   c. Both A and B
   d. Neither A nor B

4. Technician A says that the TBI system of a 1979 Chrysler operates at 14.7 psi. Technician B says that a 1981 Chrysler operates at 36 psi. Who is right?
   a. Technician A
   b. Technician B
   c. Both A and B
   d. Neither A nor B

5. A vehicle has starting or running problems. Technician A checks the battery and ignition system first. Technician B checks the fuel injection system first. Who is right?
   a. Technician A
   b. Technician B
   c. Both A and B
   d. Neither A nor B

6. What major automobile manufacturer introduced an electronic spark control system call Lean Burn, which was monitored by a microcomputer?
   a. Bosch
   b. GM
   c. Ford
   d. Chrysler

7. What tool used to service fuel injection systems is depicted in Figure 7-1?

Figure 7-1

    a. infrared analyzer
    b. digital volt-ohmmeter
    c. voltmeter
    d. STAR tester

8. Which of the following is not a test mode?
    a. circuit actuation
    b. sensor
    c. engine off
    d. switch

9. A vehicle with PFI has a problem with poor power. Technician A checks the injector spray pattern. Technician B checks the fuel pump pressure. Who is right?
    a. Technician A
    b. Technician B
    c. Both A and B
    d. Neither A nor B

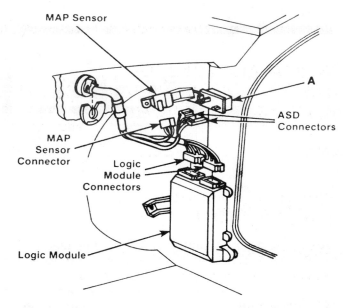

Figure 7-2

10. On a Chrysler with computer-controlled fuel injection, what component is indicated by letter A in Figure 7-2?
    a. ASD relay
    b. power module
    c. oxygen sensor
    d. TPS

JOB SHEET

SHOP ASSIGNMENT 20
RELIEVE AND CHECK THE FUEL PRESSURE ON A
CHRYSLER MODEL WITH TBI

NAME _____ **STATION** _____ **DATE** _____

Tools and Materials
Fuel pressure gauge

Protective Clothing
Safety goggles or glasses with side shields

Procedure

1. Loosen the gas cap.

Task completed _____

2. Remove the injector wiring harness.

Task completed _____

3. Ground one terminal on the connector.

Task completed _____

    a. Apply battery voltage (12 volts) to the other terminal for no longer than 10 seconds.

Task completed _____

4. Install a fuel pressure gauge between the fuel filter hose and throttle body. See Figure 7–3.

Task completed _____

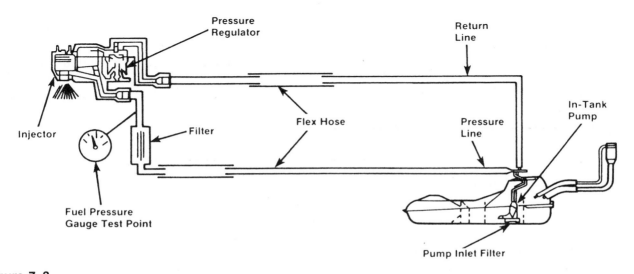

Figure 7–3

5. Start the engine.

Task completed _____

6. Check the fuel pressure gauge.

Task completed _____

   a. Is the pressure too low?

Yes _____   No _____

   b. If no, then the task is completed.

Not applicable _____   Task completed _____

   c. If yes, turn off the engine.

Not applicable _____   Task completed _____

   d. Move the gauge between the fuel filter hose and fuel line.

Not applicable _____   Task completed _____

   e. Restart the engine.

Task completed _____

   f. Check the gauge again.

Task completed _____

   g. Is the pressure now correct?

Yes _____   No _____

   h. If yes, replace the fuel filter and the task is completed.

Not applicable _____   Task completed _____

   i. If no, gently squeeze the fuel return hose.

Not applicable _____   Task completed _____

   j. Does the pressure increase?

Yes _____   No _____

   k. If yes, the pressure regulator is faulty.

Not applicable _____   Task completed _____

   l. If no, check the fuel pump.

Not applicable _____   Task completed _____

PROBLEMS ENCOUNTERED: _____

_____

_____

INSTRUCTOR'S COMMENTS: _____

_____

_____

JOB SHEET

SHOP ASSIGNMENT 21
DIAGNOSE A DRIVEABILITY PROBLEM ON A
GM VEHICLE WITH C-3

NAME _____ **STATION** _____ **DATE** _____

Tools and Materials

Hand tools
C-3 tester (scan tool), if possible
Service manual
Minor replacement parts, if necessary

Protective Clothing

Safety goggles or glasses with side shields

Procedure

1. Check for vacuum leaks.

Task completed _____

   a. Repair or replace parts if necessary.

Not applicable _____ Task completed _____

2. Check for fouled spark plugs.

Task completed _____

   a. Repair or replace parts if necessary.

Not applicable _____ Task completed _____

3. Check for arcing plug wires.

Task completed _____

   a. Repair or replace parts if necessary.

Not applicable _____ Task completed _____

4. Refer to a service manual that has complete diagnostic sequences and follow the diagnostic flow charts relating to the model line, engine design, and fuel system of the vehicle being diagnosed.

Task completed _____

5. Assuming the initial diagnosis has localized the C-3 system as being the possible suspect, continue the diagnosis with a careful visual inspection of the system inspected (ignition, emission, fuel).

Task completed _____

   a. Correct any loose, dirty connections.

Not applicable _____ Task completed _____

GM TROUBLE CODES

Code	Circuit Affected or Possible Cause	Code	Circuit Affected or Possible Cause
12	No distributor reference pulses to the ECM. This code is not stored in memory and will flash only while the fault is present. Normal code with ignition on, engine not running.	34	Vacuum sensor or Manifold Absolute Pressure (MAP) circuit. The engine must run up to 2 minutes at specified curb idle before this code will set.
13	Oxygen Sensor Circuit. The engine must run up to 4 minutes at part-throttle under road load before this code will set.	35	Idle speed control (ISC) switch circuit shorted. (Up to 70% TPS for over 5 seconds.)
14	Shorted coolant sensor circuit. The engine must run 2 minutes before this code will set.	41	No distributor reference pulses to the ECM at specified engine vacuum. This code will store in memory.
15	Open coolant sensor circuit. The engine must run 5 minutes before this code will set.	42	Electronic spark timing (EST) bypass circuit or EST circuit grounded or open.
21	Throttle Position Sensor (TPS) circuit voltage high (open circuit or misadjusted TPS). The engine must run 10 seconds at specified curb idle speed before this code will set.	43	Electronic Spark Control (ESC) retard signal for too long a time; causes retard in EST signal.
22	Throttle Position Sensor (TPS) circuit voltage low (grounded circuit or misadjsuted TPS). Engine must run 20 seconds at specified curb idle speed to set code.	44	Lean exhaust indication. The engine must run 2 minutes in closed loop and at part-throttle before this code will set.
23	M/C solenoid circuit open or grounded.	45	Rich exhaust indication. The engine must run 2 minutes in closed loop and at part throttle before this code will set.
24	Vehicle speed sensor (VSS) circuit. The vehicle must operate up to 2 minutes at road speed before this code will set.	51	Faulty or improperly installed calibration unit (PROM). It takes up to 30 seconds before this code will set.
32	Barometric pressure sensor (BARO) circuit low.	53	Exhaust Gas Recirculation (EGR) valve vacuum sensor has seen improper EGR vacuum.
		54	Shorted M/C solenoid circuit and/or faulty ECM.

Figure 7–4

b. Make sure there is a good ECM ground.

Not applicable _____ Task completed _____

c. If necessary, reroute all C-3 wires away from areas where electromagnetic induction emanating from plug wires and/or the coil could cause interference with or trigger false ECM signals.

Not applicable _____ Task completed _____

6. If possible, plug the C-3 tester (scan tool) into the ALCL connector and read the trouble codes directly. Refer to a chart such as Figure 7–4 to identify GM trouble codes.

Not applicable _____ Task completed _____

7. To read trouble codes of any C-3 system using the "check engine" light method, jump terminal A and B together in ALCL connector.

Task completed _____

a. Turn on the ignition switch.

Task completed _____

b. Does the light come on and immediately start flashing?

Yes _____ No _____

c. If no, try again.

Not applicable _____ Task completed _____

8. If yes, read and record all codes displayed until code 12 reappears and flashes three times, signaling the end of code transmission.

Not applicable _____ Task completed _____

a. If no trouble codes are found, go to step 13.

Not applicable _____ Task completed _____

9. Clear the computer's memory by pulling the ECM fuse or disconnecting the ECM feed wire at the battery for 10 seconds.

Task completed _____

10. Start and run the vehicle (road test if necessary) for at least 5 minutes to see if the codes reset.

Task completed _____

11. After verifying the validity of all the trouble codes found, use the shop manual and refer to the troubleshooting section.

Task completed _____

12. When working with multiple codes, give priority to those in the 50s range since they indicate a problem in the ECM or PROM.

Not applicable _____ Task completed _____

a. Otherwise, check and repair codes in numerical sequence.

Not applicable _____ Task completed _____

13. If no trouble codes are found, conduct a system performance test.

Not applicable _____ Task completed _____

a. Check essential sensor inputs.

Not applicable _____ Task completed _____

14. Record problems and repairs.

Task completed _____

PROBLEMS ENCOUNTERED: _____

_____

_____

INSTRUCTOR'S COMMENTS: _____

_____

_____

CHAPTER EIGHT

DIESEL FUEL INJECTION

Objectives

After reading this chapter, you should be able to:
- Describe the purpose, construction, and operation of a diesel fuel injection system and its components.
- Diagnose basic diesel fuel injection problems.
- Follow accepted general precautions prior to and during diesel fuel injection servicing.
- Describe the operation and construction of an inline injection pump.
- Describe the operation and construction of a distribution injection pump.
- Follow general corrective procedures to repair a diesel fuel injection system.

PRACTICE QUESTIONS

1. What system is not similar in both gasoline and diesel engines?
   a. starting system
   b. cooling system
   c. fuel delivery
   d. ignition system

2. What component of a diesel fuel injection system is indicated by letter A in Figure 8-1?
   a. fuel lift pump
   b. injection pump
   c. water fuel separator
   d. fuel filter

3. What component of a diesel fuel injection system is indicated by letter B in Figure 8-1?
   a. fuel lift pump
   b. injection pump
   c. water fuel separator
   d. fuel filter

4. What component of a diesel fuel injection system is indicated by letter C in Figure 8-1?
   a. fuel lift pump
   b. injection pump
   c. water fuel separator
   d. fuel filter

5. What type of diesel fuel system is the most commonly used?
   a. distributor pump system
   b. pump controlled system
   c. unit injection system
   d. common rail system

6. What type of diesel fuel feed pump is depicted in Figure 8-2?
   a. gear type
   b. piston type
   c. vane type
   d. none of the above

7. A diesel engine with fuel injection is experiencing excessive surge at light throttle when under load. Technician A adjusts the timing. Technician B checks fuel pump pressure at the inlet and outlet sides of the filter. Who is right?
   a. Technician A
   b. Technician B
   c. Both A and B
   d. Neither A nor B

8. A fuel-injected diesel engine will not crank. Technician A checks the charging system and the starter. Technician B drains the oil and refills with a recommended type. Who is right?
   a. Technician A
   b. Technician B

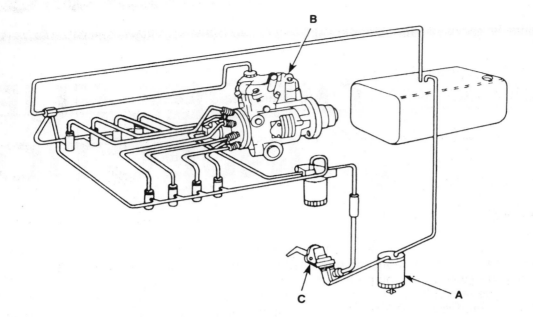

Figure 8-1

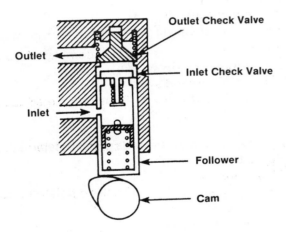

Figure 8-2

c. Both A and B
d. Neither A nor B

9. A diesel engine with fuel injection idles correctly but misfires above idle. Technician A checks the filter and replaces if necessary. Technician B checks the fuel, and flushes the fuel system and adds correct fuel if necessary. Who is right?

   a. Technician A
   b. Technician B
   c. Both A and B
   d. Neither A nor B

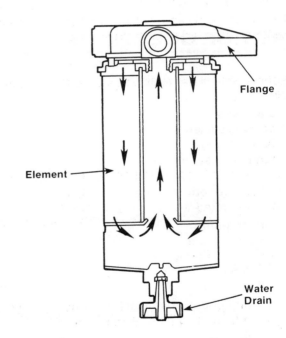

Figure 8-3

10. What component of a diesel fuel system is depicted in Figure 8-3?

    a. fuel pump
    b. fuel filter
    c. distributor
    d. water fuel separator

JOB SHEET

SHOP ASSIGNMENT 22
REPLACE A SELF-CONTAINED FUEL FILTER

NAME _____ STATION _____ DATE _____

Tools and Materials

Hand tools
New gaskets
New filter element
Clean cloth

Protective Clothing

Safety goggles or glasses with side shields

Procedure

1. Drain fuel, dirt, and water from the filter housing sump by opening the petcock or removing the drain plug.

   Task completed _____

2. Loosen the center stud bolt located at the top of the filter assembly.

   Task completed _____

   a. Discard gaskets.

   Task completed _____

   b. Remove filter housing and/or cover the element together.

   Task completed _____

3. Remove the used element from the filter housing.

   Task completed _____

   a. Discard the element.

   Task completed _____

4. Wipe the inside of the filter housing clean of fuel, residue, or other deposits.

   Task completed _____

5. Place a new gasket on the center stud bolt.

   Task completed _____

6. Place a new gasket in the filter cover assembly.

   Task completed _____

7. Place a new fuel filter element in the filter housing.

   Task completed _____

8. Assemble the housing and cover.

Task completed _____

   a. Tighten the center stud bolt to the torque indicated on the housing or in the instructions.

Task completed _____

9. Close the petcock or replace the drain plug in the filter housing.

Task completed _____

PROBLEMS ENCOUNTERED: _____

_____

_____

INSTRUCTOR'S COMMENTS: _____

_____

_____

JOB SHEET

SHOP ASSIGNMENT 23
REMOVE WATER FROM A DIESEL FUEL SYSTEM

NAME _____ STATION _____ DATE _____

Tools and Materials

Hand tools
Clear plastic line
Hand priming pump, if necessary

Protective Clothing

Safety goggles or glasses with side shields

Procedure

1. Remove the fuel return hose, usually located at the front of the engine.

   Task completed _____

2. Remove the fuel tank filler cap.

   Task completed _____

3. Pump or siphon the water from the previously disconnected line using a clear plastic line to determine when uncontaminated fuel begins to flow.

   Task completed _____

4. When the water is drained, replace the filler cap.

   Task completed _____

   a. Return the hose.

   Task completed _____

PROBLEMS ENCOUNTERED: _____

_____

_____

INSTRUCTOR'S COMMENTS: _____

_____

_____

JOB SHEET

SHOP ASSIGNMENT 24
REMOVE, REPLACE, AND TEST AN INJECTION NOZZLE

NAME _____ STATION _____ DATE _____

Tools and Materials

Nozzle stand
Hand tools
Nozzle tester
New O-rings and gaskets
Service manual

Protective Clothing

Safety goggles or glasses with side shields

Procedure

1. Clean the area around the injector(s) to be removed.

Task completed _____

2. Disconnect the fuel line(s) from the injectors to be serviced.

Task completed _____

   a. Cap all fuel lines and fittings to prevent contamination.

Task completed _____

3. Follow the procedures in the service manual and remove the injector.

Task completed _____

4. Disassemble and clean the injector.

Task completed _____

5. Inspect all parts for wear and damage.

Task completed _____

   a. Refer to specifications in the manufacturer's service manual to determine wear tolerances.

Task completed _____

6. Reassemble the injector following procedures given in the service manual.

Task completed _____

7. To test the injector, install it in an injector stand.

Task completed _____

   a. Follow the equipment manufacturer's test procedures to check for proper spray pattern.

Task completed _____

b. While rapidly moving the operating lever on the nozzle tester, listen for a creaking sound that indicates the nozzle is clean and in good condition.

Task completed _____

c. Open the gauge knob on the nozzle tester slightly.

Task completed _____

d. Observe the gauge while slowly operating the lever.

Task completed _____

e. The gauge should indicate the opening pressure of the nozzle; compare this to specifications given in the manufacturer's service manual.

Task completed _____

f. On some types of injectors the nozzle opening pressure can be adjusted with the nozzle tester.

Not applicable _____ Task completed _____

g. Again, slowly operate the lever on the nozzle tester and bring the pressure to 200 psi.

Task completed _____

h. Is the leakage at the nozzle tip acceptable? Refer to Figure 8-4 and manufacturer's service manual for proper specifications and recommendations.

Yes _____ No _____

8. Make sure no gasket material remains in the injector mounting hole.

Task completed _____

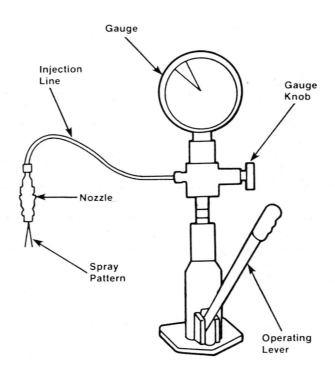

Figure 8-4

a. Clean all carbon deposits from the injector opening and the precombustion chamber.

Task completed _____

b. Remove all the O-rings; use only new O-rings and gaskets where required.

Task completed _____

9. Install the injector(s) into the cylinder head.

Task completed _____

a. Tighten the injector to the manufacturer's recommended torque.

Task completed _____

b. Make sure that there is a positive seal around the injector.

Task completed _____

10. Connect all fuel lines.

Task completed _____

a. Tighten the fuel lines to the injectors according to the manufacturer's recommended torque values.

Task completed _____

11. If necessary, bleed all air from the fuel system.

Not applicable _____ Task completed _____

12. Start the engine.

Task completed _____

a. Inspect for fuel leaks.

Task completed _____

13. Tighten all loose connections.

Task completed _____

PROBLEMS ENCOUNTERED: _____

_____

_____

INSTRUCTOR'S COMMENTS: _____

_____

_____

CHAPTER NINE

EXHAUST SYSTEMS

Objectives

After reading this chapter, you should be able to:
- Explain the three basic functions of a vehicle's exhaust system.
- Explain the function of exhaust system components, including exhaust manifold, gaskets, exhaust pipe and seal, catalytic converter, muffler, resonator, and clamps, brackets, and hangers.
- Explain the importance of proper exhaust system operations to vehicle performance and personal safety.
- Properly perform an exhaust system inspection.
- Properly service and/or replace exhaust system components.
- Explain how a turbocharger operates.
- Properly service a turbocharger.

PRACTICE QUESTIONS

1. What exhaust system component is indicated by letter A in Figure 9–1

   a. muffler
   b. tail pipe
   c. catalytic converter
   d. exhaust pipe

2. What exhaust system component is indicated by letter B in Figure 9–1?

   a. muffler
   b. tail pipe
   c. catalytic converter
   d. exhaust pipe

3. What exhaust system component is indicated by letter C in Figure 9–1?

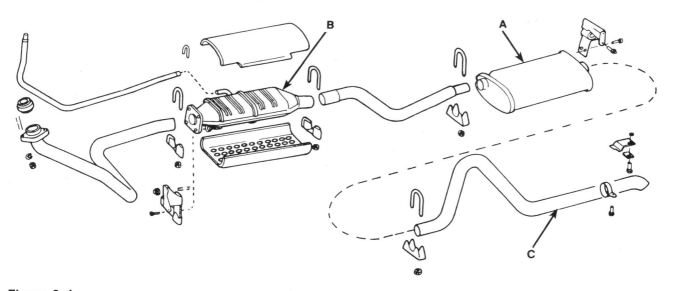

Figure 9–1

a. muffler
b. tail pipe
c. catalytic converter
d. exhaust pipe

4. Most four- and six-cylinder engines utilize what type of exhaust system?

   a. single exhaust
   b. dual exhaust
   c. double exhaust
   d. V-type exhaust

5. A customer complains because a stainless steel exhaust pipe broke without rusting or appearing to be worn out. Technician A believes the pipe is defective and attempts to get the customer a refund or replacement. Technician B explains that stainless steel pipe tends to break rather than rust when it is worn out. Who is right?

   a. Technician A
   b. Technician B
   c. Both A and B
   d. Neither A nor B

6. A customer's vehicle has a noisy heat control valve. Technician A checks the thermostat. Technician B checks the antirattle spring or shaft at the heat control assembly. Who is right?

   a. Technician A
   b. Technician B
   c. Both A and B
   d. Neither A nor B

7. What special exhaust system tool is depicted in Figure 9–2?

   a. hanger removal tool

Figure 9–2

   b. pipe cutter
   c. shaper
   d. expander

8. What is the most likely spot for leaking gaskets and seals?

   a. between the exhaust manifold and the cylinder head
   b. between the exhaust pipe and the exhaust manifold
   c. both A and B
   d. neither A nor B

9. Technician A breaks off rusted nuts with a heavy hammer. Technician B breaks off rusted nuts by tightening them instead of loosening them. Who is right?

   a. Technician A
   b. Technician B
   c. Both A and B
   d. Neither A nor B

10. On most exhaust systems, the exhaust pipe is fastened to the muffler with _____ .

    a. a slip joint and clamps
    b. a welded connection
    c. bolts
    d. neither A nor B

JOB SHEET

SHOP ASSIGNMENT 25
CONDUCT A THOROUGH EXHAUST SYSTEM INSPECTION

NAME _____ STATION _____ DATE _____

Tools and Materials

Hand tools
Flashlight or trouble light
Hammer or mallet
Vacuum gauge
Tachometer

Protective Clothing

Safety goggles or glasses with side shields

Procedure

1. Make sure the exhaust system is cool.

   Task completed _____

2. With a flashlight or trouble light, check for holes and road damage.

   Task completed _____

   a. Check for discoloration and rust.

   Task completed _____

   b. Check for carbon smudges.

   Task completed _____

   c. Check for bulging muffler seams.

   Task completed _____

   d. Check for interfering rattle points.

   Task completed _____

   e. Check for torn or broken hangers and clamps.

   Task completed _____

   f. Record any problems found.

   Not applicable _____ Task completed _____

3. Sound out the system by gently tapping the pipes and muffler with a hammer or mallet.

   Task completed _____

   a. Record a part as "good" when you hear a solid, metallic sound; record a part as "weak" when you hear a dull sound.

   Task completed _____

b. Make a note if you hear falling rust particles on the inside of the muffler.

Not applicable _____ Task completed _____

4. Grab the tail pipe and try to move it up and down and from side to side.

Task completed _____

a. If the system feels wobbly or loose, check the clamps and hangers that fasten the exhaust to the vehicle.

Not applicable _____ Task completed _____

5. Check all pipes for kinks and dents.

Task completed _____

a. Replace any pipe that is bent, kinked, or dented.

Not applicable _____ Task completed _____

6. Examine the exhaust manifold to exhaust pipe connection for any white powdery deposits.

Task completed _____

a. If any are found, check for loose bolts or leaking gasket.

Not applicable _____ Task completed _____

b. Check for loose connections at the muffler by pushing up on the muffler slightly.

Task completed _____

7. Attach a vacuum gauge to the intake manifold.

Task completed _____

a. Connect the tachometer.

Task completed _____

b. Start the engine.

Task completed _____

c. Does the vacuum gauge indicate a vacuum of 16 to 21 inches of mercury?

Yes _____ No _____

d. If no, note a probable blockage or restriction.

Not applicable _____ Task completed _____

e. Increase the engine speed to 2000 rpm.

Task completed _____

f. Does the vacuum decrease when the speed is increased rapidly then stabilize at 16 to 21 inches and remain constant?

Yes _____ No _____

g. If no, note that the exhaust system is restricted or blocked.

Not applicable _____ Task completed _____

8. Is there a bluish or brownish discoloration on the outer stainless steel shell of the catalytic converter?

Yes _____ No _____

a. If yes, note that the catalytic converter has overheated.

Not applicable _____ Task completed _____

PROBLEMS ENCOUNTERED: _____

_____

_____

INSTRUCTOR'S COMMENTS: _____

_____

_____

JOB SHEET

SHOP ASSIGNMENT 26
REPLACE A TYPICAL MUFFLER

NAME _____ STATION _____ DATE _____

Tools and Materials

Hand tools
Hammer or mallet
Muffler sealant
Shop rag

Protective Clothing

Safety goggles or glasses with side shields

Procedure

1. Raise the vehicle.

                             Task completed _____

2. Loosen the clamps and hangers from the muffler.

                             Task completed _____

3. Pull the assembly back and let it hang.

                             Task completed _____

   a. Tap the old muffler free to remove it from the system.

                             Task completed _____

4. Coat the inside of both the inlet and outlet on the new muffler with muffler sealant.

                             Task completed _____

   a. Coat the outside of the pipe coming into the muffler and the tail pipe.

                             Task completed _____

5. Make sure the muffler outlet arrow (if there is one) is pointed in the right direction.

                Not applicable _____ Task completed _____

6. Connect the muffler inlet to the exhaust or intermediate pipe with the pipe 1-1/2 to 2 inches into the muffler inlet.

                             Task completed _____

7. Rotate the muffler to its proper alignment.

                             Task completed _____

   a. Tighten the front clamp snugly.

                             Task completed _____

b. Install the rear muffler hanger.

Task completed _____

c. Place the muffler outlet into the hanger to support it.

Task completed _____

d. Insert the tail pipe into the muffler.

Task completed _____

e. Clamp snugly.

Task completed _____

8. Snug tighten the other clamps, hangers, and brackets on the remaining parts that have been loosened.

Task completed _____

9. Check that all connections are inserted fully.

Task completed _____

a. Check that all components are properly aligned.

Task completed _____

10. Make sure that all exhaust components have adequate clearance (at least 1 inch).

Task completed _____

11. Tighten all connections securely.

Task completed _____

12. With the transmission in NEUTRAL, start the engine.

Task completed _____

13. Accelerate and decelerate, listening for noise and rattles.

Task completed _____

14. If the shop has adequate ventilation, stuff a shop rag into the tail pipe and visually check for leaks.

Task completed _____

a. Remove the shop rag after inspection.

Not applicable _____ Task completed _____

15. After the exhaust system is thoroughly warmed up, retighten all connections and recheck critical clearances.

Task completed _____

PROBLEMS ENCOUNTERED: _____

_____

_____

INSTRUCTOR'S COMMENTS: _____

_____

JOB SHEET

SHOP ASSIGNMENT 27
DIAGNOSE A TURBOCHARGER GIVING OFF BLACK OR BLUE SMOKE

NAME _____ STATION _____ DATE _____

Tools and Materials

Hand tools
Shop manuals

Protective Clothing

Safety goggles or glasses with side shields

Procedure

1. If the engine is emitting blue smoke and oil consumption is high, check the air cleaner for restrictions following the engine manufacturer's shop manual.

Not applicable _____   Task completed _____

2. With the engine stopped, remove the turbo ducts.

Task completed _____

    a. Check the shaft assembly for free rotation, damage to wheels, or rubbing against housing walls.

Task completed _____

3. Check the oil drain line for restriction or damage that can cause seal flooding and leakage.

Task completed _____

4. Check for high crankcase pressure.

Task completed _____

    a. If in doubt, measure the crankcase pressure, which must be within the engine manufacturer's specifications.

Not applicable _____   Task completed _____

5. Loosen the exhaust manifold duct.

Task completed _____

    a. Check for oil in the engine exhaust.

Task completed _____

    b. If oil is present, see the engine manual for appropriate repairs.

Not applicable _____   Task completed _____

PROBLEMS ENCOUNTERED: _____

_____

_____

INSTRUCTOR'S COMMENTS: _____

_____

_____

CHAPTER TEN

EMISSION CONTROL SYSTEMS

Objectives

After reading this chapter, you should be able to:
- Explain the two types of emission control systems that are used in modern vehicles.
- Describe a PCV system, naming its components and explaining its function in emission control.
- Describe an EGR system and its role in emission control.
- Explain the operation of the fuel evaporative emission control system.
- Differentiate between carburetor fuel bowl emission control and air temperature emission control.
- Explain the operation of a basic air injection system used in most vehicles.
- Identify the importance of the catalytic converter.
- Describe the various methods of modifying engines to control vehicle emissions.
- Explain the operation of a diesel emission system.

PRACTICE QUESTIONS

1. Which EGR system component is identified by letter A in Figure 10–1?
   a. EGR control valve
   b. manifold vacuum
   c. EGR delay solenoid
   d. vacuum amplifier

2. Which EGR system component is identified by letter B in Figure 10–1?
   a. EGR control valve
   b. manifold vacuum
   c. EGR delay solenoid
   d. vacuum amplifier

3. Which EGR system component is identified by letter C in Figure 10–1?
   a. EGR control valve
   b. manifold vacuum
   c. EGR delay solenoid
   d. vacuum amplifier

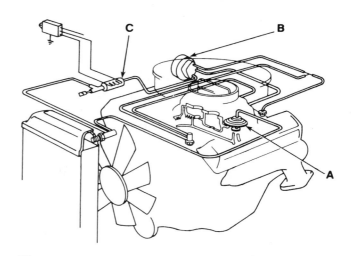

Figure 10–1

4. Which of the following contains a resistance grid that heats the mixture from the primary venturi of the carburetor?
   a. EGR
   b. EFE

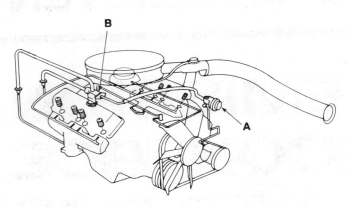

Figure 10-2

c. Thermactor Emission Control
d. None of the above

5. Which air injector system component is identified by letter A in Figure 10-2?
   a. check valve
   b. heat-operated vacuum switch
   c. diverter valve
   d. air-switching valve

6. Which air injector system component is identified by letter B in Figure 10-2?
   a. check valve
   b. heat-operated vacuum switch
   c. diverter valve
   d. air-switching valve

7. Which is the most efficient type of control system?
   a. precombustion control systems
   b. post-combustion control systems
   c. combustion control systems
   d. none of the above

8. Which catalytic converter design is used in GM and AM vehicles?
   a. monolithic
   b. dialithic
   c. pelletized
   d. oxidized

9. Which of the following is included in a closed loop fuel control system?
   a. computer
   b. sensors
   c. activators
   d. all of the above

10. Which of the following is one of the most effective spark retard delay system?
    a. CAV
    b. TVV
    c. ASV
    d. EGR

SHOP ASSIGNMENT 28: IDENTIFYING AIR TEMPERATURE EMISSION CONTROLS

Identify the type of air temperature emission control used on the vehicle and the components of that system.

SHOP ASSIGMENT 29: IDENTIFYING COMPONENTS OF AN AIR SYSTEM

Identify the components of the air injection system: vane pump, diverter valve, pressure relief valve, one-way check valve, hoses and nozzles.

SHOP ASSIGMENT 30: IDENTIFYING COMPONENTS OF EGR SYSTEM

Identify the EGR valve and other components of the EGR system used on the vehicle.

CHAPTER ELEVEN

EMISSION CONTROL SYSTEMS SERVICE

Objectives

After reading this chapter, you should be able to:
- Explain the Federal Clean Air Act and how it affects the technician.
- List the information included on a Vehicle Emission Control Information (VECI) Label.
- Explain the information coded into the Vehicle Identification Number (VIN).
- Explain the function of and the diagnosis and service procedures for the positive crankcase ventilation system.
- Explain the function of and the diagnosis and service procedures for the exhaust gas recirculating system.
- Explain the function of and the diagnosis and service procedures for the evaporative emission control system.
- Explain the function of and the diagnosis and service procedures for air temperature emission controls.
- Explain the function of and the diagnosis and service procedures for the air inspection system.
- Inspect and replace a catalytic converter.
- Explain the operation and uses of the exhaust emission analyzer.
- Explain the operation of the spark advance system.

PRACTICE QUESTIONS

1. What test on a canister purge valve does Figure 11–1 illustrate?

   a. valve open, vacuum off
   b. valve open, vacuum on
   c. valve closed, vacuum off
   d. valve closed, vacuum on

2. What is the first step in PCV servicing?

   a. Disconnect the PCV valve.

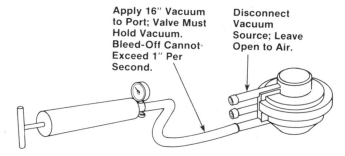

Apply 16″ Vacuum to Port; Valve Must Hold Vacuum. Bleed-Off Cannot Exceed 1″ Per Second.

Disconnect Vacuum Source; Leave Open to Air.

Figure 11–1

   b. Conduct a visual inspection.
   c. Start the engine and let it idle.
   d. None of the above

3. Which of the following is a type of back pressure EGR valve?

   a. negative transducer
   b. positive transducer
   c. both a and b
   d. none of the above

4. What does Figure 11–2 illustrate?

   a. single connector venturi vacuum amplifier system
   b. Chrysler venturi vacuum-control EGR system
   c. four input amplifier system
   d. none of the above

5. When warm driveability problems occur in cold ambient air conditions, what could be the cause?

   a. heated air duct leaking or not installed
   b. air cleaner not sealed tight

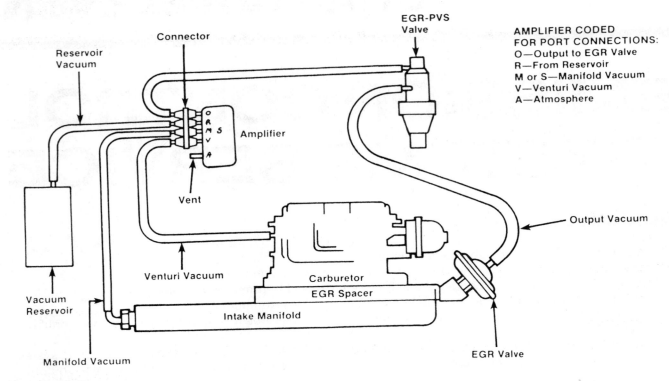

Figure 11-2

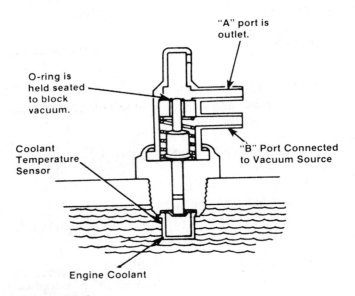

Figure 11-3

   c. broken or leaking duct or air cleaner
   d. all of the above
   e. none of the above

6. If an air injection system fails, which of the following could occur?
   a. noise
   b. poor idling
   c. poor driveability
   d. all of the above

7. Which of the following could cause high NO_x emissions?
   a. PVS fails open
   b. PVS fails closed
   c. Air bypass valve solenoid stuck open
   d. TVS fails closed

8. Which of the following could occur if the catalytic converter is clogged?
   a. loss of power at high speed
   b. drop in engine vacuum as engine rpm increases
   c. backfiring at the carburetor
   d. both a and b
   e. all of the above

9. Figure 11-3 shows the operation of the PVS under what conditions?
   a. cold engine
   b. warm engine
   c. open throttle
   d. none of the above

10. One possible spark advance problem is poor fuel economy. What could cause this problem?
   a. retard delay valve installed backward
   b. spark port plugged
   c. vacuum line leak
   d. PVS fails to open

    Chapter Eleven   **Emission Control Systems Service**

JOB SHEET

SHOP ASSIGNMENT 31
TEST A VENTURI VACUUM AMPLIFIER

NAME _____ STATION _____ DATE _____

Tools and Materials

Hand tools
Vacuum gauge
Service manual

Protective Clothing

Safety goggles or glasses with side shields

Procedure

1. Run the engine until it reaches normal operating temperatures.

   Task completed _____

2. Check all hoses and system components for good condition, proper routing, and tight connections.

   Task completed _____

3. Disconnect the hose at the EGR valve and install a vacuum gauge.

   Task completed _____

4. Disconnect the venturi vacuum hose at the carburetor.

   Task completed _____

5. Check the vacuum at curb idle.

   Task completed _____

   a. Is it within service manual specifications?

   Yes _____   No _____

   b. If so, proceed to Step 6.

   Task completed _____

   c. If not, check the manifold vacuum supply.

   Task completed _____

   d. Is the manifold vacuum supply correct?

   Yes _____   No _____

   e. If it is, replace the amplifier and repeat the test.

   Task completed _____

6. Watch the gauge as the engine accelerates to about 1500 to 2000 rpm.

Task completed _____

   a. Does the vacuum change?

Yes _____   No _____

   b. If not, proceed to Step 7.

Task completed _____

   c. If so, check the manifold vacuum supply.

Task completed _____

   d. Is the manifold vacuum supply correct?

Yes _____   No _____

   e. If it is, replace the amplifier and repeat the test.

Task completed _____

7. Connect the hose to the carburetor venturi port.

Task completed _____

8. Check for a vacuum increase.

Task completed _____

   a. Does it increase more than 1/2 inch?

Yes _____   No _____

   b. If so, check the curb idle against decal specifications.

Task completed _____

   c. Is it within specifications?

Yes _____   No _____

   d. If not, check the manifold vacuum supply.

Task completed _____

   e. Is the manifold vacuum supply correct?

Yes _____   No _____

   f. If so, replace the amplifier and repeat the test.

Task completed _____

9. Watch the gauge as the engine accelerates.

Task completed _____

   a. Does the vacuum build to 4 inches or more during acceleration and return to bias vacuum at idle?

Yes _____   No _____

   b. If not, replace the amplifier and repeat the test.

Task completed _____

PROBLEMS ENCOUNTERED: _____

_____

_____

INSTRUCTOR'S COMMENTS: _____

_____

_____

JOB SHEET

SHOP ASSIGNMENT 32
TEST AIR PUMP OUTPUT

NAME _____ STATION _____ DATE _____

Tools and Materials

Hand tools
Pressure gauge
Tachometer

Protective Clothing

Safety goggles or glasses with side shields

Procedure

1. Start the engine and warm to normal operating temperature.

   Task completed _____

2. Shut off the engine.

   Task completed _____

3. Disconnect the air outlet hose from the diverter or bypass valve.

   Task completed _____

4. Attach a pressure gauge to the air outlet hose.

   Task completed _____

   a. Is the gauge arranged so that the output air is directed away from personnel?

   Yes _____   No _____

   b. If not, reposition it.

   Task completed _____

   c. Is the gauge securely fastened in place?

   Yes _____   No _____

   d. If not, secure it.

   Task completed _____

5. Connect a tachometer

   Task completed _____

6. Restart the engine.

   Task completed _____

7. Check the reading on the gauge while increasing engine speed slowly up to 1000 rpm.

   Task completed _____

a. Is the output pressure 1 psi or more?

Yes _____   No _____

b. If the output pressure is below 1 psi or unsteady, replace the pump (see Shop Assignment 33).

Task completed _____

PROBLEMS ENCOUNTERED: _____

_____

_____

INSTRUCTOR'S COMMENTS: _____

_____

_____

JOB SHEET

SHOP ASSIGNMENT 33
INSTALL A NEW AIR PUMP

NAME _____ STATION _____ DATE _____

Tools and Materials

Hand tools
New air pump
New clamps
New gasket (if applicable)
Service manual

Protective Clothing

Safety goggles or glasses with side shields

Procedure

1. Disconnect all hoses attached to the air pump.

   Task completed _____

2. Relieve drive belt pressure by loosening the bolts holding the adjusting arm and mounting brackets.

   Task completed _____

3. Remove the pump's drive belt.

   Task completed _____

   a. If it is necessary, loosen the air conditioning compressor and drive belt to get at the air pump belt.

   Task completed _____

4. Remove the air pump pulley and mounting bolts.

   Task completed _____

5. Lift the pump away from the mounting bracket and out of the engine compartment.

   Task completed _____

6. Remove any valves that are fastened to the pump.

   Task completed _____

7. Install the valves on the new unit.

   Task completed _____

8. Was a gasket used on the rear of the old pump?

   Yes _____   No _____

   a. If yes, install a new gasket in the same location on the new unit.

   Task completed _____

Chapter Eleven   Emission Control Systems Service

9. Set the air pump on the engine and loosely install the mounting bolts.

Task completed _____

10. Install the air pump pulley.

Task completed _____

11. Torque the mounting bolts to approximately 10 foot-pounds.

Task completed _____

12. Replace the air conditioning compressor and drive belt if they were removed.

Task completed _____

13. Adjust the drive belt to the proper tension.

Task completed _____

14. Torque the mounting bolts on the pump to the manufacturer's specifications.

Task completed _____

15. Use new clamps to connect all hoses.

Task completed _____

PROBLEMS ENCOUNTERED: _____

_____

_____

INSTRUCTOR'S COMMENTS: _____

_____

_____

JOB SHEET

SHOP ASSIGNMENT 34
TEST THE OPERATION OF THE POSITIVE CRANKCASE VENTILATION SYSTEM

NAME _____ STATION _____ DATE _____

Tools and Materials

Hand tools
Infrared analyzer
PCV valve
Oil
Service manual

Protective Clothing

Safety goggles or glasses with side shields

Procedure

1. Operate the engine at specified curb idle speed.

   Task completed _____

2. Once the engine is operating at normal temperature, note the reading on the CO meter.

   CO reading _____

3. Remove the PCV valve from the cylinder head cover and allow it to hang free.

   Task completed _____

4. Does the CO reading drop more than 5 points below the CO level at curb idle speed?

   Yes _____ No _____

   a. If so, change the oil.

   Task completed _____

5. With the engine operating at specified curb idle speed, place your thumb over the end of the PCV valve.

   Task completed _____

6. Does the CO reading rise to a level higher than noted at curb idle with the PCV valve installed?

   Yes _____ No _____

   a. If not, clean or replace the PCV valve.

   Task completed _____

   b. Retest the system.

   Task completed _____

PROBLEMS ENCOUNTERED: _____

_____

_____

INSTRUCTOR'S COMMENTS: _____

_____

_____

_____

CHAPTER TWELVE

ELECTRONIC EMISSION CONTROLS

Objectives

After reading this chapter, you should be able to:
- Describe the major electronic emission control systems in use today.
- Describe the operation of the exhaust gas recirculation (EGR) system.
- Define and identify the major components/devices of the EGR system.
- Diagnose and service EGR-related malfunctions.
- Describe the operation of the secondary air system.
- Define and identify the major components/devices of the secondary air system.
- Diagnose and service secondary air system malfunctions.
- Describe the operation of an oxygen sensor.
- Diagnose the operation of an electronic spark advance system.

PRACTICE QUESTIONS

1. What secondary air system mode does Figure 12–1 illustrate?
   a. bypass mode
   b. upstream mode
   c. downstream mode
   d. midstream mode

2. The EEC/EGR valve is usually _____ .
   a. closed at idle and open during wide-open throttle conditions
   b. closed at idle and during wide-open throttle conditions
   c. open at idle and closed during wide-open throttle conditions
   d. open at idle and during wide-open throttle conditions.

3. Which of the following is a main component of the secondary air system?
   a. EVR valve
   b. SAD valve
   c. MAP valve
   d. none of the above

4. The SAB valve is _____ .
   a. normally a closed solenoid valve
   b. a component of the fuel evaporative emission system
   c. used to control the application of vacuum to the air bypass valve
   d. none of the above

5. Secondary air may be vented or bypassed for which of the following reasons?
   a. during cold engine cranking
   b. during acceleration
   c. a fuel-lean condition
   d. all of the above

6. Switching to the upstream mode allows the electronic engine control system to_____ .
   a. switch to the closed loop operation sooner
   b. postpone switching to the closed loop operation
   c. switch to the open loop operation sooner
   d. none of the above

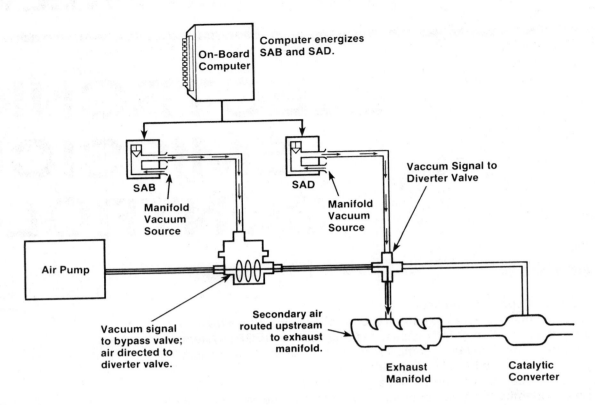

Figure 12-1

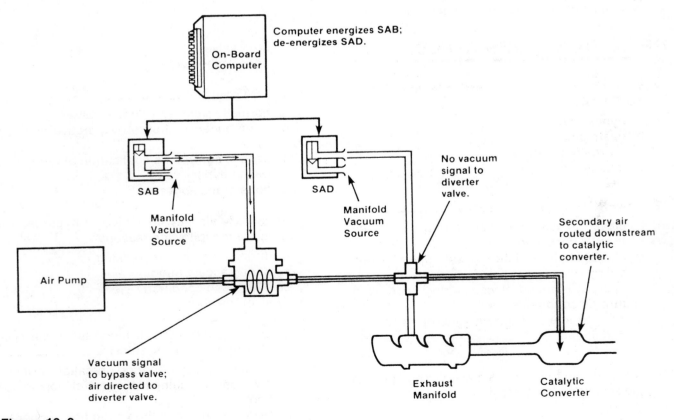

Figure 12-2

Chapter Twelve   Electronic Emission Controls

7. How much more oxygen should the secondary air system supply when it is enabled?
   a. 1 to 2 percent
   b. 2 to 5 percent
   c. 5 to 10 percent
   d. 10 to 20 percent

8. The canister of a fuel evaporative system is purged only _____ .
   a. at idle
   b. during closed throttle conditions
   c. during a part-open throttle condition
   d. during an open throttle condition

9. When troubleshooting an oxygen sensor, what type of meter should be used?
   a. ohmmeter
   b. analog voltmeter
   c. low-impedance voltmeter
   d. high-impedance voltmeter

10. What secondary air system mode does Figure 12–2 illustrate?
   a. downstream mode
   b. midstream mode
   c. upstream mode
   d. bypass mode

JOB SHEET

SHOP ASSIGNMENT 35
CONDUCT AN EGR VALVE GAS FLOW TEST

NAME _____ STATION _____ DATE _____

Tools and Materials

Hand tools
Vacuum pump

Protective Clothing

Safety goggles or glasses with side shields

Procedure

1. Run the engine to hot idle.

   Task completed _____

2. Disconnect the hose from the electronic vacuum regulator at the EGR valve diaphragm.

   Task completed _____

3. Attach a vacuum pump at the EGR valve diaphragm.

   Task completed _____

4. Slowly apply 5 to 10 inches of vacuum.

   Task completed _____

   a. Does idle speed drop 100 rpm or more?

   Yes _____   No _____

   b. Does idle speed return to normal with the vacuum removed?

   Yes _____   No _____

5. If the answer to Step 4a and Step 4b is not "yes," remove the EGR valve for a bench test/inspection.

   Task completed _____

PROBLEMS ENCOUNTERED: _____

_____

_____

INSTRUCTOR'S COMMENTS: _____

_____

_____

JOB SHEET

SHOP ASSIGNMENT 36
TEST THE VACUUM RESERVOIR
IN AN ELECTRONIC EGR SYSTEM

NAME _____ STATION _____ DATE _____

Tools and Materials

Hand tools
Vacuum pump
Reservoir

Protective Clothing

Safety goggles or glasses with side shields

Procedure

1. Remove the manifold vacuum source hose from the reservoir.

Task completed _____

2. Attach a vacuum pump.

Task completed _____

3. Apply 15 to 20 inches mercury vacuum to the reservoir.

Task completed _____

4. Allow for a time leakdown.

Task completed _____

   a. Does the reservoir leak more that 1/2 inch mercury vacuum per minute?

Yes _____   No _____

   b. If yes, replace the reservoir.

Task completed _____

PROBLEMS ENCOUNTERED: _____

_____

_____

INSTRUCTOR'S COMMENTS: _____

_____

_____

JOB SHEET

SHOP ASSIGNMENT 37
TEST A NORMALLY CLOSED (EGRC) SOLENOID VALVE

NAME _____ STATION _____ DATE _____

Tools and Materials

Hand tools
12-volt source
EGRC solenoid valve

Protective Clothing

Safety goggles or glasses with side shields

Procedure

1. Remove the hose from the solenoid.

   Task completed _____

2. Blow through the EGRC solenoid valve.

   Task completed _____

   a. Does it pass air?

   Yes _____ No _____

3. Energize the solenoid with a 12-volt source.

   Task completed _____

   a. Does the solenoid check in and the ports pass air?

   Yes _____ No _____

4. If the answer to Step 2 is "Yes" or the answer to Step 3 is "No," replace the EGRC solenoid valve.

   Task completed _____

ANSWER KEYS

ANSWERS TO PRACTICE QUESTIONS

Chapter 1

1. b	3. c	5. c	7. b	9. b
2. d	4. b	6. a	8. c	10. b

Chapter 2

1. d	3. c	5. a	7. b	9. a
2. b	4. d	6. b	8. c	10. d

Chapter 3

1. c	3. a	5. b	7. a	9. b
2. d	4. c	6. c	8. d	10. a

Chapter 4

1. b	3. c	5. b	7. c	9. d
2. a	4. d	6. b	8. a	10. b

Chapter 5

1. b	3. d	5. c	7. c	9. a
2. a	4. a	6. b	8. d	10. b

Chapter 6

1. a	3. d	5. c	7. b	9. d
2. b	4. b	6. a	8. c	10. a

Chapter 7

1. a	3. c	5. a	7. b	9. c
2. c	4. b	6. d	8. c	10. a

Chapter 8

1. d	3. b	5. a	7. c	9. c
2. c	4. a	6. b	8. a	10. d

Chapter 9

1. a	3. b	5. b	7. d	9. b
2. c	4. a	6. c	8. c	10. a

Chapter 10

1. a	3. c	5. c	7. a	9. d
2. d	4. b	6. d	8. c	10. b

Chapter 11

1. c	3. c	5. d	7. b	9. a
2. b	4. a	6. a	8. e	10. c

Chapter 12

1. b	3. b	5. a	7. b	9. d
2. b	4. c	6. a	8. c	10. a